HOW HUMANS EVOLVED

NORTON OFFERS RESOURCES ON THE WEB

To access the free How Humans Evolved Student Resource Web site,
visit: www.wwnorton.com/web/evolve

W.W. Norton & Company offers a wealth of educational resources for students and instructors
on the World Wide Web. Visit Norton's homepage —www.wwnorton.com—
To see helpful software demonstrations as well as discover exciting Web-based supplements to our texts,
including on-line study guides, Web books, and lecture aids.

FOURTH EDITION

W W NORTON & COMPANY NEW YORK • LONDON

HOW HUMANS EVOLVED

Robert Boyd • Joan B. Silk

University of California, Los Angeles

W. W. Norton & Company has been independent since its founding in 1923, when William Warder Norton and Mary D. Herter Norton first published lectures delivered at the People's Institute, the adult education division of New York City's Cooper Union. The Nortons soon expanded their program beyond the Institute, publishing books by celebrated academics from America and abroad. By mid-century, the two major pillars of Norton's publishing program—trade books and college texts—were firmly established. In the 1950s, the Norton family transferred control of them company to its employees, and today—with a staff of four hundred and a comparable number of trade, college, and professional titles published each year—W. W. Norton & Company stands as the largest and oldest publishing house owned wholly by its employees.

For Sam and Ruby

Printed in the United States of America

Manufacturing by Courier, Kendallville
Book design by Mary McDonnell
Cover design: Mary McDonnell
Page Layout: Brad Walrod/High Text Graphics, Inc

Library of Congress Cataloging-in-Publication Data
    Boyd, Robert, Ph.D.
        how humans evolved / Robert Boyd and Joan B. Silk,—4th ed.
    Includes bibliographical references and index
    ISBN 0-393-92628-1 (pbk)
    1. Human evolution. I. Silk, Joan B. II. Title.

GN281.B66 2005
599.93'8—dc22

W. W. Norton & Company, Inc., 500 Fifth Avenue, New York, N.Y. 10110
www.wwnorton.com
W. W. Norton & Company, Ltd., Castle House, 75/76 Wells Street, London W1T 3QT

ABOUT THE AUTHORS

ROBERT BOYD has written widely on evolutionary theory, focusing especially on the evolution of cooperation and role of culture in human evolution. His book, Culture and the Evolutionary Process, received the J. I. Staley Prize. He has also published numerous articles in scientific journals and edited volumes. He is currently co-director of the MacArthur Preferences Network and Professor of Anthropology at the University of California, Los Angeles.

JOAN B. SILK has conducted extensive research on the social lives of monkeys and apes, including extended field work on chimpanzees at Gombe Stream Reserve in Tanzania, and baboons in Kenya and Botswana. She is also interested in the application of evolutionary thinking to human behavior, especially adoption and friendship. She has published over 60 papers in scientific journals and scholarly edited volumes, and is currently Professor of Anthropology at the University of California, Los Angeles.

CONTENTS

PART THREE: THE HISTORY OF THE HUMAN LINEAGE

PART FOUR: EVOLUTION AND MODERN HUMANS

PREFACE

How Humans Evolved focuses on the processes that have shaped human evolution. This approach reflects our training and research interests. As anthropologists, we are interested in the evolutionary history of our own species, *Homo sapiens*. As evolutionary biologists, we study how evolution works. In this book, we integrate these two perspectives. We use current theoretical and empirical work in evolutionary theory, population genetics, and behavioral ecology to interpret human evolutionary history. We describe the changes that have occurred as the hominin lineage has evolved, and we consider why these changes may have happened. We try to give life to the creatures that left the bones and made the artifacts that paleontologists and archaeologists painstakingly excavate by focusing on the processes that generate change, create adaptations, and shape bodies and behavior. We also give serious attention to the role of evolution in shaping contemporary human behavior. There is considerable controversy about evolutionary approaches to human behavior within the social sciences, but we think it is essential to confront these issues openly and clearly. Positive responses to the first three editions of *How Humans Evolved* tell us that many of our colleagues endorse this approach.

One of the problems in writing a textbook about human evolution is that there is considerable debate on many topics. Evolutionary biologists disagree about how new species are formed and how they should be classified; primatologists argue about the adaptive significance of infanticide and the effects of dominance rank on reproductive performance; paleontologists disagree about the taxonomic relationships among early hominin species and the emergence of modern humans; and those who study modern humans disagree about the meaning and significance of race, the role of culture in shaping human behavior and psychology, the adaptive significance of many aspects of modern human behavior, and many other things. Sometimes multiple interpretations of the same data can be defended; in other cases, the facts seem contradictory. Textbook writers can handle this kind of uncertainty in two different ways. They can weigh the evidence, present the ideas that best fit the available evidence, and ignore the alternatives. Or they can present opposing ideas, evaluate the logic underlying each idea, and explain how existing data support each of the positions. We chose the second alternative, at the risk of complicating the text and frustrating readers looking for simple answers. We made this choice because we believe that this approach is essential for understanding how science works. Students need to see how theories are developed, how data are accumulated, and how theory and data interact to shape our ideas about how the world works. We hope that students remember this long after they have forgotten many of the facts that they will learn in this book.

NEW IN THE FOURTH EDITION

Many users of the book have found that it is hard to cover all of the material in a single quarter or semester. In response to this concern, we have put the book on a diet. Diets are hard because you have to give up things that you like. In this case, we eliminated the

chapter on language and the chapter on the human life cycle. We incorporated some material about the evolution of language in Part Four and some of the material about primate life history in Part Three. Readers familiar with previous editions will also notice that the readings are gone. Few of the instructors who were asked to review the third edition of the book seemed to think that the readings were essential. Removing them reduces the length of the book and helps streamline the text.

The study of human evolution is a dynamic field. No sooner do we complete one edition of this book than researchers make new discoveries that fundamentally change our view of human evolution. These kinds of discoveries include the spectacular fossil finds that reveal new chapters in human ancestry, new data that alter our interpretation of the behavioral strategies of primates, and experimental studies that reveal cross-cultural regularities in mating preferences. New developments in human evolutionary studies require regular updates of the textbook. Although we have made many changes throughout the book, readers familiar with prior editions will find significant changes in Parts II (Primate Behavior and Ecology), III (The History of the Human Lineage), and IV (Evolution and Modern Humans).

In Part I, we expanded Chapter 1 to provide more examples of the power of natural selection. We describe recent work which documents the rapid evolution of a complex placenta-like organs independently in several species of a single fish genus In Chapter 2, we provide expanded coverage of molecular genetics, and discuss more fully the role of proteins, gene regulation, and cell differentiation in development. Sometimes less is more: in Chapter 4 we have streamlined and updated the discussion of measuring genetic distance using sequence data rather than hybridization data.

In Part II, we revised a number of chapters to include recent empirical findings. For example, in Chapter 7 we discuss new evidence about the function of male-female relationships in baboons, and in Chapter 8 we describe new work on paternal kin recognition. We rewrote Chapter 9 entirely, integrating our discussion of the evolution of life history strategies (formerly covered in Part IV) with our discussion of the evolution of cognitive abilities in primates. These changes are prompted by new theoretical models that consider why evolution favored enhanced cognitive abilities in the primate order, and explore the consequences of these changes on primate life histories.

We have thoroughly revised Part III to incorporate a number of important findings that have substantially altered our understanding of human history. We rewrote Chapter 10 to include new data on the origin of primates, new fossils that tell us more about the New World primate radiation, and important new work on the the Miocene apes. In Chapter 11, we integrate information on *Sahelanthropus tchadensis* and *Orrorin tugenensis* into our discussion of the earliest hominins, and in Chapters 11 and 12, we describe new findings about the rate of development in early hominins. The tiny hominins from Flores, Indonesia—*Homo floresiensis*—make their first appearance in this edition in Chapter 13. We also discuss some new information about global climate change that may have influenced human evolution. Chapter 14 includes a new section on the earliest African *H. sapiens*, including Herto fossils and the redating of Omo Kibish I, and a new section on what genetic sequence data from modern populations (and from Neanderthals) can tell us about human evolution.

Part IV has been reorganized to provide a more integrated view of human diversity and contemporary behavior. Chapter 15 considers the sources of variation within and between human populations, and describes new findings that are relevant to understanding the processes that underlie human diversity. We have rewritten Chapter 16 to provide a more integrated view of how evolution shapes modern human behavior. We have eliminated the dichotomy between evolutionary psychology and evolutionary anthropology, focusing instead on the insights evolutionary theory provides about modern human minds and behavior. We have also included information about the evolution of language and the adaptive significance of culture in this chapter. Chapter 17 includes new data on how evolution shapes mate choice and mating tactics in men and women, a more comprehensive discussion of the factors that influence mens' investment strategies, and a broader review of evidence about the investment patterns of stepparents and grandparents.

Throughout the book, we updated the list of references for further reading, and added a number of new study questions that are related to new material in the text. With the help of our colleagues and excellent copy editor, we also corrected a number of errors that found their way into the third edition.

We wrote this book with undergraduates in mind, and we have designed a number of features to help students use the book effectively. We have retained the "key idea" statements (now printed in blue type), and we recommend that students use these key ideas to keep track of important concepts and facts, and to structure their reviews of the material. Important terms that may be unfamiliar are set in bold type and defined in the text when they first appear. Readers can also find definitions for these terms in the Glossary. Discussion questions appear at the end of each chapters. These questions are meant to help students synthesize material presented in the text. Some of the questions are designed to help students review factual material, but most are intended to help students to think about the processes or theoretical principles they have learned. Some questions are open-ended and meant to encourage students to apply their own values and judgment to the material presented in the text. Students tell us that they find these questions useful as they attempt to master the material and prepare for exams. The list of references for further reading at the end of each chapter provides a starting point for students who want to delve more deeply into the material covered in that chapter.

The book is richly illustrated with photographs, diagrams, figures, and graphs. These illustrations provide visual information to complement the text. For some subjects, a picture is clearly worth 1000 words—no amount of description can enable students to conjure up an image of an aye-aye or appreciate the how much more similar the australopithecine pelvis is to the modern human pelvis than to the chimpanee pelvis. The diagrams of evolutionary processes that appear in Part I, are designed to help students visualize how natural selection works. The figures depicting the hominin fossils are drawn to scale, so each is presented in the same orientation and to the same scale. This should help students compare one hominin specimen with another. We have often been advised that you cannot put graphs in an undergraduate textbook, but we think that the graphs help students understand the evidence more fully. For us, it is easier to remember data that is portrayed graphically, than to recall verbal descriptions of results.

In addition, students can consult the multimedia guide to the fossil record, *Human Evolution* by Philip L. Walker and Edward Hagen, that is packaged with the text. This student resource offers lesson modules that illustrate controversies, archeological discoveries, and the anatomical differences between members of the nonhuman primate and human lineage, including prosimians, monkeys, apes, and hominins. The modules contain a photographs, drawings, and artwork plus a number of three dimensional animated specimens that can be "zoomed" to half-scale. Multiple-choice quizzes conclude each of the lessons.

ANCILLARY MATERIALS

For instructors, there is an instructor's manual and test-item file, which include a detailed outline of each chapter, answers to all of the discussion questions, and a bank of exam questions. The test questions are also available on CD in MS-DOS and Macintosh formats, and will be provided free of charge upon adoption of this textbook. For instructor use we also offer a media library of full-color transparencies of the figures and a CD containing many of the figures and photographs. This media library, available to adopting instructors, is especially rich, offering images not only from *How Humans Evolved*, but also from two other fine and well-illustrated books, Dean Falk's *Primate Diversity* and Glenn Conroy's *Primate Evolution*.

ACKNOWLEDGMENTS

Over the last 10 years, many of our colleagues have provided new information, helpful comments, and critical perspectives that have enriched this book. We are grateful for all those who have responded to our requests for photographs, clarifications, references, and opinions. For the fourth edition, Laura MacClatchy provided help with the Miocene apes in Chapter 10, Dan Fessler and David Schmitt gave us access to material for Chapter 16, and Kermyt Anderson dug up original data for figures in Chapter 17. Steven Reznik reviewed our discussion of the rapid evolution of placentas in the minnows he studies and kindly provided an image. Leslie Aiello helped with our discussion of hominin developmental rates. For help with the third edition, we thank Carola Borries, Colin Chapman, Richard Klein, Cheryl Knott, Sally McBrearty, Ryne Palombit, Steve Pinker, Karin Stronswold, and Bernard Wood. For help with the second edition, we also thank Tom Plummer, Daniel Povinelli, Beverly Strassman, and Patricia Wright. We remain grateful for the help we received for the first edition from Leslie Aiello, Monique Borgerhoff Mulder, Scott Carroll, Dorothy Cheney, Glenn Conroy, Martin Daly, Robin Dunbar, Lynn Fairbanks, Sandy Harcourt, Kristin Hawkes, Richard Klein, Phyllis Lee, Nancy Levine, Jeff Long, Joseph Manson, Henry McHenry, John Mitani, Jocelyn Peccei, Susan Perry, Steve Pinker, Tom Plummer, Tab Rasmussen, Mark Ridley, Alan Rogers, Robert Seyfarth, Frank Sulloway, Don Symons, Alan Walker, Tim White, and Margo Wilson.

A number of people reviewed all or parts of the first three editions. We thank the following: Thad Bartlett, Barry Bogin, Margaret Clarke, Douglas Crews, Sharon Gursky, Mark Griffin, Andrew Irvine, Richard Klein, Darrell La Lone, Clark Larsen, Lynette Leidy, Marilyn Norconk, Ann Palkovich, James Paterson, Eric Smith, Craig Stanford, Horst Steklis, Joan Stevenson, Mark Stoneking, Rebecca Storey, and Patricia Wright. Several anonymous reviewers read previous editions and provided suggestions. Although we are certain that we have not satisfied all those who read and commented on parts of the book, we found all of the comments to be very helpful as we revised the text.

Richard Klein provided us with many exceptional drawings of fossils that appear in Part Three—an act of generosity that we continue to appreciate. We also give special thanks to Neville Agnew and the Getty Conservation Institute, which granted us permission to use images of the Laetoli conservation project for the cover of the second edition.

Many users of the book have commented on the quality of the illustrations. For this we must thank the many friends and colleagues who allowed us to use their photographs: Bob Bailey, Carola Borries, Colin Chapman, Nick Blurton Jones, Sue Boinski, Monique Borgerhoff Mulder, Richard Byrne, Scott Carroll, Marina Cords, Diane Doran, Robert Gibson, Peter Grant, Kim Hill, Kevin Hunt, Lynne Isbell, Charles Janson, Alex Kacelnik and the Behavioral Ecology Research Group, Nancy Levine, Carlão Limeira, Joe Manson, Frank Marlowe, Laura MacLatchy, Bill McGrew, John Mitani, Claudio Nogueira, Ryne Palombit, Susan Perry, Craig Stanford, Karen Strier, Alan Walker, Katherine West, and John Yellen. The National Museums of Kenya kindly allowed us to reprint a number of photographs.

We also acknowledge the thousands of students and dozens of teaching assistants at UCLA who have used various versions of this material over the years. Student evaluations of the original lecture notes, the first draft of the text, and the first three editions were helpful as we revised and rewrote various parts of the book. The teaching assistants helped us identify many parts of the text that needed to be clarified, corrected, or reconsidered.

We thank all the people at Norton who helped us produce this book, particularly our outstanding editor Leo Wiegman. We are grateful to our excellent copy editor, Stephanie Hiebert, for her relentless pursuit of clarity, accuracy, and typographical errors. We also thank Christopher Granville, who saw the book through production; and Elizabeth Erhart at Texas State University, San Marcos for preparing the revised instructor's manual.

WHY STUDY HUMAN EVOLUTION?

Origin of man now proved—Metaphysics must flourish—He who
understand baboon would do more toward metaphysics than Locke.

—Charles Darwin, *M Notebook*, August 1838

In 1838, Charles Darwin discovered the principle of evolution by natural selection and revolutionized our understanding of the living world. Darwin was 28 years old, and it was just two years since he had returned from a five-year voyage around the world as a naturalist on the HMS *Beagle* (Figure 1). Darwin's observations and experiences during the journey had convinced him that biological species change through time and that new species arise by the transformation of existing ones, and he was avidly searching for an explanation of how these processes worked. In late September of the same year, Darwin read Thomas Malthus's *Essay on the Principle of Population*, in which Malthus (Figure 2) argued that human populations invariably grow until they are limited by starvation, poverty, and death. Darwin realized that Malthus's logic also applied to the natural world, and this intuition inspired the conception of his theory of evolution by natural selection. In the intervening century and a half, Darwin's theory has been augmented by discoveries in genetics and amplified by studies of the evolution of many types of organisms. It is now the foundation of our understanding of life on Earth.

This book is about human evolution, and we will spend a lot of time explaining how natural selection and other evolutionary processes have shaped the human species. Before we begin, it is important to consider why you should care about this topic. Many of you will be working through this book as a requirement for an undergraduate class in biological anthropology and will read the book in order to earn a good grade. As instructors of a class like this ourselves, we approve of this motive. However, there is a much better reason to care about the processes that have shaped human evolution: understanding how humans evolved is the key to understanding why people look and behave the way they do.

The profound implications of evolution for our understanding of humankind were apparent to Darwin from the beginning. We know this today because he kept notebooks in which he recorded his private thoughts about various topics. The quotation that begins this prologue is from the *M Notebook*, begun in July 1838, in which Darwin jotted down his ideas about humans, psychology, and the philosophy of science. In the nineteenth century, metaphysics involved the study of the human mind. Thus Darwin was saying that because he believed humans evolved from a creature something like a baboon, it followed that an understanding of the mind of a baboon would contribute more to an understanding of the human mind than would all of the works of the great English philosopher John Locke.

Darwin's reasoning was simple. Every species on this planet has arisen through the same evolutionary processes. These processes determine why organisms are the way they are by shaping their morphology, physiology, and behavior. The traits that characterize the human species are the result of the same evolutionary processes that created all other species. If we understand these processes, and the conditions under which the human species evolved, then we will have the basis for a scientific understanding of human nature. Trying to comprehend the human mind without an understanding of human evolution is, as Darwin wrote in another notebook that October, "like puzzling at astronomy without mechanics." By this, Darwin meant that his theory of evolution could play the same role in biology and psychology that Isaac Newton's laws of motion had played in astronomy. For thousands of years, stargazers, priests, philosophers, and mathematicians had struggled to understand the motions of the planets without success. Then, in the late 1600s, Newton discovered the laws of mechanics and showed how all of the intricacies in the dance of the planets could be explained by the action of a few simple processes (Figure 3).

In the same way, understanding the processes of evolution enables us to account for the stunning sophistication of organic design and the diversity of life, and to understand why people are the way they are. As a consequence, understanding how natural selection and other evolutionary processes shaped the human species is relevant to all of the academic disciplines that are concerned with human beings. This vast intellectual domain includes medicine, psychology, the social sciences, and even the humanities. Beyond academia, understanding our own evolutionary history can help

FIGURE 1 When this portrait of Charles Darwin was painted, he was about 30 years old. He had just returned from his voyage on the HMS *Beagle* and was still busy organizing his notes, drawings, and vast collections of plants and animals.

FIGURE 2 Thomas Malthus was the author of *An Essay on the Principle of Population*, a book Charles Darwin read in 1838 that profoundly influenced the development of his theory of evolution by natural selection.

FIGURE 3 Sir Isaac Newton discovered the laws of celestial mechanics, a body of theory that resolved age-old mysteries about the movements of the planets.

us answer many questions that confront us in everyday life. Some of these questions are relatively trivial: Why do we sweat when hot or nervous? Why do we crave salt, sugar, and fat, even though large amounts of these substances cause disease (Figure 4)? Why are we better marathon runners than mountain climbers? Other questions are more profound: Why do only women nurse their babies? Why do we grow old and eventually die? Why do people look so different around the world? As you will see, evolutionary theory provides answers or insights about all of these questions. Aging, which eventually leads to death, is an evolved characteristic of humans and most other creatures. Understanding how natural selection shapes the life histories of organisms tells us why we are mortal, why our life span is about 70 years, and why other species live shorter lives. In an age of horrific ethnic conflicts and growing respect for multi-cultural diversity, we are constantly reminded of the variation within the human species. Evolutionary analyses tell us that genetic differences between human groups are relatively minor, and that our notions of race and ethnicity are culturally constructed categories, not biological realities.

All of these questions deal with the evolution of the human body. However, understanding evolution is also an important part of our understanding of human behavior and the human mind. The claim that understanding evolution will help us understand contemporary human behavior is much more controversial than the claim that it will help us understand how human bodies work. But it should not be. The human brain is an evolved organ of great complexity, just like the endocrine system, the nervous system, and all of the other components of the human body that regulate our behavior. Understanding evolution helps us understand our minds and behavior because evolutionary processes forged the brain that controls human behavior, just as they forged the brain of the chimpanzee and the salamander.

One of the great debates in Western thought centers on the essence of human nature. One view is that people are basically honest, generous, and cooperative creatures who are corrupted by an immoral economic and social order. The opposing view is that we are fundamentally amoral, egocentric beings whose antisocial impulses are held in check by social pressures. This question turns up everywhere. Some people believe that children are little barbarians who are civilized only through sustained parental effort; others think that children are gentle beings who are socialized into competitiveness and violence by exposure to negative influences like toy guns and violent TV programs (Figure 5). The same dichotomy underpins much political and economic thought. Economists believe that people are rational and selfish, but other social scientists, particularly anthropologists and sociologists, question and sometimes reject this assumption. We can raise an endless list of interesting questions about human nature: Does the fact that, in most societies, women rear children and men make war mean that men and women differ in their innate predispositions? Why do men typically find younger women attractive? Why do some people neglect and abuse their children, while others adopt and lovingly raise children who are not their own?

FIGURE 4 A strong appetite for sugar, fat, and salt may have been adaptive for our ancestors, who had little access to sweet, fatty, and salty foods. We have inherited these appetites and have easy access to these foods. As a consequence, many of us suffer from obesity, high blood pressure, diabetes, and heart disease.

Understanding human evolution does not reveal the answers to all of these questions, or even provide a complete answer to any one of them. As we will see, however, it can provide useful insights about all of them. An evolutionary approach does not imply that behavior is "genetically determined," or that learning and culture are unimportant. In fact, we will argue that learning and culture play crucial roles in human behavior. Behavioral differences among peoples living in different times and places result mainly from flexible adjustments to different social and environmental conditions. Understanding evolution is useful precisely because it helps us understand why humans respond in different ways to different conditions.

OVERVIEW OF THE BOOK

Humans are the product of organic evolution. By this we mean that there is an unbroken chain of descent that connects every living human being to a bipedal, apelike creature that walked through tall grasses of the African savanna 3 million years ago (mya); to a monkeylike animal that clambered through the canopy of great tropical forests covering much of the world 35 mya; and finally to a small, egg-laying, insect-eating mammal that scurried about at night during the age of the dinosaurs, 100 mya. To understand what we are now, you have to understand how this transformation took place. We tell this story in four parts.

FIGURE 5 One of the great debates in Western thought focuses on the essential elements of human nature. Are people basically moral beings corrupted by society, or fundamentally amoral creatures socialized by cultural conventions, social strictures, and religious beliefs?

Part One: How Evolution Works

More than a century of hard work has given us a good understanding of how evolution works. The transformation of apes into humans involved the assembly of many new, complex adaptations. For example, in order for early humans to walk upright on two legs, there had to be coordinated changes in many parts of their bodies, including their feet, legs, pelvis, backbone, and inner ear. Understanding how natural selection gives rise to such complex structures, and why the genetic system plays a crucial role in this process, is essential for understanding how new species arise. An understanding of these processes also allows us to reconstruct the history of life from the characteristics of contemporary organisms.

Part Two: Primate Behavior and Ecology

In the second part of the book, we consider how evolution has shaped the behavior of nonhuman primates—an exercise that helps us understand human evolution in two ways. First, humans are members of the primate order: we are more similar to other primates, particularly the great apes, than we are to wolves, raccoons, or other mammals. Studying how primate morphology and behavior are affected by ecological conditions helps us determine what our ancestors might have been like and how they may have been transformed by natural selection. Second, we study primates because they are an extremely diverse order and are particularly variable in their social behavior. Some are solitary, others live in monogamous pairs, and some live in large groups that contain many adult females and males. Data derived from studies of these species help us understand how social behavior is molded by natural selection. We can then use these insights to interpret the hominin fossil record and the behavior of contemporary people (Figure 6).

FIGURE 6 We will draw on information about the behavior of living primates, such as this chimpanzee, to understand how behavior is molded by evolutionary processes, to interpret the hominin fossil record, and to draw insights about the behavior of contemporary humans.

Part Three: The History of the Human Lineage

General theoretical principles are not sufficient to understand the history of any lineage, including our own. The transformation of a shrewlike creature into the human species involved many small steps, and each step was affected by specific environmental and biological circumstances. To understand human evolution, we have to reconstruct the actual history of the human lineage and the environmental context in which these events occurred. Much of this history is chronicled in the fossil record. These bits of mineralized bone, painstakingly collected and reassembled by paleontologists, document the sequence of organisms that link early mammals to modern humans. Complementary work by geologists, biologists, and archaeologists allows us to reconstruct the environments in which the human lineage evolved (Figure 7).

Part Four: Evolution and Modern Humans

Finally, we turn our attention to modern humans and ask why we are the way we are. Why is the human species so variable? How do we acquire our behavior? How has evolution shaped human psychology and behavior? How do we choose our mates? Why do people commit infanticide? Why have humans succeeded in inhabiting every corner of the Earth when other species have more limited ranges? We will explain how an understanding of evolutionary theory and a knowledge of human evolutionary history provide a basis for addressing questions like these.

The history of the human lineage is a great story, but it is not a simple one. The relevant knowledge is drawn from many disciplines in the natural sciences, such as physics, chemistry, biology, and geology; and from the social sciences, mainly anthropology, psychology, and economics. Learning this material is an ambitious task, but it offers a very satisfying reward. The better you understand the processes that have shaped human evolution and the historical events that took place in the human lineage, the better you will understand how we came to be and why we are the way we are.

FIGURE 7   Fossils painstakingly excavated from many sites in Africa, Europe, and Asia provide us with a record of our history as a species. Two million years ago in Africa, there were a number of apelike species that walked bipedally but still had ape-sized brains and apelike developmental patterns. These are the fossilized remains of *Australopithecus habilis*, a species that some think is ancestral to modern humans.

HOW EVOLUTION WORKS

ADAPTATION BY NATURAL SELECTION

EXPLAINING ADAPTATION BEFORE DARWIN

Animals and plants are adapted to their conditions in subtle and marvelous ways.

Even the casual observer can see that organisms are well suited to their circumstances. For example, fish are clearly designed for life under water, and certain flowers are designed to be pollinated by particular species of insects. More careful study reveals that organisms are more than just suited to their environments: they are complex machines, made up of many exquisitely constructed components, or **adaptations**, that interact to help the organism survive and reproduce.

2

The human eye provides a good example of an adaptation. Eyes are amazingly useful: they allow us to move confidently through the environment, to locate critical resources like food and mates, and to avoid dangers like predators and cliffs. Eyes are extremely complex structures made up of many interdependent parts (Figure 1.1). Light enters the eye through a transparent opening, then passes through a diaphragm called the "iris," which regulates the amount of light entering the eye and allows the eye to function in a wide range of lighting conditions. The light then passes through a lens that projects a focused image on the retina on the back surface of the eye. Several different kinds of light-sensitive cells then convert the image into nerve impulses that encode information about spatial patterns of color and intensity. These cells are more sensitive to light than the best pho-

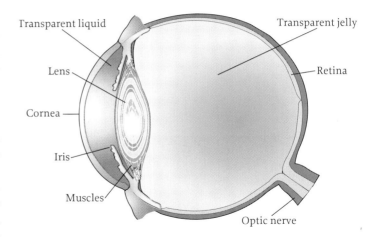

FIGURE 1.1 A cross section of the human eye.

tographic film is. The detailed construction of each of these parts of the eye makes sense in terms of the eye's function: seeing. If we probed into any of these parts, we would see that they, too, are made of complicated, interacting components whose structure is understandable in terms of their function.

Differences between human eyes and the eyes of other animals make sense in terms of the types of problems each creature faces. Consider, for example, the eyes of fish and humans (Figure 1.2). The lens in the eyes of humans and other terrestrial mammals is much like a camera lens; it is shaped like a squashed football and has the same index of refraction (a measure of light-bending capacity) throughout. In contrast, the lens in fish eyes is a sphere located at the center of the curvature of the retina, and the index of refraction of the lens increases smoothly from the surface of the lens to the center. It turns out that this kind of lens, called a "spherical gradient lens," provides a sharp image over a full 180° visual field, a very short focal length, and high light-gathering power—all desirable properties. Terrestrial creatures like us cannot use this design because light is bent when it passes from the air through the cornea (the transparent cover of the pupil), and this fact constrains the design of the remaining lens elements. In contrast, light is not bent when it passes from water through the cornea of aquatic animals, and the design of their eyes takes advantage of this fact.

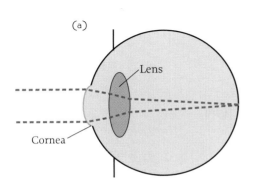

(a)

Human eye

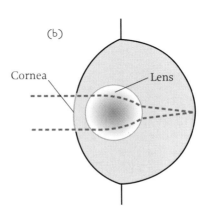

(b)

Fish eye

FIGURE 1.2 (a) Like those of other terrestrial mammals, human eyes have more than one light-bending element. A ray of light entering the eye (dashed lines) is bent first as it moves from the air to the cornea and then again as it enters and leaves the lens. (b) In contrast, fish eyes have a single lens that bends the light throughout its volume. As a result, fish eyes have a short focal length and high light-gathering power.

Before Darwin there was no scientific explanation for the fact that organisms are well adapted to their circumstances.

As many nineteenth-century thinkers were keenly aware, complex adaptations like the eye demand a different kind of explanation than other natural objects do. This is not simply because adaptations are complex, since many other complicated objects exist in nature. Adaptations require a special kind of explanation because they are complex in a particular, highly improbable way. For example, the Grand Canyon, with its maze of delicate towers intricately painted in shades of pink and gold, is byzantine in its complexity (Figure 1.3). Given a different geological history, however, the Grand Canyon might be quite different—different towers, in different hues—yet we would still recognize it as a canyon. The particular arrangement of painted towers of the Grand Canyon is improbable, but the existence of a spectacular canyon with a complex array of colorful cliffs in the dry sandstone country of the American Southwest is not unexpected at all; and in fact, wind and water produced several other canyons in this region. In contrast, any substantial changes in the structure of the eye would prevent the eye from functioning, and then we would no longer recognize it as an eye. If the cornea were opaque, or the lens were on the wrong side of the retina, then the eye would not transmit visual images to the brain. It is highly improbable that natural processes would randomly bring together bits of matter having the detailed structure of the eye, because only an infinitesimal fraction of all arrangements of matter would be recognizable as a functioning eye.

In Darwin's day, most people were not troubled by this problem, because they believed that adaptations were the result of divine creation. In fact, the theologian William Paley used a discussion of the human eye to argue for the existence of God in his book *Natural Theology*, published in 1802. Paley argued that the eye is clearly *designed* for seeing; and where there is design in the natural world, there certainly must be a heavenly designer.

Although most scientists of the day were satisfied with this reasoning, a few, including Charles Darwin, sought other explanations.

FIGURE 1.3 Although an impressive geological feature, the Grand Canyon is much less remarkable in its complexity than the eye is.

DARWIN'S THEORY OF ADAPTATION

Charles Darwin was expected to become a doctor or clergyman, but instead he revolutionized science.

Charles Darwin was born into a well-to-do, intellectual, and politically liberal family in England. Like many prosperous men of his time, Darwin's father wanted his son to

FIGURE 1.4 The HMS *Beagle* in Beagle Channel on the southern coast of Tierra del Fuego.

become a doctor. But after failing at the prestigious medical school at the University of Edinburgh, Charles went on to Cambridge University, resigned to becoming a country parson. He was, for the most part, an undistinguished student—much more interested in tramping through the fields around Cambridge in search of beetles than in studying Greek and mathematics. After graduation, one of Darwin's biology professors, William Henslow, provided him with a chance to pursue his passion for natural history as a naturalist on the HMS *Beagle*.

The *Beagle* was a Royal Navy vessel whose charter was to spend two to three years mapping the coast of South America and then to return to London, perhaps by circling the globe (Figure 1.4). Darwin's father forbade him to go, preferring that he get serious about his career in the clergy, but Darwin's uncle (and future father-in-law) Josiah Wedgwood intervened. The voyage turned out to be the turning point in Darwin's life. His work during the voyage established his reputation as a skilled naturalist. His observations of living and fossil animals ultimately convinced him that plants and animals sometimes change slowly through time, and that such evolutionary change is the key to understanding how new species come into existence. This view was rejected by most scientists of the time and was considered heretical by the general public.

Darwin's Postulates

Darwin's theory of adaptation follows from three postulates: (1) the struggle for existence, (2) variation in fitness, and (3) the inheritance of variation.

In 1838, shortly after the *Beagle* returned to London, Darwin formulated a simple mechanistic explanation for *how* species change through time. His theory follows from three postulates:

1. The ability of a population to expand is infinite, but the ability of any environment to support populations is always finite.

2. Organisms within populations vary, and this variation affects the ability of individuals to survive and reproduce.
3. This variation is transmitted from parents to offspring.

Darwin's first postulate means that populations grow until they are checked by the dwindling supply of resources in the environment. Darwin referred to the resulting competition for resources as "the struggle for existence." For example, animals require food to grow and reproduce. When food is plentiful, animal populations grow until their numbers exceed the local food supply. Because resources are always finite, it follows that not all individuals in a population will be able to survive and reproduce. According to the second postulate, some individuals will possess traits that enable them to survive and reproduce more successfully (producing more offspring) than others in the same environment. The third postulate holds that if the advantageous traits are inherited by offspring, then these traits will become more common in succeeding generations. Thus, traits that confer advantages in survival and reproduction are retained in the population, and traits that are disadvantageous disappear. When Darwin coined the term **natural selection** for this process, he was making a deliberate analogy to the artificial selection practiced by animal and plant breeders of his day. A much more apt term would be "evolution by variation and selective retention."

An Example of Adaptation by Natural Selection

Contemporary observations of Darwin's finches provide a particularly good example of how natural selection produces adaptations.

In his autobiography, first published in 1887, Darwin claimed that the curious pattern of adaptations he observed among the several species of finches that live on the Galápagos Islands off the coast of Ecuador—now referred to as "Darwin's finches"—was crucial in the development of his ideas about evolution (Figure 1.5). Recently discovered documents suggest that Darwin was actually quite confused about the Galápa-

(a)

(b)

FIGURE 1.5   (a) The islands of the Galápagos, which are located off the coast of Ecuador, house a variety of unique species of plants and animals. (b) Cactus finches from Charles Darwin's *The Zoology of the Voyage of H.M.S. Beagle.*

gos finches during his visit, and they played little role in his discovery of natural selection. Nonetheless, Darwin's finches hold a special place in the minds of most biologists.

Peter and Rosemary Grant, biologists at Princeton University, conducted a landmark study of the ecology and evolution of one particular species of Darwin's finches on one of the Galápagos Islands. The study is remarkable because the Grants were able to directly document how Darwin's three postulates lead to evolutionary change. The island, Daphne Major, is home to the medium ground finch (*Geospiza fortis*), a small bird that subsists mainly by eating seeds (Figure 1.6). The Grants and their colleagues caught, measured, weighed, and banded nearly every finch on the island each year of their study—some 1500 birds in all. They also kept track of critical features of the birds' environment, such as the distribution of seeds of various sizes, and they observed the birds' behavior.

FIGURE 1.6 The medium ground finch, *Geospiza fortis*, uses its beak to crack open seeds. (Photograph courtesy of Peter Grant.)

A few years into the Grants' study, a severe drought struck Daphne Major (Figure 1.7). During the drought, plants produced many fewer seeds, and the finches soon depleted the stock of small, soft, easily processed seeds, leaving only large, hard, difficult-to-process seeds (Figure 1.8). The bands on the birds' legs enabled the Grants to track the fates of individual birds during the drought, and the regular measurements that they had made of the birds allowed them to compare the traits of birds that survived the drought with the traits of those that perished. The Grants also kept detailed records of the environmental conditions, which allowed them to determine how the drought affected the finch's habitat. It was this vast body of data that enabled the Grants to document the action of natural selection among the finches of Daphne Major.

The Grants' data show how the processes identified in Darwin's postulates lead to adaptation.

The events on Daphne Major embodied all three of Darwin's postulates. First, the supply of food on the island was not sufficient to feed the entire population, and many finches did not survive the drought. From the beginning of the drought in 1976 until

(a)

(b)

FIGURE 1.7 Daphne Major in March 1976 (a), after a year of good rains; and in March 1977 (b), after a year of very little rain.

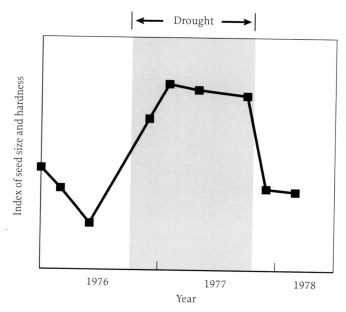

FIGURE 1.8 During the two-year drought, the size and hardness of seeds available on Daphne Major increased because birds consumed all of the desirable small, soft seeds, leaving mainly larger and harder seeds. Each point on this plot represents an index of seed size and hardness at a given time.

the rains came nearly two years later, the population of medium ground finches on Daphne Major declined from 1200 birds to only 180.

Second, beak depth (the top-to-bottom dimension of the beak) varied among the birds on the island, and this variation affected the birds' survival. Before the drought began, the Grants and their colleagues had observed that birds with deeper beaks were able to process large, hard seeds more easily than birds with shallower beaks were. Deep-beaked birds usually concentrated on large seeds, while shallow-beaked birds normally focused their efforts on small seeds. The open bars in the histogram in Figure 1.9a show what the distribution of beak sizes in the population was like before the drought. The height of each open bar represents the number of birds with beaks in a given range of depths—for example, 8.8 to 9.0 mm, or 9.0 to 9.2 mm. During the drought, the relative abundance of small seeds decreased, forcing shallow-beaked birds to shift to larger and harder seeds. Shallow-beaked birds were then at a distinct disadvantage because it was harder for them to crack these seeds. The distribution of individuals within the population changed during the drought because finches with deeper beaks were more likely to survive than finches with shallow beaks (Figure 1.9b). The shaded portion of the histogram in Figure 1.9a shows what the distribution of beak depths would have been like among the survivors. Because many birds died, there were fewer remaining in each category. However, mortality was quite specific. The proportion of shallow-

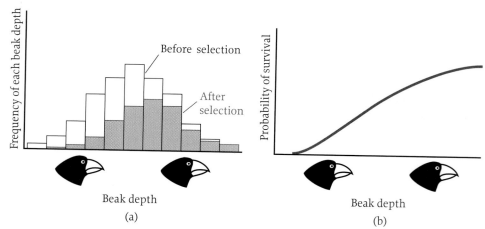

FIGURE 1.9 A schematic diagram of how directional selection increased mean beak depth among medium ground finches on Daphne Major. (a) The heights of each bar represent the numbers of birds whose beak depths fall within each of the intervals plotted on the x axis, with beak depth increasing to the right. The histogram with open bars shows the distribution of beak depths before the drought began. The histogram with shaded bars shows the distribution of beak depths after a year of drought. Notice that the number of birds in each category has decreased. Because birds with deep beaks were less likely to die than birds with shallow beaks, the peak of the distribution shifted to the right, indicating that the mean beak depth had increased. (b) The probability of survival for birds of different beak depths is plotted. Birds with shallow beaks are less likely to survive than are birds with deep beaks.

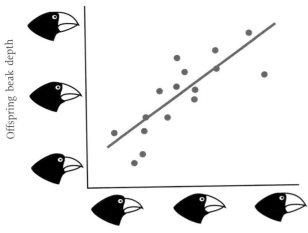

FIGURE 1.10 Parents with deeper-than-average beaks tend to have offspring with deeper-than-average beaks. Each point represents one offspring. Offspring beak depth is plotted on the vertical axis (deeper beaks farther up the axis), and the average of the two parents' beak depths is plotted on the horizontal axis (deeper beaks farther to the right).

beaked birds that died greatly exceeded the proportion of deep-beaked birds that died. As a result, the shaded portion of the histogram shows a shift to the right, which means that the average beak depth in the population increased. Thus the average beak depth among the survivors of the drought was greater than the average beak depth in the same population before the drought.

Third, parents and offspring had similar beak depths. The Grants discovered this by capturing and banding nestlings and recording the identity of the nestlings' parents. When the nestlings became adults, the Grants recaptured and measured them. The Grants found that, on average, parents with deep beaks produced offspring with deep beaks (Figure 1.10). Because parents were drawn from the pool of individuals who survived the drought, their beaks were, on average, deeper than those of the original residents of the island; and because offspring resemble their parents, the average beak depth of the survivors' offspring was greater than the average beak depth before the drought. This means that, through natural selection, the average **morphology** (an organism's size, shape, and composition) of the bird population changed so that birds became better adapted to their environment. This process, operating over approximately two years, led to a 4% increase in the mean beak depth in this population (Figure 1.11).

Selection preserves the status quo when the most common type is the best adapted.

So far, we have seen how natural selection led to adaptation as the population of finches on Daphne Major evolved in response to changes in their environment. Will this process continue forever? If it did, eventually all the finches would have deep enough beaks to efficiently process the largest seeds available. However, large beaks have disadvantages as well as benefits. The Grants showed, for instance, that birds with large beaks are less likely to survive the juvenile period than are birds with small beaks, probably because large-beaked birds require more food

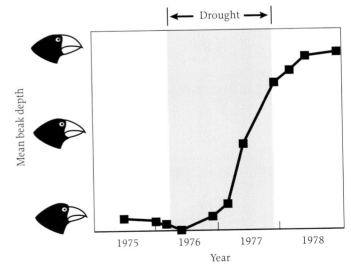

FIGURE 1.11 The average beak depth in the population of medium ground finches on Daphne Major increased during the drought of 1975–1978. Each point plots an index of average beak depth of the population in a particular year. Deeper beaks are plotted higher on the y axis.

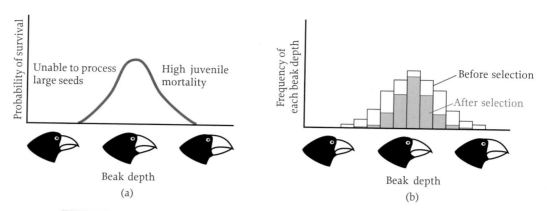

FIGURE 1.12 When birds with the most common beak depth are most likely to survive and reproduce, natural selection keeps the mean beak depth constant. (a) Birds with deep or shallow beaks are less likely to survive than birds with average beaks. Birds with shallow beaks cannot process large, hard seeds; and birds with deep beaks are less likely to survive to adulthood. (b) The open bars represent the distribution of beak depths before selection, and the shaded bars represent the distribution after selection. As in Figure 1.9, notice that there are fewer birds in the population after selection. Because birds with average beaks are most likely to survive, however, the peak of the distribution of beak depths is not shifted and mean beak depth remains unchanged.

(Figure 1.12). Evolutionary theory predicts that, over time, selection will increase the average beak depth in the population until the costs of larger-than-average beak size exceed the benefits. At this point, finches with the average beak size in the population will be the most likely to survive and reproduce, and finches with deeper or shallower beaks than the new average will be at a disadvantage. When this is true, beak size does not change, and we say that an **equilibrium** exists in the population with regard to beak size. The process that produces this equilibrium state is called **stabilizing selection**. Notice that even though the average characteristics of the beak in the population will not change in this situation, selection is still going on. Selection is required to change a population, and selection is also required to keep a population the same.

It might seem that beak depth would also remain unchanged if this trait had no effect on survival (or put another way, if there were no selection favoring one type of beak over another). Then all types of birds would be equally likely to survive from one generation to the next, and beak depth would remain constant. This logic would be valid if selection were the only process affecting beak size. However, real populations are also affected by other processes that cause **traits**, or **characters**, to change in unpredictable ways. We will discuss these processes further in Chapter 3. The point to remember here is that populations do not remain static over the long run unless selection is operating.

Evolution need not always lead to change in the same direction.

Natural selection has no foresight; it simply causes organisms to change so that they are better adapted to their current environment. Often environments fluctuate over time; and when they do, selection may track these fluctuations. We see this kind of pattern in the finches on the Galápagos over the last 25 years. During this time there have been dry periods like 1976–1978, but there have also been wet periods like

1983–1985, when small, soft, easy-to-process seeds were exceedingly abundant. During wet years, selection favors smaller beaks, reversing the changes in beak size and shape wrought by natural selection during the drought years. As Figure 1.13 shows, beak size has wobbled up and down during the Grants' long study of the medium ground finch on Daphne Island.

Species are populations of varied individuals that may or may not change through time.

As the Grants' work on Daphne Major makes clear, a species is not a fixed type or entity. Species change in their general characteristics from generation to generation according to the postulates Darwin described. Before Darwin, however, people thought of species as fixed, unchanging categories, much the same way that we think of geometrical figures: A finch could no more change its properties than a triangle could. If a triangle acquired another side, it would not be a modified triangle, but rather a rectangle. In much the same way, to biologists before Darwin, a changed finch was not a finch at all. The late Ernst Mayr, a distinguished evolutionary biologist, called this pre-Darwinian view of immutable species "essentialism." According to Darwin's theory, a **species** is a dynamic *population* of individuals. The characteristics of a particular species will be static over a long period of time only if the most common type of individual is consistently favored by stabilizing selection. Both stasis (staying the same) and change result from natural selection, and both require explanation in terms of natural selection. Stasis is not the natural state of species.

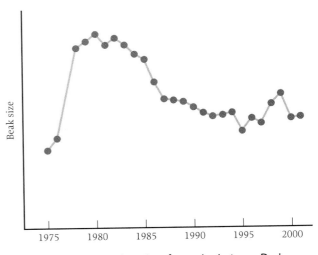

FIGURE 1.13 An index of mean beak size on Daphne Major for the period 1975–2001.

Individual Selection

Adaptation results from the competition among individuals, not between entire populations or species.

It is important to notice that selection produces adaptations that benefit *individuals*. Such adaptation may or may not benefit the population or species. In the case of simple morphological characters such as beak depth, selection probably does allow the population of finches to compete more effectively with other populations of seed predators. However, this need not be the case. Selection often leads to changes in behavior or morphology that increase the reproductive success of individuals but decrease the average reproductive success of the group, population, and species.

The fact that almost all organisms produce many more offspring than are necessary to maintain the species provides an example of the conflict between individual and group interests. A female monkey may, on average, produce 10 offspring during her lifetime (Figure 1.14). In a stable population, perhaps only two of these offspring will survive and reproduce. From the point of view of the species, the other eight are a waste of resources. They compete with other members of their species for food, water, and sleeping sites. The demands of a growing population can lead to serious overexploitation of the environment, and the species as a whole might be more likely to survive if

FIGURE 1.14 A female blue monkey holds her infant. (Photograph courtesy of Marina Cords.)

all females produced fewer offspring. This does not happen, however, because natural selection among individuals favors females who produce many offspring.

To see why selection on individuals will lead to this result, let's consider a simple, hypothetical case. Suppose the females of a particular species of monkey are maximizing individual reproductive success when they produce 10 offspring. Females that produce more than or less than 10 offspring will tend to leave fewer descendants in the next generation. Further suppose that the likelihood of the species becoming extinct would be lowest if females produced only two offspring apiece. Now suppose that there are two kinds of females. Most of the population is composed of low-fecundity females that produce just two offspring each, but a few high-fecundity females produce 10 offspring each. (**Fecundity** is the term demographers use for the ability to produce offspring.) High-fecundity females have high-fecundity daughters, and low-fecundity females have low-fecundity daughters. The proportion of high-fecundity females will increase in the next generation because such females produce more offspring than do low-fecundity females. Over time, the proportion of high-fecundity females in the population will increase rapidly. As fecundity increases, the population will grow rapidly and may deplete available resources. The depletion of resources, in turn, will increase the chance that the species becomes extinct. However, this fact is irrelevant to the evolution of fecundity before the extinction, because natural selection results from competition among individuals, not competition among species.

The idea that natural selection operates at the level of the individual is a key element in understanding adaptation. In discussing the evolution of social behavior in Chapter 8, we will encounter several additional examples of situations in which selection increases individual success but decreases the competitive ability of the population.

THE EVOLUTION OF COMPLEX ADAPTATIONS

The example of the evolution of beak depth in the medium ground finch illustrates how natural selection can cause adaptive change to occur rapidly in a population. Deeper beaks enabled the birds to survive better, and deeper beaks soon came to predominate in the population. Beak depth is a fairly simple character, lacking the intricate complexity of an eye. As we will see, however, the accumulation of small variations by natural selection can also give rise to complex adaptations.

Why Small Variations Are Important

There are two categories of variation: continuous and discontinuous.

It was known in Darwin's day that most variation is continuous. An example of **continuous variation** is the distribution of heights in people. Humans grade smoothly from one extreme to the other (short to tall), with all the intermediate types (in this case, heights) represented. However, Darwin's contemporaries also knew about **discontinuous variation**, in which a number of distinct types exist with no intermediates. In humans, height is also subject to discontinuous variation. For example, there is a genetic condition, called **achondroplasia**, that causes affected individuals to be much shorter than other people, have proportionately shorter arms and legs, and bear a variety of other distinctive features. Discontinuous variants are usually quite rare in nature. Nonetheless, many of Darwin's contemporaries who were convinced of the reality of evolution believed that new species arise as discontinuous variants.

Discontinuous variation is not important for the evolution of complex adaptations, because complex adaptations are extremely unlikely to arise in a single jump.

Unlike most of his contemporaries, Darwin thought that discontinuous variation did not play an important role in evolution. A hypothetical example, described by Oxford University biologist Richard Dawkins in his book *The Blind Watchmaker*, illustrates Darwin's reasoning. Dawkins recalls an old story in which an imaginary collection of monkeys sit at typewriters happily typing away. Lacking the ability to read or write, they strike keys at random. Given enough time, the story goes, the monkeys will reproduce all the great works of Shakespeare. Dawkins points out that this is not likely to happen in the lifetime of the universe, let alone the lifetime of one of the monkey typists. To illustrate why it would take so long, Dawkins presents these illiterate monkeys with a much simpler problem: reproducing a single line from *Hamlet*, "Methinks it is like a weasel." To make the problem even simpler for the monkeys, Dawkins ignores the difference between uppercase and lowercase letters and omits all punctuation except spaces. There are 28 characters (including spaces) in the phrase. Because there are 26 characters in the alphabet and Dawkins is keeping track of spaces, each time a monkey types a character, there is only a 1-in-27 chance that it will type the right character. There is also only a 1-in-27 chance that the second character will be correct. Again, there is a 1-in-27 chance that the third character will be right, and so on up to the twenty-eighth character. Thus the chance that a monkey will type the correct sequence at random is $\frac{1}{27}$ multiplied by itself 28 times, or

$$\underbrace{\frac{1}{27} \times \frac{1}{27} \times \frac{1}{27} \times \cdots \times \frac{1}{27}}_{28 \text{ times}} \approx 10^{-40}$$

This is a *very* small number. To get a feeling for how small a chance there is of the monkeys typing the sentence correctly, suppose a very fast computer (faster than any currently in existence) could generate 100 billion (10^{11}) characters per second and run for the lifetime of the Earth, about 4 billion years, or 10^{17} seconds. Then the chance of randomly typing the line "Methinks it is like a weasel" even once during the whole of Earth's history would be about one in a trillion! Typing the whole play is obviously astronomically less likely, and although *Hamlet* is a very complicated thing, it is much less complicated than a human eye. There's no chance that a structure like the human eye would arise by chance in a single trial. If it did, it would be, as the astrophysicist Sir Fred Hoyle said, "like a hurricane blowing through a junk yard and chancing to assemble a Boeing 747."

Complex adaptations can arise through the accumulation of small random variations by natural selection.

Darwin argued that continuous variation is essential for the evolution of complex adaptations. Once again, Richard Dawkins provides an example that makes clear

Darwin's reasoning. Again imagine a room full of monkeys and typewriters, but now the rules of the game are different. The monkeys type the first 28 characters at random, and then during the next round they attempt to copy the same initial string of letters and spaces. Most of the sentences are just copies of the previous string, but because monkeys sometimes make mistakes, some strings have small variations, usually in only a single letter. During each trial, the monkey trainer selects the string that most resembles Shakespeare's phrase "Methinks it is like a weasel" as the string to be copied by all the monkeys in the next trial. This process is repeated until the monkeys come up with the correct string. Calculating the exact number of trials required to generate the correct sequence of characters is quite difficult, but it is easy to simulate the process on a computer. Here's what happened when Dawkins performed the simulation. The initial random string was

WDLMNLT DTJBKWIRZREZLMQCO P

After one trial Dawkins got

WDLMNLT DTJBSWIRZREZLMQCO P

After 10 trials:

MDLDMNLS ITJISWHRZREZ MECS P

After 20 trials:

MELDINLS IT ISWPRKE Z WECSEL

After 30 trials:

METHINGS IT ISWLIKE B WECSEL

After 40 trials:

METHINKS IT IS LIKE I WEASEL

The exact phrase was reached after 43 trials. Dawkins reports that it took his 1985-vintage Macintosh only 11 seconds to complete this task.

Selection can give rise to great complexity starting with small random variations because it is a *cumulative* process. As the typing monkeys show us, it is spectacularly unlikely that a single random combination of keystrokes will produce the correct sentence. However, there is a much greater chance that some of the many *small* random changes will be advantageous. The combination of reproduction and selection allows the typing monkeys to accumulate these small changes until the desired sentence is reached.

Why Intermediate Steps Are Favored by Selection

The evolution of complex adaptations requires that all of the intermediate steps be favored by selection.

There is a potent objection to the example of the typing monkeys. Natural selection, acting over time, can lead to complex adaptations, but it can do so only if each small change along the way is itself adaptive. Although it is easy to assume that this is true in a hypothetical example of character strings, many people have argued that it is unlikely for every one of the changes necessary to assemble a complex organ like the eye to be adaptive. An eye is useful, it is claimed, only after all parts of the complexity have been assembled; until then, it is worse than no eye at all. After all, what good is 5% of an eye?

Darwin's answer, based on the many adaptations for seeing or sensing light that exist in the natural world, was that 5% of an eye *is* often better than no eye at all. It is quite possible to imagine that a very large number of small changes—each favored by selection—led cumulatively to the wonderful complexity of the eye. Living mollusks, which display a broad range of light-sensitive organs, provide examples of many of the likely stages in this process (Figure 1.15):

1. Many invertebrates have a simple light-sensitive spot (Figure 1.15a). Photoreceptors of this kind have evolved many times from ordinary epidermal (surface) cells —usually ciliated cells whose biochemical machinery is light-sensitive. Those individuals whose cells are more sensitive to light are favored when information about changes in light intensity is useful. For example, a drop in light intensity may often be an indicator that a predator is in the vicinity.

2. By having the light-sensitive cells in a depression (see Figure 1.15a), the organism will get some additional information about the direction of the change in light intensity. The surface of organisms is variable, and those individuals whose photoreceptors are in depressions will be favored by selection in environments where such information is useful. For example, mobile organisms may need better information about what is happening in front of them than immobile ones need.

3. Through a series of small steps, the depression could deepen (Figure 1.15b), and each step could be favored by selection because better directional information would be available.

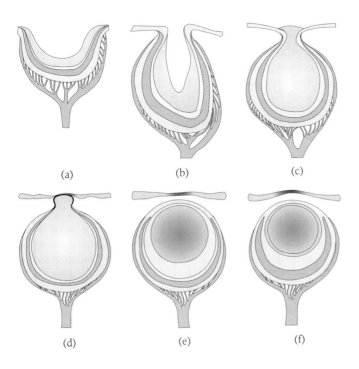

FIGURE 1.15 Living gastropod mollusks illustrate all of the intermediate steps between a simple eye cup and a camera-type eye: (a) The eye pit of a limpet, *Patella* sp.; (b) the eye cup of Beyrich's slit shell, *Pleurotomaria beyrichi*; (c) the pinhole eye of a California abalone, *Haliotis* sp.; (d) the closed eye of a turban shell, *Turbo creniferus*; (e) the lens eye of the spiny dye murex, *Murex brandaris*; (f) the lens eye of the Atlantic dog whelk, *Nucella lapillus*. (Lens is shaded in e and f.)

(a)   (b)   (c)

(d)   (e)   (f)

4. If the depression got deep enough (Figure 1.15c), it could form images on the light-sensitive tissue, much the way pinhole cameras form images on photographic film. In settings in which detailed images are useful, selection could then favor the elaboration of the neural machinery necessary to interpret the image.

5. The next step is the formation of a transparent cover (Figure 1.15d). This might be favored because it protects the interior of the eye from parasites and mechanical damage.

6. A lens could evolve through gradual modification of either the transparent cover or the internal structures within the eye (Figure 1.15e, f).

Notice that evolution produces adaptations like a tinkerer, not an engineer. New organisms are created by small modifications of existing organisms, not by starting with a clean slate. Clearly many beneficial adaptations will not arise because they are blocked at some step along the way when a particular variation is not favored by selection. Darwin's theory explains how complex adaptations can arise through natural processes, but it does not predict that every possible adaptation, or even most, will occur. This is not the best of all possible worlds; it is just one of many possible worlds.

Sometimes unrelated species have independently evolved the same complex adaptation, suggesting that the evolution of complex adaptations by natural selection is not a matter of mere chance.

The fact that natural selection constructs complex adaptations like a tinkerer might lead you to think that the assembly of complex adaptations is a chancy business. If even a single step were not favored by selection, the adaptation could not arise. Such reasoning suggests that complex adaptations are mere coincidence. Although chance does play a very important role in evolution, the power of cumulative natural selection should not be underestimated. The best evidence that selection is a powerful process for generating complex adaptations comes from a phenomenon called **convergence**, the evolution of similar adaptations in unrelated groups of animals.

The similarity between the marsupial faunas of Australia and South America and the placental faunas of the rest of the world provides a good example of convergence. In most of the world, the mammalian fauna is dominated by **placental mammals**, which nourish their young in the uterus during long pregnancies. Both Australia and South America, however, became separated from an ancestral supercontinent, known as Pangaea, long before placental mammals evolved. In Australia and South America, **marsupials** (nonplacental mammals, like kangaroos, that rear their young in external pouches) came to dominate the mammalian fauna, filling all available mammalian niches. Some of these marsupial mammals were quite similar to the placental mammals on the other continents. For example, there was a marsupial wolf in Australia that looked very much like placental wolves of Eurasia, even sharing subtle features of their feet and teeth (Figure 1.16). These marsupial wolves became extinct in the 1930s. Similarly, in South America a marsupial saber-toothed cat independently evolved many of the same adaptations as the placental saber-toothed cat that stalked North America 10,000 years ago. These similarities are more impressive when you consider that the last common ancestor of marsupial and placental mammals was a small, nocturnal, insectivorous creature, something like a shrew, that lived about 120 million years ago (mya). Thus, selection transformed a shrew step by small step, each step favored by selection, into a saber-toothed cat—and it did it twice. This cannot be coincidence.

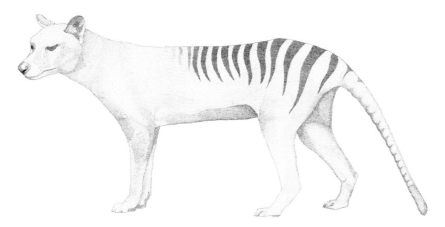

FIGURE 1.16 The marsupial wolf that lived in Tasmania until early in the twentieth century (drawn from a photograph of one of the last living animals). Similarities with placental wolves of North America and Eurasia illustrate the power of natural selection to create complex adaptations. Their last common ancestor was probably a small insectivorous creature like the shrew.

The evolution of eyes provides another good example of convergence. Remember that the spherical gradient lens is a good lens design for aquatic organisms because it has good light-gathering ability and provides a sharp image over the full 180° visual field. Complex eyes with lenses have evolved independently eight different times in distantly related aquatic organisms: once in fish, once in cephalopod mollusks like squid, several times among gastropod mollusks like the Atlantic dog whelk, once in annelid worms, and once in crustaceans (Figure 1.17). These are very diverse creatures whose last common ancestor was a simple creature that did not have a complex eye. Nonetheless, in every case they have evolved very similar spherical gradient lenses. Moreover, no other lens design is found in aquatic animals. Despite the seeming chanciness of assembling complex adaptations, natural selection has achieved the optimal design in every case.

(a)

(b)

FIGURE 1.17 Complex eyes with lenses have evolved independently in a number of different kinds of aquatic animals, including this slingjaw wrasse (a) and squid (b).

RATES OF EVOLUTIONARY CHANGE

Natural selection can cause evolutionary change that is much more rapid than we commonly observe in the fossil record.

In Darwin's day, the idea that natural selection could change a chimpanzee into a human, much less that it might do so in just a few million years, was unthinkable. Though people are generally more accepting of evolution today, many still think of evolution by natural selection as a glacially slow process that requires millions of years to accomplish noticeable change. Such people often doubt that there has been enough time for selection to accomplish the evolutionary changes observed in the fossil record. And yet, as we will see in subsequent chapters, most scientists now believe that humans evolved from an apelike creature in only 5 million to 10 million years. In fact, some of the rates of selective change observed in contemporary populations are far faster than necessary for such a transition. The puzzle is not whether there has been enough

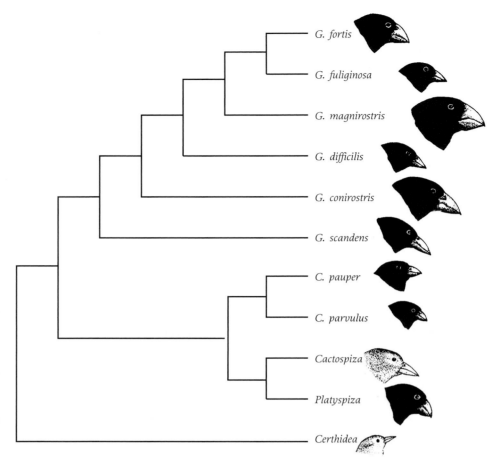

FIGURE 1.18 We can trace the relationship among various species of Darwin's finches by analyzing their protein polymorphisms. Species that are closely linked in the phylogenetic tree are more similar to one another genetically than to other species because they share a more recent common ancestor. The tree does not include 3 of the 14 species of Darwin's finches.

time for natural selection to produce the adaptations that we observe. The real puzzle is why the change observed in the fossil record was so slow.

The Grants' observation of the evolution of beak morphology in Darwin's finches provides one example of rapid evolutionary change. The medium ground finch of Daphne Major is one of 14 species of finches that live in the Galápagos. Evidence suggests that all 14 are descended from a single South American species that migrated to the newly emerged islands about half a million years ago (Figure 1.18). This doesn't seem like a very long time. Is it possible that natural selection created 14 species in only half a million years?

To start to answer the question, let's calculate how long it would take for the medium ground finch to come to resemble its closest relative, the large ground finch (*Geospiza magnirostris*), in beak size and weight (Figure 1.19). The large ground finch is 75% heavier than the medium ground finch, and its beak is about 20% deeper. Remember that beak size increased about 4% in two years during the 1977 drought. The Grants' data indicate that body size also increased by a similar amount. At this rate, Peter Grant calculated that it would take between 30 and 46 years for selection to increase the beak size and body weight of the medium ground finch to match those of the large ground finch. But these changes occurred in response to an extraordinary environmental crisis. The data suggest that selection doesn't generally push consistently in one direction. Instead, in the Galápagos, evolutionary change seems to go in fits and starts, moving traits one way and then another. So let's suppose that a net change in beak size like the one that occurred during 1977 occurs only once every century. Then it would take about 2000 years to transform the medium ground finch into the large ground finch—still a very rapid process.

Similar rates of evolutionary change have been observed elsewhere when species invade new habitats. For example, about 100,000 years ago a population of elk (called "red deer" in Great Britain) colonized and then became isolated on the island of Jersey, off the French coast, presumably by rising sea levels. By the time the island was reconnected with the mainland approximately 6000 years later, the red deer had shrunk to about the size of a large dog. University of Michigan paleontologist Philip Gingerich compiled data on the rate of evolutionary change in 104 cases in which species invaded new habitats. These rates ranged from a low of zero (that is, no change) to a high of about 22% per year, with an average of about 0.1% per year.

The changes the Grants observed in the medium ground finch are relatively simple: birds and their beaks just got bigger. More complex changes usually take longer to evolve, but several kinds of evidence suggest that selection can produce big changes in remarkably short periods of time.

One line of evidence comes from artificial selection. Humans have performed selection on domesticated plants and animals for thousands of years, and for most of this period this selection was not deliberate. There are many familiar examples. All domesticated dogs, for instance, are believed to be descendants of wolves. Scientists are not sure when dogs were domesticated, but 15,000 years ago is a good guess, which means that in a few thousand generations, selection changed wolves into Pekingese, beagles, greyhounds, and Saint Bernards. In reality, most of these breeds were created fairly recently, and most were the products of directed breeding. Darwin's favorite example of artificial selection was the domestication of pigeons. In the nineteenth century, pigeon breeding was a popular hobby, especially among working people who competed to produce showy birds (Figure 1.20). They created a menagerie of wildly different forms, all descended from the rather plain-looking rock pigeon. Darwin pointed out that these breeds are so different that if they were discovered in nature, biologists would surely classify them as members of different species. Yet they were produced by artificial selection within a few hundred years.

Rapid evolution of a complex feature has also been documented in a recent study of a group of very closely related species of fish from the genus *Poeciliopsis* (Figure 1.21). These small minnows can be found in tropical lowland streams, high-altitude lakes, and desert springs and streams in Mexico and Central America. All the species in this genus bear live young, but the sequence of events between fertilization and birth varies. In most species, females endow the eggs with nutrients before fertilization. As the young develop, they consume this endowment. As a consequence, the emerging offspring are smaller than the egg they came from. In a few species, however, females continue to provide nutrients to their unborn offspring throughout development using highly differentiated tissues that are analogous to mammalian placentas. When they are born, these offspring can be more than 100 times the mass of the egg at fertilization. Biologist David Reznick of the University of California Riverside and his colleagues have shown that these placental tissues evolved independently in three different groups of species within the genus *Poeciliopsis*. Genetic data indicate that one of these species groups diverged from ancestors lacking placental tissue only 0.75 mya, and the other two diverged less than 2.4 millions of years ago (mya). These time estimates actually represent the time since these species shared a common ancestor, and they set an upper bound on the amount of time required for the placenta to evolve. The generation time for these fish ranges

FIGURE 1.19 The large ground finch (*Geospiza magnirostris*) has a beak that is nearly 20% deeper than the beak of its close relative, the medium ground finch (*G. fortis*). At the rate of evolution observed during the drought of 1977, Peter Grant calculated that selection could transform the medium ground finch into the large ground finch in less than 46 years. (Photograph courtesy of Peter and Rosemary Grant.)

(a)

(b)

(c)

(d)

FIGURE 1.20 In Darwin's day, pigeon fanciers created many new breeds of pigeons—including (a) pouters, (b) fantails, and (c) carriers—from (d) the common rock pigeon.

from six months to a year, so this complex adaptation evolved in fewer than a million generations.

A third line of evidence comes from theoretical studies of the evolution of complex characters. Dan-Eric Nilsson and Susanne Pleger of Lund University, in Sweden, have built a mathematical model of the evolution of the eye in an aquatic organism. They start with a population of organisms, each with a simple eyespot, a flat patch of light-sensitive tissue sandwiched between a transparent protective layer and a layer of dark pigment. They then consider the effect of every possible small (1%) deformation of the shape of the eyespot on the resolving power of the eye. They determine which 1% change has the greatest positive effect on the eye's resolving power and then repeat the process again and again, deforming the new structure by 1% in every possible way at each step. The

FIGURE 1.21 Fish in the genus *Poeciliopsis* include small minnows like *P. occidentalis*, shown here. (Photograph courtesy of David Reznik.)

results are shown in Figure 1.22. After 538 changes of 1% each, a simple concave eye cup evolves; after 1033 changes of 1%, crude pinhole eyes emerge; after 1225 changes of 1%, an eye with an elliptical lens is created; and after 1829 steps the process finally comes to a halt because no small changes increase resolving power. The end result is an eye with a spherical gradient lens just like those in fish and other aquatic organisms. As Nilsson and Pleger point out, 1829 changes of 1% add up to a substantial amount of change. For instance, 1829 changes of 1% would lengthen a 10-cm (4-in.) human finger to a length of 8000 km (5500 miles)—about the distance from Los Angeles to New York and back. Nonetheless, making very conservative assumptions about the strength of selection, Nilsson and Pleger calculate that this would take only about 364,000 generations. For organisms with short generations, the complete structure of the eye can evolve from a simple eyespot in less than a million years, a brief moment in evolutionary time.

By comparison, most changes observed in the fossil record are much slower. Human brain size has roughly doubled in the last 2 million years—a rate of change of 0.00005% per year. This is 10,000 times slower than the rate of change that the Grants observed in the Galápagos. Moreover, such slow rates of change typify what can be observed from the fossil record. As we will see, however, the fossil record is incomplete. It is quite likely that some evolutionary changes in the past were rapid, but the sparseness of the fossil record prevents us from detecting them.

DARWIN'S DIFFICULTIES EXPLAINING VARIATION

Darwin's *On the Origin of Species*, published in 1859, was a best-seller during his day, but his proposal that new species and other major evolutionary changes arise by the accumulation of small variations through natural selection was not widely embraced. Most educated people accepted the idea that new species arise through the transformation of existing species, and many scientists accepted the idea that natural selection is the most important cause of organic change (although by the turn of the twentieth century even this consensus had broken down, particularly in the United States). But only a minority endorsed Darwin's view that major changes occur through the accumulation of small variations.

Darwin couldn't convince his contemporaries that evolution occurred through the accumulation of small variations because he couldn't explain how variation is maintained.

Darwin's critics raised a telling objection to his theory: the actions of blending inheritance (described in the next paragraph) and selection would both inevitably deplete variation in populations and make it impossible for natural selection to continue. These were potent objections that Darwin was unable to resolve in his lifetime because he and his contemporaries did not yet understand the mechanics of inheritance.

Everyone could readily observe that many of the characteristics of offspring are an average of the characteristics of their parents. Most people, including Darwin, believed this phenomenon to be caused by the action of **blending inheritance,** a model of inheritance that assumes the mother and father each contribute a hereditary substance that mixes, or "blends," to determine the characteristics of the offspring. Shortly after publication of *On the Origin of Species*, a Scottish engineer named Fleeming

Stage

1 176 steps (1%)

2 362 steps (1%)

3 270 steps (1%)

4 225 steps (1%)

5 192 steps (1%)

6 308 steps (1%)

7 296 steps (1%)

8

FIGURE 1.22 A computer simulation of the evolution of the eye generates this sequence of forms. In the first step, the eye is a flat patch of light-sensitive cells that lies between a transparent protective layer and a layer of dark pigment. Eventually the eyespot deepens, the lens curves outward, and the radius lengthens. This process involves about 1800 changes of 1% each.

Jenkin published a paper in which he clearly showed that, with blending inheritance, there could be little or no variation available for selection to act on. The following example shows why Jenkin's argument was so compelling. Suppose a population of one species of Darwin's finches displays two forms: tall and short. Further suppose that a biologist controls mating so that every mating is between a tall individual and a short individual. Then, with blending inheritance, all of the offspring will be the same intermediate height, and their offspring will be the same height as they are. All of the variation for height in the population will disappear in a single generation. With random mating, the same thing will occur, though it will take longer. If inheritance were purely a matter of blending parental traits, then Jenkin would have been right about its effect on variation. However, as we will see in Chapter 3, genetics accounts for the fact that offspring are intermediate between their parents and does not assume any kind of blending.

Another problem arose because selection works by removing variants from populations. For example, if finches with small beaks are more likely to die than finches with larger beaks over the course of many generations, eventually all that will be left are birds with large beaks. There will be no variation for beak size, and Darwin's second postulate holds that without variation there can be no evolution by natural selection. For example, suppose the environment changes so that individuals with small beaks are less likely to die than those with large beaks are. The average beak size in the population will not decrease, because there are no small-beaked individuals. Natural selection destroys the variation required to create adaptations.

Even worse, as Jenkin also pointed out, there was no explanation of how a population might evolve beyond its original range of variation. The cumulative evolution of complex adaptations requires that populations move far outside their original range of variation. Selection can cull away some characters from a population, but how can it lead to new types not present in the original population? This apparent contradiction was a serious impediment to explaining the logic of evolution. How could elephants, moles, bats, and whales all descend from an ancient shrewlike insectivore unless there were a mechanism for creating new variants not present at the beginning? For that matter, how could all the different breeds of dogs have descended from their one common ancestor, the wolf (Figure 1.23)?

Remember that Darwin and his contemporaries knew there were two kinds of variation: continuous and discontinuous. Because Darwin believed that complex adaptations could arise only through the accumulation of small variations, he thought discontinuous variants were unimportant. However, many biologists thought that the discontinuous variants, called "sports" by nineteenth-century animal breeders, were the key to evolution because they solved the problem of the blending effect. The following hypothetical example illustrates why. Suppose that a population of green birds has entered a new environment in which red birds are better adapted. Some of Darwin's critics believed that any new variant that was slightly more red would have only a small advantage and would be rapidly swamped by blending. In contrast, an all-red bird would have a large enough selective advantage to overcome the effects of blending, and could increase its frequency in the population.

Darwin's letters show that these criticisms worried him greatly, and although he tried a variety of counterarguments, he never found one that was satisfactory. The solution to these problems required an understanding of genetics, which was not available for another half century. As we will see, it was not until well into the twentieth century that geneticists came to understand how variation is maintained and Darwin's theory of evolution was generally accepted.

(a)

(b)

FIGURE 1.23 (a) The wolf is the ancestor of all domestic dogs, including (b) the poodle and the Saint Bernard. These transformations were accomplished in several thousand generations of artificial selection.

FURTHER READING

Browne, J. 1995. *Charles Darwin: A Biography,* Vol. 1: *Voyaging.* New York: Knopf.
Dawkins, R. 1996. *The Blind Watchmaker.* Norton, New York.
Dennett, D. 1995. *Darwin's Dangerous Idea.* Touchstone, New York.
Ridley, M. 1996. *Evolution* (2nd ed.). Blackwell Scientific, Oxford, England.
Weiner, J. 1994. *The Beak of the Finch.* Knopf, New York.

STUDY QUESTIONS

1. It is sometimes observed that offspring do not resemble their parents for a particular character, even though the character varies in the population. Suppose this were the case for beak depth in the medium ground finch.
   (a) What would the plot of offspring beak depth against parental beak depth look like?
   (b) Plot the mean depth in the population among (i) adults before a drought, (ii) the adults that survived a year of drought, and (iii) the offspring of the survivors.
2. Many species of animals engage in cannibalism. This practice certainly reduces the ability of the species to survive. Is it possible that cannibalism could arise by natural selection? If so, with what adaptive advantage?
3. Some insects mimic dung. Ever since Darwin, biologists have explained this behavior as a form of camouflage: selection favors individuals who most resemble dung because they are less likely to be eaten. The late Harvard paleontologist Stephen Jay Gould objected to this explanation. He argued that although selection could perfect such mimicry once it evolved, it could not cause the resemblance to arise in the first place. "Can there be any edge," Gould asked, "to looking 5% like a turd?" (R. Dawkins, 1996, *The Blind Watchmaker*, Norton, p. 81). Can you think of a reason why looking 5% like a turd would be better than not looking at all like a turd?
4. In the late 1800s an American biologist named Hermon Bumpus collected a large number of sparrows that had been killed in a severe ice storm. He found that birds whose wings were about average length were rare among the dead birds. What kind of selection is this? What effect would this episode of selection have on the mean wing length in the population?

KEY TERMS

adaptations

equilibrium

stabilizing selection

traits

characters

species

fecundity

continuous variation

discontinuous variation

achondroplasia

placental mammals

marsupials

blending inheritance

GENETICS

MENDELIAN GENETICS

Although none of the main participants in the nineteenth-century debate about evolution knew it, the key experiments necessary to understand how genetic inheritance really worked had already been performed by an obscure Silesian monk, Gregor Mendel, living in what is now Slovakia (Figure 2.1). The son of peasant farmers, Mendel was recognized by his teachers as an extremely bright student, and he enrolled in the University of Vienna to study the natural sciences. While he was there, Mendel received a first-class education from some of the scientific luminaries of Europe. Unfortunately, Mendel had an extremely nervous disposition:

every time he was faced with an examination, he became physically ill, taking months to recover. As a result, he was forced to leave the university, after which he joined a monastery in the city of Brno, more or less because he needed a job. Once there, Mendel continued to study inheritance, an interest he had developed in Vienna.

By conducting careful experiments with plants, Mendel discovered how inheritance works.

FIGURE 2.1   Gregor Mendel, about 1884, 15 years after abandoning his botanical experiments.

Between 1856 and 1863, using the common edible garden pea plant (Figure 2.2), Mendel isolated a number of traits with only two forms, or **variants**. For example, one of the traits he studied was pea color. This trait had two variants: yellow and green. He also studied pea texture, a trait that also had two variants: wrinkled and smooth. Mendel cultivated populations of plants in which these traits bred true, meaning that the traits did not change from one generation to the next. For example, **crosses** (matings) between plants that bore green peas always produced offspring with green peas, and crosses between plants that bore yellow peas consistently produced offspring with yellow peas. Mendel performed a large number of crosses between these different kinds of true-breeding peas.

Before going further, we need to establish a way to keep track of the results of the matings. Geneticists refer to the original founding population as the F_0 **generation**, the offspring of the original founders as the F_1 **generation**, and so on. In this case, the original true-breeding plants constitute the F_0 generation and the plants created by crossing true-breeding parents constitute the F_1 generation. The offspring of the F_1 generation will be the F_2 **generation**.

In one set of Mendel's experiments with garden peas, a series of crosses between green and yellow variants yielded offspring that all bore yellow peas, matching only one of the parent plants (Figure 2.3). Mendel's next step was to perform crosses among the offspring of these crosses. When members of the F_1 generation (all of which bore yellow peas) were crossed, some of the offspring produced yellow seeds and some produced green seeds. Unlike most of the people who had experimented with plant crosses before, Mendel performed many of these kinds of crosses and kept careful count of the numbers of each kind of individual that resulted. These data showed that, in the F_2 generation, there were three individuals with yellow seeds for every one with green seeds.

Mendel was able to formulate two principles that accounted for his experimental results.

Mendel derived two insightful conclusions from his experimental results:

1. The observed characteristics of organisms are determined jointly by two particles, one inherited from the mother and one from the father. The American geneticist T. H. Morgan later named these particles **genes**.
2. Each of these two particles, or genes, is equally likely to be transmitted when **gametes** (eggs and sperm) are formed. Modern scientists call this **independent assortment**.

These two principles account for the pattern of results in Mendel's breeding experiments, and as we shall see, they are the key to understanding how variation is preserved.

FIGURE 2.2   Mendel's genetic experiments were conducted on the common garden pea.

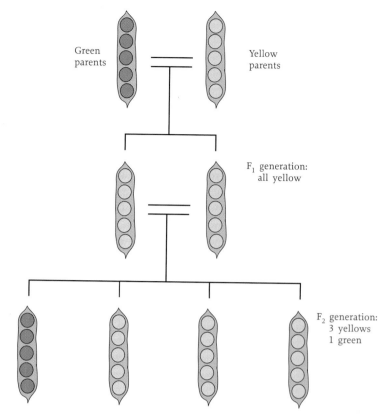

Green parents

Yellow parents

F₁ generation: all yellow

F₂ generation: 3 yellows 1 green

FIGURE 2.3 In one of Mendel's experiments, crossing true-breeding lines of green and yellow peas led to all yellow offspring. Crossing these F₁ individuals led to an F₂ generation with a 3:1 ratio of yellow to green individuals.

CELL DIVISION AND THE ROLE OF CHROMOSOMES IN INHERITANCE

Nobody paid any attention to Mendel's results for almost 40 years.

Mendel thought his findings were important, so he published them in 1866 and sent a copy of the paper to Karl Nägeli, a very prominent botanist. Nägeli was studying inheritance and should have understood the importance of Mendel's experiments. Instead, Nägeli dismissed Mendel's work, perhaps because it contradicted his own results or because Mendel was an obscure monk. Soon after this, Mendel was elected abbot of his monastery and was forced to give up his experiments. His ideas did not resurface until the turn of the twentieth century, when several botanists independently replicated Mendel's experiments and rediscovered the laws of inheritance.

In 1896, the Dutch botanist Hugo de Vries unknowingly repeated Mendel's experiments with poppies. Instead of publishing his results immediately, however, he cautiously waited until he had replicated his results with more than 30 plant species. Then in 1900, just as de Vries was ready to send off a manuscript describing his experiments, a colleague sent him a copy of Mendel's paper. Poor de Vries; his hot new results were already 30 years old! About the same time, two other European botanists, Carl Correns and Erich von Tschermak, also duplicated Mendel's breeding experiments, derived similar conclusions, and discovered that they, too, had been scooped. Correns and von Tschermak graciously acknowledged Mendel's primacy in discovering the laws of inheritance, but de Vries was less magnanimous. He did not cite Mendel in his

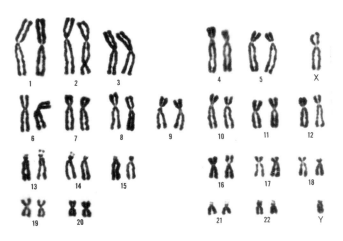

FIGURE 2.4 When a human cell divides, 23 pairs of chromosomes appear in its nucleus, including a pair of sex chromosomes (X and Y for a male, as shown here). Different chromosomes can be distinguished by their shape and by the banding patterns created by dyes that stain the chromosomes.

treatise on plant genetics, and he refused to sign a petition advocating the construction of a memorial in Brno commemorating Mendel's achievements.

When Mendel's results were rediscovered, they were widely accepted because scientists now understood the role of chromosomes in the formation of gametes.

By the time Mendel's experiments were rediscovered in 1900, it was well known that virtually all living organisms are built out of cells. Moreover, careful embryological work had shown that all the cells in complex organisms arise from a single cell through the process of cell division. Between the time of Mendel's initial discovery of the nature of inheritance and its rediscovery at the turn of the twentieth century, a crucial feature of cellular anatomy was discovered: the **chromosome**. Chromosomes are small, linear bodies contained in every cell and replicated during cell division (Figure 2.4). Moreover, scientists had also learned that chromosomes are replicated in a special kind of cell division that creates gametes. As we will see in subsequent sections, this research provides a simple material explanation for Mendel's results. Our current model of cell division, which was developed in small steps by a number of different scientists, is summarized in the sections that follow.

Mitosis and Meiosis

Ordinary cell division, called "mitosis," creates two copies of the chromosomes present in the nucleus.

When plants and animals grow, their cells divide. Every cell contains within it a body called the **nucleus** (plural *nuclei*; Figure 2.5); when cells divide, their nuclei also divide. This process of ordinary cell division is called **mitosis**. As mitosis begins, a cloud of material begins to form in the nucleus, and gradually this cloud condenses into a number of linear chromosomes. The chromosomes can be distinguished under the microscope by their shape and by how they stain. (Stains are dyes added to cells in the laboratory that allow researchers to distinguish different parts of a cell.) Different organisms have different numbers of chromosomes, but in **diploid** organisms,

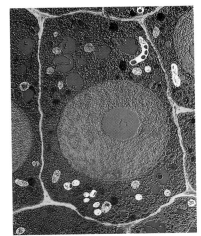

FIGURE 2.5 In all plants and animals, every cell contains a body called the nucleus (the dark circle in the center of this magnified image). The nucleus contains the chromosomes.

chromosomes come in **homologous pairs** (pairs whose members have similar shapes and staining patterns). All primates are diploid, but other organisms have a variety of arrangements. Diploid organisms also vary in the number of chromosome pairs their cells have. The fruit fly *Drosophila* has four pairs of chromosomes, humans have 23, and some organisms have many more.

Two features of mitosis suggest that the chromosomes play an important role in determining the properties of organisms. First, the original set of chromosomes is duplicated so that each new daughter cell has an exact copy of the chromosomes present in its parent. This means that, as an organism grows and develops through a sequence of mitotic divisions, every cell will have the same chromosomes that were present when the egg and sperm united. Second, the material that makes up the chromosome is present even when cells are not dividing. Cells spend little of their time dividing; most of the time they are in a "resting" period, doing what they are supposed to do as liver cells, muscle cells, bone cells, and so on. During the resting period, chromosomes are not visible. However, the material that makes up the chromosomes is always present in the cell.

In meiosis, the special cell division process that produces gametes, only half of the chromosomes are transmitted from the parent cell to the gamete.

The sequence of events that occur during mitosis is quite different from the sequence of events during **meiosis**, the special form of cell division leading to the production of gametes. The key feature of meiosis is that each gamete contains only *one* copy of each chromosome, whereas cells that undergo mitosis contain a *pair* of homologous chromosomes. Cells that contain only one copy of each chromosome are said to be **haploid** (Figure 2.6). When a new individual is conceived, a haploid sperm from the father unites with a haploid egg from the mother to produce a diploid **zygote**. The zygote is a single cell that then divides mitotically over and over to produce the millions and millions of cells that make up an individual's body.

Chromosomes and Mendel's Experimental Results

Mendel's two principles can be deduced from the assumption that genes are carried on chromosomes.

In 1902, less than two years after the rediscovery of Mendel's findings, Walter Sutton, a young graduate student at Columbia University, made the connection between chromosomes and the properties of inheritance revealed by Mendel's principles. Recall that the first of Mendel's two principles states that an organism's observed characteristics are determined by particles acquired from each of the parents. This concept fits with the idea that genes reside on chromosomes, because individuals inherit one copy of each chromosome from each parent. The idea that observed characteristics are determined by genes from both parents is consistent with the observation that mitosis transmits a copy of both chromosomes to every daughter cell, so every cell contains copies of both the maternal and the paternal chromosomes. Mendel's second principle states that genes segregate independently. The observation that meiosis involves the creation of gametes with only one of the two possible chromosomes from each homologous pair is consistent with two notions: (1) that one gene is inherited from each parent, and (2) that each of these genes is equally likely to be transmitted to gametes. Not every-

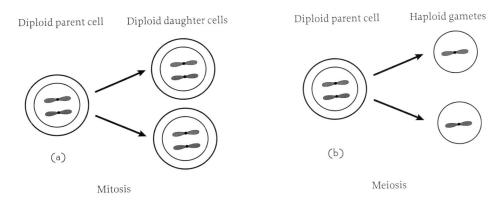

Diploid parent cell   Diploid daughter cells   Diploid parent cell   Haploid gametes

(a)

Mitosis

(b)

Meiosis

FIGURE 2.6 Diploid cells have n pairs of homologous chromosomes; n varies widely among species, but here n = 1. Members of homologous pairs may differ in the alleles they carry at specific sites along the chromosome. (a) Mitosis duplicates the chromosomes. (b) Meiosis creates gametes that carry only one member of each homologous pair of chromosomes.

one agreed with Sutton, but over the next 15 years, T. H. Morgan and his colleagues at Columbia performed many experiments that proved Sutton right.

Different varieties of a particular gene are called "alleles." Individuals with two copies of the same allele are "homozygous"; individuals with different alleles are "heterozygous."

To see more clearly the connection between chromosomes and the results of Mendel's experiments with the peas, we need to introduce some new terms. The word *gene* is used to refer to the material particles carried on chromosomes. Later you will learn that genes are made of a molecule called "DNA." **Alleles** are varieties of genes with the same effect on the organism. Individuals with two copies of the same allele are **homozygous** for that allele and are called "homozygotes." When individuals carry copies of two different alleles, they are said to be **heterozygous** for those alleles and are called "heterozygotes."

Consider the case in which all the yellow individuals in the parental generation carry two genes for yellow pea color, one on each chromosome. We will use the symbol *A* for this allele. Thus, these plants are homozygous (*AA*) for yellow pea color. All of the individual plants with green peas are homozygous for a different allele, which we denote *a*, so green-pea plants are *aa*. As we will see, this is the only pattern that is consistent with Mendel's model. What happens if we cross two yellow-pea plants with each other, or two green-pea plants with each other? Because they are homozygous, all of the gametes produced by the yellow parents will carry the *A* allele. This means that all of the offspring produced by the crossing of two *AA* parents will also be homozygous for that *A* allele, and therefore will also produce yellow peas. Similarly, all of the gametes produced by parents that are homozygous for the *a* allele will carry the *a* allele; when the gametes of two *aa* parents unite, they will produce only *aa* individuals, with green peas. Thus, we can explain why each type breeds true.

A cross between a homozygous dominant parent and a homozygous recessive parent produces all heterozygotes in the F₁ generation.

Next let's consider the offspring of a mating between a true-breeding green parent and a true-breeding yellow parent (Figure 2.7). The green parent produces only *a* gametes, and the yellow parent produces only *A* gametes. Thus, every one of their offspring inherits an *a* gamete from one parent and an *A* gamete from the other parent. According to Mendel's model, all the individuals in the F_1 generation will be *Aa*.

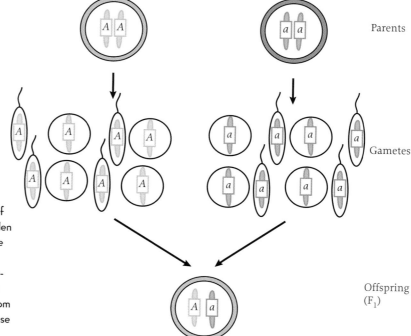

Parents

Gametes

Offspring
(F₁)

FIGURE 2.7 In Mendel's experiments, crosses of two true-breeding lines of the garden pea produced offspring that all had yellow peas; All of the gametes produced by homozygous *AA* parents carry the *A* allele. Similarly, all of the gametes produced by homozygous *aa* parents carry the *a* allele. All of the zygotes from an *AA* × *aa* mating get an *A* from one parent and an *a* from the other parent. Thus, all F₁ offspring are *Aa*, and because *A* is dominant, they all produce yellow seeds.

Given that Mendel discovered that all offspring of such crosses have yellow peas, it must be that heterozygotes bear yellow peas. To describe these effects, geneticists use the following four terms:

1. *Genotype.* **Genotype** refers to the particular combination of genes or alleles that an individual carries.
2. *Phenotype.* **Phenotype** refers to the observable characteristics of the organism, such as the color of the peas in Mendel's experiments.
3. *Dominant.* The *A* allele is **dominant** because individuals with only one copy of the dominant allele have the same phenotype, yellow peas, that individuals with two copies of that allele have.
4. *Recessive.* The *a* allele is **recessive** because it has *no* effect on phenotype in heterozygotes.

As you can see in Table 2.1, *AA* and *Aa* individuals have the same phenotype but different genotypes; knowing an individual's observable characteristics, or phenotype, does not necessarily tell you its genetic composition, or genotype.

TABLE 2.1 The relationship between genotype and phenotype in Mendel's experiment on pea color.

Genotype	Phenotype
AA	Yellow
Aa	Yellow
aa	Green

A cross between heterozygous parents produces a predictable mixture of all three genotypes.

Now consider the second stage of Mendel's experiment: crossing members of the F₁ generation with each other to create an F₂ generation (Figure 2.8). We have seen that every individual in the F₁ generation is heterozygous, *Aa*. This means that half of their gametes contain a chromosome with an *A* allele and half with an *a* allele. (Remember, meiosis produces haploid gametes.) On average, if we draw pairs of

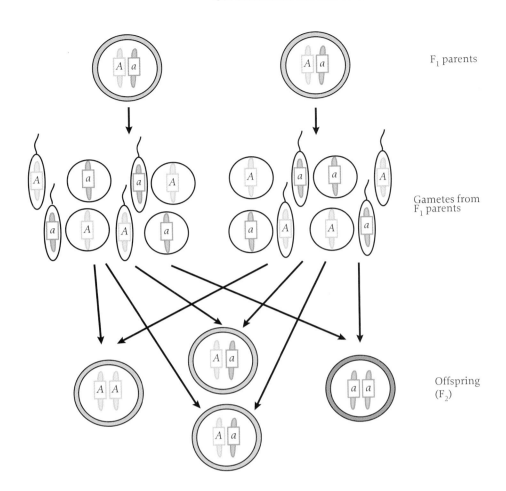

F₁ parents

Gametes from
F₁ parents

Offspring
(F₂)

FIGURE 2.8 Crosses of the F₁ heterozygotes yielded a 3:1 ratio of phenotypes in Mendel's experiments. All the parents are heterozygous, or *Aa*. This means that half of the gametes produced by each parent carry the *A* allele, and the other half produce the *a* allele. Thus, ¼ of the zygotes will be *AA*, ½ will be *Aa*, and the remaining ¼ will be *aa*. Because *A* is dominant, ¾ of the offspring (¼ *AA* + ½ *Aa*) will produce yellow seeds.

gametes at random from the gametes produced by the F₁ generation, ¼ of the individuals will be *AA*, ½ will be *Aa*, and ¼ will be *aa*.

To see why there is a 1:2:1 ratio in the F₂ generation, it is helpful to construct an event tree, like the one shown in Figure 2.9. First pick the paternal gamete. Every male in the F₁ generation is heterozygous: he has one chromosome with an *A* allele and one with an *a* allele. Thus there is a probability of ½ of getting a sperm that carries an *A*, and a probability of ½ of getting a sperm that carries an *a*. Suppose you select an *A* by chance. Now pick the maternal gamete. Once again you have a ½ chance of getting an *A* and a ½ chance of getting an *a*. The probability of getting two *A*s, one from the father and one from the mother, is

$$\text{Pr}(AA) = \text{Pr}(A \text{ from dad}) \times \text{Pr}(A \text{ from mom}) = \text{½} \times \text{½} = \text{¼}$$

If you repeated this process a large number of times—first picking male gametes and then picking female gametes—roughly ¼ of the F₂ individuals produced would be *AA*. Similar reasoning shows that ¼ would be *aa*. One-half would be *Aa* because there are two ways of combining the *A* and *a* alleles: the *A* could come from the mother and the *a* from the father, or vice versa. Because both *AA* and *Aa* individuals have yellow peas, there will be three yellow individuals for every green one among the offspring of the F₁ parents. Another way to visualize this result is to construct a diagram called a **Punnett square**, like the one shown in Figure 2.10.

| FIGURE 2.9 | This event tree shows why there is a 1:2:1 genotypic ratio among offspring |

in the F$_2$ generation. Imagine forming a zygote by first choosing a sperm and then an egg. The numbers along each branch give the probability that a choice at the previous node ends on that branch. The circles at the end of each branch represent the resulting zygote, and the color of the ring denotes the phenotype: yellow or green peas. The first node represents the choice of a sperm. There is a 50% chance of selecting a sperm that carries the *A* allele, and a 50% chance of selecting a sperm that carries the *a* allele. Now choose an egg. Once again, there is a 50% chance of selecting an egg with each allele. Thus, the probability of getting both an *A* sperm and an *A* egg is 0.5 × 0.5 = 0.25. Similarly, the chance of getting an *a* egg and an *a* sperm is 0.5 × 0.5 = 0.25. The same kind of calculation shows that there is a 25% chance of getting an *a* sperm and an *A* egg, and a 25% chance of getting an *A* sperm and an *a* egg. Thus, the probability of getting an *Aa* zygote is 50%.

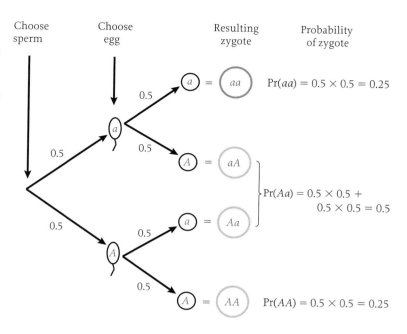

Choose sperm | Choose egg | Resulting zygote | Probability of zygote

Pr(*aa*) = 0.5 × 0.5 = 0.25

Pr(*Aa*) = 0.5 × 0.5 + 0.5 × 0.5 = 0.5

Pr(*AA*) = 0.5 × 0.5 = 0.25

Linkage and Recombination

Mendel also performed experiments involving two traits that he believed showed that separate characters segregate independently.

Mendel also performed experiments that involved two characters. For example, he crossed individuals that bred true for two different traits: pea color and seed texture. He crossed plants with smooth yellow seeds and plants with wrinkled green seeds. All of the F$_1$ individuals were smooth-yellow, but the F$_2$ individuals occurred in the following ratio:

9 smooth-yellow:3 smooth-green:3 wrinkled-yellow:1 wrinkled-green

This experiment is important because it demonstrates that sexual reproduction shuffles genes that affect different traits and thereby produces new combinations of traits —a phenomenon called **recombination**. This process is extremely important for maintaining variation in natural populations. We will return to this issue in Chapter 3.

To understand Mendel's experiment, recall that each of the parental characters (pea color and seed texture) breeds true, which means that each is homozygous for a different allele controlling seed color: the yellow line is *AA*, and the green line is *aa*. Because seed texture is also a true-breeding character, the parental lines must also be homozygous for genes affecting seed texture. Let's suppose that the parents with smooth seeds are *BB* and the parents with wrinkled seeds are *bb*. Then in the parental generation there are only two genotypes: *AABB* and *aabb*. By the F$_2$ generation, sexual reproduction has generated all 16 possible genotypes and two new phenotypes: smooth-green and wrinkled-yellow. The details are given in Box 2.1.

The 9:3:3:1 ratio of phenotypes tells us that the genes determining seed color and the genes determining seed texture each segregate independently, as Mendel's second principle predicts. Thus, knowing that a gamete has the *A* allele tells us nothing about

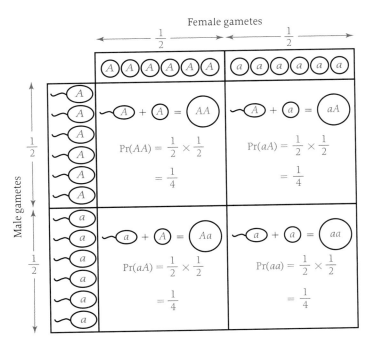

Female gametes

FIGURE 2.10 This diagram, called a Punnett square, provides another way to see why there is a 1:2:1 genotypic ratio among offspring in the F_2 generation. The horizontal axis is divided in half to reflect the equal proportion of A and a eggs. The vertical axis is divided according to the proportion of sperm carrying each allele, and again it is divided in half. The areas of squares formed by the intersection of the vertical and horizontal dividing lines give the proportion of zygotes that result from each of the four possible fertilization events: AA = 0.25 (one of the four squares), Aa = 0.25 + 0.25 = 0.50 (two of the four squares), and aa = 0.25 (one of the four squares). Thus, zygotes will have genotypes in the ratio 1:2:1.

whether it will have B or b; it is equally likely to carry either of these alleles. Mendel's experiments convinced him that all traits segregate independently. Today, however, we know that independent segregation occurs only when the traits measured are controlled by genes that reside on different chromosomes.

Genes are arranged on chromosomes like beads on a string.

It turns out that the genes for a particular character occur at a particular site on a particular chromosome. Such a site is called a **locus** (plural *loci*). Loci are arranged on the chromosomes in a line, like beads on a string. A particular locus may hold any of several alleles. The gene for seed color is always in the same position on a particular chromosome, whether it codes for green or for yellow seeds. Keep in mind that because organisms have more than one pair of chromosomes, the locus for seed color may be located on one chromosome and the locus for seed texture located on another chromosome. All of the genes carried on all of the chromosomes are referred to as the **genome**.

Traits may not segregate independently if they are affected by genes on the same chromosome.

Mendel's conclusion that traits segregate independently is true only if the loci that affect the traits are on different chromosomes, because links between loci on the same chromosome alter the patterns of segregation. When loci for different traits occur on the same chromosome, they are said to be **linked**; loci on different chromosomes are said to be **unlinked**. You might think that genes at two loci on the same chromosome would always segregate together as if they were a single gene. If this were true, then a gamete receiving a particular chromosome from one of its parents would get all of the same genes that occurred on that chromosome in its parent. There would be no

BOX 2.1

More on Recombination

The first step to a deeper understanding of recombination is to see the connections between Mendel's two-trait experiment, independent segregation, and chromosomes. Mendel crossed a smooth-yellow (*AABB*) parent and a wrinkled-green (*aabb*) parent to produce members of an F$_1$ generation. Smooth-yellow parents produce only *AB* gametes, and wrinkled-green parents produce only *ab* gametes; all members of the F$_1$ generation are therefore *AaBb*. If we assume that the genes for seed color and seed texture enter gametes independently, then each of the four possible types—*AB*, *Ab*, *aB*, and *ab*—will be represented by ¼ of the gametes produced by members of the F$_1$ generation (Figure 2.11).

With this information we can construct a Punnett square predicting the proportions of each genotype in the F$_2$ generation (Figure 2.12). Once again we divide the vertical and horizontal axes in proportion to the frequency of each type of gamete, in this case dividing each axis into four equal parts. The areas of the rectangles formed by the intersection of the vertical and horizontal dividing lines give the proportions of zygotes that result from each of the 16 possible fertilization events. Because the area of each cell in this matrix is the same, we can determine the phenotypic ratios of zygotes that result from each of the 16 possible fertilization events in this example by simply counting the number

of squares that contain each phenotype. There are nine smooth-yellow squares, three wrinkled-yellow squares, three smooth-green squares, and one wrinkled-green square. Thus the 9:3:3:1 ratio of phenotypes that Mendel observed was consistent with the assumption that genes on different chromosomes segregate independently. If the genes controlling these traits did not segregate independently, the ratio of phenotypes would be different, as we will see next.

During meiosis, chromosomes frequently become damaged, break, and recombine. This process, called "crossing over," creates chromosomes with combinations of genes not present in the parent (see text). In the preceding example, remember that all members of the F$_1$ generation were *AaBb*. Now suppose that the locus controlling seed color and the locus controlling seed texture are carried on the same chromosome. Further assume that when the chromosomes are duplicated during meiosis, a fraction *r* of the time there is crossing over (Figure 2.13), and a fraction 1 − *r* of the time there is no crossing over. Next, chromosomes segregate independently into gametes. The types of chromosomes present in the parental generation, *Ab* and *ab*, each occur in a fraction (1 − *r*)/2 of the gametes; the novel, recombinant types *Ab* and *aB* each occur in a fraction *r*/2 of the gametes (Figure 2.14).

FIGURE 2.11 This event tree shows why an F$_1$ parent is equally likely to produce all four possible gametes when genes for two traits are carried on different chromosomes. The first node represents the choice of the chromosome carrying the gene for seed color (*A* or *a*), and the second node represents the choice of the chromosome that carries the gene for seed texture (*B* or *b*). The number along each branch gives the probability that the choice at the previous node ends on that branch. The circles at the end of each branch represent the resulting gametes.

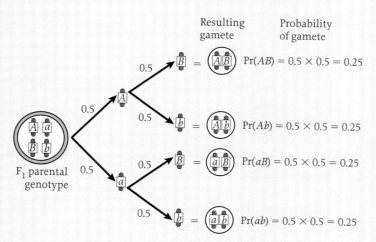

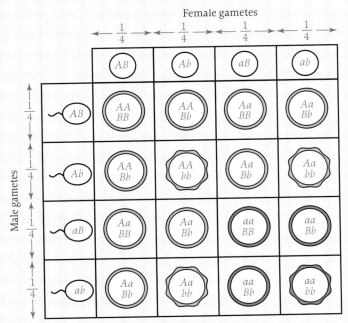

Female gametes

Male gametes

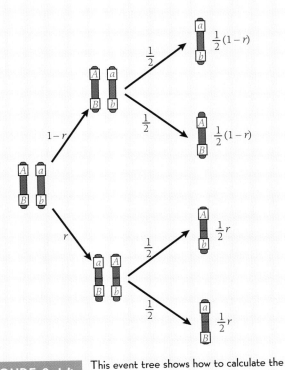

This Punnett square shows why there is a 9:3:3:1 phenotypic ratio among offspring of the F$_2$ generation when the genes for the two traits, seed color and seed texture, are carried on different chromosomes. The rings in each square show the color (green or yellow) and seed texture (wrinkled or smooth) of the phenotype associated with each genotype.

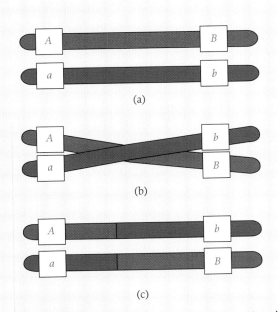

FIGURE 2.13 Crossing over during meiosis sometimes leads to recombination and produces novel combinations of traits. Suppose that the A allele leads to yellow seeds and the a allele to green seeds, while the B allele leads to smooth seeds and the b allele to wrinkled seeds. (a) Here an individual carries one AB chromosome and one ab chromosome. (b) During meiosis, the chromosomes are damaged and crossing over occurs. (c) Now the A allele is paired with b, and the a allele is paired with B.

FIGURE 2.14 This event tree shows how to calculate the fraction of each type of gamete that will be produced when genes are carried on the same chromosome. The first node represents whether or not crossing over takes place. There is a probability r that crossing over takes place and produces novel trait combinations, and a probability 1 – r that there is no crossing over. At the second node, chromosomes are randomly assigned to gametes. The value given above each branch represents the probability of reaching that branch from the previous node. The likelihood of forming each type of genotype is the product of the probabilities along each pathway.

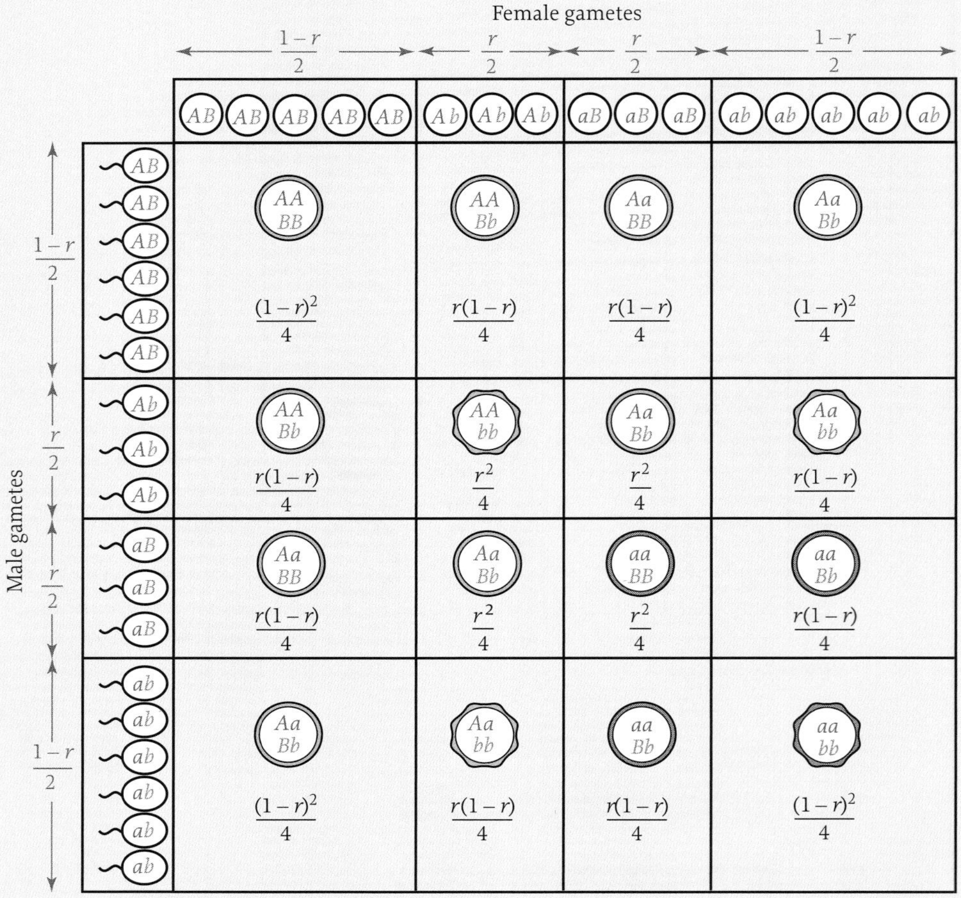

Female gametes

$$\frac{1-r}{2} \qquad \frac{r}{2} \qquad \frac{r}{2} \qquad \frac{1-r}{2}$$

		AB AB AB AB AB	A b A b A b	aB aB aB	ab ab ab ab ab
Male gametes	$\frac{1-r}{2}$ AB AB AB AB AB AB		$\begin{array}{c}AA\\BB\end{array}$ $\dfrac{(1-r)^2}{4}$	$\begin{array}{c}AA\\Bb\end{array}$ $\dfrac{r(1-r)}{4}$	$\begin{array}{c}Aa\\BB\end{array}$ $\dfrac{r(1-r)}{4}$	$\begin{array}{c}Aa\\Bb\end{array}$ $\dfrac{(1-r)^2}{4}$
	$\frac{r}{2}$ Ab Ab Ab		$\begin{array}{c}AA\\Bb\end{array}$ $\dfrac{r(1-r)}{4}$	$\begin{array}{c}AA\\bb\end{array}$ $\dfrac{r^2}{4}$	$\begin{array}{c}Aa\\Bb\end{array}$ $\dfrac{r^2}{4}$	$\begin{array}{c}Aa\\bb\end{array}$ $\dfrac{r(1-r)}{4}$
	$\frac{r}{2}$ aB aB aB		$\begin{array}{c}Aa\\BB\end{array}$ $\dfrac{r(1-r)}{4}$	$\begin{array}{c}Aa\\Bb\end{array}$ $\dfrac{r^2}{4}$	$\begin{array}{c}aa\\BB\end{array}$ $\dfrac{r^2}{4}$	$\begin{array}{c}aa\\Bb\end{array}$ $\dfrac{r(1-r)}{4}$
	$\frac{1-r}{2}$ ab ab ab ab ab ab		$\begin{array}{c}Aa\\Bb\end{array}$ $\dfrac{(1-r)^2}{4}$	$\begin{array}{c}Aa\\bb\end{array}$ $\dfrac{r(1-r)}{4}$	$\begin{array}{c}aa\\Bb\end{array}$ $\dfrac{r(1-r)}{4}$	$\begin{array}{c}aa\\bb\end{array}$ $\dfrac{(1-r)^2}{4}$

FIGURE 2.15 This Punnett square shows how to calculate the frequency of each phenotype among offspring in the F_2 generation if the genes for seed color and seed texture are carried on the same chromosome. The horizontal axis is divided according to the proportion of each type of egg: *AB*, *Ab*, *aB*, and *ab*. When genes are carried on the same chromosome, the frequency of each type of gamete is calculated as shown in Figure 2.14. The vertical axis is divided according to the proportion of each type of sperm, and the horizontal axis is divided according to the proportion of each type of egg. The areas of the rectangles formed by the intersection of the vertical and horizontal dividing lines give the proportions of zygotes that result from each of the 16 possible fertilization events.

Now we can use a Punnett square to calculate the frequency of each of the 16 possible genotypes in the F_2 generation (Figure 2.15). As before, we divide the vertical and horizontal axes in proportion to the relative frequency of each type of gamete, and the area of the rectangles formed by the intersection of these grid lines gives the frequency of each genotype. If recombination rates are low, most members of the F_2 generation will be of three genotypes—*AABB*, *AaBb*, and *aabb*—just as if there were only two alleles, *AB* and *ab*. If the recombination rate is higher, however, more of the novel recombinant genotypes will be produced.

recombination. However, chromosomes frequently tangle and break as they are replicated during meiosis. Thus, chromosomes are not always preserved intact, and genes on one chromosome are sometimes shifted from one member of a homologous pair to the other (see Figure 2.11). We call this process **crossing over**. Linkage reduces the rate of recombination but does not eliminate it altogether. The rate at which recombination generates novel combinations of genes at two loci on the same chromosome depends on the likelihood that a crossing-over event will occur. If two loci are located close together on the chromosome, crossing over will be rare and the rate of recombination will be low. If the two loci are located far apart, then crossing over will be common and the rate of recombination will approach the rate for genes on different chromosomes. This process is discussed more fully in Box 2.1.

MOLECULAR GENETICS

Genes are segments of a long molecule called DNA, which is contained in chromosomes.

In the first half of the 1900s, geneticists made substantial progress in describing the cellular events that took place during meiosis and mitosis, and in understanding the chemistry of reproduction. For instance, by the middle of the twentieth century it was known that chromosomes contain two structurally complex molecules: protein and **deoxyribonucleic acid**, or **DNA**. It had also been determined that the particle of heredity postulated by Mendel was DNA, not protein, though exactly how DNA might contain and convey the information essential to life was still a mystery. But in the early 1950s, two young biologists at Cambridge University, Francis Crick and James Watson, made a discovery that revolutionized biology: they deduced the structure of DNA. Watson and Crick's elucidation of the structure of DNA was the wellspring of a great flood of research that continues to provide a deep and powerful understanding of how life works at the molecular level. We now know how DNA stores information and how this information controls the chemistry of life, and this knowledge explains why heredity leads to the patterns Mendel described in pea plants, and why there are sometimes new variations.

Understanding the chemical nature of the gene is critical to the study of human evolution: (1) molecular genetics links biology to chemistry and physics, and (2) molecular methods help us reconstruct the evolutionary history of the human lineage.

Modern molecular genetics, the product of Watson and Crick's discovery, is a field of great intellectual excitement producing many new discoveries of practical value in medicine and agriculture. However, progress in molecular genetics has not yet fundamentally changed our understanding of how morphology and behavior evolve. Morphological and behavioral traits are usually affected by many genes that interact in ways that are, for the most part, still too complex to understand at the molecular level. There are several important exceptions to this generalization. For example, in recent years there has been rapid progress in our understanding of the genes that control the development of body size and shape, and in our understanding of the genetic basis of certain forms of behavior. Nonetheless, having a grasp of molecular genetics is crucially important to comprehending the evolution of the human phenotype, for two reasons:

1. Molecular genetics links biology to chemistry and physics. One of the grandest goals of science is to provide a single, consistent explanatory framework for the way the world works. We want to place evolution in this grand scheme of scientific explanation. It is important to be able to explain not only how new species of plants and animals arise, but also how a wide range of phenomena—from the origin of stars and galaxies to the rise of complex societies—have evolved. Modern molecular biology is of profound importance because it explains how life and evolution work at the level of physics and chemistry, and it thereby connects physical and geochemical evolution to Darwinian processes.

2. Molecular genetics provides important data for reconstructing evolutionary history. In recent years, molecular geneticists have used information about varia-

tion in DNA sequences to reconstruct the history of particular lineages. To see how this is done, you have to understand a little about the molecular biology of the gene.

Genes Are DNA

DNA is unusually well suited to be the chemical basis of inheritance.

The discovery of the structure of DNA was fundamental to genetics because the structure itself implied how inheritance must work. Each chromosome contains a single DNA molecule roughly 2 m (about 6 ft) long that is folded up to fit in the nucleus. DNA molecules consist of two long strands, and each strand has a "backbone" of alternating sequences of phosphate and sugar molecules. Attached to each sugar is one of four molecules, collectively called **bases: adenine, guanine, cytosine,** or **thymine.** The two strands of DNA are held together by very weak chemical bonds, called "hydrogen bonds," which connect some of the bases on different strands. Thymine bonds only with adenine, and guanine bonds only with cytosine (Figure 2.16).

The repeating four-base structure of DNA allows the molecule to assume a vast number of distinct forms. Furthermore, the staggering number of DNA molecules that exist in nature are equally stable chemically. DNA is not the only complex molecule with many alternative forms, but other molecules have some forms that are less stable than others. Such molecules would be unsuitable for carrying information because the messages would degrade (become garbled) as the molecules changed toward a more stable form. What makes DNA unusual is that all of its nearly infinite number of forms are equally stable.

Each DNA configuration is exactly like a message written in an alphabet with letters that stand for each of the four bases (T for thymine, A for adenine, G for guanine, and C for cytosine). Thus,

TCGGTAGTAGTTACGG

is one message, and

ATCCGGATGCAATCCA

is another message. Because the DNA in a single chromosome is millions of bases long, there is room for a nearly infinite variety of messages.

In addition to preserving a message faithfully, hereditary material must be replicable. Without the ability to make copies of itself, the genetic message that directs the activities of living cells could not be spread to offspring, and natural selection would be impossible. DNA is replicated within cells by a highly efficient cellular machinery: it first unzips the two strands, and then, with the help of other specialized molecular machinery, it adds complementary bases to each of the strands until two identical sugar and phosphate backbones are built (Figure 2.17). There are also mechanisms that "proofread" the copies and correct errors that crop up. These proofreading mechanisms are astoundingly good; they miss only one error in every *billion* bases replicated.

The message encoded in DNA affects phenotype in several different ways.

For DNA to to play an important role in evolution, it is also important for variation in DNA coding to lead to variation in phenotypes. Here the story gets more

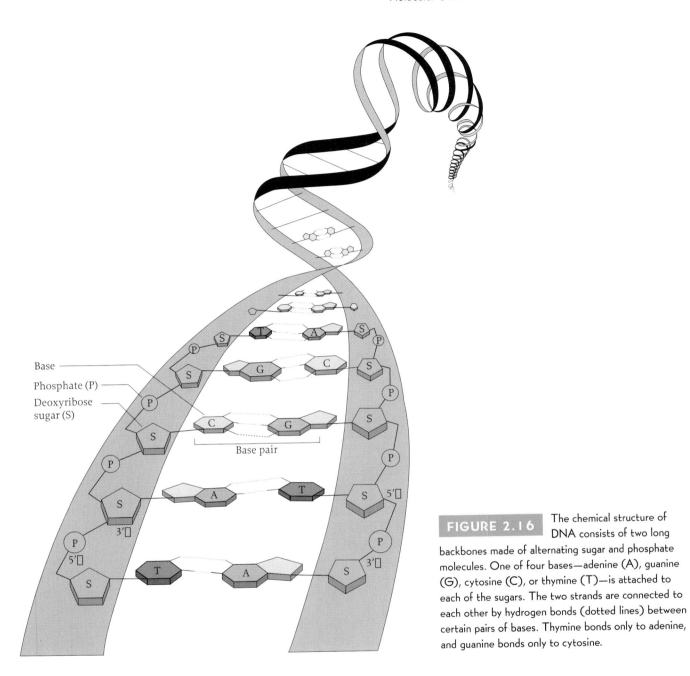

Base

Phosphate (P)

Deoxyribose
sugar (S)

Base pair

FIGURE 2.16 The chemical structure of DNA consists of two long backbones made of alternating sugar and phosphate molecules. One of four bases—adenine (A), guanine (G), cytosine (C), or thymine (T)—is attached to each of the sugars. The two strands are connected to each other by hydrogen bonds (dotted lines) between certain pairs of bases. Thymine bonds only to adenine, and guanine bonds only to cytosine.

complicated because DNA affects phenotypes in a number of different ways. The two most important ways that DNA affects phenotypes are the following:

1. DNA in **coding sequences** specifies the structure of proteins. Proteins play many important roles in the machinery of life. In particular, many proteins are **enzymes**, which regulate much of the biochemical machinery of organisms. Coding sequences are sometimes referred to as **structural genes**.
2. DNA in **regulatory sequences** (or **regulatory genes**) determines the conditions under which the message encoded in a coding sequence gene will be expressed. Regulatory genes play a crucial role in shaping the differentiation of cells during development.

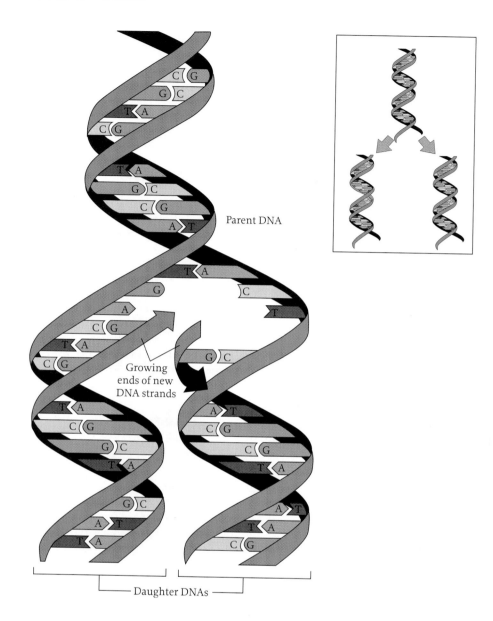

Parent DNA

Growing
ends of new
DNA strands

Daughter DNAs

FIGURE 2.17 When DNA is replicated, the two strands of the parent DNA are separated and two daughter DNA strands are formed.

Structural Genes Code for Proteins

Proteins called enzymes influence an organism's biochemistry.

The cells and organs of living things are made up a very large number of chemical compounds, and it is this combination of compounds that gives each organism its characteristic form and structure. All organisms use the same raw materials, but they achieve different end results. How does this happen?

The answer is that enzymes present in the cell determine what the raw materials are transformed into when cells are built (Figure 2.18). The best way to understand how enzymes determine the characteristics of organisms is to think of an organism's biochemical machinery as a branching tree. Enzymes act as switches to determine what will happen at each node and thus what chemicals will be present in the cell. For

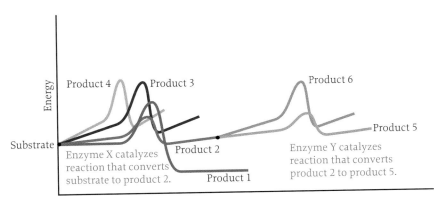

FIGURE 2.18 Enzymes control the chemical composition of cells by catalyzing some chemical reactions but not others. In this hypothetical example, the molecules that provide the initial raw material (called a "substrate") of the pathway could undergo four different reactions yielding different molecules, labeled products 1 through 4. However, because enzyme X is present, the reaction that yields product 2 proceeds much more rapidly than the other reactions (note that this reaction has the lowest activation energy), and all of the substrate is converted to that product. Product 2 could then undergo two different reactions yielding products 5 and 6. When enzyme Y is present, it lowers the activation energy, thus causing the reaction yielding product 5 to proceed much more rapidly, and only product 5 is produced. In this way, enzymes link products and reactants into pathways that satisfy particular chemical objectives, such as the extraction of energy from glucose.

example, glucose serves as a food source for many cells, meaning that it provides energy and a source of raw materials for the construction of cellular structures. Glucose might initially undergo any one of an extremely large number of slow-moving reactions. The presence of particular catalytic enzymes will determine which reactions occur rapidly enough to alter the chemistry of the cell. For example, some enzymes lead to the metabolism of glucose and the release of its stored energy. At the end of the first branch there is another node representing all of the reactions that could involve the product(s) of the first branch. Again, one or more enzymes will determine what happens next.

This picture has been greatly simplified. Real organisms take in many different kinds of compounds, and each compound is involved in a complicated tangle of branches that biochemists call "pathways." Real **biochemical pathways** are very complex, as you can see in Figure 2.19, which shows the pathway for the conversion of glucose to energy in animal cells. One set of enzymes causes glucose to be shunted to a pathway that yields energy. A different set of enzymes causes the glucose to be shunted to a pathway that binds glucose molecules together to form glycogen, a starch that functions as energy storage. The presence of a different set of enzymes would lead to the synthesis of cellulose, a complex molecule that provides the structural material in plants. Enzymes play roles in virtually all cellular processes—from the replication of DNA and division of cells to the contraction and movement of muscles.

Proteins play a number of other important roles in the machinery of life.

Some proteins have crucial structural functions in living things. For example, your hair is made up mainly of a protein called "keratin," and your ligaments and tendons are strengthened by another protein, called "collagen." Other proteins act as tiny mechanical contraptions that accomplish many important functions. Some are tiny

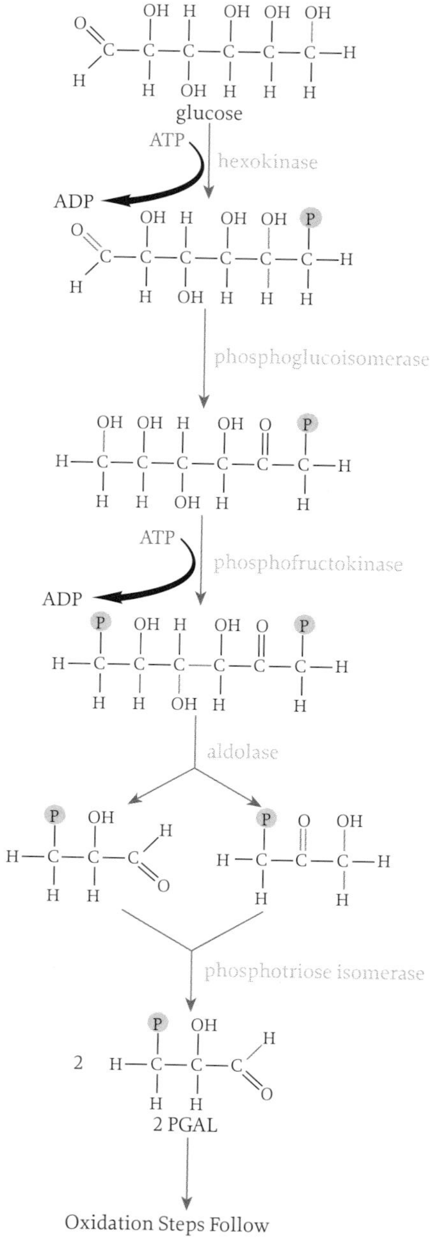

FIGURE 2.19 This is the glycolysis pathway, a biochemical pathway found in animals that converts glucose to energy, which can then be used to power cellular processes. Each step involves a chemical reaction controlled by a different enzyme.

valves for regulating what goes into and out of cells. Still others, like insulin, convey chemical signals from one part of the body to another, or act as receptors that respond to these messages.

The sequence of amino acids in proteins determines their properties.

Proteins are constructed of amino acids. There are 20 different amino acid molecules. All **amino acids** have the same chemical backbone, but they differ in the chemical composition of the side chain connected to this backbone (Figure 2.20). The sequence of amino acid side chains, called the **primary structure** of the protein, is what makes one protein different from others. You can think of a protein as a very long railroad train in which there are 20 different kinds of cars, each representing a different amino acid. The primary structure is a list of the types of cars in the order that they occur.

When proteins are actually doing their business, they are folded in complex ways. The three-dimensional shape of the folded protein, called the **tertiary structure**, is crucial to its catalytic function. The way the protein folds depends on the sequence of amino acid molecules that make up its primary sequence. This means that the function of enzymes depends on the sequence of the amino acids that make them up. (Proteins also have secondary structure, and sometimes also quaternary structure, but to keep things simple, we will ignore those levels here.)

These ideas are illustrated in Figure 2.21, which shows the folded shape of part of a **hemoglobin** molecule, a protein that transports oxygen from the lungs to the tissues via red blood cells. As you can see, the protein folds into a roughly spherical glob, and oxygen is bound to the protein near the center of the glob. **Sickle-cell anemia**, a condition common among people in West Africa and among African Americans, is caused by a single change in the primary sequence of amino acids in the hemoglobin molecule. Glutamic acid is the sixth amino acid in normal hemoglobin molecules, but in people afflicted with sickle-cell anemia, valine is substituted for glutamic acid. This single substitution changes the way that the molecule folds, and reduces its ability to bind oxygen.

DNA specifies the primary structure of protein.

Now we return to our original question: how does the information contained in DNA—its sequence of bases—determine the structure of proteins? Remember that DNA encodes messages in a four-letter alphabet. Researchers determined that these letters are combined into three-letter "words" called **codons**, each of which specifies a particular amino acid. Because there are four bases, there are 64 possible three-letter combinations for codons (4 possibilities for the first base times 4 for the second times 4 for the third, or $4 \times 4 \times 4 = 64$). Of these codons, 61 are used to code for the 20 amino acids that make up proteins. For example, the codons GCT, GCC, GCA, and GCG all code for alanine; GAT and GAC code for asparagine; and so on. The remaining three codons are "punctuation marks" that mean either "start, this is the beginning of the protein" or "stop, this is the end of the protein." Thus, if you can identify the base pairs, it is a simple matter to determine what proteins are encoded on the DNA.

You might be wondering why several different codons code for the same amino acid. This redundancy serves an important function. Because a number of processes can damage DNA and cause one base to be substituted for another, redundancy decreases the chance that a random change will alter the primary

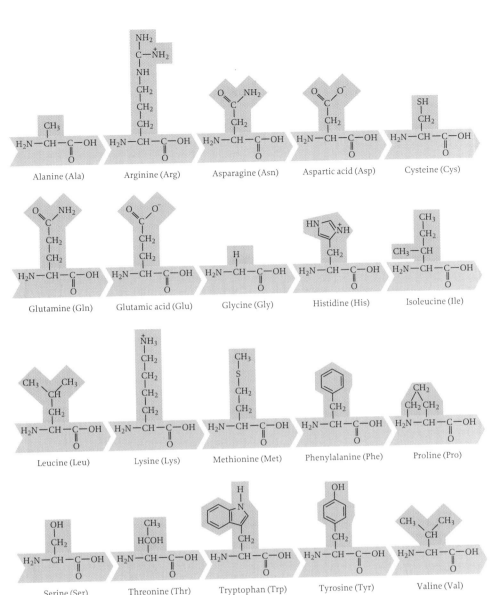

Alanine (Ala)

Arginine (Arg)

Asparagine (Asn)

Aspartic acid (Asp)

Cysteine (Cys)

Glutamine (Gln)

Glutamic acid (Glu)

Glycine (Gly)

Histidine (His)

Isoleucine (Ile)

Leucine (Leu)

Lysine (Lys)

Methionine (Met)

Phenylalanine (Phe)

Proline (Pro)

Serine (Ser)

Threonine (Thr)

Tryptophan (Trp)

Tyrosine (Tyr)

Valine (Val)

FIGURE 2.20 All amino acids share the same chemical backbone, here colored blue. They differ according to the chemical structure of the side group attached to the backbone, here colored pink.

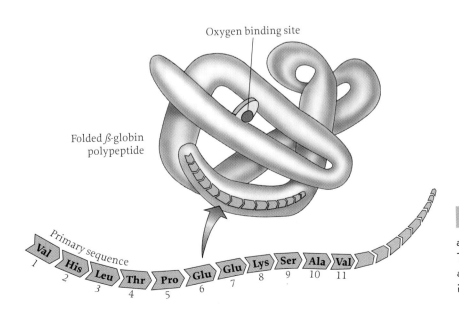

Oxygen binding site

Folded ß-globin polypeptide

Primary sequence

Val 1 His 2 Leu 3 Thr 4 Pro 5 Glu 6 Glu 7 Lys 8 Ser 9 Ala 10 Val 11

FIGURE 2.21 The primary and tertiary structures of hemoglobin, a protein that transports oxygen on red blood cells. The primary structure is the sequence of amino acids making up the protein. The tertiary structure is the way the protein folds into three dimensions.

sequence of the protein produced. Proteins represent complex adaptations, so we would expect most changes to be deleterious (harmful). But because the code is redundant, many substitutions have no effect on the message of a particular stretch of DNA. The importance of this redundancy is underscored by the fact that the most common amino acids are the ones with the greatest number of codon variants.

Before DNA is translated into proteins, its message is first transcribed into messenger RNA.

DNA can be thought of as a set of instructions for building proteins, but the real work of synthesizing proteins is performed by other molecules. The first step in the translation of DNA into protein occurs when a facsimile of one of the strands of DNA, which will serve as a messenger or chemical intermediary, is made, usually in the cell's nucleus. This copy is **ribonucleic acid, or RNA**. RNA is similar to DNA, except that it has a slightly different chemical backbone, and the base **uracil** (denoted U) is substituted for thymine. RNA comes in several forms, all of which aid in protein synthesis. The form of RNA used in this first step is **messenger RNA (mRNA)**. Messenger RNA then leaves the nucleus and migrates to the **cytoplasm**, the part of the cell outside the nucleus.

Each kind of amino acid is attached to a transfer RNA molecule that bears the anticodon for that amino acid.

Meanwhile, another essential part of protein synthesis is going on in the cytoplasm. Amino acid molecules are bound to a different kind of RNA, called **transfer RNA (tRNA)**. Each tRNA molecule has a triplet of bases, called an **anticodon**, at a particular site (Figure 2.22). Different tRNA molecules have different anticodon sequences and also differ in other ways. Each type of tRNA is bound to the amino acid whose codon binds to the anticodon on the tRNA. For example, one of the codons for the amino acid alanine is the base sequence GCU, and GCU binds only to the anticodon CGA. Thus the tRNA with the anticodon CGA binds only to the amino acid alanine. Because several different codons correspond to each amino acid, there are more types of tRNA than there are kinds of amino acids.

The tRNAs are bound to the appropriate amino acid by a type of enzyme with the tongue-twisting name **aminoacyl-tRNA synthetase**. For each type of tRNA there is a distinctive aminoacyl-tRNA synthetase that recognizes the tRNA and binds it to the appropriate amino acid. Although the anticodon appears to play a role in this crucial task, other distinctive features of the appropriate tRNA seem to be more important.

The ribosome then synthesizes a particular protein by reading the mRNA copy of the gene.

The next step in the process involves the **ribosomes**. Ribosomes are small cellular **organelles** composed of protein and nucleic acid. Organelles are membrane-bounded cellular components that perform a particular function, analogous to the way organs such as the liver perform a function for the body as a whole. The mRNA first binds to ribosomes at a binding site and then moves through the binding site one codon at a time. As each codon of mRNA enters the binding site, a tRNA with a complementary anticodon is drawn from the complex soup of chemicals inside the cell and

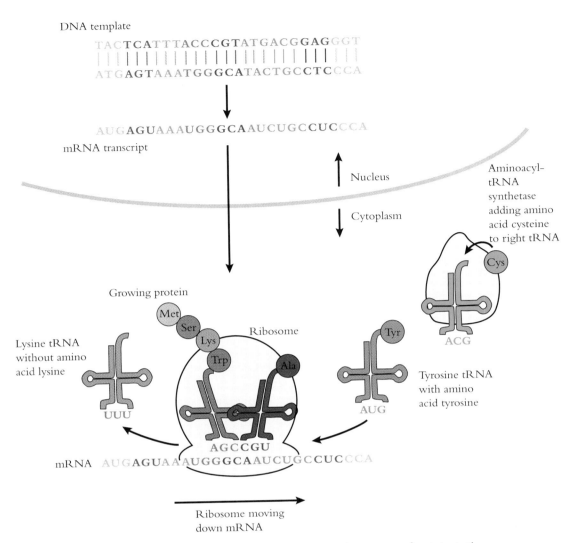

DNA template

TACTCATTTACCCGTATGACGGAGGGT
||||||||||||||||||||||| |||| |||
ATGAGTAAATGGGCATACTGCCTCCCA

AUGAGUAAAUGGGCAAUCUGCCUCCCA

mRNA transcript

Growing protein

Lysine tRNA
without amino
acid lysine

Tyrosine tRNA
with amino
acid tyrosine

Aminoacyl-
tRNA
synthetase
adding amino
acid cysteine
to right tRNA

mRNA   AUGAGUAAAUGGGCAAUCUGCCUCCCA

Ribosome moving
down mRNA

FIGURE 2.22 The information encoded in DNA determines the structure of proteins in the following way. Inside the nucleus, an mRNA copy is made of the original DNA template. The mRNA is coded in three-base codons. Here each codon is given a different color. For example, the sequence AUG codes for the start of a protein and the amino acid methionine, the sequence AGU codes for serine, and the sequence AAA codes for lysine. The mRNA then migrates to the cytoplasm. In the cytoplasm, special enzymes called aminoacyl-tRNA synthetases locate a specific kind of tRNA and attach the amino acid whose mRNA codon will bind to the anticodon on the tRNA. For example, the mRNA codon for cysteine is UGG, and the appropriate anticodon is ACC because U binds to A and C binds to G. In this diagram, matching mRNA codons and tRNA anticodons are given the same color. The initiation of protein assembly is complicated and involves specialized enzymes. Once the process is started, each codon of the mRNA binds to the ribosome. Then the matching tRNA is bound to the mRNA, the amino acid is transferred to the growing protein, the tRNA is released, the ribosome shifts to the next codon, and the process is repeated.

bound to the mRNA. The amino acid bound to the other end of the tRNA is then detached from the tRNA and added to one end of the growing protein chain. The process repeats for each codon, continuing until the end of the mRNA molecule passes through the ribosome. Voilà! A new protein is ready for action.

Regulatory Sequences Control Gene Expression

The DNA sequence in regulatory genes determines when structural genes are expressed.

Gene regulation in *Escherichia coli*, a bacterium that lives in the human gut, provides a good example of how the DNA sequence in regulatory genes controls gene expression. *E. coli* use the sugar glucose as their primary source of energy. When glucose runs short, these bacteria can switch to other sugars, such as lactose, but this switch requires a number of enzymes that allow lactose to be metabolized. The genes for these lactose-specific enzymes are always present, but they are not expressed when there is plenty of glucose available. This is efficient because it would be wasteful to produce these enzymes if they were not needed. The genes for making enzymes that allow the metabolism of lactose to glucose are expressed only when glucose is in short supply *and* sufficient lactose is present. Two regulatory sequences are located near the structural genes that encode the amino acid sequence for three enzymes necessary for lactose metabolism. When there is glucose in the environment, a **repressor** protein binds to one of the two regulatory sequences, thereby preventing the structural genes from being transcribed (Figure 2.23a). When glucose is absent, the repressor protein changes shape and does not bind to the DNA in the regulatory sequence (Figure 2.23b). The second regulatory sequence is an **activator**. In the presence of lactose, an activator protein binds to this DNA sequence, greatly increasing the rate at which the structural genes are transcribed (Figure 2.23c). The specific DNA sequences of the regulatory genes control whether or not the repressor and activator proteins that control DNA transcription bind to the DNA. This means that the sequence of DNA in regulatory genes affects phenotype and creates variation. Therefore, regulatory genes are subject to natural selection, just as structural genes are.

In humans and other eukaryotes, the expression of a given structural gene is often affected by many regulatory sequences, which are sometimes located quite far from the structural gene that they regulate. The proteins that bind to these regulatory genes can interact in complex ways, so that multiple proteins bound to sequences of DNA at widely separate sites may act to activate or repress a particular structural gene. However, the basic principle is usually the same: under some circumstances the DNA sequence in the regulatory gene binds to a protein that in turn affects the expression of a particular structural gene.

The existence of multiple regulatory sequences allows for **combinatorial control** of gene expression. In *E. coli* the combination of a repressor and an activator means that the genes necessary to metabolize lactose are synthesized in a particular combination of circumstances: the presence of lactose *and* the absence of glucose. This is a simple example. Combinatorial control of gene expression is often more complex in eukaryotes: there may be dozens of activators and multiple different combinations that influence gene expression.

Gene regulation allows cell differentiation in complex multicellular organisms like humans.

Complex, multicellular organisms are made of many different kinds of cells, each with its own specific chemical machinery: cells in the liver and pancreas secrete digestive enzymes, nerve cells carry electrical signals throughout the body, muscle cells respond to these signals by performing mechanical work, and so on. Nonetheless, in

(a)

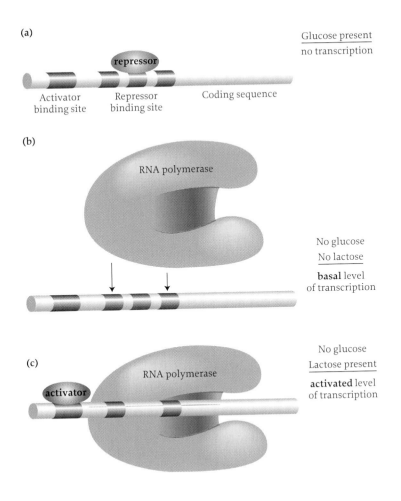

Glucose present
no transcription

Activator   Repressor   Coding sequence
binding site   binding site

(b)

RNA polymerase

No glucose
No lactose

basal level
of transcription

(c)

RNA polymerase

activator

No glucose
Lactose present

activated level
of transcription

FIGURE 2.23 This diagram shows how regulatory sequences binding activator and repressor proteins control the expression of the enzymes necessary to digest lactose in *E. coli*. (a) When glucose is present, the repressor protein binds to the regulatory sequence, thus preventing the enzyme RNA polymerase from creating an mRNA from the DNA template. (b) When glucose is absent and there is no lactose that the RNA polymerase can bind, the genes necessary for metabolism of lactose are expressed, but at a low (basal) level. (c) When lactose is present, the activator protein can bind to its regulatory sequence, thereby greatly increasing the rate at which the RNA polymerase binds and increasing the level of gene expression.

a single individual, all cells have exactly the same genetic composition. The cells differ in their function because different genes are activated in different cell types.

The development of the vertebrate nervous system provides an example of how this works. At a certain point in the development of all vertebrate embryos, cells that are going to give rise to the spinal cord differentiate to form the neural tube. Special cells at one end of this structure secrete a molecule called "Sonic Hedgehog" (after the video game character of the same name). Cells close to the source of this signaling molecule experience a high concentration of Sonic Hedgehog; those more distant experience a lower concentration. The concentration of Sonic Hedgehog affect the expression of genes in these future nerve cells. Low concentrations lead to the expression of genes that destine the cells to become motor neurons, which control muscles; higher concentrations lead to the expression of genes that cause cells to become the neurons of the brain and spinal cord.

A single signal can trigger the complex series of events that transform a cell into a liver cell or a nerve cell because the expression of one gene causes a different set of genes to be expressed, and this in turn causes additional sets of genes to be expressed. For example, a gene called *Pax6* codes for a regulatory protein that is important in the differentiation of the cells in the developing eye. Artificially activating this gene in one cell in a fruit fly antenna leads to the synthesis of a regulatory protein that sets off a cascade of gene expression, which eventually involves the expression of about 2500 other genes and the development of an extra eye on the fly's antenna.

Not All DNA Codes for Proteins

In eukaryotes, the DNA that codes for proteins is interrupted by noncoding sequences called introns.

So far, our description of protein synthesis applies to almost all organisms. However, most of this information was learned through the study of *E. coli*. Like other bacteria, *E. coli* belongs to a group of organisms called the **prokaryotes** because it does not have chromosomes or a cell nucleus. In prokaryotes the DNA sequence that codes for a particular protein is uninterrupted. A stretch of DNA is copied to RNA and then translated into a protein. For many years, biologists thought that the same would prove to be true of **eukaryotes** (organisms like plants, birds, and humans that have chromosomes and a cell nucleus).

Beginning in the 1970s, new recombinant DNA technology allowed molecular geneticists to study eukaryotes. These studies revealed that in eukaryotes, the segment of DNA that codes for a protein is almost always interrupted by at least one, and sometimes many, noncoding sequences called **introns.** (The coding sequences are called **exons.**) Protein synthesis in eukaryotes includes one additional step not mentioned in our discussion so far: after the entire DNA sequence is copied to make an mRNA molecule, and while the mRNA is still in the nucleus, the introns are snipped out and the mRNA molecule is spliced back together. Only then is the mRNA exported outside of the nucleus, where protein synthesis takes place.

The function of introns is unclear. Some biologists think that introns facilitate the evolution of novel, useful proteins. According to this view, complex proteins are constructed from a number of standardized components—analogous to pillars, beams, and arches in buildings. Each component has its own exon because this makes it easier for exons to recombine to form novel proteins. Adherents of this hypothesis think that introns pre-date the origin of eukaryotes, which arose about 2 billion years ago, and that most of the modern proteins in prokaryotes and eukaryotes result from the recombination of exons. Prokaryotes have subsequently lost their introns, perhaps because DNA synthesis is more costly for such small organisms. Other biologists think that introns provide no benefit to the organism as a whole but are bits of "selfish" DNA that replicate themselves by jumping from one place in the genome to another. According to this view, introns arose after the origin of the eukaryotes.

Chromosomes also contain long strings of simple repeated sequences.

Introns are not the only kind of DNA that is not involved in the synthesis of proteins. Chromosomes also contain a lot of DNA that is composed of simple repeated patterns. For example, in fruit flies there are long segments of DNA composed of adenine and thymine in the following monotonous five-base pattern:

... ATAATATAATATAATATAATATAATATAATATAATATAAT ...

In all eukaryotes, simple repeated sequences of DNA are found in particular sites on particular chromosomes.

Only DNA that affects phenotype is subject to natural selection.

DNA found in introns and in simple repeated sequences is not translated into proteins and has no direct effect on phenotype. Consequently, this DNA is not subject to natural selection. Fruit flies with the repeated sequence ATATT would be no more likely to survive and reproduce than those with ATAAT. It turns out that the bulk of the DNA in eukaryotes is found in introns and simple repeated sequences. According to one recent estimate, only 1.5% of human DNA codes for regulatory or structural genes. Thus the evolution of 98.5% of human DNA is controlled not directly by natural selection, but by random processes like mutation. As we will see, this fact is very important because it allows us to date distant evolutionary events using genetic information.

At this point there is some danger of losing sight of the genes amid the discussion of introns, exons, and repeat sequences. A brief reprise may be helpful.

In summary, chromosomes contain an enormously long molecule of DNA. Genes are short segments of this DNA. After suitable editing, a gene's DNA is transcribed into mRNA, which in turn is translated into a protein whose structure is determined by the gene's DNA sequence. Proteins determine the properties of living organisms by selectively catalyzing some chemical reactions and not others, and by forming some of the structural components of cells, organs, and tissues. Genes with different DNA sequences lead to the synthesis of proteins with different catalytic behavior and structural characteristics. Many changes in the morphology or behavior of organisms can be traced back to variations in the proteins and genes that build them. It is important to remember that evolution has a molecular basis. The changes in the genetic constitution of populations that we will explore in Chapter 3 are grounded in the physical and chemical properties of molecules and genes discussed here.

FURTHER READING

Maynard Smith, J. 1989. *Evolutionary Genetics*. Oxford University Press, New York.
Olby, R. C. 1966. *Origins of Mendelism*. Schocken, New York.
Ridley, M. 1996. *Evolution* (2nd ed.). Blackwell Scientific, Oxford, England.
Watson, J. D., Baker, T. A., Bell, S. P., Gann, A., Levine, M, and Losick, R. 2004. *Molecular Biology of the Gene* (5th ed). Benjamin Cummings, San Francisco.

STUDY QUESTIONS

1. Explain why Mendel's principles follow from the mechanics of meiosis.
2. Mendel did experiments in which he kept track of the inheritance of seed texture (wrinkled or smooth). First he created true-breeding lines: parents with smooth seeds produced offspring with smooth seeds, and parents with wrinkled seeds produced offspring with wrinkled seeds. When he crossed wrinkled and smooth peas from these true-breeding lines, all of the offspring were smooth. Which trait is dominant? What happened when he crossed members of the F_1 generation? What would have happened had he backcrossed members of the F_1 generation with individuals from a true-breeding line (like their parents) with smooth seeds? What about crosses between the F_1 generation and the wrinkled-seed true-breeding line?

3. Mendel also did experiments in which he kept track of two traits. For example, he created true-breeding lines of smooth-green and wrinkled-yellow, and then crossed these lines to produce an F_1 generation. What were the F_1 individuals like? He then crossed the members of the F_1 generation to form an F_2 generation. Assuming that the seed color locus and the seed texture locus are on different chromosomes, calculate the ratio of each of the four phenotypes in the F_2 generation. Calculate the ratios (approximately), assuming that the two loci are very close together on the same chromosome.

4. Many animals, such as birds and mammals, have physiological mechanisms that enable them to maintain their body temperature well above the air temperature. They are called "homeotherms." "Poikilotherms," like snakes and other reptiles, regulate body temperature by moving closer to or farther away from sources of heat. Use what you have learned about chemical reactions to develop a hypothesis that explains why animals have mechanisms for controlling body temperature.

5. Why is DNA so well suited for carrying information?

6. Recall the discussion about crossing over, and try to explain why introns might increase the rate of recombination between surrounding exons.

KEY TERMS

variants
crosses
F_0 generation
F_1 generation
F_2 generation
genes
gametes
independent assortment
chromosome
nucleus
mitosis
diploid
homologous pairs
meiosis
haploid
zygote
alleles
homozygous
heterozygous
genotype
phenotype
dominant
recessive

Punnett square
recombination
locus
genome
linked
unlinked
crossing over
deoxyribonucleic acid/
    DNA
bases:
    adenine
    guanine
    cytosine
    thymine
coding sequences
enzymes
structural genes
regulatory sequences/
    regulatory genes
biochemical pathways
proteins
amino acids
primary structure

tertiary structure
hemoglobin
sickle-cell anemia
codons
ribonucleic acid/RNA
uracil
messenger RNA/mRNA
cytoplasm
transfer RNA/tRNA
anticodon
aminoacyl-tRNA
    synthetase
ribosomes
organelles
repressor
activator
combinatorial control
prokaryotes
eukaryotes
introns
exons

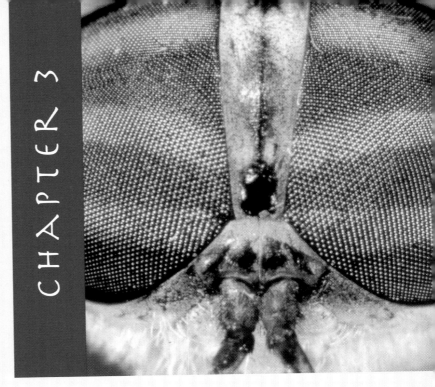

CHAPTER 3

THE MODERN SYNTHESIS

POPULATION GENETICS

Evolutionary change in a phenotype reflects change in the underlying genetic composition of the population.

When we discussed Mendel's experiments in Chapter 2, we briefly introduced the distinction between genotype and phenotype: phenotypes are the observable characteristics of organisms, and genotypes are the underlying genetic compositions. There need not be a one-to-one correspondence between genotype and phenotype. For example, there were only two seed color phenotypes in Mendel's pea populations (yellow and green), but there were three genotypes (*AA*, *Aa*, and *aa*; Figure 3.1).

FIGURE 3.1 Peas were a useful subject for Mendel's botanical experiments because they have a number of dichotomous traits. For example, pea seeds are either yellow or green, but not an intermediate color.

From what we now know about the genetic nature of inheritance, it is also clear that evolutionary processes must entail changes in the genetic composition of populations. When evolution alters the morphology of a character like the finch's beak, there must be a corresponding change in the distribution of genes that control beak development within the population. To understand how Mendelian genetics solves each of Darwin's difficulties, we need to look more closely at what happens to genes in populations that are undergoing natural selection. This is the domain of **population genetics**.

Genes in Populations

Biologists describe the genetic composition of a population by specifying the frequency of alternative genotypes.

It's easier to see what happens to genes in populations if we consider a trait that is controlled by one gene operating at a single locus on a chromosome. Phenylketonuria (PKU), a potentially debilitating, genetically inherited disease in humans, is determined by the substitution of one allele for another at a single locus. Individuals who are homozygous for the PKU allele are missing a crucial enzyme in the biochemical pathway that allows people to metabolize the amino acid phenylalanine. If the disease is not treated, phenylalanine builds up in the bloodstream of children with PKU and leads to severe mental retardation. Fortunately, treatment is possible: people with PKU raised on a special low-phenylalanine diet can develop normally and lead normal lives as adults.

How do evolutionary processes control the distribution of the PKU allele in a population? The first step in answering this question is to characterize the distribution of the harmful allele. Population geneticists do this by specifying the **genotypic frequency**, which is simply the fraction of the population that carries that genotype. Let's label the normal allele *A* and the deleterious PKU allele *a*. Suppose we perform a census of a population of 10,000 individuals and determine the number of individuals with each genotype. Geneticists can do this using established biochemical methods for determining an individual's genotype for a specific genetic locus. Table 3.1 shows the number of individuals with each genotype and the frequencies of each genotype in this hypothetical population. (In most real populations, the frequency of individuals homozygous for the PKU allele is only about 1 in 10,000. We have used larger numbers here to make the calculations simpler.)

Genotypic frequencies must add up to 1.0 because every individual in the popu-

TABLE 3.1   The distribution of individuals with each of the three genotypes in a population of 10,000.

Genotype	Number of Individuals	Frequency of Genotype
aa	2000	freq(*aa*) = 2000/10,000 = 0.2
Aa	4000	freq(*Aa*) = 4000/10,000 = 0.4
AA	4000	freq(*AA*) = 4000/10,000 = 0.4

lation has to have a genotype. We keep track of the frequencies of genotypes, rather than the numbers of individuals with each genotype, because the frequencies provide a description of the genetic composition of populations that is independent of population size. This makes it easy to compare populations of different sizes.

One goal of evolutionary theory is to determine how genotypic frequencies change through time.

A variety of events in the lives of plants and animals act to change the frequency of alternative genotypes in populations from generation to generation. Population geneticists categorize these processes into a number of evolutionary mechanisms, or "forces." The most important mechanisms are sexual reproduction, natural selection, mutation, and genetic drift. In the remainder of this section we will see how sexual reproduction and natural selection alter the frequencies of genes and genotypes, and later in the chapter we will return to consider the effects of mutation and genetic drift.

How Random Mating and Sexual Reproduction Change Genotypic Frequencies

The events that occur during sexual reproduction can lead to changes in genotypic frequencies in a population.

First let's consider the effects of the patterns of inheritance that Mendel observed. Imagine that men and women do not choose their mates according to whether or not they are afflicted with PKU, but mate randomly with respect to the individual's genotype for PKU. It is important to study the effects of random mating because, for most genetic loci, mating is random. Even though humans may choose their mates with care, and might even avoid mates with particular genetic characteristics, such as PKU, they cannot choose mates with a particular allele at each locus because there are about 30,000 genetic loci in humans. Random mating between individuals is equivalent to the random union of gametes. In this sense it's not really different from oysters shedding their eggs and sperm into the ocean, where chance dictates which gametes will form zygotes.

The first step in determining the effects of sexual reproduction on genotypic frequencies is to calculate the frequency of the PKU allele in the pool of gametes.

We can best understand how segregation affects genotypic frequencies by breaking the process into two steps. In the first step we determine the frequency of the PKU allele among all the gametes in the mating population. Remember that the a allele is the PKU allele and the A allele is the "normal" allele. Table 3.1 gives the frequency of each of the three genotypes among the parental generation. We want to use this information to determine the genotypic frequencies among the F_1 generation. First we calculate the frequencies of the two types of alleles in the pool of gametes. (The frequency of an allele is also referred to as its **gene frequency**.) Let's label the frequency of A as

p and the frequency of a as q. (Because there are only two alleles, $p + q = 1$.) If all individuals produce the same number of gametes, then we can calculate q as follows:

$$q = \frac{\text{no. of } a \text{ gametes}}{\text{total no. of gametes}}$$

Note that this is simply the definition of a frequency. We can calculate the values of the numerator and denominator of this fraction from information we already know. Because a gametes can be produced only by aa and Aa individuals, the total number of a gametes is simply the sum of the number of Aa and aa parents multiplied by the number of a gametes that each parent produces. The denominator is the number of gametes per parent multiplied by the total number of parents. Hence,

$$q = \frac{\left(\begin{array}{c}\text{no. of } a \\ \text{gametes per} \\ aa \text{ parent}\end{array}\right)\left(\begin{array}{c}\text{no. of } aa \\ \text{parents}\end{array}\right) + \left(\begin{array}{c}\text{no. of } a \\ \text{gametes per} \\ Aa \text{ parent}\end{array}\right)\left(\begin{array}{c}\text{no. of } Aa \\ \text{parents}\end{array}\right)}{(\text{no. of gametes per parent})(\text{total no. of parents})} \tag{3.1}$$

To simplify this equation, let's first examine the terms that involve numbers of individuals. Remember that the population size is 10,000 individuals. This means that the number of aa parents is equal to the frequency of aa parents multiplied by 10,000. Similarly, the number of Aa parents is equal to the frequency of Aa parents multiplied by 10,000. Now we examine the terms that involve numbers of gametes. Suppose each parent produces two gametes. For aa individuals, both gametes contain the a allele, so the number of a gametes per aa parent is 2. For Aa individuals, half of the gametes will carry the a allele and half will carry the A allele, so here the number of a gametes per Aa parent is 0.5×2. Now we substitute all these values into Equation (3.1) to get

$$q = \frac{2[\text{freq}(aa) \times 10{,}000] + (0.5 \times 2)[\text{freq}(Aa) \times 10{,}000]}{2 \times 10{,}000}$$

We can reduce this fraction by dividing the top and bottom by $2 \times 10{,}000$, which yields the following formula for the frequency of the a allele in the pool of gametes:

$$q = \text{freq}(aa) + 0.5 \times \text{freq}(Aa) \tag{3.2}$$

Notice that this form of the formula contains neither the population size nor the average number of gametes per individual in the population. Under normal circumstances the population size and the average number of gametes per individual do not matter. This means that you can use this expression as a general formula for calculating gene frequencies among the gametes produced by any population of individuals, as long as the genetic locus of interest has only two alleles. It is important to keep in mind that the formula results from applying Mendel's laws and from counting the number of a gametes produced.

By using Equation (3.2) and values from Table 3.1, for this particular population we get $q = 0.2 + (0.5 \times 0.4) = 0.4$. Because $p + q$ must sum to 1.0, p (the frequency of the A gametes) must be 0.6. Notice that the frequency of each allele in the pool of gametes is the same as the frequency of the same allele among parents.

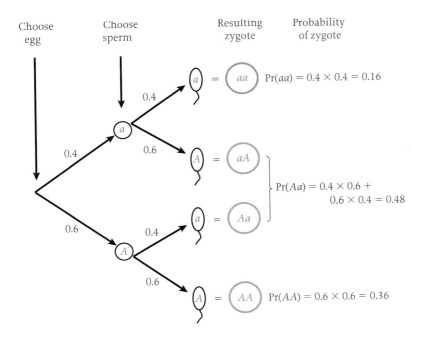

Choose egg   Choose sperm   Resulting zygote   Probability of zygote

a = aa $\text{Pr}(aa) = 0.4 \times 0.4 = 0.16$

0.4

a

0.4   0.6

A = aA

$\text{Pr}(Aa) = 0.4 \times 0.6 + 0.6 \times 0.4 = 0.48$

a = Aa

0.6   0.4

A

0.6

A = AA $\text{Pr}(AA) = 0.6 \times 0.6 = 0.36$

FIGURE 3.2 This event tree shows how to calculate the frequency of each genotype among the zygotes created by the random union of gametes. In this pool of gametes, the frequency of a is 0.4 and the frequency of A is 0.6. The first node represents the choice of an egg. There is a 40% chance that the egg will carry the a allele and a 60% chance it will carry the A allele. The second node represents the choice of the sperm. Once again there is a 40% chance of drawing an a-bearing sperm and a 60% chance of drawing an A-bearing sperm. We can calculate the probability of each genotype by computing the probability of taking a path through the tree. For example, the probability of getting an aa zygote is $0.4 \times 0.4 = 0.16$.

The next step is to calculate the frequencies of all the genotypes among the zygotes.

Now that we have calculated the frequency of the PKU *allele* among the pool of gametes, we can determine the frequencies of the *genotypes* among the zygotes. If each zygote is the product of the random union of two gametes, the process of zygote formation can be schematically represented in an event tree (Figure 3.2) similar to the one we used to represent Mendel's crosses (for example, Figure 2.9). First we select a gamete—say, an egg. The probability of selecting an egg carrying the a allele is 0.4 because this is the frequency of the a allele in the population. Now we randomly draw a second gamete, the sperm. Again, the probability of getting a sperm carrying an a allele is 0.4. The chance that these two randomly chosen gametes are both a is 0.4×0.4, or 0.16. Figure 3.2 also shows how to compute the probabilities of the other two genotypes. Note that the sum of the three genotypes always equals 1.0, because all individuals must have a genotype.

If we form a large number of zygotes by randomly drawing gametes from the gamete pool, we will obtain the genotypic frequencies shown in Table 3.2. Compare the frequencies of each genotype in the parental population (Table 3.1) with the frequencies of each genotype in the F_1 generation: the genotypic frequencies have changed because the processes of independent segregation of alleles into gametes and

TABLE 3.2 The distribution of genotypes in the population of zygotes in the F_1 generation.

freq(aa)	=	0.4×0.4	=	0.16
freq(Aa)	=	$(0.4 \times 0.6) + (0.4 \times 0.6)$	=	0.48
freq(AA)	=	0.6×0.6	=	0.36

random mating alter the distribution of alleles in zygotes. As a result, genotypic frequencies between the F_0 generation and the F_1 generation are altered. [Note, however, that the frequencies of the two *alleles*, *a* and *A*, have not changed; you can check this out using Equation (3.2).]

When no other forces (such as natural selection) are operating, genotypic frequencies reach stable proportions in just one generation. These proportions are called the Hardy–Weinberg equilibrium.

If no other processes act to change the distribution of genotypes, the set of genotypic frequencies in Table 3.2 will remain unchanged in subsequent generations. That is, if members of the F_1 generation mate at random, the distribution of genotypes in the F_2 generation will be exactly the same as the distribution of genotypes in the F_1 generation. The fact that genotypic frequencies remain constant was recognized independently by British mathematician Godfrey Harold Hardy and German physician Wilhelm Weinberg in 1908, and these constant frequencies are now called the **Hardy–Weinberg equilibrium**. As we will see later in this chapter, the realization that it's not just sexual reproduction alone that alters phenotypic and genotypic frequencies was the key to understanding how variation is maintained.

In general, the Hardy–Weinberg proportions for a genetic locus with two alleles are

$$\text{freq}(aa) = q^2$$
$$\text{freq}(Aa) = 2pq$$
$$\text{freq}(AA) = p^2 \qquad\qquad (3.3)$$

where q is the frequency of allele a, and p is the frequency of allele A. Figure 3.3 shows how to calculate these frequencies using a Punnett square.

If no other processes act to change genotypic frequencies, the Hardy–Weinberg equilibrium frequencies will be reached after only one generation and will remain unchanged thereafter. Moreover, if the Hardy–Weinberg proportions are altered by

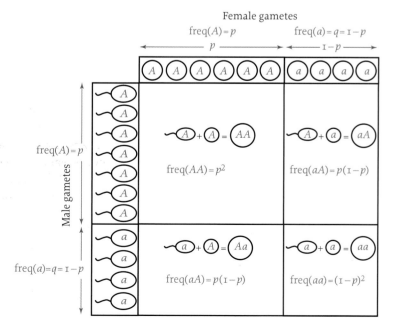

FIGURE 3.3 This Punnett square shows a second way to calculate the frequency of each type of zygote when there is random mating. The horizontal axis represents the proportion of *A* and *a* eggs, and is thus divided into fractions *p* and *q*, where *q* = 1 – *p*. The vertical axis is divided according to the proportion of sperm carrying each allele, and again it is divided into fractions *p* and *q*. The areas of the rectangles formed by the intersection of the vertical and horizontal dividing lines give the proportion of zygotes that result from each of the four possible fertilization events. The area of the square containing *aa* zygotes is q^2, the area of the square containing *AA* zygotes is p^2, and the total area of the two rectangles containing *Aa* zygotes is $2pq$.

BOX 3.1

Genotypic Frequencies after Two Generations of Random Mating

From Equation (3.2) given in the text, we know that the frequency of the gametes produced by adults in the second generation carrying the a allele will be

$$q = \text{freq}(aa) + 0.5 \times \text{freq}(Aa)$$
$$= 0.16 + 0.5 \times 0.48$$
$$= 0.16 + 0.24$$
$$= 0.4$$

Because there are only two alleles, the frequency of A is 0.6. As we asserted in the text, the frequency of the two alleles remains unchanged, and the frequencies of the gametes produced by random mating among members of this generation are the same as in the previous generation. The calculations for these frequencies are the same as those shown in Table 3.2.

chance, the population will return to Hardy–Weinberg proportions in one generation. If this seems like an unlikely conclusion, work through the calculations in Box 3.1, which shows how to calculate the genotypic frequencies after a second episode of segregation and random mating. As you can see, genotypic frequencies remain constant.

We have seen that sexual reproduction and random mating can change the distribution of genotypes, which will reach equilibrium after one generation. We have also seen that neither process changes the frequencies of alleles. Clearly, sexual reproduction and random mating alone cannot lead to evolution over the long run. We now turn to a process that can produce changes in the frequencies of alleles: natural selection.

How Natural Selection Changes Gene Frequencies

If different genotypes are associated with different phenotypes and those phenotypes differ in their ability to reproduce, then the alleles that lead to the development of the favored phenotype will increase in frequency.

Genotypic frequencies in the population will remain at the Hardy–Weinberg proportions [Equation (3.3)] as long as all genotypes are equally likely to survive and produce gametes. For PKU, this is approximately correct in wealthy countries where the disease can be treated. However, this assumption will not be valid in environments in which PKU is not treated. Suppose a catastrophe interrupted the supply of modern medical care to our hypothetical population just after a new generation of zygotes was formed. In this case, virtually none of the PKU individuals (genotype aa) would survive to reproduce. Suppose the frequency of the PKU allele was 0.4 among zygotes and the population was in Hardy–Weinberg equilibrium. Let's compute q', the frequency of the PKU allele among adults, which is given by

$$q' = \frac{\text{no. of } a \text{ gametes produced by adults in the next generation}}{\text{total no. of gametes produced by adults in the next generation}} \quad (3.4)$$

We have already calculated the frequency of each of the genotypes among zygotes just after conception (see Table 3.2), but if PKU is a lethal disease without treatment, not all of these individuals will survive. Now we need to calculate the frequency of a gametes after selection, which we can do by expanding Equation (3.4) as follows:

$$q' = \frac{\begin{pmatrix} \text{no. of } a \\ \text{gametes per} \\ aa \text{ parent after} \\ \text{selection} \end{pmatrix}\begin{pmatrix} \text{no. of } aa \\ \text{parents} \\ \text{after} \\ \text{selection} \end{pmatrix} + \begin{pmatrix} \text{no. of } a \\ \text{gametes per} \\ Aa \text{ parent after} \\ \text{selection} \end{pmatrix}\begin{pmatrix} \text{no. of } Aa \\ \text{parents} \\ \text{after} \\ \text{selection} \end{pmatrix}}{(\text{no. of gametes per parent})(\text{total no. of parents after selection})} \quad (3.5)$$

Because none of the aa individuals will survive, the first term in the numerator will be equal to zero. If we assume that all of the AA and Aa parents survive, then the number of parents after selection is $10{,}000 \times [\text{freq}(AA) + \text{freq}(Aa)]$ and thus Equation (3.5) can be simplified to

$$q' = \frac{(0.5 \times 2)(0.48 \times 10{,}000)}{2 \times (0.36 + 0.48) \times 10{,}000} = 0.2857$$

This calculation shows that the frequency of the PKU allele in adults is 0.2857, a big decrease from 0.4.

Several important lessons can be drawn from this example:

- Selection cannot produce change unless there is variation in the population. If all the individuals were homozygous for the normal allele, gene frequencies would not change from one generation to the next.
- Selection does not operate directly on genes and does not change gene frequencies directly. Instead, natural selection changes the frequency of different phenotypes. In this case, individuals with PKU cannot survive without treatment. Selection decreases the frequency of the PKU allele because it is more likely to be associated with the lethal phenotype.
- The strength and direction of selection depend on the environment. In an environment with medical care, the strength of selection against the PKU allele is negligible.

It is also important to see that although this example shows how selection can change gene frequencies, it does not yet show how selection can lead to the evolution of new adaptations. Here, all phenotypes were present at the outset; all selection did was change their relative frequency.

THE MODERN SYNTHESIS

The Genetics of Continuous Variation

When Mendelian genetics was rediscovered, biologists at first thought it was incompatible with Darwin's theory of evolution by natural selection.

Darwin believed that evolution proceeded by the gradual accumulation of small changes. But Mendel and the biologists who elucidated the structure of the genetic system around the turn of the twentieth century were dealing with genes that had a noticeable effect on the phenotype. The substitution of one allele for another in

Mendel's peas changed pea color. Genetic substitutions at other loci had visible effects on the shape of the pea seeds and the height of the pea plants. Genetics seemed to prove that inheritance was fundamentally discontinuous, and early-twentieth-century geneticists like Hugo de Vries and William Bateson argued that this fact could not be reconciled with Darwin's idea that adaptation occurs through the accumulation of small variations. If one genotype produces short plants and the other two genotypes produce tall plants, then there will be no intermediate types, and the size of pea plants cannot change in small steps. In a population of short plants, tall ones must be created all at once by mutation, not gradually lengthened over time by selection. Most biologists of the time found these arguments convincing, and consequently Darwinism was in decline during the early part of the twentieth century.

Mendelian genetics and Darwinism were eventually reconciled, resulting in a body of theory that solved the problem of explaining how variation is maintained.

In the early 1930s, the British biologists Ronald A. Fisher and J. B. S. Haldane and the American biologist Sewall Wright showed how Mendelian genetics could be used to explain continuous variation (Figure 3.4). We will see how their insights led to the resolution of the two main objections to Darwin's theory: (1) the absence of a theory of inheritance, and (2) the problem of accounting for how variation is maintained in populations. When the theory of Wright, Fisher, and Haldane was combined with Darwin's theory of natural selection, and with modern field studies by biologists such as Theodosius Dobzhansky, Ernst Mayr, and George Gaylord Simpson, a powerful explanation of organic evolution emerged. This body of theory, and the supporting empirical evidence, is now called the **modern synthesis.**

Continuously varying characters are affected by genes at many loci, each locus having only a small effect on phenotype.

To see how the theory of Wright, Fisher, and Haldane works, let's start with an unrealistic, but instructive, case. Suppose that there is a measurable, continuously varying character such as beak depth, and suppose that two alleles, + and –, operating at

(a)

(b)

(c)

FIGURE 3.4 Shown here are the three architects of the modern synthesis, which showed how Mendelian genetics could be used to account for continuous variation: (a) Ronald A. Fisher, (b) J. B. S. Haldane, and (c) Sewall Wright.

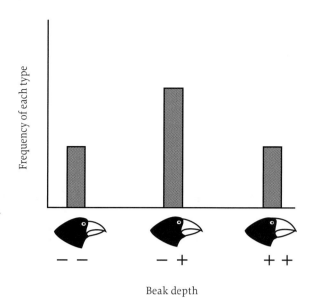

FIGURE 3.5 The hypothetical distribution of beak depth, assuming that beak depth is controlled by a single genetic locus with two alleles that occur at equal frequency. The height of each bar represents the fraction of birds in the population with a given beak depth. The heights of the bars are computed by means of the Hardy–Weinberg formula. The birds with the smallest beaks are homozygous for the – allele and have a frequency of 0.5 × 0.5 = 0.25. The birds with intermediate beaks are heterozygotes and have a frequency of 2(0.5 × 0.5) = 0.5. The birds with the deepest beaks are homozygous for the + allele and thus have a frequency of 0.5 × 0.5 = 0.25.

a single genetic locus control the character. We'll assume that the gene at this locus influences the production of a hormone that stimulates beak growth, and that each allele leads to production of a different amount of the growth hormone. Let's say that each "dose" of the + allele increases the beak depth, while a – dose decreases it. Thus, + + individuals have the deepest beaks, – – individuals have the shallowest beaks, and + – individuals have intermediate beaks. In addition, suppose the frequency of the + allele in the population is 0.5. Now we use the Hardy–Weinberg rule [Equation (3.3)] to calculate the frequencies of different beak depths in the population. A quarter of the population will have deep beaks (+ +), half will have intermediate beaks (+ –), and the remaining quarter will have shallow beaks (– –; Figure 3.5).

This does not look like the smooth, bell-shaped distribution of beak depths that the Grants observed on Daphne Major (see Chapter 1). If beak depth were controlled by a single locus, natural selection could not increase beak depth in small increments. But imagine what would happen if other genes at a second locus on a different chromosome also affected beak depth, perhaps because they controlled the synthesis of the receptors for the growth hormone. As before, we assume there is a + allele that leads to larger beaks and a – allele that leads to smaller beaks. Using the Hardy–Weinberg proportions and assuming the independent segregation of chromosomes, we can show that there are now more types of genotypes, and the distribution of phenotypes begins to look somewhat smoother (Figure 3.6). Now imagine that a third locus controls the calcium supply to the growing beak, and that once again there is a + allele leading to larger beaks and a – allele leading to smaller beaks. As Figure 3.7 shows, the distribution of beak depths is now even more like the bell-shaped distribution seen in nature. However, small gaps still exist in this distribution of beaks. If genes were the only influence on beak depth, we might expect to see a more broken distribution than the Grants actually observed in the Galápagos.

The observed distribution of phenotypic values is a smooth, bell-shaped curve because of the effect of **environmental variation**. The phenotypic expression of all characters, whether affected by one locus or by many loci, will depend, at least to some extent, on the environment in which the organism develops. For example, the size of a bird's beak will depend on how well the bird is nourished during its development.

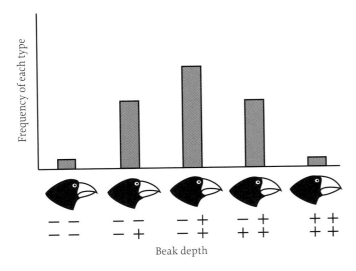

FIGURE 3.6 When beak depth is controlled by two loci, each locus having two alleles that occur with equal frequency, intermediate types are observed. The frequencies are calculated by means of a more advanced form of the Hardy–Weinberg formula, on the assumption that the genotypes at each locus are independent.

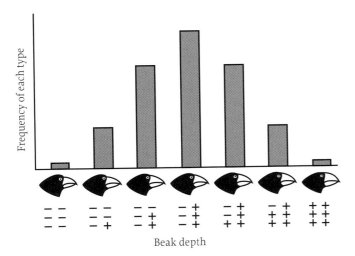

FIGURE 3.7 When beak depth is controlled by three loci, each locus having two alleles that occur with equal frequency, the distribution of phenotypes begins to resemble a bell-shaped curve.

When there is only one locus and the effect of an allelic substitution at that locus is large, environmental variation is not important; it is easy to distinguish the phenotypes associated with different genotypes. This was the case for seed color in Mendel's peas. But when there are many loci and each has a small effect on the phenotype, environmental variation tends to blur together the phenotypes associated with different genotypes. You can't be sure whether you are measuring a + + + − − − bird that matured in a good environment or a + + + + − − bird that grew up in a poor environment. The result is that we observe a smooth curve of variation, as shown in Figure 3.8.

Darwin's view of natural selection is easily incorporated into the genetic view that evolution typically results from changes in gene frequencies.

Darwin knew nothing about genetics, and his theory of adaptation by natural selection was framed in purely phenotypic terms: there is a struggle for existence, there is phenotypic variation that affects survival and reproduction, and this phenotypic variation is heritable. In Chapter 1 we saw how this theory could explain the adaptive changes in beak depth in a population of Darwin's finches on the Galápagos

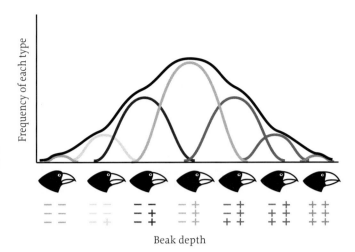

FIGURE 3.8 Environmental variation influences the distribution of beak depths. Again, as in Figure 3.7, three loci affect beak depth. Now, however, we assume that environmental conditions cause individuals with the same genotype to develop beaks of different depths. Each colored curve gives the distribution of phenotypes associated with the genotype of the same color, and the black curve gives the overall distribution of phenotypes.

Islands. As we saw earlier in this chapter, however, population geneticists take the seemingly different view that evolution means changes in allelic frequencies by natural selection. But these two views of evolution are easily reconciled. Suppose that Figure 3.8 gave the distribution of beak depths before the drought on Daphne Major. Remember that individuals with deep beaks were more likely to survive the drought and reproduce than individuals with smaller beaks were. Figure 3.8 illustrates that individuals with deep beaks are more likely to have + alleles at the three loci assumed to affect beak depth. Thus, at each locus, + + individuals have deeper beaks on average than + − individuals, which in turn have deeper beaks on average than − − individuals. Because individuals with deeper beaks had higher fitness, natural selection would favor the + alleles at each of the three loci affecting beak depth, and thus the + alleles would increase in frequency.

How Variation Is Maintained

Genetics provides a ready explanation for why the phenotypes of offspring tend to be <u>intermediate</u> between those of their parents.

Recall from Chapter 1 that the blending model of inheritance appealed to nineteenth-century thinkers because it explained the fact that, for most continuously varying characters, offspring are intermediate between their parents. However, the genetic model developed by Fisher, Wright, and Haldane is also consistent with this fact. To see why, consider a cross between the individuals with the biggest and smallest beaks: (+ + + + + +) × (− − − − − −). All of the offspring will be (+ − + − + −), intermediate between the parents because during development the effects of + and − alleles are averaged. Other matings will produce a distribution of different kinds of offspring, but intermediate types will tend to be the most common.

There is <u>no blending of genes</u> during sexual reproduction. only phenotypes

We know from population genetics that, in the absence of selection (and other factors to be discussed later in this chapter), genotypic frequencies reach equilibrium after

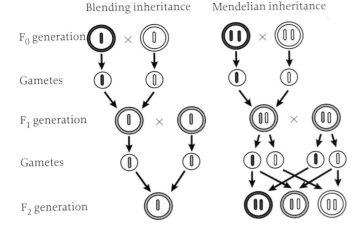

Blending inheritance   Mendelian inheritance

F₀ generation

Gametes

F₁ generation

Gametes

F₂ generation

FIGURE 3.9 The blending model of inheritance assumes that the hereditary material is changed by mating. When red and white parents are crossed to produce a pink offspring, the blending model posits that the hereditary material has mixed, so that when two pink individuals mate, they produce only pink offspring. According to Mendelian genetics, however, the effects of genes are blended in their *expression* to produce a pink phenotype, but the genes themselves remain unchanged. Thus, when two pink parents mate, they can produce white, pink, or red offspring.

one generation and the distribution of phenotypes does not change. Furthermore, we know that sexual reproduction produces no blending in the genes, even though offspring may appear to be intermediate between their parents. This is because genetic transmission involves faithful copying of the genes themselves and their reassembly in different combinations in zygotes. The only blending that occurs takes place at the level of the expression of genes in phenotypes. The genes themselves remain distinct physical entities (Figure 3.9).

These facts do not completely solve the problem of the maintenance of variation, because selection tends to deplete variation. When selection favors birds with deeper beaks, we might expect – alleles to be replaced at all three loci affecting the trait, leaving a population in which every individual has the genotype + + + + + +. There would still be phenotypic variation due to environmental effects, but without genetic variation there can be no further adaptation.

Mutation slowly adds new variation.

Genes are copied with amazing fidelity, and their messages are protected from random degradation by a number of molecular repair mechanisms. Every once in a while, however, a mistake in copying is made and goes unrepaired, and a new allele is introduced into the population. In Chapter 2 we learned that genes are pieces of DNA. Certain forms of ionizing radiation (such as X-rays) and certain kinds of chemicals damage the DNA and alter the message that it carries. These changes are called **mutations**, and they add variation to a population by continuously introducing new alleles, some of which may produce novel phenotypic effects that selection can assemble into adaptations. Although rates of mutation are very low—ranging from 1 in 100,000 to 1 in 10 million per locus per gamete in each generation—this process plays an important role in generating variation.

Low mutation rates can maintain variation because a lot of variation is protected from selection.

For characters that are affected by genes at many different loci, low rates of mutation can maintain variation in populations. This is possible because many different genotypes generate intermediate phenotypes that are favored by stabilizing selection. If individuals with a variety of genotypes are equally likely to survive and reproduce, a considerable amount of variation is protected (or hidden) from selection. To make this concept clearer, let's consider the action of stabilizing selection on beak depth in the medium ground finch. Remember that stabilizing selection occurs when birds with intermediate-sized beaks have higher fitness than birds with deeper or shallower beaks. If beak depth is affected by genes at three loci, as shown in Figure 3.8, individuals with several different genotypes may develop very similar, intermediate-sized beaks. For example, individuals that are + + at one locus affecting the trait may be − − at another locus, and thus may have the same phenotype as an individual that is + − at both loci. Thus, when a large number of loci affect a single trait, only a small fraction of the genetic variability present in the population is expressed phenotypically. As a result, selection removes variation from the population very slowly. The processes of segregation and recombination slowly shuffle and reshuffle the genome, and thereby expose the hidden variation to natural selection in subsequent generations (Figure 3.10). This process provides the solution to Darwin's other dilemma; because a considerable amount of variation is protected from selection, a very low mutation rate can maintain variation despite the depleting action of selection.

Hidden variation explains why selection can move populations far beyond their initial range of variation.

Remember from Chapter 1 Fleeming Jenkin's argument that Darwin's theory could not explain cumulative evolutionary change because it provided no account of how a population could evolve beyond the initial range of variation present. Selection would cull away all of the small-beaked finches, he would have argued, but it could never make the average beak bigger than the biggest beak initially present. If this argument were correct, then selection could never lead to cumulative, long-term change.

Jenkin's argument was wrong because it failed to take hidden variation into account. Hidden variation is always present in continuously varying traits. Let's suppose that environmental conditions favor larger beaks. When beak depth is affected by genes at many loci, the birds in a population with the deepest beaks do not carry

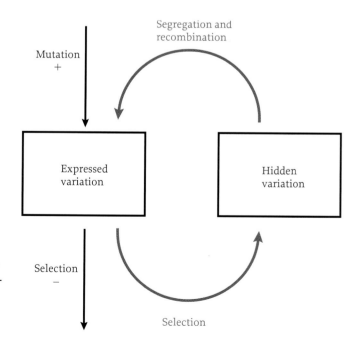

FIGURE 3.10 There are two pools of genetic variation: hidden and expressed. Mutation adds new genetic variation to the pool of expressed variation, and selection removes it. Segregation and recombination shuffle variation back and forth between the two pools within each generation.

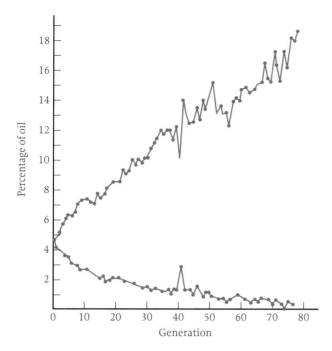

FIGURE 3.11 Selection can carry a population far beyond the original range of variation because at any given time a lot of genetic variation is not expressed as phenotypic variation. The range of variation in the oil content of corn at the beginning of this experiment was 4% to 6%. After about 80 generations, oil content in the line selected for high oil content had increased to 19%, and oil content in the low-oil line had decreased to less than 1%.

all + alleles. They carry a lot of + alleles and some − alleles. When the finches with the shallowest beaks die, alleles leading to small beaks are removed from the breeding population. As a result, the frequency of + alleles at every locus increases, but because even the deepest-beaked individuals had some − alleles, a huge amount of variation remains. This variation is shuffled through the process of sexual reproduction. Because + alleles become more common, more of these alleles are likely to be combined in the genotype of a single individual. The greater the proportion of + alleles in an individual, the larger the beak will be. Thus the biggest beak will be larger than the biggest beak in the previous generation. In the next generation the same thing happens again: the deepest-beaked individuals still carry − alleles, but fewer than before. As a result, the biggest beaks can be bigger than any beaks in the previous generation.

This process can go on for many generations. An experiment on oil content in corn begun at the Illinois Experiment Station in 1896 provides a good example. Researchers selected for high and low oil content in corn, and as Figure 3.11 shows, each generation showed significant change. In the initial population of 163 ears of corn, oil content ranged from 4% to 6%. After nearly 80 generations of selection, both the high and low values for oil content far exceeded the initial range of variation. Researchers were then even able to reverse the direction of adaptation by taking plants from the high-oil line and selecting for low oil content.

An even more dramatic example is the Pekingese, whose very existence is a testament to the power of hidden variation (Figure 3.12). In a few thousand generations, a mere instant on the evolutionary timescale, wolves were transformed into Pekingese, as well as into bulldogs, Great Danes, dachshunds, Chihuahuas, Irish wolfhounds, and toy poodles. This is far too little time for a large number of new mutations to accumulate. Instead, most of the genes necessary for the development of Irish wolfhounds and Pekingese must have been present in the ancestral population of wolves from which dogs were selected, their effects hidden by other genes. Dog breeders, acting as agents of selection, chose the genes leading to small size, long hair, and funny little faces, and

FIGURE 3.12 This Pekingese is smaller than the smallest wolf, although wolves are the ancestors of all domestic breeds of dogs. Some of Darwin's critics argued that natural selection could not produce this kind of transformation because it could not lead to changes that fell outside the range of variation found in the original population. However, genes for very small size are present in wolves, but their effects are hidden by other genes.

made them more common in some populations; segregation and recombination brought these features together over time in the Pekingese. The same process created dogs that are well adapted for herding sheep, ferreting out badgers, retrieving game, chasing mechanical rabbits around a track, and warming the laps of Chinese emperors.

NATURAL SELECTION AND BEHAVIOR

The evolution of mate guarding in the soapberry bug illustrates how flexible behavior can evolve.

So far we have considered the evolution of morphological characters, like beak depth and eye morphology, that do not change once an individual has reached adulthood. In much of this book we will be interested in the evolution of the behavior of humans and other primates. Behavior is different from morphology in an important way: it is flexible, and individuals adjust their behavior in response to their circumstances. Some people think natural selection cannot account for flexible responses to environmental contingencies because natural selection acts only on phenotypic variation that results from genetic differences. Although this view is very common (particularly among social scientists), it is incorrect. To see why, let's consider an elegant empirical study that illustrates exactly how natural selection can shape flexible behavioral responses.

The soapberry bug (*Jadera haematoloma*), a seed-eating insect found in the southeastern part of the United States, has been studied by biologist Scott Carroll of the University of New Mexico. (It's OK to call them "bugs" because they are members of the insect order Hemiptera, the true bugs.) Adult soapberry bugs are bright red and black, and about 1 to 1.5 cm (0.5 in.) long (Figure 3.13). They gather in huge groups near the plants they eat. During mating, males mount females and copulate. The transfer of sperm typically takes about 10 minutes. However, males remain in the copulatory position, sometimes for hours, securely anchored to females by large genital hooks. This behavior is called **mate guarding**. Biologists believe that the function of mate guarding is to prevent other males from copulating with the female before she lays her eggs. When a female mates with several males, they share in the paternity of the eggs that she lays. By guarding his mate and preventing her from mating with other males, a male can increase his reproductive success. However, mate guarding also has a cost: a male cannot find and copulate with other females while guarding a mate. The relative magnitude of the costs and benefits of mate guarding depends on the **sex ratio** (the relative numbers of males and females). When the ratio of males to females is high, males have little chance of finding another female, and guarding is the best strategy. When females are more common than males, the chance of finding an unguarded female increases, so males may benefit more from looking for additional mates than from guarding.

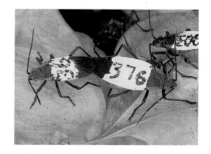

FIGURE 3.13 The male soapberry bug on the left is guarding the (larger) female below him. Carroll painted numbers on the bugs in order to identify individuals. (Photograph courtesy of Scott Carroll.)

Behavioral plasticity allows male soapberry bugs to vary their mate-guarding behavior adaptively in response to variation in the local abundance of females.

In populations of soapberry bugs in western Oklahoma, the sex ratio is quite variable. In some places there are equal numbers of males and females; in others, there are twice as many males as females. Males guard their mates more often where females

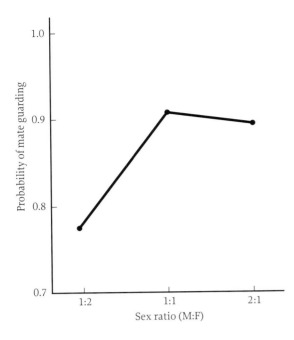

FIGURE 3.14 The probability of mate guarding is a function of sex ratio in this population of the soapberry bug. Males are less likely to guard their mates when females are relatively more common than when they are relatively rare.

are rare than where they are common (Figure 3.14). Two possible mechanisms might produce this pattern: (1) males in populations with high sex ratios might differ genetically from males in populations with low sex ratios, or (2) males might adjust their behavior in response to the local sex ratio. To distinguish between these two possibilities, Carroll brought soapberry bugs into the laboratory and created populations with different sex ratios. Then he watched males mate in each of these populations. If the mate-guarding trait were **canalized** (that is, showing the same phenotype in a wide range of environments), then a male would behave the same way in each population; if the trait were **plastic**, then a male would adjust his behavior in relation to the local sex ratio. Carroll's data confirmed that soapberry bug males in Oklahoma have a plastic behavioral strategy that causes them to modify their mating behavior in response to current social conditions.

Evidence suggests that the soapberry bug's plasticity has evolved in response to the variability in conditions in Oklahoma.

Most soapberry bugs live south of Oklahoma, in warmer, more stable habitats like the Florida Keys, and the sex ratio in these areas is always close to even. (Hence, females are relatively rare.) Carroll subjected male soapberry bugs from the Florida Keys to the same experimental protocol as the bugs from Oklahoma. As Figure 3.15 shows, the Florida males did not change their behavior in response to changes in sex ratio. They guarded their mates about 90% of the time, regardless of the abundance of females. In a stable environment like the Florida Keys, sex ratios do not vary much from time to time, and the ability to adjust mate-guarding behavior provides no advantage. Behavioral flexibility is costly in a number of ways. For example, flexible males must spend time and energy assessing the sex ratio before they mate, males sometimes make mistakes about the local sex ratio and behave inappropriately, and flexibility probably requires a more complex nervous system. Thus a simple fixed behavioral rule is likely to be best in stable environments. In the variable climate of Oklahoma, however, the

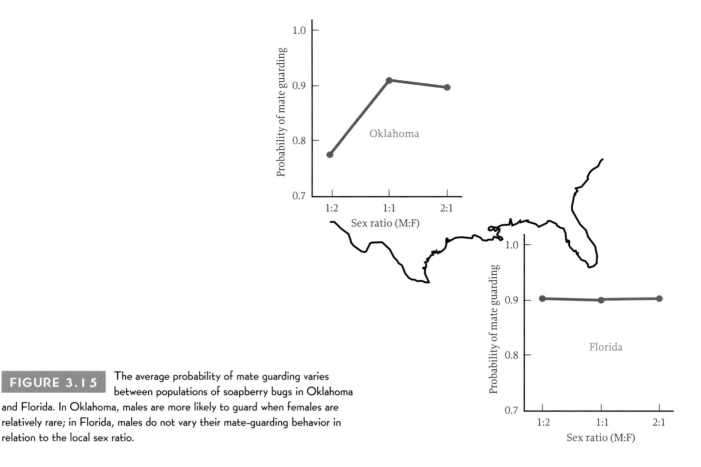

FIGURE 3.15 The average probability of mate guarding varies between populations of soapberry bugs in Oklahoma and Florida. In Oklahoma, males are more likely to guard when females are relatively rare; in Florida, males do not vary their mate-guarding behavior in relation to the local sex ratio.

ability to adjust mate-guarding behavior provides enough fitness benefits to compensate for the costs of maintaining behavioral flexibility.

A comparison of the mate-guarding behavior in two different populations of soapberry bugs indicates that behavioral plasticity evolves when the nature of the behavioral response to the environment is genetically variable.

How did behavioral flexibility evolve in the Oklahoma bugs? It evolved like any other adaptation: by the selective retention of beneficial genetic variants. For any character to evolve, (1) the character must vary, (2) the variation must affect reproductive success, and (3) the variation must be heritable. Mate guarding in soapberry bugs satisfies each of these conditions: First, there is variation, and this variation affects fitness. By bringing individual bugs into the laboratory and observing their behavior at different sex ratios, Carroll showed that individual males in Oklahoma have different behavioral strategies. Figure 3.16 plots the probability of mate guarding for four representative individuals numbered 1 to 4. Males 1 and 4 both have fixed behavioral strategies. Male 1 guards about 90% of the time, and male 4 guards about 80% of the time. Males 2 and 3 have variable behavioral strategies. Male 2 is very sensitive to changes in sex ratio; male 3 is less sensitive. Notice that both the amount of mate guarding and the amount of flexibility vary among the bugs in Oklahoma.

Second, it seems likely that this variation would affect male reproductive success. In Oklahoma, males would experience a range of sex ratios, so it seems likely that

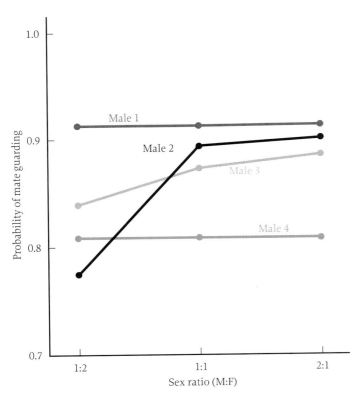

FIGURE 3.16 There is genetic variation in the behavioral rules of individual soapberry bugs, illustrated here by four representative individuals. There is variation in both the level of mate guarding (for example, male 1 spends more time guarding mates than male 4 does for all sex ratios) and in the amount of plasticity (for example, the behavior of males 1 and 4 does not change, the behavior of male 3 changes a bit, and the behavior of male 2 changes a lot). If this variation is heritable, the rule that works best, averaged over all of the environments of the population, will tend to increase.

males with flexible strategies, like male 2, would tend to have the most offspring. In Florida, males with inflexible strategies, like male 1, would have the most offspring.

Third, the character is heritable. By controlling matings in the laboratory, Carroll was able to show that males tended to have the same strategies as their fathers. Bugs like male 1 had sons with fixed strategies, and bugs like male 2 had sons with flexible strategies. Thus the Oklahoma bugs would come to have a variable strategy, while the Florida bugs would come to have a fixed strategy.

Behavior in the soapberry bug is relatively simple. Mate guarding depends on the sex ratio in the local population. The behavior of humans and other primates is much more complex, but the principles that govern the evolution of complex forms of behavior are the same as the principles that govern the evolution of simpler forms of behavior. That is, individuals must differ in the ways they respond to the environment, these differences must affect their ability to survive and reproduce, and at least some of these differences must be heritable. Then individual responses to environmental circumstances will evolve in much the same way that finch beaks and male soapberry bugs' mate-guarding behavior evolve.

CONSTRAINTS ON ADAPTATION

Natural selection plays a central role in our understanding of evolution because it is the only mechanism that can explain adaptation. However, evolution does not always lead to the best possible phenotype. In this section we consider five reasons why this is the case.

Correlated Characters

When individuals that have particular variants of one character also tend to have particular variants of a second character, the two characters are said to be correlated.

So far, we have considered the action of natural selection on only one character at a time. This approach is misleading if natural selection acts on more than one character simultaneously and the characters are nonrandomly associated, or **correlated**. It's easiest to grasp the meaning and importance of correlated characters in the context of a now familiar example: Darwin's finches. When the Grants and their colleagues captured medium ground finches on Daphne Major, they measured beak depth, beak width, and a number of other morphological characters. Beak depth is measured as the top-to-bottom dimension of the beak; beak width is the side-to-side dimension. As is common for morphological characters like these, the Grants found that beak depth and beak width are positively correlated: birds with deep beaks also have wide beaks (Figure 3.17). Each point in Figure 3.17 represents one individual. Beak width is plotted on the vertical axis, and beak depth is plotted along the horizontal axis. If the cloud of points were round or the points were randomly scattered in the graph, it would mean these characters were uncorrelated. Then information about an individual's beak depth would tell us nothing about its beak width. However, the cloud of points forms an ellipse with the long axis oriented from the lower left to the upper right, so we know that the two characters are **positively correlated**: deep beaks also tend to be wide. If birds with deep beaks tended to have narrow beaks, and birds with shallow beaks tended to have wide beaks, the two characters would be **negatively correlated**, and the long axis of the cloud of points would run from the upper left to the lower right.

Correlated characters occur because some genes affect more than one character.

Genes that affect more than one character are said to have **pleiotropic effects**, and probably most genes fall into this category. Often genes that are expressed early in

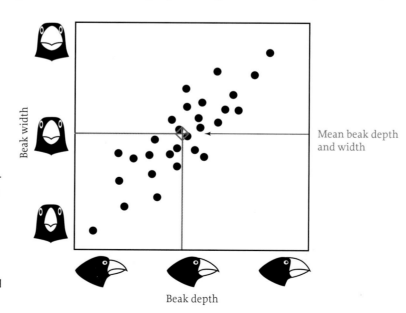

FIGURE 3.17 Beak depth and beak width are correlated in the medium ground finch on Daphne Major. The vertical axis gives the difference between the individual's beak width and the mean width in the population, and the horizontal axis gives the difference between the individual's beak depth and the population mean. Each point represents one individual. The data show that birds with deep beaks are likely to have wide beaks, and birds with shallow beaks generally have narrow beaks.

Mean beak depth and width

Beak width

Beak depth

development and affect overall size also influence a number of other discrete morphological traits. In Darwin's finches, individuals that have alleles for bigness at loci affecting overall size will have deeper beaks and wider beaks than individuals that have alleles for smallness. The PKU locus provides another good example of pleiotropy. Untreated PKU homozygotes are phenotypically different from heterozygotes or normal homozygotes in several ways. For example, they have lower IQs and different hair color from individuals with other genotypes. Thus, the substitution of another PKU allele for the normal allele in a heterozygote would affect a wide variety of phenotypic characters.

When two characters are correlated, selection that changes the mean value of one character in the population also changes the mean value of the correlated character.

Returning to the finches, suppose that there is selection for individuals with deep beaks, and that beak width has no effect on survival (Figure 3.18). As we would expect, the mean value of beak depth increases. Notice, however, that the mean value of beak width also increases, even though beak width has no effect on the probability that an individual will survive. Selection on beak depth affects the mean value of beak width because the two traits are correlated. This effect is called the **correlated response** to selection. It results from the fact that selection increases the frequency of genes that increase both beak depth and beak width.

A correlated response to selection can cause other characters to change in a maladaptive direction.

To understand how selection on one character can cause other characters to change in a **maladaptive** (less fit) direction, let's continue with our example. It turns out that beak width did affect survival during the Galápagos drought. By holding beak depth

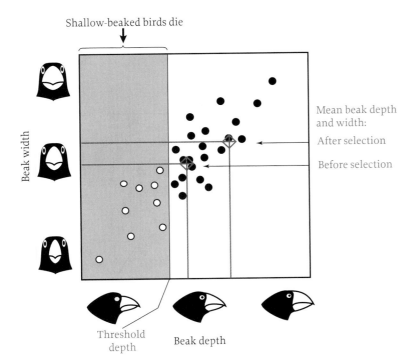

FIGURE 3.18 Selection on one character affects the mean value of correlated characters, even if the other characters have no effect on fitness. The distribution of beak depth and beak width in the population is plotted here, assuming that only birds whose beaks are greater than a threshold depth survive. Beak width is assumed to have no effect on survival. Selection leads to an increase in both average beak depth and average beak width. Mean beak width is increased by the correlated response of selection on beak depth.

constant, the Grants showed that individuals with *thinner* beaks were more likely to survive during the drought, probably because birds with thinner beaks were able to generate more pressure on the tough seeds that predominated during the drought. If selection were acting only on beak width, then we would expect the mean beak width in the population to decrease. However, the correlated response to selection on beak depth also acts to increase mean beak width. If the correlated response to selection on beak depth were stronger than the effect of selection directly on beak width, mean beak width would increase, even though selection favors thinner beaks. This is exactly what happened on Daphne Major (Figure 3.19).

Disequilibrium

Selection produces optimal adaptations only at equilibrium.

In Chapter 1 we saw how natural selection could gradually increase beak depth, generation by generation, until an equilibrium was reached—a point where stabilizing selection maintained the average beak depth in the population at the optimum size. This example illustrates the principle that selection keeps changing a population until an adaptive equilibrium is reached. It is easy to forget that the populations being observed have not necessarily reached equilibrium. If the environment has changed recently, there is every reason to suspect that the morphology or behavior of residents is not adaptive under current conditions. How long does it take for populations to adapt to environmental change? This depends on how quickly selection acts. As we have seen, enormous changes can be created by artificial selection in a few dozen generations.

Disequilibrium is particularly important for some human characters because there have been big changes in the lives of humans during the last 10,000 years. It seems likely that many aspects of the human phenotype have not had time to catch up with recent changes in our subsistence strategies and living conditions. Our diet provides

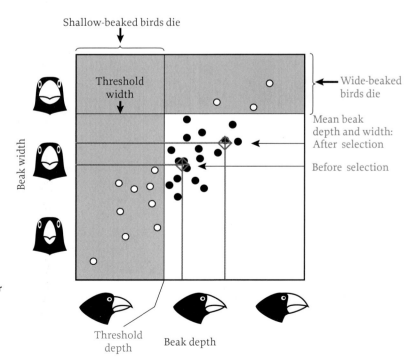

FIGURE 3.19 The correlated response to selection can cause less fit phenotypes to become more common. The distribution of beak depth and beak width in the population is plotted here, assuming that all birds whose beaks are less than a threshold depth or greater than a threshold width die. All others survive. Even though there is selection favoring narrower beaks, beak width increases as a result of the correlated response to selection on beak depth.

a good example. Five thousand years ago, most people hunted wild game and gathered wild plants for food, and typically people had little access to sugar, fat, and salt (Figure 3.20). These are essential dietary requirements for the proper functioning of the human body, and so it was probably always good for people to eat as much of these things as they could find. In an adaptive response to human dietary needs, evolution equipped people with a nearly insatiable appetite for fat, salt, and sugar. With the advent of agriculture and trade, however, these substances became readily available and our evolved appetites for them became more problematic. Today, eating too much fat, salt, and sugar is associated with a variety of health problems, including bad teeth, obesity, diabetes, and high blood pressure.

Genetic Drift

When populations are small, genetic drift may cause random changes in gene frequencies.

FIGURE 3.20 For most of our evolutionary history, humans have subsisted on wild game and plant foods. Sugar, salt, and fat were in short supply. Here a Hadza woman digs up a tuber. (Photograph courtesy of Nick Blurton Jones.)

So far, we have assumed that evolving populations are always very large. When populations are small, however, random effects caused by sampling variation can be important. To see what this means, consider a statistical analogy. Suppose that we have a huge urn like the one in Figure 3.21. The urn contains 10,000 balls—half of them black and half red. Suppose we also have a collection of small urns, each of which holds 10 balls. We draw 10 balls at random from the big urn to put in each small urn. Not all of the little urns will have five red balls and five black balls. Some will have four red balls, some three red balls, and a few may even have no red balls. The fact that the distribution of black and red balls among the small urns varies is called **sampling variation**.

The same thing happens during genetic transmission in small populations (Figure 3.22). Suppose there is an organism that has only one pair of chromosomes with two possible alleles, *A* and *a*, at a particular locus and that selection does not act on this trait. In addition, a population of five individuals of this species is newly isolated from the rest. In this small population (generation 1), each allele has a frequency of 0.5, so five chromosomes carry *A* and five carry *a*. These five individuals mate at random and produce five surviving offspring (generation 2). To keep things simple, assume that the gamete pool is large, so there is no sampling variation. This means that half of the gametes produced by generation 1 will carry *A* and half will carry *a*. However, only 10 gametes will be drawn from this gamete pool to form the next generation of five individuals. These gametes will be sampled in the same way the balls were sampled from the urn (see Figure 3.21). The most likely result is that there are five *A* alleles and five *a* alleles in generation 2, as there were in generation 1. However, just as there is some chance of drawing three, four, six, seven, or even zero black balls from the urn, there is some chance that generation 2 will not carry equal numbers of *A* and *a* alleles. Suppose the five individuals in generation 2 have six *A* alleles and four *a* alleles. The frequency of the *A* allele is now 0.6, so when these individuals mate, the frequency of the *A* allele in the pool of gametes will be 0.6 as well. This means that the frequency

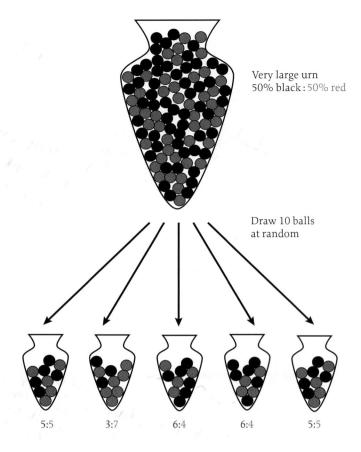

Very large urn
50% black : 50% red

Draw 10 balls
at random

FIGURE 3.21 A large urn contains 10,000 balls, half of them black and half red. Ten balls are drawn from the large urn at random and placed in each of five small urns. Not all of the small urns will contain five red balls and five black balls. This phenomenon is called sampling variation.

5:5   3:7   6:4   6:4   5:5

of the allele in the population has changed by chance alone. In the third generation the gene frequencies change again, this time in the opposite direction. This phenomenon is called **genetic drift**. In small populations, genetic drift causes random fluctuations in genetic frequencies.

The way alleles are sampled in real populations depends on the biology of the evolving species and the nature of the selective forces that the species faces. For instance, population size in some bird species might be limited by nesting sites, and success in obtaining nesting sites might be related to body size: perhaps the larger the bird, the more chance there would be that it would have success in battling others for a prime nesting spot. Although body size is influenced partly by genes, and selection would favor genes for large body size, many genes don't affect body size, and they may be sampled randomly in this particular population.

The rate of genetic drift depends on population size.

Genetic drift causes more rapid change in small populations than in large ones because sampling variation is more pronounced in small samples. To see why this is true, consider a simple example. Suppose a company is trying to predict whether a new product—say, low-fat peanut butter—will succeed in the marketplace. The company commissions two different survey firms to find out how many peanut butter fans would switch to the new product. One firm conducts interviews with five people, four of whom say that they would switch immediately to the low-fat variety. The other company polls 1000 people, and 50% of them say they would prefer low-fat peanut butter. Which survey should the company trust? Clearly, the second survey is more

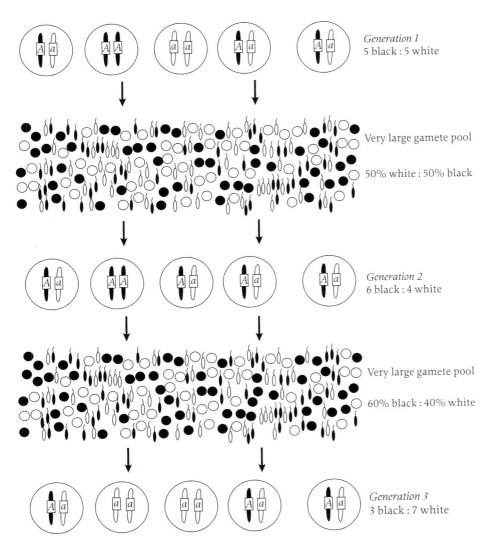

Generation 1
5 black : 5 white

Very large gamete pool

50% white : 50% black

Generation 2
6 black : 4 white

Very large gamete pool

60% black : 40% white

Generation 3
3 black : 7 white

FIGURE 3.22 Sampling variation leads to changes in gene frequency in small populations. Suppose that a population produces half *A* gametes (black) and half *a* gametes (white), and the next generation consists of just five individuals. It is not unlikely that, by chance, these individuals will carry six *A* gametes and four *a* gametes. This new population will produce 60% *A* gametes. Thus, sampling variation can change gene frequencies. Notice here that gene frequencies change again in generation 3, this time in the reverse direction, to 30% *A*.

credible than the first one. It is easy to imagine that by chance we might find four fans of low-fat peanut butter in a sample of five people, even if the actual frequency of such preferences in the population were only 50%. In contrast, in a sample of 1000 people, we would be very unlikely to find such a large discrepancy between our sample and the population.

Exactly the same principle applies to genetic drift. In a population of five individuals (with 10 chromosomes) in which two alleles are equally common (frequency = 0.5), there is a good chance that in the next generation the frequency of one of the alleles will be greater than or equal to 0.8. But in a population of 1000 individuals, there is virtually no chance that such a large deviation from the initial frequency will occur.

Genetic drift causes isolated populations to become genetically different from each other.

Genetic drift leads to unpredictable evolution because the changes in gene frequency caused by sampling variation are random. As a result, drift causes isolated populations to become genetically different from one another over time. Results from a computer simulation illustrate how this works for a single genetic locus with two

FIGURE 3.23 This computer simulation demonstrates that genetic drift causes isolated populations to become genetically different from one another. Four populations are initially formed by 20 individuals being drawn from a single large population. In the original population, one gene has two alleles that each have a frequency of 0.5. The graph tracks the frequency of one of these alleles in each of these populations over time. Population 1 is in red, population 2 in blue, population 3 in green, and population 4 in purple. The dotted line shows the global frequency of the *A* allele. In each population, the frequency of the allele fluctuates randomly under the influence of genetic drift. Notice that populations 2 and 4 eventually reach fixation, meaning that one of the two alleles is lost.

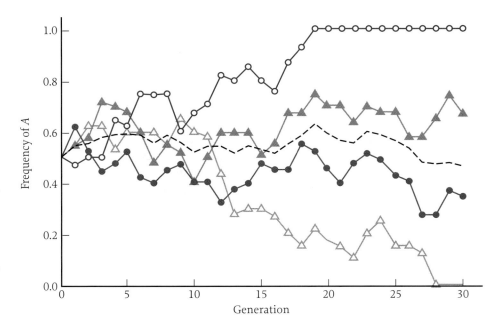

alleles (Figure 3.23). Initially there is a large population in which each allele has a frequency of 0.5. Then four separate populations of 20 individuals (each individual carrying two sets of chromosomes) are created. These four populations are maintained at a constant size and remain isolated from each other during all subsequent generations. During the first generation the frequency of *A* in the gamete pool is 0.5 in each population. However, the unpredictable effects of sampling variation alter the frequencies of *A* among adults in each population. From the first generation we sample 40 gametes from each of the four populations. In population 1 (marked in red) the frequency of *A* increases dramatically, in populations 2 (blue) and 3 (green) it increases a little, and in population 4 (purple) it decreases. In each case, the change is created by chance alone.

During the next generation the same process is repeated, except that the gamete pools of each of the four populations now differ. Once again, 40 gametes are sampled from each population. This time the frequency of *A* in populations 2, 3, and 4 increases, while the frequency of *A* in population 1 decreases substantially. As these random and unpredictable changes continue, the four populations become more and more different from one another. Eventually, one of the alleles is lost entirely in two of our populations (*A* in population 2 and *a* in population 4), making all individuals in these populations identical at the locus in question. Such populations are said to have reached **fixation**. In general, the smaller the population, the sooner it reaches fixation. Theoretically, if genetic drift goes on long enough, all populations eventually reach fixation. When populations are large, however, this may take such a long time that the species becomes extinct before fixation occurs. Populations remain at fixation until mutation introduces a new allele.

As you might expect, the rate at which populations become genetically different is strongly affected by their size: small populations differentiate rapidly; larger populations differentiate more slowly. We will see in Chapter 14 that the relationship between population size and the rate of genetic drift allows us to make interesting and important inferences about the size of human populations tens of thousands of years ago.

Populations must be quite small for drift to lead to significant maladaptation.

Random changes produced by genetic drift will create adaptations only by chance, and we have seen that the probability of assembling complex adaptations in this way is small indeed. This means that genetic drift usually leads to maladaptation. The importance of genetic drift in evolution is a controversial topic. Nearly everyone agrees that populations must be fairly small (say, less than 100 individuals) for drift to have an important effect when it is opposed by natural selection. There is a lot of debate about whether populations in nature are usually that small. Most scientists agree, however, that genetic drift is not likely to generate significant maladaptation in traits that vary continuously and are affected by many genetic loci. Thus a trait such as beak size in Darwin's finches is unlikely to be influenced much by genetic drift.

Local versus Optimal Adaptations

Natural selection may lead to an evolutionary equilibrium at which the most common phenotype is not the best possible phenotype.

Natural selection acts to increase the adaptedness of populations, but it does not necessarily lead to the best possible phenotypes. The reason is that natural selection is myopic: it favors small improvements to the existing phenotype, but it does not take into account the long-run consequences of these alterations. Selection is like a mountaineer who tries to climb a peak cloaked in dense cloud; she cannot see the surrounding country, but she figures she will reach her goal if she keeps climbing uphill. The mountain climber will eventually reach the top of something if she continues climbing, but if the topography is the least bit complicated, she won't necessarily reach the summit. Instead, when the fog clears, she is likely to find herself on the top of a small, subsidiary peak. In the same way, natural selection keeps changing the population until no more small improvements are possible, but there is no reason to believe that the end product is the optimal phenotype. The phenotype arrived at in such cases is called a "local adaptation"; it is analogous to the subsidiary peak reached by the mountaineer.

As an example of this effect, let's consider the evolution of eyes (Figure 3.24). Humans, other vertebrates, and some invertebrates such as octopi have **camera-type eyes**. In this type of eye there is a single opening in front of a lens, which projects an image on photoreceptive tissue. Insects, by contrast, have **compound eyes**, in which many very small separate photoreceptors build up an image composed of a grid of dots, something like a television image. Compound eyes are inferior to camera-type eyes in most ways. For example, they have lower resolution and less light-gathering power than similarly sized camera-type eyes have. If this is so, why haven't insects evolved camera-type eyes?

The most likely answer is that once a species has evolved complex, compound eyes, selection cannot favor intermediate types, even though this might eventually allow superior camera-type eyes to evolve. Consider the early lineages in which eyes were first evolving. There may have come a time when selection favored greater light sensitivity. In the vertebrates and mollusks, greater light sensitivity was achieved by increases in the area of sensitive tissue within each eye. In insects, it seems likely that increased sensitivity was achieved by multiplication of the number of small photoreceptors, each in its own cup. At this early stage, these alternatives yielded equally useful eyes. Once the insect lineage had evolved a visual system based on many small eyes,

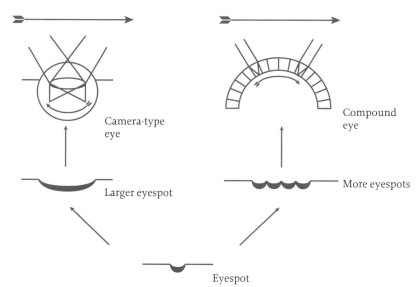

FIGURE 3.24 Researchers believe that different evolutionary pathways led to the development of camera-type and compound eyes. Selection for greater light sensitivity in a simple organism could favor the development of a larger eyespot or of multiple eyespots. The first pathway could lead to camera-type eyes, and the second to compound eyes. Images are formed very differently in compound and camera-type eyes, making it virtually impossible for compound eyes to evolve into camera-type eyes.

however, images could not be formed the same way as in camera-type eyes because whereas the camera-type eye inverts images, the compound eye does not. It does not seem structurally possible to evolve from a compound eye to a camera-type eye in a series of small steps, each favored by selection.

Some local adaptations are called "developmental constraints."

Most kinds of organisms begin life as a single fertilized cell, a zygote. As an organism grows, this cell divides many times, giving rise to specialized nerve cells, liver cells, and so on. This process of growth and differentiation is called **development**, and the development of complex structures like eyes involves many different interdependent processes. Developmental changes that would produce desirable modifications are often selected against because these alterations have many other negative effects. For example, as we will see later in the text, there is good reason to believe that it would be adaptive for males in some primate species to be able to **lactate** (produce milk for their young). However, no primate males can lactate, and most biologists believe that the developmental changes that would allow males to lactate would also make them sterile. Thus, developmental processes constrain evolution, but such constraints are not absolute. They result from the particular phylogenetic history of the lineage. If primate reproductive biology had evolved along a different pathway so that the development of mammary glands and other elements of the primate reproductive system were independent, then there would be no constraints on the evolution of male lactation.

Other Constraints on Evolution

Evolutionary processes are also constrained by the laws of physics and chemistry.

The laws of physics and chemistry place additional constraints on the kinds of adaptations that are possible. For example, the laws of mechanics predict that the strength of bones is proportional to their cross-sectional area. This fact constrains

The Geometry of Area/ Volume Ratios

BOX 3.2

When an animal becomes larger without otherwise changing shape, the ratio of any fixed area measurement to its volume decreases. This is easiest to understand if we compute how the ratio of an animal's entire surface area to its volume changes as the animal becomes larger. To see why, suppose that the animal is a cube x centimeters on a side. Then its volume is x^3 and its surface area is $6x^2$. Thus the ratio of its surface area to its volume is

$$\frac{\text{surface area}}{\text{volume}} = \frac{6x^2}{x^3} = \frac{6}{x}$$

This means that an animal measuring 1 cm on a side has 6 cm² of surface area for each cubic centimeter of volume. An animal 2 cm on a side has only 3 cm² of surface area for each cubic centimeter of volume. Of course, there aren't any cubic animals, but it turns out that the shape doesn't alter this relationship. When an animal's linear dimension is doubled without a change in shape, the ratio of surface area

to volume is halved. We will see later that this fact has consequences for how temperature affects the size of animals.

The same geometric principle governs the relationship between any area measure and volume. Suppose the cubic animal has a vertical bone running through its center that supports the weight of the rest of the animal. For simplicity, we make the bone square in cross section and assume that it has the dimensions $\frac{1}{2}\,x$ by $\frac{1}{2}\,x$, as shown in Figure 3.25. The cross-sectional area of this bone, then, is $\frac{1}{2}\,x \times \frac{1}{2}\,x = \frac{1}{4}\,x^2$. Thus the ratio of the volume of the animal to the cross-sectional area of the bone is

$$\frac{\frac{1}{4}x^2}{x^3} = \frac{1}{4x}$$

So if the linear dimensions of an animal are doubled, its weight will increase eightfold, but the cross-sectional area of its bones—and therefore their strength—will increase only fourfold.

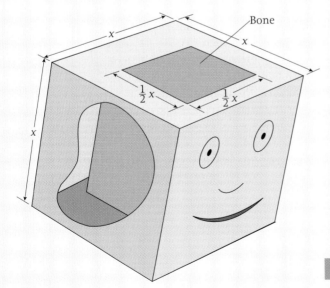

FIGURE 3.25 A cutaway diagram of the cubic animal discussed here.

FIGURE 3.26 It might be advantageous for animals to be large enough to be relatively invulnerable to predators but agile enough to leap considerable distances. But physical constraints limit evolution; not all things are possible.

the evolution of morphology. To see why, suppose that an animal's size (that is, its *linear* dimensions) doubles as the result of natural selection. This means that the animal's weight (which is proportional to its *volume*) will increase by a factor of 8 (see Box 3.2). The strength to bear an animal's weight comes from its muscles and bones and is determined largely by their cross-sectional areas. If the linear dimensions of bones and muscles double, however, their cross-sectional areas will increase by a factor of only 4. This means that the bones and muscles will be only half as strong relative to the animal's weight as they were before. Thus, if the bones are to be as strong as they were before, they must be more than four times greater in cross section, which means that they must become proportionally thicker. Of course, thicker bones are heavier, and this greater weight imposes other constraints on the animal. If it is heavier, it may not be able to move as fast or suspend itself from branches. This constraint (and a closely related one that governs the strength of muscles) explains why big animals like elephants are typically slow and ponderous, whereas smaller animals like squirrels move quickly and are often agile leapers. It also explains why all flying animals are relatively light. Natural selection might well favor an elephant that could run like a cheetah, vault obstacles like an impala, and climb like a monkey. But because of the trade-offs imposed by the laws of physics, there are no mutant pachyderms whose bones are both light and strong enough to permit these behaviors (Figure 3.26).

Such constraints are closely related to genetic correlations. Selection eliminates mutants with heavier, weaker bones; and the trade-offs inherent in the design of structures mean that mutant animals that have lighter, stronger bones are also not possible. The genotypes that remain in the population lead to the development of either lighter bones or stronger bones, but not both. Thus, bone weight and bone strength typically show a positive genetic correlation.

FURTHER READING

Crow, J. F. 1986. *Basic Concepts in Population, Quantitative, and Evolutionary Genetics.* Freeman, New York.

Dawkins, M. S. 1986. *Unraveling Animal Behavior.* Longham, Essex, England.

Dawkins, R. 1996. *The Blind Watchmaker.* Norton, New York.

Falconer, D. S. 1960. *Quantitative Genetics.* Ronald, New York.

Hedrick, P. 1983. *Genetics of Populations.* Science Books International, Boston.

Maynard Smith, J. 1989. *Evolutionary Genetics.* Oxford University Press, Oxford, England.

Ridley, M. 1985. *The Problems of Evolution.* Oxford University Press, New York.

Ridley, M. 1996. *Evolution* (2nd ed.). Blackwell Scientific, Oxford, England.

STUDY QUESTIONS

1. In human populations in Africa, two common alleles affect the structure of hemoglobin, the protein that carries oxygen on red blood cells. What is regarded as the normal allele (the allele common in European populations) is usually labeled *A*, and the sickle-cell allele is designated *S*. There are three hemoglobin genotypes,

and the phenotypes associated with these genotypes can be distinguished in a variety of ways. In one African population of 10,000 adults, for example, there are 3000 AS individuals, 7000 AA individuals, and 0 SS individuals.

(a) Suppose that these individuals were to mate at random. What would be the frequency of the A and S alleles among the gametes that they produced?

(b) What would be the frequency of the three genotypes that resulted from random mating?

(c) Is the original population of adults in Hardy–Weinberg equilibrium?

2. The three common genotypes at the hemoglobin locus have very different phenotypes: SS individuals suffer from severe anemia, AS individuals have a relatively mild form of anemia but are resistant to malaria, and AA individuals have no anemia but are susceptible to malaria. The frequency of the S allele among the gametes produced by the first generation of a central African population is 0.2.

(a) Assuming that mating occurs at random, what are the frequencies of the three genotypes among zygotes produced by this population?

(b) In this area, no SS individuals survive to adulthood, 70% of the AA individuals survive, and all of the AS individuals survive. What is the frequency of each of the three genotypes among the second generation of adults?

(c) What is the frequency of the S allele among gametes produced by these adults?

3. Tay-Sachs disease is a lethal genetic disease controlled by two alleles: the Tay-Sachs allele T and the normal allele N. Children who have the TT genotype suffer from mental deterioration, blindness, paralysis, and convulsions; and they die sometime between the ages of three and five years. The proportion of infants afflicted with Tay-Sachs disease varies in different ethnic groups. The frequency of the Tay-Sachs *allele* is highest among descendants of central European Jews. Suppose that in a population of 5000 people descended from central European Jews the frequency of the Tay-Sachs allele is 0.02.

(a) Assuming that members of this population mate at random, and each adult has four children on average, how many gametes from each person are successful at being included in a zygote?

(b) Calculate the frequency of each of the three genotypes among the population of zygotes.

(c) Assume that no TT zygotes survive to become adults, and that 50% of the TN and NN individuals survive to become adults. What is the frequency of the TT and TN genotypes among these adults?

(d) When the adult individuals resulting from zygotes in part 3b produce gametes, what is the frequency of the T allele among their gametes? Compare your answer to the frequency of the T allele among gametes produced by their parents. Think about your result, given that these numbers are roughly realistic and that there have been no substantial medical breakthroughs in the treatment of Tay-Sachs disease. What is the paradox here?

4. Consider a hypothetical allele that is lethal, like the Tay-Sachs allele, but is dominant rather than recessive. Assume that this allele has a frequency of 0.02, like the Tay-Sachs allele. Recalculate the answers to parts a, b, and c in question 3. Is selection stronger against the recessive allele or the dominant one? Why? Assuming that mutation to the deleterious allele occurs at the same rate at both loci, which allele will occur at a higher frequency? Why?

5. Consider the Rh blood group. For this example we will assume that there are only two alleles, R and r, of which r is recessive. Homozygous rr individuals produce certain nonfunctional proteins and are said to be Rh-negative (Rh⁻), while Rr and

RR individuals are Rh-positive (Rh⁺). When an Rh⁻ female mates with an Rh⁺ male, their offspring may suffer serious anemia while still in the uterus. The frequency of the *r* allele in a hypothetical population is 0.25. Assuming that the population is in Hardy–Weinberg equilibrium, what fraction of the offspring in each generation will be at risk for this form of anemia?

6. The blending model of inheritance was attractive to nineteenth-century biologists because it explained why offspring tend to be intermediate in appearance between their parents. This model fails, however, because even though it predicts a loss in variation, variation is maintained. How does Mendelian genetics explain the intermediate appearance of offspring without the loss of variation? Be sure to explain why the properties of the Hardy–Weinberg equilibrium are an important part of this explanation.

7. In a particular species of fish, egg size and egg number are negatively correlated. Draw a graph that illustrates this fact. What will happen to egg size if selection favors larger numbers of eggs?

8. A plant breeder trying to increase the yield of wheat selects plants with larger kernels. After several generations of selection, the size of kernels produced has increased substantially. However, the number of kernels per plant has decreased, so the total yield remains constant. What two sources of maladaptation best explain this result?

9. The compound eyes of insects provide poor image clarity compared with the images produced by the camera-type eyes of vertebrates. Explain why insects don't evolve camera-type eyes.

10. Explain why genetic drift has no effect on genetic loci that are not variable. (When all individuals in the population are homozygous for the same allele, that genetic locus is not variable.) What does this mean about the long-term outcome of genetic drift?

11. Why does natural selection produce adaptations only at equilibrium?

12. Two dog breeders are using artificial selection to create new breeds. Both begin with dachshunds. One breeder wants to create a breed with longer forelimbs and hindlimbs. The second breeder wants to create a breed with shorter forelimbs and longer hindlimbs. Which breeder will make more rapid progress? Why?

13. When selection makes animals larger overall, their bones also get thicker. The increase in bone thickness could be due to the correlated response to selection, or to independent selection for thicker bones, or to some combination of the two factors. How could you determine the relative importance of these two processes?

KEY TERMS

population genetics	pleiotropic effects
genotypic frequency	correlated response (positive and
gene frequency	negative correlation)
Hardy–Weinberg equilibrium	maladaptive
modern synthesis	sampling variation
environmental variation	fixation
mate guarding	camera-type eyes
sex ratio	compound eyes
canalized	development
plastic	lactate

CHAPTER 4

SPECIATION AND PHYLOGENY

WHAT ARE SPECIES?

Microevolution refers to how populations change under the influence of natural selection and other evolutionary forces; macroevolution refers to how new species and higher taxa are created.

So far, we have focused on how natural selection, mutation, and genetic drift cause populations to change through time. These are mechanisms of **microevolution**, and they affect the morphology, physiology, and behavior of particular species in particular environments. For example, microevolutionary processes are responsible for variation in the size and shape of the beaks of medium ground finches on Daphne Major.

However, this is not all there is to the process of evolution. Darwin's major work was entitled *On the Origin of Species* because he was interested in how new species are created, as well as how natural selection operates within populations. Evolutionary theory tells us how new species, genera, families, and higher groupings come into existence. These processes are mechanisms of **macroevolution**. Macroevolutionary processes play an important role in the story of human evolution. To properly interpret the fossil record and reconstruct the history of the human lineage, we need to understand how new species and higher groupings are created and transformed over time.

Species can usually be distinguished by their behavior and morphology.

Organisms cluster into distinct types called **species**. The individual organisms that belong to a species are similar to each other and are usually quite distinct from the members of other species. For example, in Africa some tropical forests house two species of apes: chimpanzees and gorillas. These two species are similar in many ways: both are tailless, both bear weight on their knuckles when they walk, and both defend territories. Nonetheless, these two species can easily be distinguished on the basis of their morphology: chimpanzees are smaller than gorillas; male gorillas have tiny **testes** (singular *testis*; the organs that produce sperm), while male chimpanzees have quite large ones; and gorillas have a fin of bone on their skull, while chimpanzees have more rounded skulls. Gorillas and chimpanzees also differ in their behavior: chimpanzees make and use tools in foraging, while gorillas do not; male gorillas beat their chests when they perform displays, while male chimpanzees flail branches and charge about; and gorillas live in smaller groups than chimpanzees do. These two species are easy to distinguish because no animals are intermediate between them; there are no "gimps" or "chorillas" (Figure 4.1).

Species are not abstractions created by scientists; they are real biological categories. People all over the world name the plants and animals around them, and biologists use the same kinds of phenotypic characteristics to sort animals into species that other people use. For the most part there is little problem in identifying any particular specimen from its phenotype.

Although nearly everyone agrees that species exist and can recognize species in nature, biologists are much less certain about how species should be defined. This

FIGURE 4.1 Chimpanzees (a) and gorillas (b) sometimes occupy the same forests and share certain traits. However, these two species are readily distinguished because no animals are intermediate between them.

(a)   (b) uncertainty arises from the fact that evolutionary biologists do not agree about *why* species exist. There is now a considerable amount of controversy about the processes that give rise to new species and the processes that maintain established ones. Although there are many different views on these topics, we will concentrate on two of the most widely held points of view: the biological species concept and the ecological species concept.

The Biological Species Concept

The biological species concept defines a species as a group of interbreeding organisms that are reproductively isolated from other organisms.

Most zoologists believe in the **biological species concept**, which defines a biological species as a **group of organisms that interbreed in nature and are reproductively isolated**. **Reproductive isolation** means that members of a given group of organisms do not mate successfully with organisms outside the group. For example, there is just one species of gorilla, *Gorilla gorilla*, and this means that all gorillas are capable of mating with one another and they do not breed with any other kinds of animals in nature. According to adherents of the biological species concept, reproductive isolation is the reason why there are no gimps or chorillas.

The biological species concept defines a species in terms of the ability to interbreed because successful mating leads to **gene flow**, the movement of genetic material within parts of a population or from one population to another. Gene flow tends to maintain similarities among members of the same species. To see how gene flow preserves homogeneity within species, consider the hypothetical situation diagrammed in Figure 4.2a. Imagine a population of finches living on a small island in which there are two habitats: wet and dry. Natural selection favors different-sized beaks in each habitat. Large beaks are favored in the dry habitat, and small beaks are favored in the wet habitat. Because the island is small, however, birds fly back and forth between the two environments and mate at random, so there is a lot of gene flow between habitats. Unless selection is very strong, gene flow will swamp its effects. On average, birds in both habitats will have medium-sized beaks, a compromise between the optimal phenotypes for each habitat. In this way, gene flow tends to make the members of a species evolve as a unit.

Now suppose there are finches living on two different islands, one wet and one dry, and that the islands are far enough apart that the birds are unable to fly from one island to the other (Figure 4.2b). This means that there will be no interbreeding and no gene flow will occur. With no genetic exchange between the two independent groups to counter the effects of selection, the two populations will diverge genetically and become less similar phenotypically. Birds on the dry island will develop large beaks, and birds on the wet island will develop small beaks.

Reproductive isolation prevents species from genetically blending.

Reproductive isolation is the flip side of interbreeding. Suppose chimpanzees and gorillas could and did interbreed successfully in nature. The result would be gene flow between the two kinds of apes, and this phenomenon would produce animals that were genetically intermediate between chimpanzees and gorillas. Eventually the two

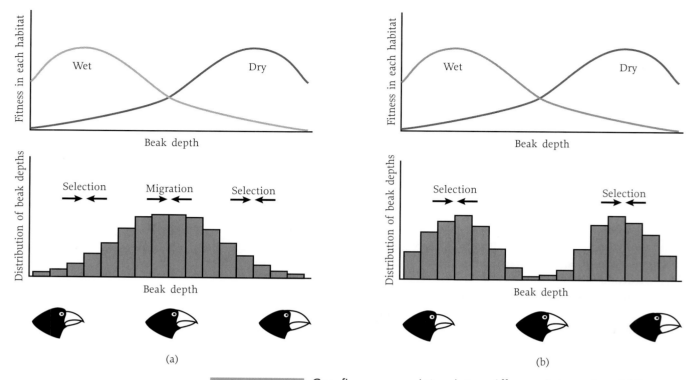

FIGURE 4.2 Gene flow among populations destroys differences between them. (a) Suppose a population of finches lives on an island with both wet and dry habitats. In the dry habitat, selection favors deep beaks, but selection in the wet habitat favors shallow beaks. Because interbreeding leads to extensive gene flow between populations in the two habitats, the population beak size responds to an average of the two environments. (b) Now suppose two populations of finches live on different islands, one wet and one dry, with no gene flow between the two islands. In this scenario, deep beaks are common on the dry island, and shallow beaks are common on the wet island.

species would merge into one. There are no chorillas or gimps in nature because chimpanzees and gorillas are reproductively isolated.

Reproduction is a complicated process, and anything that alters the process can act as an isolating mechanism. Even subtle differences in activity patterns, courtship behavior, or appearance may prevent individuals of different types from mating. Moreover, even if a mating among individuals of different types does take place, the egg may not be fertilized or the zygote may not survive.

The Ecological Species Concept

The ecological species concept emphasizes the role of selection in maintaining species boundaries.

Critics of the biological species concept point out that gene flow is neither necessary nor sufficient to maintain species boundaries in every case, and they argue that selection plays an important role in preserving the boundaries between species. The view that emphasizes the role of natural selection in creating and maintaining species is called the **ecological species concept.**

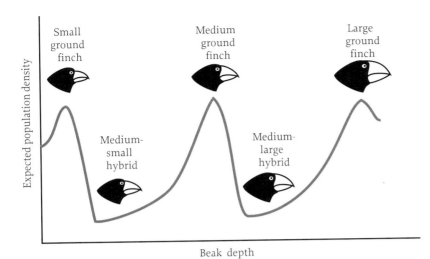

FIGURE 4.3 In the Galápagos, selection maintains three species of ground finches, even though there is substantial gene flow among them. The red line represents the amount of food available in the environment for birds with different-sized beaks. Each peak in this curve represents a different species.

In nature, species boundaries are often maintained even when there are substantial amounts of gene flow between species. For example, the medium ground finch readily breeds with the large ground finch on islands where they coexist. Peter Grant and his colleagues have estimated that approximately 10% of the time, medium ground finches mate with large ground finches, leading to considerable gene flow between these two species. Yet they have not merged into a single species. Grant and his colleagues have concluded that the medium ground finch and the large ground finch have remained distinct because these two species represent two of the three optimal beak sizes for ground finches (Figure 4.3). These three optimal sizes are based on the availability of seeds of different size and hardness and the ability of birds with different-sized beaks to harvest these seeds. According to the calculations of Grant's group, the three optimal beak sizes for ground finches correspond to the average beak sizes of the small (*Geospiza fuliginosa*), medium (*G. fortis*), and large (*G. magnirostris*) ground finch species (Figure 4.4). These researchers suggest that hybrids are selected against because their beaks fall in the "valleys" between these selective "peaks." The kind of interbreeding seen in the Galápagos finches is not particularly unusual; interspecific matings occur in a sizable minority of bird and mammal species.

(a)

(b)

(c)

FIGURE 4.4 (a) The small ground finch, *Geospiza fuliginosa*; (b) the medium ground finch, *G. fortis*; (c) the large ground finch, *G. magnirostris*.

FIGURE 4.5   Checkerspot butterflies living in localized populations throughout California are all members of the same species, but there is probably very little gene flow among different populations.

In addition, a number of species have maintained their coherence with no gene flow between isolated subpopulations. For example, the checkerspot butterfly (Figure 4.5) is found in scattered populations throughout California. Members of different populations are very similar morphologically and are all classified as members of the species *Euphydryas editha*. However, careful studies by Stanford University biologist Paul Ehrlich have shown that these butterflies rarely disperse more than 100 m (about 110 yd) from their place of birth. Given that populations are often separated by several kilometers and sometimes by as much as 200 km (about 120 miles), it seems unlikely that there is enough gene flow to unify the species.

The existence of strictly asexual organisms provides additional evidence that species can be maintained without gene flow. Asexual species, which reproduce by budding or fission, simply replicate their own genetic material and produce exact copies of themselves. Thus, gene flow cannot occur in asexual organisms, and it does not make sense to think of them as being reproductively isolated. Nonetheless, many biologists who study purely asexual organisms contend that it is just as easy to classify them into species as to classify organisms that reproduce sexually.

What maintains the coherence of species in these cases? Most biologists think the answer is natural selection. To see why, let's return to the Galápagos once more. Imagine a situation in which selection favored finches with small beaks in one habitat and finches with large beaks in another habitat, but did not favor birds with medium-sized beaks in either habitat. If birds with medium-sized beaks consistently failed to survive or reproduce successfully, natural selection could maintain the difference between the two bird species even if they interbred freely. In the case of the checkerspot butterflies and asexual species, we can imagine the opposite scenario: selection favors organisms with the same morphology and physiology and thus maintains their similarity even in the absence of gene flow.

Today most biologists concede that selection maintains species boundaries in a few odd cases like Darwin's finches, but they generally insist that reproductive isolation plays the major role in most instances. However, a growing minority of researchers contend that selection plays an important role in maintaining virtually all species boundaries.

THE ORIGIN OF SPECIES

Speciation is difficult to study empirically.

The species is one of the most important concepts in biology, so it would be helpful to know how new species come into existence. Despite huge amounts of hard work and many heated arguments, however, there is still uncertainty about what Darwin called the "mystery of mysteries": the origin of species. The mystery persists because the process of speciation is very difficult to study empirically. Unlike microevolutionary change within populations, which researchers are sometimes able to study in the field or laboratory, new species usually evolve too slowly for any single individual to study the entire process. And on the flip side, speciation usually occurs much too rapidly to be detected in the fossil record. Nonetheless, biologists have compiled a substantial body of evidence that provides important clues about the processes that give rise to new species.

Allopatric Speciation

If geographic or environmental barriers isolate part of a population, and selection favors different phenotypes in these regions, then a new species may evolve.

Allopatric speciation occurs when a population is divided by some type of barrier and different parts of the population adapt to different environments. The following hypothetical scenario captures the essential elements of this process. Mountainous islands in the Galápagos contain dry habitats at low elevations and wet habitats at higher elevations. Suppose that a finch species that lives on mountainous islands has a medium-sized beak, which represents a compromise between the best beak for survival in the wet areas and the best beak for survival in the dry areas. Further suppose that a number of birds are carried on the winds of a severe storm to another island that is uniformly dry, like the small, low-lying islands in the Galápagos archipelago. On this new dry island, only large-beaked birds are favored because large beaks are best for processing the large, hard seeds that predominate there. As long as there is no competition from some other small bird adapted to the dry habitat and there is very little movement between the two islands, the population of finches on the dry island will rapidly adapt to their new habitat and the birds' average beak size will increase (Figure 4.6).

Now let's suppose that after some time, finches from the small, dry island are blown back to the large island from which their ancestors came. If the large-beaked newcomers successfully mate with the medium-beaked residents, then gene flow between the two populations will rapidly eliminate the differences in beak size between them, and the recently created large-beaked variety will disappear. If, on the other hand, large-beaked immigrants and small-beaked residents cannot successfully interbreed, then the differences between the two populations will persist. As we noted earlier, a variety of mechanisms can prevent successful interbreeding, but it seems that the most common obstacle to interbreeding is that hybrid progeny are less viable than other offspring. In this case, we might imagine that when the two populations of finches became isolated, they diverged genetically because of natural selection and genetic drift. The longer they remain isolated, the greater the genetic difference between the populations becomes. When the two distinct types then come into contact, hybrids may have reduced viability, either because genetic incompatibilities have arisen during their isolation or because hybrid birds are unable to compete successfully for food. If these processes cause complete reproductive isolation, then a new species has been formed.

Even if there is some gene flow after the members of the two populations initially come back into contact, two additional processes may increase the degree of reproductive isolation and facilitate the formation of a new species. The first process, **character displacement**, may occur if competition over food, mates, or other resources increases the morphological differences between the immigrants and the residents.

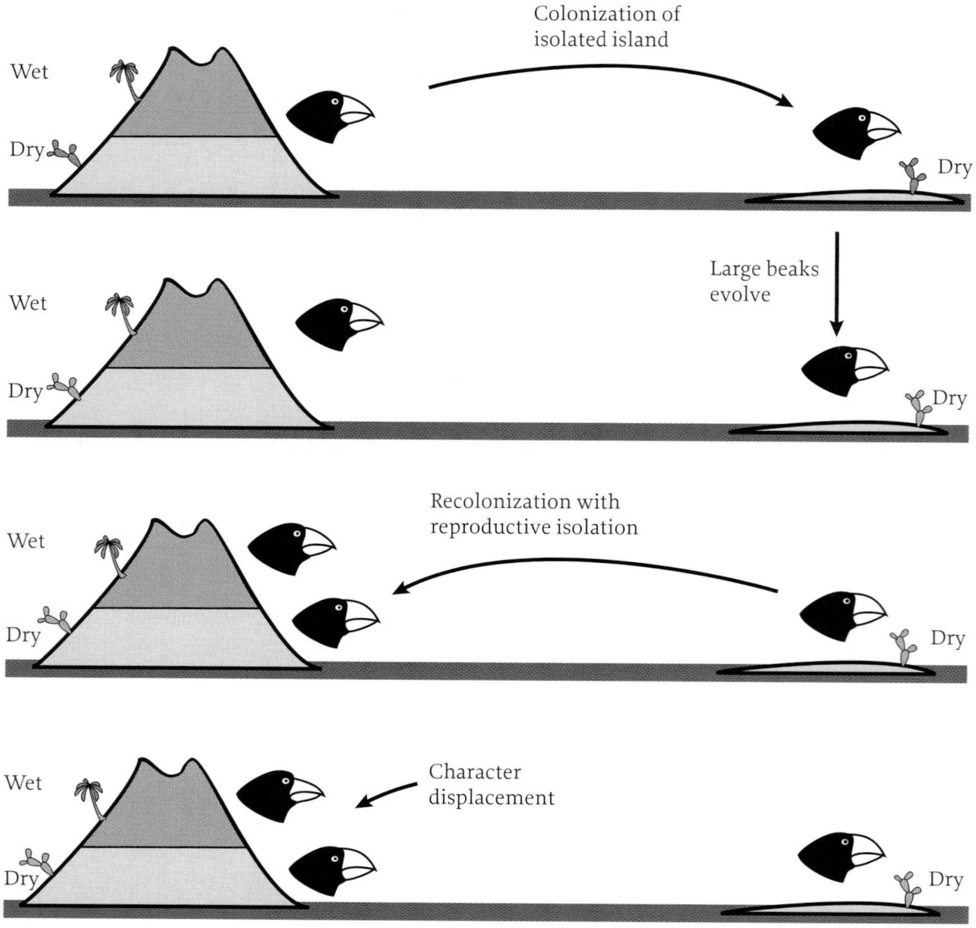

FIGURE 4.6 A likely sequence of events in a hypothetical allopatric speciation event in the Galápagos. Initially, one population of finches occupies an island with both wet and dry habitats. Because there is extensive gene flow between habitats, these finches have intermediate-sized beaks. By chance, some finches disperse to a dry island, where they evolve larger beaks. Then, again by chance, some of these birds are reintroduced to the original island. If the two populations are reproductively isolated, then a new species has been formed. Even if there is some gene flow, competition between the two populations may cause the beaks of the residents and immigrants to diverge further, a process called character displacement.

In our example, the large-beaked immigrants will be better suited to the dry parts of the large island and will be able to outcompete the original residents of these areas. Resident birds will be better off in the wet habitats, where they face less competition from the immigrants. Because small beaks are advantageous in wet habitats, residents with smaller-than-average beaks will be favored by selection. At the same time, because large beaks are advantageous in dry habitats, natural selection will favor increased beak size among the immigrants. This process will cause the beaks of the competing populations to diverge. There is good evidence that character displacement has played an important role in shaping the morphology of Darwin's finches (Figure 4.7).

A second process, called **reinforcement**, may act to reduce the extent of gene flow between the populations. Because hybrids have reduced viability, selection will favor behavioral or morphological adaptations that prevent matings between members of the two populations. This process will further increase the reproductive isolation between the two populations (Figure 4.8). Thus, character displacement and reinforcement may amplify the initial differences between the populations and lead to two new species.

Allopatric speciation requires a physical barrier that initially isolates part of a population, interrupts gene flow, and allows the isolated subpopulation to diverge from the original population under the influence of natural selection. In our example, the physical barrier is the sea, but mountains, rivers, and deserts can also restrict move-

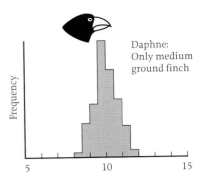

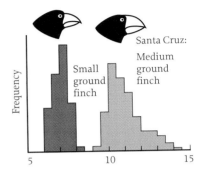

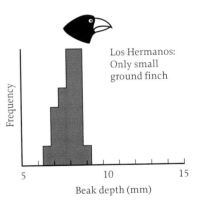

FIGURE 4.7 The distribution of beak sizes of the small ground finch (*G. fuliginosa*) and the medium ground finch (*G. fortis*) on three of the Galápagos islands illustrates the effects of character displacement. (b) The distribution of beak depth for each species on Santa Cruz Island, where the two species compete. (a, c) The distributions of beak depth for the small and medium ground finches on islands where they do not compete with each other (*G. fortis* on Daphne Major and *G. fuliginosa* on Los Hermanos). The beaks of the two species are the most different where there is direct competition. Careful measurements of seed density and other environmental conditions suggest that competition, not environmental differences, is responsible for the difference between the populations.

ment and interrupt gene flow. Character displacement and reinforcement may work to increase differences when members of the two populations renew contact. However, these processes are not necessary elements in allopatric speciation. Many species become completely isolated while they are separated by a physical barrier.

Parapatric and Sympatric Speciation

New species may also form if there is strong selection that favors two different phenotypes.

Biologists who endorse the biological species concept usually argue that allopatric speciation is the most important mechanism for creating new species in nature. For these scientists, gene flow "welds" a species together, and species can be split only if gene flow is interrupted. Biologists who think that natural selection plays an important role in maintaining species often contend that selection can lead to speciation even when there is interbreeding. There are strong and weak versions of this hypothesis. The weak version, called **parapatric speciation**, holds that selection alone is not sufficient to produce a new species, but new species can be formed if selection is combined with partial genetic isolation. For example, baboons range from Saudi Arabia to the Cape of

FIGURE 4.8 Finch courtship displays can be quite elaborate. Here a male finch displays to attract the female's attention. Subtle variations in courtship behavior among members of different populations may prevent mating and increase reproductive isolation.

Good Hope and occupy an extremely broad range of environments. Some baboons live in moist tropical forests, some live in arid deserts, and some live in high-altitude grasslands (Figure 4.9). Different behaviors and morphological traits may be favored in each of these environments, and this variation may cause baboons in different regions to vary. At habitat boundaries, animals that come from different habitats and have different characteristics may mate and create a **hybrid zone**. Study of hybrid zones in a wide variety of species suggests that hybrids are usually less fit than nonhybrids.

(a)

(b)

(c)

(d)

FIGURE 4.9 Baboons are distributed all over Africa and occupy a very diverse range of habitats. (a) In the Drakensberg Mountains in South Africa, winter temperatures drop below freezing and snow sometimes falls. (b) In the Matopo Hills of Zimbabwe, woodlands are interspersed with open savanna areas. (c) Amboseli National Park lies at the foot of Mount Kilimanjaro just inside the Kenyan border. (d) Gombe Stream National Park lies on the hilly shores of Lake Tanganyika in Tanzania.

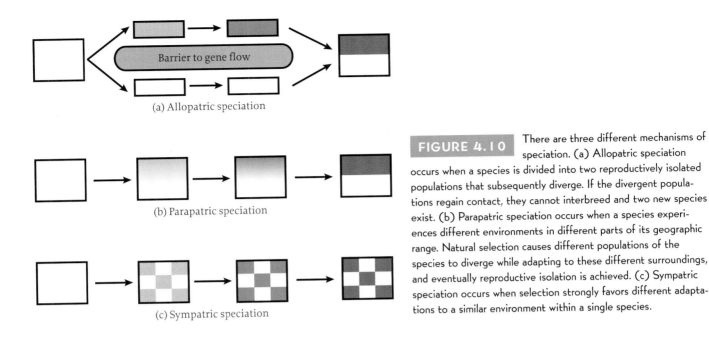

(a) Allopatric speciation

(b) Parapatric speciation

(c) Sympatric speciation

FIGURE 4.10 There are three different mechanisms of speciation. (a) Allopatric speciation occurs when a species is divided into two reproductively isolated populations that subsequently diverge. If the divergent populations regain contact, they cannot interbreed and two new species exist. (b) Parapatric speciation occurs when a species experiences different environments in different parts of its geographic range. Natural selection causes different populations of the species to diverge while adapting to these different surroundings, and eventually reproductive isolation is achieved. (c) Sympatric speciation occurs when selection strongly favors different adaptations to a similar environment within a single species.

When this is the case, selection should favor behavior or morphology that prevents mating between members of individuals from different habitats. If such reinforcement does occur, gene flow will be reduced, and eventually two new, reproductively isolated species will evolve.

The strong version of the hypothesis, called **sympatric speciation**, contends that strong selection favoring different phenotypes can lead to speciation even when there is no geographic separation, and therefore initially there is extensive gene flow among individuals in the population. Sympatric speciation is theoretically possible, and this form of speciation has been induced in laboratory populations, but it is uncertain whether it occurs in nature. The three speciation mechanisms are diagrammed in Figure 4.10.

Adaptive radiation occurs when there are many empty niches.

Ecologists use the term **niche** to refer to a particular way of "making a living," which includes the kinds of food eaten, as well as when, how, and where the food is acquired. One consequence of all three models of speciation is that the rate of speciation depends on the number of available ecological niches. Once again, Darwin's finches provide a good example. When the first finches migrated to the Galápagos from the mainland of South America (or perhaps from the Cocos Islands) about half a million years ago, all of the niches for small birds in the Galápagos were empty. There were opportunities to make a living as a seed eater, a cactus eater, and so on. The finches' ancestors diversified to fill all of these ecological niches, and eventually they became 14 distinct species (Figure 4.11). An even more spectacular example of this process occurred at the end of the Cretaceous era. Dinosaurs dominated the Earth during the Cretaceous era but disappeared suddenly 65 mya (million years ago). The mammals that coexisted with the dinosaurs were mostly small, nocturnal, and insectivorous. But when the dinosaurs became extinct, these small creatures diversified to fill a broad range of ecological niches, evolving into elephants, killer whales, buffalo,

(a)

(b)

(c)

FIGURE 4.11   When the first finches arrived in the Galápagos 500,000 years ago, there were many empty niches. Some finches became cactus eaters (a), others became seed eaters (b), and some became predators on insects and other arthropods (c).

FIGURE 4.12   Darwin's finches are not the only example of adaptive radiations that occur when immigrants encounter a range of empty ecological niches. In the Hawaiian archipelago, the adaptive radiation of honeycreeper finches, shown here, produced an even greater diversity of species.

wolves, bats, gorillas, humans, and other kinds of mammals. When a single kind of animal or plant diversifies to fill many available niches, the process is called **adaptive radiation** (Figure 4.12).

THE TREE OF LIFE

Organisms can be classified hierarchically on the basis of similarities. Many such similarities are unrelated to adaptation.

In Chapter 1 we saw how Darwin's theory of evolution explains the existence of adaptation. Now we can show how the same theory explains why organisms can be classified into a hierarchy on the basis of their similarities, a fact that puzzled nineteenth-century biologists much more than the existence of adaptations did. Richard Owen, a nineteenth-century anatomist who was one of Darwin's principal opponents in the debates following the publication of *On the Origin of Species*, used dugongs (aquatic mammals much like manatees), bats, and moles as an example of this phenomenon. All three of these creatures have the same kind and number of bones in their forelimbs, even though the shapes of these bones are quite different (Figure 4.13). The forelimb of the bat is adapted to flying, and that of the dugong to paddling (Figure 4.14). Nonetheless, the basic structure of a bat's forelimb is much more similar to that of a dugong than to the forelimb of a swift, even though the swift's forelimb is also designed for flight.

Such patterns of similarity make it possible to cluster species hierarchically—like a series of nested boxes (Figure 4.15). This remarkable property of life is the basis of the system for classifying plants and animals devised by the eighteenth-century Swedish biologist Carolus Linnaeus. All species of bats share many similarities and are grouped together in one box. Bats can be clustered with dugongs and moles in a larger box that contains all of the mammals. Mammals in turn are classified together with birds, reptiles, and amphibians in an even larger box. Sometimes the similarities that lead us to classify animals together are functional similarities. Many of the features shared by different species of bats are related to the fact that they fly at night. How-

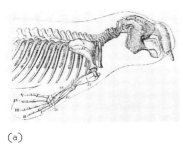

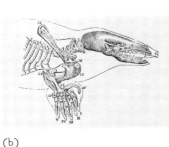

(a)

(b)

(c)

FIGURE 4.13 Organisms show patterns of similarity that have nothing to do with adaptation. Richard Owen's drawings of the forelimbs of three mammalian species illustrate this fact. The dugong (a), the mole (b), and the bat (c) all have the same basic bone structure in their forelimbs; dugongs use their forelimbs to swim in the sea, moles use theirs to burrow into the earth, and bats use their forelimbs to fly.

ever, many shared features bear little relation to adaptation; bats, tiny aerial acrobats, for example, are grouped with the enormous placid dugong, and not with the small acrobatic swift.

Speciation explains why organisms can be classified hierarchically.

The fact that new species derive from existing species accounts for the existence of the patterns of nonadaptive similarity that allow organisms to be classified hierarchically. Clearly, new species originate by splitting off from older ones, and we can arrange a group of species that share a common ancestor into a family tree, or **phylogeny**. Figure 4.16 shows the family tree of the **hominoids**—the superfamily that includes apes and humans—and Figure 4.17 shows what some of these present-day apes look like. At the root of the tree is an unknown ancestral species from which all hominoids evolved. Each branch represents a speciation event in which one species split into two daughter species, and one or both of the new daughter species diverged in morphology or behavior from the parent species.

It is important to realize that when two daughter species diverge from each other, they don't differ in all phenotypic details. A few traits may differ while most others retain their original form. For example, the hominoids share many common features, including an unspecialized digestive system, five toes on each foot, and no tail. At the same time, these animals have diverged in a number of ways. Gorillas live in groups that contain one adult male and several adult females; orangutans are mainly solitary. Male and female gibbons are about the same size, but among the other apes, males are larger than females. In general, we expect to see the greatest divergence in those traits that are related to making a living in different habitats and to choosing mates.

FIGURE 4.14 The dugong is an aquatic animal. Its forelimb is adapted for paddling through the water.

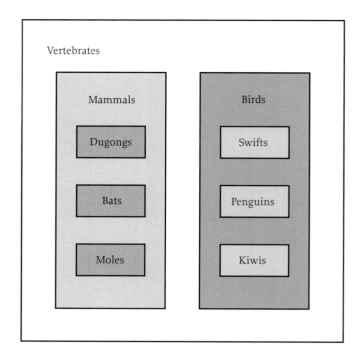

FIGURE 4.15 Patterns of similarity allow organisms to be classified hierarchically into a series of nested boxes. All mammals share more similarities with each other than they do with birds, even though some mammals, such as bats, must solve the same adaptive problems that face birds.

Now let's return to the family tree. Each time one species splits to become two new species, the new daughter species will differ in some way. They will continue to diverge through time because once speciation has occurred, the two lineages evolve independently. Some differences will arise because the two species adapt to different habitats; other differences may result from random processes like genetic drift. In general, species that have recently diverged will have more characters in common with one another than with species that diverged in the more distant past. For example, chimpanzees and gorillas share more traits with one another than they do with orangutans or gibbons because they share a more recent common ancestor. This pattern of sharing traits is the source of the hierarchical nature of life.

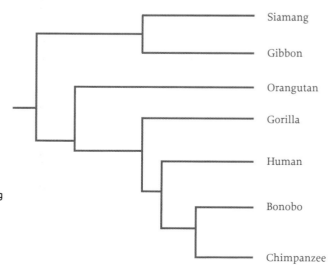

FIGURE 4.16 A phylogeny for the hominoids. The names of living species are given at the ends of the branches. The branching pattern of the phylogenetic tree reflects the ancestry of extant (still-living) lineages. For example, chimpanzees and bonobos have a more recent common ancestor than chimpanzees and orangutans do.

(a)   (b)   (c)   (d)

Gibbons (a) and siamangs are called "lesser apes" and are classified in the family Hylobatidae. Orangutans (b), gorillas (c), and chimpanzees (d) are called "great apes" and are classified in the family Pongidae.

WHY RECONSTRUCT PHYLOGENIES?

Phylogenetic reconstruction plays three important roles in the study of organic evolution.

We have seen that descent with modification explains the hierarchical structure of the living world. Because new species always evolve from existing species and species are reproductively isolated, all living organisms can be placed on a single phylogenetic tree, which we can then use to trace the ancestry of all living species. In the remainder of this chapter, we will see how the pattern of similarities and differences observed in living things can be used to construct phylogenies and to help establish the evolutionary history of life.

Reconstructing phylogenies plays an important role in the study of evolution for three reasons:

1. *Phylogeny is the basis for the identification and classification of organisms.* In the latter part of this chapter we will see how scientists use phylogenetic relationships to name organisms and arrange them into hierarchies. This endeavor is called **taxonomy.**

2. *Knowing phylogenetic relationships often helps explain why a species evolved certain adaptations and not others.* Natural selection creates new species by modifying existing body structures to perform new functions. To understand why a new organism evolved a particular trait, it helps to know what kind of organism it evolved from, and this is what phylogenetic trees tell us. The phylogenetic relationships among the apes provide a good example of this point. Most scientists used to believe that chimpanzees and gorillas shared a more recent common ancestor than either of them shared with humans. This view influenced their interpretation of the evolution of locomotion, or forms of movement, among the apes. All of the great apes are **quadrupedal**, which means that

they walk on their hands and feet. However, whereas gorillas and chimpanzees curl their fingers over their palms and bear weight on their knuckles, a form of locomotion called **knuckle walking**, orangutans bear weight on their palms (Figure 4.18). Humans, of course, stand upright on two legs. Knuckle walking involves distinctive modifications of the anatomy of the hand, and because human hands show none of these anatomical features, most scientists believed that humans did not evolve from a knuckle-walking species. Because both chimpanzees and gorillas are knuckle walkers, it was generally assumed that this trait evolved in their common ancestor (Figure 4.19). However, recent measurements of genetic similarity now have led most scientists to believe that humans and chimpanzees are more closely related to one another than either species is to the gorilla. If this is correct, then the old account of the evolution of locomotion in apes must be wrong. Two accounts are consistent with the new phylogeny. It is possible that the common ancestor of humans, chimpanzees, and gorillas was a knuckle walker and that knuckle walking was retained in the common ancestor of humans and chimpanzees. This would mean that knuckle walking evolved only once and that humans are descended from a knuckle-walking species (Figure 4.20a). Alternatively, the common ancestor of humans, chimpanzees, and other apes may not have been a knuckle walker. If this was the case, then knuckle walking evolved independently in chimpanzees and gorillas (Figure 4.20b). Each of these scenarios raises interesting questions about the evolution of locomotion in humans and apes. If chimpanzees and gorillas evolved knuckle walking independently, then a close examination should reveal subtle differences in their morphology. If humans evolved from a knuckle walker, then perhaps more careful study will reveal the traces of our former mode of locomotion. As we will see in Chapter 11, such traces can be found in the wrists of one of our ancestors.

3. *We can deduce the function of morphological features or behaviors by comparing the traits of different species.* This technique is called the **comparative method**. As we will see in Part Two, most primates live in groups. Some scientists have argued that **terrestrial** (ground-dwelling) primates live in larger groups than **arboreal** (tree-dwelling) primates because terrestrial species are more vulnerable to predators and animals are safer in larger groups. To test the relationship between group size and terrestriality using the comparative method, we would collect data on group size and lifestyle (arboreal/terrestrial) for many different primate species. However, most biologists believe that only independently evolved cases should be counted in comparative analyses, so we must take the phylogenetic relationships among species into account. Box 4.1 provides a hypothetical example of how phylogenetic information can alter our interpretations of comparative data.

For many years, scientists constructed phylogenies only for the purpose of classification. The terms *taxonomy* and *systematics* were used interchangeably to refer to the construction of phylogenies and to the use of such phylogenies for naming and classifying organisms. With the recent realization that phylogenies have other important uses like the ones just described, there is a need for terms that distinguish phylogenetic construction from classification. Here we adopt the suggestion of University of Chicago anthropologist Robert Martin to employ the term **systematics** to refer to the construction of phylogenies, and the term **taxonomy** to mean the use of phylogenies in naming and classification. Although this distinction may not seem important now, it will become more relevant as we proceed.

(a)

(b)

FIGURE 4.18 Knuckle walking among the great apes. (a) Chimpanzees and gorillas are knuckle walkers; this means that when they walk, they bend their fingers under their palms and bear weight on their knuckles. (b) In contrast, orangutans are not knuckle walkers; when they walk, they bear their weight on their palms.

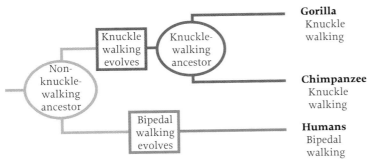

Knuckle walking evolved in the common ancestor of chimpanzees and gorillas

FIGURE 4.19 If chimpanzees are more closely related to gorillas than to humans, as morphological evidence suggests, then this is the most plausible scenario for the evolution of locomotion in gorillas, chimpanzees, and humans. The fossil record suggests that the ancestor of chimpanzees, gorillas, and humans was neither knuckle-walking nor bipedal. The simplest account of the current distribution of locomotion is that knuckle walking evolved once in the common ancestor of chimpanzees and gorillas.

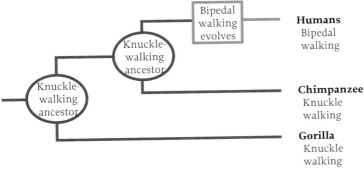

(a) Bipedal locomotion in humans evolved from knuckle walking

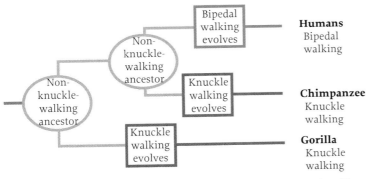

(b) Knuckle walking evolved independently in chimpanzees and gorillas

FIGURE 4.20 If humans are more closely related to chimpanzees than to gorillas, as the genetic data suggest, then there are two possible scenarios for the evolution of locomotion in these three species. (a) If the common ancestor of chimpanzees and gorillas was a knuckle walker, then it follows that human bipedal locomotion evolved from knuckle walking. (b) If the common ancestor of all three species was not a knuckle walker, then knuckle walking must have evolved independently in chimpanzees and gorillas.

HOW TO RECONSTRUCT PHYLOGENIES

We reconstruct phylogenies on the assumption that species with many phenotypic similarities are more closely related than species with fewer phenotypic similarities are.

To see how systematists reconstruct a phylogeny, consider the following example. We begin with three species, named for the moment A, B, and C, whose myoglobin (a cellular protein involved in oxygen metabolism) differs in the ways shown in Table 4.1. Remember from Chapter 2 that proteins are long chains of amino acids; the letters in Table 4.1 stand for the amino acids present at various positions along the protein chain.

BOX 4.1

The Role of Phylogeny in the Comparative Method

To understand why it is important to take phylogeny into account when using the comparative method, consider the phylogeny shown in Figure 4.21, which shows the pattern of relationships for eight hypothetical primate species.

As you can see, three terrestrial species live in large groups and three arboreal species live in small groups. Only one terrestrial species lives in small groups, and only one arboreal species lives in large groups. Thus, if we based our determinations on living species, we would conclude that there is a statistical relationship between group size and lifestyle. If we count independent evolutionary events, however, we get a very different answer. Large group size and terrestriality are found together only in species B and its descendants: B1, B2, and B3. This combination evolved only once, though we now observe this combination in three living species. There is also only one case of selection creating an arboreal species that lives in small groups (species C and its descendants: C1, C2, and C3). Note that each of the other possible combinations has also evolved once. When we tabulate independent evolutionary events, we find no consistent relationship between lifestyle and group size. Clearly, phylogenetic information is crucial to making sense of the patterns we see in nature.

FIGURE 4.21 The phylogenetic relationships among eight hypothetical primate species are shown here. Living species lie at the ends of branches, and their ancestors are identified at the branching points of the tree. These species vary in group size (small or large) and lifestyle (arboreal or terrestrial). The lineages in which novel associations between group size and lifestyle first evolved are marked with a double red bar. The left-hand matrix below tallies the association between group size and lifestyle among living species, and suggests that whereas arboreal species live in small groups, terrestrial species live in large ones. The right-hand matrix tallies the number of times the association between group size and lifestyle changed in the course of the evolution of these species. In this case, no relationship between group size and lifestyle is evident. This example demonstrates why it is important to keep track of independent evolutionary events in comparative analyses.

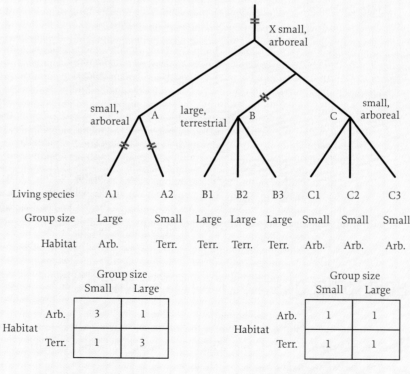

Living species	A1	A2	B1	B2	B3	C1	C2	C3
Group size	Large	Small	Large	Large	Large	Small	Small	Small
Habitat	Arb.	Terr.	Terr.	Terr.	Terr.	Arb.	Arb.	Arb.

		Group size	
		Small	Large
Habitat	Arb.	3	1
	Terr.	1	3

Counting living species implies that habitat predicts group size.

		Group size	
		Small	Large
Habitat	Arb.	1	1
	Terr.	1	1

Counting evolutionary events implies that habitat does *not* predict group size.

TABLE 4.1 The amino acid sequence for the protein myoglobin is shown for three different species. The numbers refer to positions on the myoglobin chain, and the letter in each cell stands for the particular amino acid found at that position in each kind of myoglobin. All three species have the same amino acids at all of the positions not shown here, as well as at positions 1 to 3. At eight positions, however, there is at least one discrepancy among the three species (shaded).

AMINO ACID NUMBER

Species	1	2	3	5	9	13	30	34	48	59	66
A	G	L	S	G	L	V	I	K	H	E	A
B	G	L	S	G	L	V	I	K	G	A	I
C	G	L	S	Q	Q	I	M	H	G	E	Q

A systematist wants to find the pattern of descent that is most likely to have produced these data. The first thing she would notice is that all three types of myoglobin have the same amino acid at many positions, exemplified here by the positions 1 (all G), 2 (all L), and 3 (all S). At positions 5, 9, 13, 30, and 34, species A and B have the same amino acids, but species C has a different set. At position 48, species B and C have the same amino acid; and at position 59, A and C have the same amino acid. At position 66, all three species have different amino acids. The systematist would see that there are fewer differences between species A and B than between A and C or between B and C. She would infer that fewer genetic changes have accumulated since A and B shared a common ancestor sometime in the past than since A and C, or B and C, shared a common ancestor. Because fewer evolutionary changes are necessary to convert A to B than A to C or B to C, A and B are assumed to have a more recent common ancestor than either species shares with C (Figure 4.22). In other words, species that are more similar are assumed to be more closely related.

Let's end the suspense and reveal the identities of the three species. Species A is our own species, *Homo sapiens*; species B is the duck-billed platypus (Figure 4.23); and species C is the domestic chicken. The phylogeny shown in Figure 4.22 suggests that humans and duck-billed platypuses (A and B) are more closely related to each other than either of them is to chickens (C). However, this phylogeny is not based on nearly enough data to be conclusive. To be convinced of the relationships among these three organisms, we would need to gather and analyze data on many more characters, and generate the same tree each time. In fact, many other characters show the same pattern as myoglobin. Humans and duck-billed platypuses, for example, have hair and mammary glands, but chickens lack these structures. The phylogeny in Figure 4.22 is the currently accepted tree for these three species.

Problems Due to Convergence

In constructing phylogenies, we must avoid basing decisions on characters that are similar because of convergent evolution.

Despite the evidence just discussed. not everything about the phylogeny for humans, platypuses, and chickens is hunky-dory. Two amino acid positions, 48 and 59 (see Table 4.1), are not consistent with the phylogeny shown in Figure 4.22. Moreover,

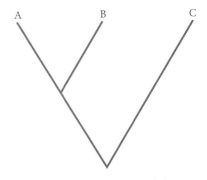

FIGURE 4.22 The phylogenetic tree for species A, B, and C derived from their myoglobin amino acid sequences, portions of which are shown in Table 4.1.

FIGURE 4.23 Duck-billed platypuses illustrate the importance of distinguishing between derived and ancestral traits. These creatures lay eggs and have horny bills like birds, but they lactate like mammals.

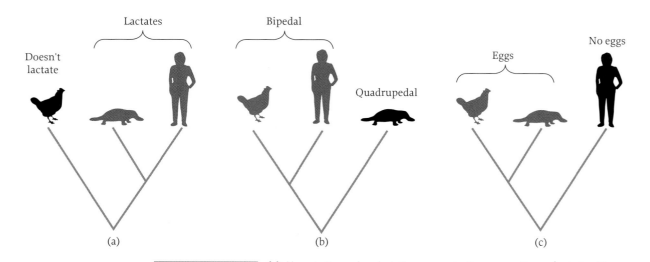

FIGURE 4.24   (a) Many traits, such as lactation, generate the same pattern of relationships among humans, chickens, and platypuses that myoglobin does (see Figure 4.22). (b) Other characters, such as bipedal locomotion, suggest a closer relationship between humans and chickens than between humans and platypuses or between platypuses and chickens. (c) Some traits, like egg laying, suggest a closer relationship between chickens and platypuses than between chickens and humans or between platypuses and humans.

certain other characters don't fit this tree. For example, both platypuses and humans lactate, but chickens don't (Figure 4.24a); both humans and chickens are bipedal, but platypuses are not (Figure 4.24b); and both platypuses and chickens lay eggs (Figure 4.24c), have a feature of the gut called a cloaca, and sport horny bills, but humans don't. Why don't these characters fit neatly into our phylogeny?

One reason for these anomalies is convergent evolution. Sometimes traits shared by two species are not the result of common ancestry. Instead, they are separate adaptations independently produced by natural selection. Chickens and humans are bipedal *not* because they are descended from the same bipedal ancestor, but rather because they *each* evolved this mode of locomotion independently. Similarly, the horny bills of chickens and platypuses are not similarities due to descent; they are independently derived characters. Systematists say that characters similar because of convergence are **analogous**, while characters whose similarity is due to descent from a common ancestor are **homologous**. It is important to avoid using convergent traits in reconstructing phylogenetic relationships. Although it is fairly obvious that bipedal locomotion in humans and chickens is not homologous, convergence is sometimes very difficult to detect.

Problems Due to Ancestral Characters

It is also important to ignore similarity based on ancestral characters, traits that also characterized the common ancestor of the species being classified.

There are also homologous traits that do not fit neatly into correct phylogenies. For example, chickens and platypuses reproduce by laying eggs, while humans do not. It seems likely that both chickens and platypuses lay eggs because they are descended from a common egg-laying ancestor. But if these characters are homologous, why don't they allow us to generate the correct phylogeny (see Figure 4.24c)?

Red-eyed   Blue-necked   Orange-spotted
cootie   cootie   cootie

FIGURE 4.25   The red-eyed cootie and the blue-necked
cootie share more traits with one another
(such as green legs, blue body, and green antennae) than either of
them shares with the orange-spotted cootie. If phylogenetic reconstruction were based on overall similarity, we would conclude that the
blue-necked cootie and the red-eyed cootie are more closely related to
each other than either is to the orange-spotted cootie.

Egg laying is an example of what systematists call an **ancestral trait**, one that characterized the common ancestor of chickens, platypuses, and humans. Egg laying has
been retained in the chicken and the platypus but lost in humans. It is important to
avoid using ancestral characters when constructing phylogenies. Only **derived traits**—
features that have evolved since the time of the last common ancestor of the species
under consideration—can be used in constructing phylogenies.

To see why we need to distinguish between ancestral and derived traits, consider
the three hypothetical species of "cooties" pictured in Figure 4.25. At first glance the
red-eyed cootie and the blue-necked cootie seem more similar to one another than
either is to the orange-spotted cootie, and a careful count of characters will show that
they do share more traits with one another than with the orange-spotted variety. But
consider Figure 4.26, which shows the phylogeny for the three species. You can see
that the blue-necked cootie is actually more closely related to the orange-spotted cootie
because they share a more recent common ancestor. The blue-necked cootie seems
more similar to the red-eyed cootie because they share many ancestral characters, but
the orange-spotted cootie has undergone a period of rapid evolution that has eliminated most ancestral characters. Looking at ancestral traits will not generate the correct phylogeny if rates of evolution differ among species.

Red-eyed   Blue-necked   Orange-spotted
cootie   cootie   cootie

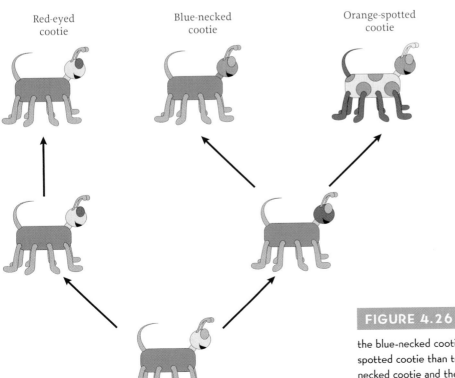

FIGURE 4.26   The phylogenetic history of the three cootie
species shown in Figure 4.25 indicates that
the blue-necked cootie is actually more closely related to the orange-
spotted cootie than to the red-eyed cootie. This is because the blue-
necked cootie and the orange-spotted cootie have a more recent
common ancestor with each other than with the red-eyed cootie.

If we base our assessment of similarity only on the number of derived characters that each species displays (as shown in Figure 4.27), then the most similar species are the ones most closely related. The blue-necked cootie and orange-spotted cootie share one derived character (an orange face) with each other, but they share no derived characters with the red-eyed cootie. Thus, if we avoid ancestral characters, we can construct the correct phylogeny.

Systematists distinguish ancestral from derived characters by the following criteria: ancestral characters (1) appear earlier in organismal development, (2) appear earlier in the fossil record, and (3) are seen in out-groups.

It is easy to see that distinguishing ancestral characters from derived characters is important, but hard to see how to do it in practice. If you can observe only living organisms, how can you tell whether a particular character in two species is ancestral or derived? There is no surefire solution to this problem, but biologists use three rules of thumb:

1. The development of a multicellular organism is a complex process, and it seems logical that modifications early in the process are likely to be more disruptive than modifications occurring later. As a result, evolution often (but not always) proceeds by modifying the ends of existing developmental pathways. To the extent that this generalization is true, characters that occur early in development are ancestral. For example, humans and other apes do not have tails, but the fact that a tail appears in the development of the human embryo and then disappears

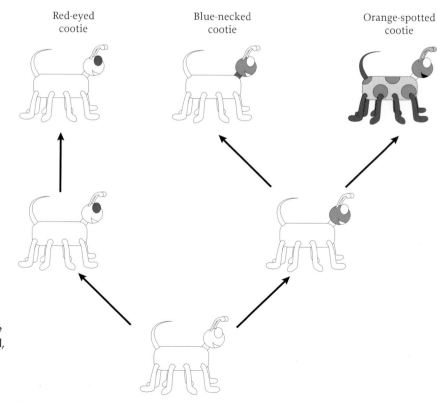

FIGURE 4.27 In this phylogeny, only derived characters are shown: red eyes in the red-eyed cootie; orange face in the blue-necked and orange-spotted cooties; blue neck in the blue-necked cootie; and red legs and tail, yellow antennae, and orange spots in the orange-spotted species. The correct phylogeny is based on similarity in shared, derived characters.

is evidence that the tail is an ancestral character (Figure 4.28). We conclude from this that the absence of a tail in humans is derived. This reasoning will not be helpful if ancestral traits have been completely lost during development or if traits (like egg laying) are expressed only in adults.

2. In many cases, fossils provide information about the ancestors of modern species. If we see in the fossil record that all of the earliest likely ancestors of apes had tails, and that primates without tails appear only later in the fossil record, then it is reasonable to infer that having a tail is an ancestral character. This criterion may fail if the fossil record is incomplete, and therefore a derived character appears in the fossil record before an ancestral one. As we will see in Chapter 11, new fossil finds can sometimes lead to radical revisions of existing phylogenies.

3. Finally, we can determine which characters are ancestral in a particular group by looking at neighboring groups, or **out-groups**. Suppose we are trying to determine whether having a tail is ancestral in the primates. We know that monkeys have tails and that apes do not, but we don't know which state is ancestral. To find out, we look at neighboring mammalian groups, such as insectivores or carnivores. Because the members of these out-groups typically have tails, it is reasonable to infer that the common ancestor of all primates also had a tail.

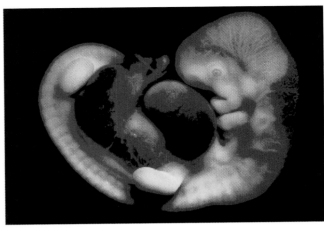

FIGURE 4.28 At an early stage of development, the human embryo has a tail. This feature disappears as development continues.

Using Genetic-Distance Data to Date Phylogenetic Events

Genetic distance measures the overall genetic similarity of two species.

Biologists and anthropologists often find it useful to compute a measure of overall genetic similarity, called **genetic distance**, between pairs of species. There are a number of different ways to measure genetic distance, but here we will focus on genetic-distance estimates based on DNA sequence data. For example, to compute the genetic distance between humans and chimpanzees, the biologist first identifies homologous DNA segments in the two species. This means that the segments are descended from the same DNA sequence in the common ancestor of humans and chimpanzees. These DNA segments are then sequenced. The number of nucleotide sites at which the two sequences differ is used to compute the genetic distance between the two species. The mathematical formulas used for these calculations are beyond the scope of this book, but suffice it to say that the larger the number of nucleotide differences, the bigger the genetic distance.

Genetic-distance data are often consistent with the hypothesis that genetic distance changes at an approximately constant rate.

The genetic distances among noncoding sequences of the DNA of humans, chimpanzees, gorillas, and orangutans are shown in Table 4.2. Notice that gorillas,

| TABLE 4.2 | Human | Chimpanzee | Gorilla | Orangutan |

Genetic distances among humans and the three great ape species, based on sequence divergence in noncoding regions of DNA. (Data from M. A. Jobling, M. Hurles, and C. Tyler-Smith, 2004, *Human Evolutionary Genetics: Origins, Peoples & Disease,* Garland Science, New York, p. 214.)

	Human	Chimpanzee	Gorilla	Orangutan
Human	—	1.24	1.62	3.08
Chimpanzee		—	1.63	3.12
Gorilla			—	3.09
Orangutan				—

humans, and chimpanzees are essentially the same genetic distance from orangutans. This close similarity is evidence that genetic distance changes at an approximately constant rate. To see why, think of genetic distance accumulating along the path leading from the last common ancestor of all of the hominoids to each living species. The total distance between any pair of living species is the sum of their paths from the last common ancestor. For example, the genetic distance between humans and orangutans is the accumulated distance between the last common ancestor of all four of these species and orangutans plus the accumulated distance between the last common ancestor and humans. Now notice that the distance between orangutans and humans is the same as the distance between chimpanzees and orangutans. This means that the genetic distance between the last common ancestor of orangutans and humans is the same as the distance between the last common ancestor of chimpanzees and orangutans. Because the time that elapsed since the branching off from the last common ancestor is the same for all these species, it follows that the rate of change of genetic distance along both of these paths through the phylogeny must be approximately the same, and that the genetic distance between two species is a measure of the time elapsed since both had a common ancestor. Evolutionists refer to genetic distances with a constant rate of change as **molecular clocks** because genetic change acts like a clock that measures the time since two species shared a common ancestor. Data from many other groups of organisms suggest that genetic distances often have this clocklike property, but there are also important exceptions.

A majority of biologists agree that as long as genetic distances are not too big or too small, the molecular-clock assumption is a useful approximation. However, there is some controversy about *why* genetic distance changes in a clocklike way. Advocates of the **neutral theory** believe that most changes in DNA sequences have little or no effect on fitness, so the evolution of this neutral DNA must be controlled by drift and mutation. The neutral theory suggests that, under the right circumstances, mutation and drift will produce clocklike change. Other biologists argue that the molecular clock is the product of natural selection in variable environments.

If the molecular-clock hypothesis is correct, then knowing the genetic distance between two living species allows us to estimate how long ago the two lineages diverged.

The data from contemporary species indicate that genetic distance changes at a constant rate, but the data don't provide a clue as to what that rate might be. However, dated fossils allow us to estimate when the splits between lineages occurred. By dividing the known genetic distance between a pair of species by the time since the last common ancestor, we can estimate the rate at which genetic distance changes through time. For example, fossil evidence (see Chapter 10) indicates that the last common ancestor of orangutans and humans lived about 14 mya and that the genetic distance between these two species is 3.1. Dividing the genetic distance between humans and orangutans by the time since their last common ancestor indicates that genetic distance accumulates at a rate of 0.22 unit per million years.

Once we have an estimate of the rate at which genetic distance changes through time, the molecular-clock hypothesis can be used to date the divergence times for lineages, even when we don't have any fossils. For example, to estimate the last common ancestor of humans and chimpanzees, we divide the genetic distance, 1.24, by the estimated rate, 0.22. This calculation indicates that the last common ancestor of these two species lived about 5.6 mya.

In practice, scientists use a number of different divergence dates to calibrate the rate at which genetic distance changes through time. Each divergence date produces a slightly different estimate of the rate of change of genetic distance, and this estimate in turn generates different estimates of the divergence times. Thus, divergence dates for humans and chimpanzees range from 7 to 5 mya.

TAXONOMY: NAMING NAMES

The hierarchical pattern of similarity created by evolution provides the basis for the way science classifies and names organisms.

Putting names on things is, to some extent, arbitrary. We could give species names like Sam or Ruby, or perhaps use numbers like the Social Security system does. This is, in fact, the way common names work. The word *lion* is an arbitrary label, as is "Charles" or "550-72-9928." The problem for scientists is that the number of animals and plants is very large, encompassing far too many species for any individual to keep track of. One way to cope with this massive complexity is to devise a system in which organisms are grouped together in a hierarchical system of classification. Once again there are many possible systems. For example, we could categorize organisms alphabetically, grouping alligators and apricots with the *A*s, barnacles and baboons with the *B*s, and so on. This approach has little to recommend it, because knowing how an animal is classified tells us nothing about the organism besides its location in the alphabet. An alternative approach would be to adopt a system of classification that groups together organisms with similar characteristics, analogous to the Library of Congress system used by libraries to classify books. The *Q*s might be predators, the *QH*s aquatic predators, the *QP*s aerial predators, and so on. Thus, knowing that the scientific name of the red-tailed hawk is QP604.4 might tell you that the hawk is a small aerial predator that lives in North America. The problem with such a system is that not all organisms would fall into a single category. Where would you classify animals that eat both animals and plants, such as bears, or amphibious predators, such as frogs?

The scientific system for naming animals is based on the hierarchy of descent: species that are closely related are classified together. Closely related species are grouped together in the same **genus** (plural *genera*; Figure 4.29). For example, the genus *Pan* contains two closely related species of chimpanzees: the common chimpanzee, *Pan troglodytes*; and the bonobo, *Pan paniscus*. Closely related genera are

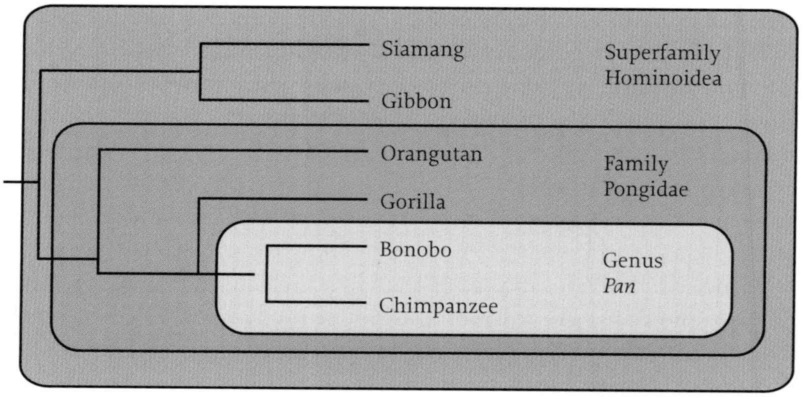

FIGURE 4.29 Bonobos and chimpanzees are classified together in the genus *Pan*, both of these species are classified with gorillas and orangutans in the family Pongidae, and all of the apes are grouped together in the superfamily Hominoidea. Note that humans are missing from this phylogenetic tree.

usually grouped together in a higher unit, often the **family**. Chimpanzees are in the family Pongidae along with the other great apes: the orangutan (*Pongo pygmaeus*) and the gorilla (*Gorilla gorilla*). Closely related families are then grouped together in a more inclusive unit, often a **superfamily**. The Pongidae belong to the superfamily Hominoidea along with the gibbons, siamangs, and humans.

Taxonomists disagree about whether overall similarity should also be used in classifying organisms.

The majority of taxonomists agree that descent should play a major role in classifying organisms. However, they vehemently disagree about whether descent should be the *only* factor used to classify organisms. The members of a relatively new school of thought, called **cladistic taxonomy** (or sometimes "cladistic systematics"), argue that only descent should matter. Adherents to an older school taxonomy, called **evolutionary taxonomy** (or "systematics"), believe that classification should be based both on descent *and* on overall similarity. To understand the difference between these two philosophies, consider Figure 4.30, in which humans have been added to the phylogeny of the apes shown in Figure 4.29. Evolutionary taxonomists would say that humans are qualitatively different from other apes and so deserve to be distinguished at a higher taxonomic level (Figure 4.30a). Accordingly, these taxonomists classify humans in a family of their own, the Hominidae. For a cladist, this is unacceptable because humans are descended from the same common ancestor as other members of the family Pongidae. This means that humans *must* be classified in the same family as chimpanzees, bonobos, and gorillas (Figure 4.30b). It is not just chauvinism about our own place in the primate phylogeny that causes discrepancies between these classification schemes. The same problem arises in many other taxa. For example, it turns out that crocodiles and birds share a more recent common ancestor than either does with lizards. For a cladist, this means that birds and crocodiles must be classified together, and lizards must be classified separately. Evolutionary taxonomists argue that birds are obviously distinctive and deserve a separate taxonomic grouping.

In theory, cladistic taxonomy is both informative and unambiguous. It is informative because knowing an organism's name and its position in the hierarchy of life tells us how it is related to other organisms. It is unambiguous because the position of each organism is given by the actual pattern of descent. Once you are confident that you understand the phylogenetic relationships within a group, there is no doubt about how any organism in that group should be classified. Cladists believe that evolutionary tax-

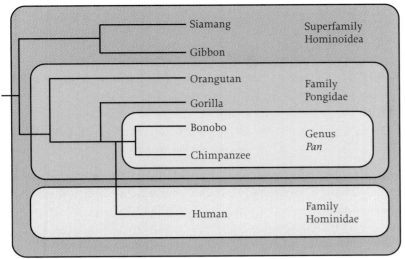

(a) Evolutionary classification

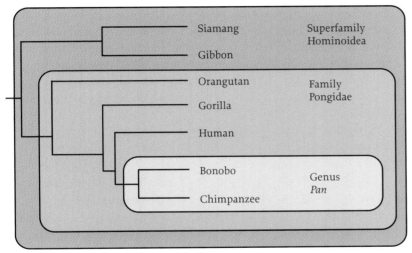

(b) Cladistic classification

FIGURE 4.30 Cladistic and evolutionary taxonomic schemes generate two different phylogenies for the hominoids. (a) The evolutionary classification classifies humans in a different family from other apes because apes are more similar to each other than they are to humans. (b) According to the cladistic classification, humans must be classified in the same family as the other great apes because they share a common ancestor with these creatures.

onomy is ambiguous because judgments of overall similarity are necessarily subjective. On the other hand, evolutionary taxonomists complain that the advantages of cladistics are mainly theoretical. In real life, they argue, uncertainty about phylogenetic relationships introduces far more ambiguity and instability in classification than do judgments about overall similarity. For example, we are not completely certain whether chimpanzees are more closely related to humans or to gorillas. Morphological data suggest that chimpanzees are more closely related to gorillas. Although most measures of genetic distance indicate that chimpanzees are more closely related to humans, some genetic-distance data suggest the opposite. Given such uncertainty, how would cladists name and classify these species?

It is important to keep in mind that this controversy is not about what the world is like, or even about how evolution works. Instead it is a debate about how we should name and classify organisms. Thus, no experiment or observation can prove either school right or wrong. Instead, scientists must determine which system is more useful in practice.

FURTHER READING

Dawkins, R. 1996. *The Blind Watchmaker*. Norton, New York, chap. 11.
Ridley, M. 1986. *Evolution and Classification*. Longman, London.
Ridley, M. 1996. *Evolution* (2nd ed.). Blackwell Scientific, Oxford, England.

STUDY QUESTIONS

1. There are two large plants in the authors' yard: a yucca tree and an enormous prickly pear cactus. Both have rough scaly bark, but it is known that these plants are not closely related.
   (a) What are two different explanations for the similarity of the bark?
   (b) It is also known that yuccas are descended from a grass species instead of a tree species. Which of the two explanations given in part 1a is consistent with this fact?

2. Chimpanzees and gorillas more closely resemble each other anatomically than either resembles humans. For example, the hands of chimpanzees and gorillas are structurally similar and quite different from human hands. Genetic-distance data suggest, however, that humans and chimpanzees are more closely related to each other than either is to gorillas. Assuming that the genetic-distance data are correct, give two different explanations for the observed anatomical similarity between chimpanzees and gorillas.

3. Use the genetic-distance matrix below to establish the taxonomic relationships between the species listed. (*Hint:* Draw a phylogenetic tree to illustrate these taxonomic relationships.)

	A	B	C	D
B	4.8	—		
C	0.7	5.0	—	
D	3.6	4.7	3.6	—

4. According to the biological species concept, what is a species? Why do some biologists define species in this way?

5. What is the ecological species concept? Why have some biologists questioned the biological species concept?

6. Molecular methods allow biologists to measure the amount of gene flow among populations that make up a species. When such methods first became available, systematists were surprised to find that many morphologically indistinguishable populations seem to be reproductively isolated from each other and thus, according to the biological species concept, are considered entirely different species. Is it possible to account for the existence of such cryptic species by allopatric speciation? by parapatric speciation? by sympatric speciation?

7. New plant species are sometimes formed by the hybridization of existing species. A new species retains all of the genes of each parent. For example, the variety of wheat used to make bread is a hybrid of three different grass species. Explain how such hybridization affects the family tree of these plants.

KEY TERMS

microevolution

macroevolution

biological species concept

reproductive isolation

gene flow

ecological species concept

allopatric speciation

character displacement

reinforcement

parapatric speciation

hybrid zone

sympatric speciation

niche

adaptive radiation

phylogeny

hominoids

taxonomy

quadrupedal

knuckle walking

comparative method

terrestrial

arboreal

systematics

analogous

homologous

ancestral trait

derived trait

out-groups

genetic distance

molecular clocks

neutral theory

cladistic taxonomy

evolutionary taxonomy

PRIMATE ECOLOGY AND BEHAVIOR

INTRODUCTION TO THE PRIMATES

TWO REASONS TO STUDY PRIMATES

The chapters in Part Two focus on the behavior of living nonhuman primates. Studies of nonhuman primates help us understand human evolution for two complementary but distinct reasons. First, closely related species tend to be similar morphologically. As we saw in Chapter 4, this similarity is due to the fact that closely related species retain and share traits acquired through descent from a common ancestor. For example, **viviparity** (bearing live young) and lactation are traits that all placental and marsupial mammals share, and these traits distinguish mammals from other taxa, such as reptiles. The existence of such similarities means that studies of living primates often give us more insight about the behavior of our ancestors than do studies of other organisms. This approach is called "reasoning by homology." The second reason we study primates is based on the idea that natural selection leads to similar organisms in similar environments. By assessing the patterns of diversity in the behavior and morphology of organisms in relation to their environments, we can see how evolution shapes adaptation in response to different selective pressures. This approach is called "reasoning by analogy."

Primates Are Our Closest Relatives

The fact that humans and other primates share many characteristics means that other primates provide valuable insights about early humans.

We humans are more closely related to nonhuman primates than we are to any other animal species. The anatomical similarities among monkeys, apes, and humans led the Swedish naturalist Carolus Linnaeus to place humans in the order Primates in the first scientific taxonomy, *Systema Naturae*, published in 1735. Later, naturalists such as Georges Cuvier and Johann Blumenbach placed humans in their own order because of our distinctive mental capacities and upright posture. In *The Descent of Man*, however, Charles Darwin firmly advocated reinstating humans in the order Primates; he cited biologist Thomas Henry Huxley's essay enumerating the many anatomical similarities between us and apes, and he suggested that "if man had not been his own classifier, he would never have thought of founding a separate order for his own reception." Modern systematics unambiguously confirms that humans are more closely related to other primates than to any other living creatures.

Because we are closely related to other primates, we share with them many aspects of morphology, physiology, and development. For example, like other primates, we have well-developed visual abilities and grasping hands and feet. We share certain features of our life history with other primates as well, including an extended period of juvenile development, and primates as a whole have larger brains in relation to body size than the members of most other taxonomic groups have. Homologies between humans and other primates also extend to behavior, since the physiological and cognitive structures that underlie human behavior are more similar to those of other primates than to members of other taxonomic groups. The existence of this extensive array of homologous traits, the product of the common evolutionary history of the primates, means that nonhuman primates provide useful models for understanding the evolutionary roots of human morphology and for unraveling the origins of human nature.

Primates Are a Diverse Order

Diversity within the primate order helps us to understand how natural selection shapes behavior.

During the last 30 years, hundreds of researchers from a variety of academic disciplines have spent thousands of hours observing many different species of nonhuman primates in the wild, in captive colonies, and in laboratories. All primate species have evolved adaptations that enable them to meet the basic challenges of life, such as finding food, avoiding predators, obtaining mates, rearing young, and coping with competitors. At the same time, there is great morphological, ecological, and behavioral diversity among species within the primate order. For example, primates range in size from the tiny mouse lemur, who weighs less than 30 g (about 1 oz) to male gorillas weighing about 160 kg (350 lb). Some species live in dense tropical forests; others are at home in open woodlands and savannas. Some subsist almost entirely on leaves; others rely on an omnivorous diet that includes fruits, leaves, flowers, seeds, gum, nectar, insects, and small animal prey. Some species are solitary, and others are highly gregarious. Some are active at night (**nocturnal**); others are active during daylight hours (**diurnal**). One primate, the fat-tailed dwarf lemur, enters a torpid state and sleeps for six months each year. Some species actively defend territories from incursions by other members of their own species (**conspecifics**); others do not. In some species, only females provide care of their young; in others, males participate actively in this process.

This variety is inherently interesting. Researchers who study primates tend to be so absorbed in the lives of their subjects that they are motivated to endure the hardships of fieldwork, the frustrations of attempting to obtain a share of ever-shrinking research funds, and the puzzlement of family and friends who wonder why they have chosen such an odd occupation. However, evidence of diversity among closely related organisms living under somewhat different ecological and social conditions also helps researchers to understand how evolution shapes behavior. Animals that are closely related to one another phylogenetically tend to be very similar in morphology, physiology, life history, and behavior. Thus, differences observed among closely related species are likely to represent adaptive responses to specific ecological conditions. At the same time, similarities among more distantly related creatures living under similar ecological conditions are likely to be the product of convergence.

This approach, sometimes called the "comparative method," has become an important form of analysis as researchers attempt to explain the patterns of variation in morphology and behavior observed in nature. The same principles have been borrowed to reconstruct the behavior of extinct hominins, early members of the human lineage. Because behavior leaves virtually no trace in the fossil record, the comparative method provides one of our only objective means of testing hypotheses about the lives of our hominin ancestors. For example, the observation that there are substantial differences in male and female body size, a phenomenon called **sexual dimorphism**, in species that form nonmonogamous groups suggests that highly dimorphic hominins in the past were not monogamous. In Part Three we will see how the data and theories about behavior produced by primatologists have played an important role in reshaping our ideas about human origins.

FEATURES THAT DEFINE THE PRIMATES

The primate order is generally defined by a number of shared, derived characters, but not all primates share all of these traits.

The animals pictured in Figure 5.1 are all members of the primate order. These animals are similar in many ways: they are covered with a thick coat of hair, they have four limbs, and they have five fingers on each hand. However, they share these ancestral features with all mammals. Beyond these ancestral features, it is hard to see what this group of animals have in common that makes them distinct from other mammals. What distinguishes a ring-tailed lemur from a mongoose or a raccoon? What features link the langur and the aye-aye?

(a)

(b)

(c)

(d)

(e)

FIGURE 5.1 All of these animals are primates: (a) aye-aye, (b) ring-tailed lemur, (c) langur, (d) howler, (e) gelada baboon. Primates are a diverse order and do not possess a suite of traits that unambiguously distinguish them from other animals.

In fact, primates are a rather nondescript mammalian order that cannot be unambiguously characterized by a single derived feature shared by all members. In his extensive treatise on primate evolution, however, University of Chicago biologist Robert Martin defines the primate order in terms of the derived features listed in Table 5.1.

The first three traits in Table 5.1 are related to the flexible movement of hands and feet. Primates can grasp with their hands and feet (Figure 5.2a), and most monkeys and apes can oppose their thumb and forefinger in a precision grip (Figure 5.2b). The flat nails, distinct from the claws of many animals, and the tactile pads on the tips of primate fingers and toes further enhance their dexterity (Figure 5.2c). These traits enable primates to use their hands and feet differently than most other animals do.

> **TABLE 5.1**   Definition of the primate order. See the text for more complete descriptions of these features.

1. The big toe on the foot is **opposable**, and hands are **prehensile.** This means that primates can use their feet and hands for grasping. The opposable big toe has been lost in humans.

2. There are flat nails on the hands and feet in most species, instead of claws, and there are sensitive tactile pads with "fingerprints" on fingers and toes.

3. Locomotion is **hindlimb-dominated**, meaning the hindlimbs do most of the work, and the center of gravity is nearer the hindlimbs than the forelimbs.

4. There is an unspecialized **olfactory** (smelling) apparatus that is reduced in diurnal primates.

5. The visual sense is highly developed. The eyes are large and moved forward in the head, providing stereoscopic vision.

6. Females have small litters, and gestation and juvenile periods are longer than in other mammals of similar size.

7. The brain is large compared with the brains of similarly sized mammals, and it has a number of unique anatomical features.

8. The **molars** are relatively unspecialized, and there is a maximum of two **incisors**, one **canine**, three **premolars**, and three molars on each half of the upper and lower jaw.

9. There are a number of other subtle anatomical characteristics that are useful to systematists but are hard to interpret functionally.

(a)

(b)

(c)

FIGURE 5.2 (a) Primates have grasping feet, which they use to climb, cling to branches, hold food, and scratch themselves. (b) Primates can oppose the thumb and forefinger in a precision grip —a feature that enables them to hold food in one hand while they are feeding, to pick small ticks and bits of debris from their hair while grooming, and (in some species) to use tools. (c) Most primates have flat nails on their hands and sensitive tactile pads on the tips of their fingers.

Primates are able to grasp fruit, squirming insects, and other small items in their hands and feet, and they can grip branches with their fingers and toes. During grooming sessions, they delicately part their partner's hair and use their thumb and forefinger to remove small bits of debris from the skin.

Traits 4 and 5 in Table 5.1 are related to a shift in emphasis among the sense organs. Primates are generally characterized by a greater reliance on visual stimuli and a reduced reliance on olfactory stimuli than other mammals are. Many primate species can perceive color, and their eyes are set forward in the head, providing them with binocular, stereoscopic vision (Figure 5.3). **Binocular vision** means that the fields of vision of the two eyes overlap so that both eyes perceive the same image. **Stereoscopic vision** means that each eye sends a signal of the visual image to both hemispheres in the brain to create an image with depth. These trends are not uniformly expressed within the primate order; for example, olfactory cues play a more important role in the lives of prosimian primates than in the lives of anthropoid primates. As we will explain shortly, the **prosimian** primates include the lorises and lemurs, and the **anthropoid** primates include the monkeys and apes.

Features 6 and 7 in Table 5.1 result from the distinctive life history of primates. As a group, primates have longer pregnancies, mature at later ages, live longer, and have larger brains than other animals of similar body size do. These features reflect a progressive trend toward increased dependence on complex behavior, learning, and behavioral flexibility within the primate order. As the noted primatologist Alison Jolly points out, "If there is an essence of being a primate, it is the progressive evolution of intelligence as a way of life." As we will see in the chapters that follow, these traits have a profound impact on the mating and parenting strategies of males and females and the patterns of social interaction among members of the order Primates.

The eighth feature in Table 5.1 concerns primate dentition. Teeth play a very important role in the lives of primates and in our understanding of their evolution. The utility of teeth to primates themselves is straightforward: teeth are necessary for processing food and are also used as weapons in conflicts with other animals. Teeth are also useful features for those who study living and fossil primates. Primatologists sometimes rely on tooth wear to gauge the age of individuals, and they use features of the teeth to assess the phylogenetic relationships among species. As we will see, paleontologists often rely on teeth, which are hard and preserve well, to identify the phylogenetic relationships of extinct creatures and to make inferences about their developmental patterns, their dietary preferences, and their social structure. Box 5.1 describes primate dentition in greater detail.

Although these traits are generally characteristic of primates, you should keep two points in mind. First, none of them makes primates unique. Dolphins, for example, have large brains and extended periods of juvenile dependence, and their social behavior may be just as complicated and flexible as that of any nonhuman primate (Figure 5.4). Second, not every primate possesses all of these traits.

FIGURE 5.3 In most primates, the eyes are moved forward in the head. The field of vision of the two eyes overlaps, creating binocular, stereoscopic vision. (Photograph courtesy of Carola Borries.)

FIGURE 5.4 A high degree of intelligence characterizes some animals besides primates. Dolphins, for example, have very large brains in relation to their body size, and their behavior is quite complex.

BOX 5.1

What's in a Tooth?

To appreciate the basic features of primate dentition, you can consult Figure 5.12, or you can simply look in a mirror, since your teeth are much like those of other primates. Teeth are rooted in the jaw. The jaw holds four different kinds of teeth; in order, they are first the incisors at the front, then the canines, premolars, and the molars in the rear. To understand what each kind of tooth does, imagine yourself eating a sandwich. You bite into the sandwich with your incisors and canines and use your front teeth to detach a piece. The incisors are relatively small, peg-shaped teeth. The canines are dagger-shaped. The upper canines are usually considerably longer than the other teeth, and the upper canine is sharpened on the lower premolar. The incisors and canines are involved mainly in getting food into the mouth and preparing food for further processing by the molars and premolars. The molars and premolars have broad surfaces, covered by a series of enamel bumps, or **cusps**, connected by crests or ridges. The molars and premolars are mainly used to crush, shred, and chew food before it is swallowed. In Chapter 6 we will show how the size and shape of the teeth are related to the types of foods primates eat.

Although all primates have the same kinds of teeth, species vary in how many of each kind of tooth they have. For convenience, these combinations are expressed in a standard format called the **dental formula**, which is commonly written in the following form:

$$\frac{2.1.3.3}{2.1.3.3}$$

The numerals separated by periods tell us how many of each of the four types of teeth a particular species has (or had) on one side of its jaw. Left to right, the four types are given for the front of the mouth (incisors) to the rear (molars). The top line of numbers represents the teeth on one side of the upper jaw (**maxilla**), and the bottom line represents the teeth on the corresponding side of the lower jaw (**mandible**). Hence this species—which happens to be the common ancestor of all primates—had two incisors, one canine, three premolars, and three molars each on one side of the upper and lower jaws. Usually, but not always, the formula is the same for both upper and lower jaws. Like most other parts of the body, our dentition is **bilaterally symmetrical**, which means that the left side is identical to the right side. The ancestral pattern shown here has been modified in

various primate taxa, as the total number of teeth has been reduced (Table 5.2).

The dental formulas among living primates vary. Prosimians have the most variable dentition. The lorises, pottos, galagos, and a number of lemurids have retained the primitive dental formula, but other groups have lost incisors, canines, or premolars. Tarsiers have lost one incisor on the mandible but retained two on the maxilla. Dentition is generally less variable among the anthropoid primates than among the prosimians. All of the New World monkeys, except the marmosets and tamarins, have retained the primitive dental formula; the marmosets and tamarins have lost one molar. The Old World monkeys, apes, and humans have reduced the number of premolars from three to two.

TABLE 5.2 Primates vary in the numbers of each type of tooth that they have. The dental formulas here give the number of incisors, canines, premolars, and molars on each side of the upper and lower jaws. For example, lorises have two incisors, one canine, three premolars, and three molars on each side of their upper and lower jaws.

Primate Group	Dental Formula
Prosimians	
Lorises, pottos, and galagos	$\frac{2.1.3.3}{2.1.3.3}$
Dwarf lemurs, mouse lemurs, and true lemurs	$\frac{2.1.3.3}{2.1.3.3}$
Indris	$\frac{2.1.2.3}{2.0.3.3}$
Aye-ayes	$\frac{1.0.1.3}{1.0.0.3}$
Tarsiers	$\frac{2.1.3.3}{1.1.3.3}$
New World monkeys	
Most species	$\frac{2.1.3.3}{2.1.3.3}$
Marmosets and tamarins	$\frac{2.1.3.2}{2.1.3.2}$
All Old World monkeys, apes, and humans	$\frac{2.1.2.3}{2.1.2.3}$

Humans have lost the grasping big toe that characterizes other primates, and some prosimians have claws on some of their fingers and toes.

PRIMATE BIOGEOGRAPHY

Primates are generally restricted to tropical regions of the world.

The continents of Asia, Africa, and South America and the islands that lie near their coasts are home to most of the world's primates (Figure 5.5). A few species remain in Mexico and Central America. Primates were once found in southern Europe, but no natural populations survive there now. There are no natural populations of primates in Australia or Antarctica, and none occupied these continents in the past.

Primates are found mainly in tropical regions, where the fluctuations in temperature from day to night greatly exceed fluctuations in temperature over the course of the year. In the tropics the distribution of resources that primates rely on for subsistence is affected more strongly by seasonal changes in rainfall than by seasonal changes in temperature. Some primate species extend their ranges into temperate areas of Africa and Asia, where they manage to cope with substantial seasonal fluctuations in environmental conditions.

Within their ranges, primates occupy an extremely diverse set of habitats that includes all types of tropical forests, savanna woodlands, mangrove swamps, grasslands, high-altitude plateaus, and deserts. The vast majority of primates, however, are found in forested areas, where they travel, feed, socialize, and sleep in a largely arboreal world.

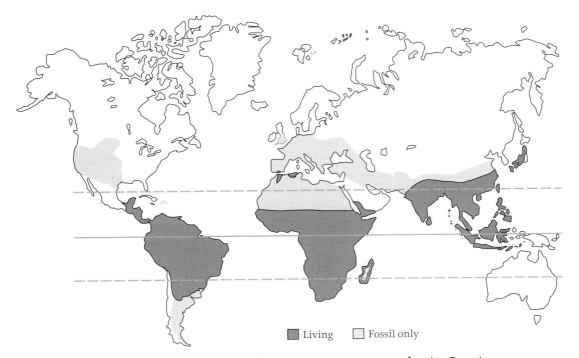

Living   Fossil only

FIGURE 5.5 The distribution of living and fossil primates. Primates are now found in Central America, South America, Africa, and Asia. They are found mainly in tropical regions of the world. Primates were formerly found in southern Europe and northern Africa. There have never been indigenous populations of primates in Australia or Antarctica.

A TAXONOMY OF LIVING PRIMATES

The primates are divided into two groups: the prosimians and the anthropoids.

Scientists classify primates into two suborders: Prosimii and Anthropoidea (Table 5.3). Many of the primates included in the suborder Prosimii are nocturnal, and like some of the earliest primates who lived 50 mya, they have many adaptations to living in darkness, including a well-developed sense of smell, large eyes, and independently movable ears. By contrast, monkeys and apes, who make up the suborder Anthropoidea, evolved adaptations more suited to a diurnal lifestyle early in their evolutionary history. In the Anthropoidea, the traits related to increased complexity of behavior are most fully developed.

The classification of the primates into prosimians and anthropoids does not strictly reflect the patterns of genetic relationships among the animals in the suborders. Tarsiers are included in the prosimians because, like the lorises and lemurs, they are nocturnal creatures that have retained many ancestral characters. However, both genetic and morphological data suggest that tarsiers are more closely related to monkeys and apes than to prosimians. Thus a purely cladistic classification would place tarsiers in the same **infraorder** (the taxonomic level immediately below suborder) as the monkeys. In fact, many primate taxonomists advocate a taxonomy in which the lemurs and lorises are classified together as **strepsirhines** and the rest of the primates are classified together as **haplorhines**. The more traditional division into prosimians and anthropoids is an example of evolutionary taxonomy in which overall similarity and relatedness are used to classify species.

The Prosimians

The prosimians are divided into three infraorders: Lemuriformes, Lorisiformes, and Tarsiiformes.

The infraorder Lemuriformes includes lemurs, which are found only on Madagascar and the Comoro Islands, off the southeastern coast of Africa. These islands have been separated from Africa for 120 million years. The primitive prosimians who reached Madagascar evolved in total isolation from primates elsewhere in the world, as well as from many of the predators and competitors that primates confront in other places. Faced with a diverse set of available ecological niches, the lemurs underwent a spectacular adaptive radiation. When humans first colonized Madagascar about 2000 years ago, there were approximately 44 species of lemurs, some as small as mouse lemurs and others as big as gorillas. In the next few centuries, all of the larger lemur species became extinct, probably the victims of human hunters. The extant lemurs are mainly small or medium-sized arboreal residents of forested areas; they travel quadrupedally or by jumping in an upright posture from one tree to another, a form of locomotion known as **vertical clinging and leaping**. Activity patterns of lemurs are quite variable: about half are primarily diurnal, others are nocturnal, and some are active during both day and night. One of the most interesting aspects of lemur behavior is that females routinely dominate males. In most lemur species, females are able to supplant males from desirable feeding sites; and in some lemur species, females regularly defeat males in aggressive encounters. Alison Jolly, one of the first observers of

TABLE 5.3	A taxonomy of the living primates. (From R. Martin, 1992, Classification of primates, pp. 20–21 in S. Jones, R. Martin, and D. Pilbeam, The Cambridge Encyclopedia of Human Evolution, Cambridge University Press, Cambridge.)

Suborder	Infraorder	Superfamily	Family	Subfamily
Prosimii (prosimians)	Lemuriformes	Lemuroidea (lemurs)	Cheirogaleidae (dwarf and mouse lemurs)	
			Lemuridae	Lemurinae (true lemurs)
				Lepilemurinae (sportive lemurs)
			Indridae (indris)	
			Daubentoniidae (aye-ayes)	
	Lorisiformes	Lorisoidea (loris group)	Lorisidae	Lorisinae (lorises)
				Galaginae (galagos)
	Tarsiiformes	Tarsioidea	Tarsiidae (tarsiers)	
Anthropoidea (anthropoids)	Platyrrhini	Ceboidea (New World monkeys)	Cebidae	Cebinae (e.g., capuchins)
				Aotinae (e.g., owl monkeys)
				Atelinae (e.g., spider monkeys)
				Alouattinae (howling monkeys)
				Pithecinae (e.g., sakis)
				Callimiconinae (Goeldi's monkeys)
			Callitrichidae (marmosets and tamarins)	
	Catarrhini	Cercopithecoidea (Old World monkeys)	Cercopithecidae	Cercopithecinae (e.g., macaques and vervets)
				Colobinae (e.g., langurs)
		Hominoidea (apes and humans)	Hylobatidae	Hylobatinae (gibbons and siamangs)
			Pongidae	Ponginae (great apes)
			Hominidae	Homininae (humans)

FIGURE 5.6 Galagos are small, arboreal, nocturnal animals who can leap great distances. They are mainly solitary, though residents of neighboring territories sometimes rest together during the day.

FIGURE 5.7 Tarsiers are small, insectivorous primates who live in Asia. Some tarsiers form monogamous pairs.

free-ranging lemurs, noted, "At any time a female may casually supplant any male or irritably cuff him over the nose and take a tamarind pod from his hand." Although such behavior may seem unremarkable in our own liberated times, female dominance is very rare in other primate species.

The infraorder Lorisiformes comprises small, nocturnal, arboreal residents of the forests of Africa and Asia. These animals include two subfamilies with different types of locomotion and activity patterns. Galagos are active and agile, leaping through the trees and running quickly along branches (Figure 5.6). The lorises move with ponderous deliberation, and their wrists and ankles have a specialized network of blood vessels that allows them to remain immobile for long periods of time. These traits may be adaptations that help them avoid detection by predators. Traveling alone, the lorisiforms generally feed on fruit, gum, and insect prey. The lorisiforms leave their dependent offspring in nests built in the hollows of trees or hidden in masses of tangled vegetation. During the day, females sleep, nurse their young, and groom, sometimes in the company of mature offspring or familiar neighbors.

The infraorder Tarsiiformes includes tarsiers, which are enigmatic primates who live in the rain forests of Borneo, Sulawesi, and the Philippines (Figure 5.7). Like many other prosimians, tarsiers are small, nocturnal, and arboreal, and they move by vertical clinging and leaping. Some tarsiers live in monogamous family groups, but many groups have more than one breeding female. Female tarsiers give birth to infants who weigh 25% of their own weight; mothers leave their bulky infants behind in safe hiding places when they forage for insects. Tarsiers are unique among primates because they are the only primates who rely exclusively on animal matter, feeding on insects and small vertebrate prey.

The Anthropoids

The suborder Anthropoidea contains the infraorders Platyrrhini and Catarrhini.

The two infraorders Platyrrhini and Catarrhini are commonly referred to as the New World monkeys and the Old World monkeys and apes, respectively, because platyrrhine monkeys are found in South and Central America, while catarrhine monkeys and apes are found in Africa and Asia. This geographic dichotomy breaks down with humans, however: we are catarrhine primates, but we are spread over the globe.

The infraorder Platyrrhini (New World monkeys) is divided into two families: Callitrichidae and Cebidae.

Although the New World monkeys encompass considerable diversity in size, diet, and social organization, they do share some basic features. All but those in one genus

(*Aotus*) are diurnal, all live in forested areas, and all are mainly arboreal. Most New World monkeys are quadrupedal, moving along the tops of branches and jumping between adjacent trees. Some species in the family Cebidae can suspend themselves by their hands, feet, or tail and can move by swinging by their arms beneath branches.

The family Callitrichidae is composed of the marmosets and tamarins. These species share several morphological features that distinguish them from other anthropoid primate species: they are extremely small, the largest weighing less than 1 kg (2.2 lb); they have claws instead of nails; they have only two molars, while all other monkeys have three; and they frequently give birth to twins and sometimes triplets (Figure 5.8). Marmosets and tamarins are also notable for their domestic arrangements: most species seem to be monogamous, and some may be polyandrous. **Polyandry,** which occurs when two or more males simultaneously form pair-bonds with a single female, is extremely uncommon among mammalian species. In caring for their young, marmoset and tamarin mothers receive a considerable amount of assistance from their mates and older offspring.

The cebid monkeys are generally larger than the marmosets and tamarins, ranging in size from the 600-g (21-oz) squirrel monkey to the 9.5-kg (21-lb) muriqui (Figure 5.9). Although many people think that all monkeys can swing by their tails, prehensile tails are actually restricted to only the largest species of the Cebidae.

The family Cebidae is divided into six subfamilies that encompass considerable diversity in social organization, feeding behavior, and ecology. The subfamily Alouattinae is composed of several species of howlers, named for the long-distance roars they give in intergroup interactions. Howlers live in small one-male or multimale groups, defend their home ranges, and feed mainly on leaves. The subfamily Atelinae includes spider monkeys and woolly monkeys (Figure 5.9a). These species subsist mainly on fruit and leaves, and they live in multimale, multifemale groups of 15 to 25. Spider monkeys, who rely heavily on ripe fruit, typically break up into small parties for feeding (Figure 5.9b). The subfamily Cebinae includes capuchins and squirrel monkeys. Capuchins are best known to the public as the clever creatures dressed in red caps and jackets who retrieve coins for organ grinders (Figure 5.9c). To primatologists, capuchins are notable in part because they have very large brains in relation to their body size (see Chapter 9). Squirrel monkeys and capuchins live in multimale, multifemale groups of 10 to 50 individuals and forage for fruit, leaves, and insects (Figure 5.9d). The subfamily Aotinae includes the diurnal titi monkey and the only nocturnal anthropoid primate, the owl monkey. The subfamily Callimiconinae is composed of just one species, Goeldi's monkey. Aotinae and Callimiconinae monkeys are small-bodied fruit eaters and live in monogamous groups. Finally, Pithecinae is a subfamily composed of the uakaris and sakis. Uakaris have little hair around their faces, making them look like wizened old men.

The infraorder Catarrhini contains the monkeys and apes of the Old World and humans.

As a group, the catarrhine primates share a number of anatomical and behavioral features that distinguish them from the New World primates. For example, most Old World monkeys and apes have narrow nostrils that face downward, while New World monkeys have round nostrils. Old World monkeys have two premolars on each side of the upper and lower jaws; New World monkeys have three. Most Old World primates are larger than most New World species, and Old World monkeys and apes occupy a wider range of habitats than New World species do.

FIGURE 5.8 Marmosets are small-bodied South American monkeys who form monogamous or polyandrous social groups. Males and older offspring participate actively in the care of infants.

(a)

(b)

(c)

(d)

FIGURE 5.9 Portraits of some cebid monkeys. (a) Muriquis, or woolly spider monkeys, are large-bodied and arboreal. They are extremely peaceful creatures, rarely fighting or competing over access to resources. (b) Spider monkeys rely heavily on ripe fruit and travel in small parties. They have prehensile tails that they can use much like an extra hand or foot. (c) Capuchin monkeys have larger brains in relation to their body sizes than any of the other nonhuman primates. (d) Squirrel monkeys form large multimale, multifemale groups. In the mating season, males gain weight and become "fatted," and then compete actively for access to receptive females. [Photographs courtesy of Sue Boinski (a), Susan Perry (c), and Carlão Limeira (d).]

The catarrhine primates are divided into two superfamilies: Cercopithecoidea (Old World monkeys) and Hominoidea (apes and humans). Cercopithecoidea contains one extant (still living) family, which is further divided into two subfamilies of monkeys: Cercopithecinae and Colobinae.

The superfamily Cercopithecoidea encompasses great diversity in social organization, ecological specializations, and biogeography.

Colobine monkeys are found in the forests of Africa and Asia and are collectively perhaps the most elegant of the primates (Figure 5.10). They have slender bodies, long legs, long tails, and often beautifully colored coats. The black-and-white colobus, for example, has a white ring around its black face, a striking white cape on its black back, and a bushy white tail that flies out behind as it leaps from tree to tree. Colobines are mainly leaf and seed eaters, and most species spend the majority of their time in trees. They have complex stomachs, almost like the chambered stomachs of cows, which allow them to maintain bacterial colonies that facilitate the digestion of cellulose.

(a)

(b)

FIGURE 5.10 (a) African colobines, like this black-and-white colobus monkey, are arboreal and feed mainly on leaves. These animals are sometimes hunted for their spectacular coats. (b) Hanuman langurs are native to India and have been the subject of extensive study during the last three decades. In some areas, hanuman langurs form one-male, multifemale groups, and males engage in fierce fights over membership in bisexual groups. In these groups, infanticide often follows when a new male takes over the group. (b, Photograph courtesy of Carola Borries.)

Colobines are most often found in groups composed of one adult male and a number of adult females. As in many other vertebrate taxa, the replacement of resident males in one-male groups is often accompanied by lethal attacks on infants by new males. Infanticide under such circumstances is believed to be favored by selection because it improves the relative reproductive success of infanticidal males. This issue is discussed more fully in Chapter 7.

Most cercopithecine monkeys are found in Africa, though one successful genus (*Macaca*) is widely distributed through Asia and part of Europe (Figure 5.11). The cercopithecines are more variable in size and diet than the colobines are. The social behavior, reproductive behavior, life history, and ecology of a number of cercopithecine species (particularly baboons, macaques, and vervets) have been studied extensively. Cercopithecines typically live in medium or large bisexual (multimale, multifemale) groups. Females typically remain in their **natal groups** (the groups into which they are born) throughout their lives and establish close and enduring relationships with their maternal kin; males leave their natal groups and join new groups when they reach sexual maturity.

The superfamily Hominoidea includes three families of apes: Hylobatidae (gibbons), Pongidae (orangutans, gorillas, and chimpanzees), and Hominidae (humans).

The hominoids are different from the cercopithecoids in a number of ways. The most readily observed difference between apes and monkeys is that apes lack tails. But there are many other more subtle differences between apes and monkeys. For example, the apes share some derived traits, including broader noses, broader palates, and

(a)

(b)

(c)

FIGURE 5.11 Some representative cercopithecines: (a) Bonnet macaques are one of several species of macaques that are found throughout Asia and North Africa. Like other macaques, bonnet macaques form multimale, multifemale groups, and females spend their entire lives in their natal (birth) groups. (b) Vervet monkeys are found throughout Africa. Like macaques and baboons, females live among their mothers, daughters, and other maternal kin. Males transfer to nonnatal groups when they reach maturity. Vervets defend their ranges against incursions by members of other groups. (c) Blue monkeys live in one-male, multifemale groups. During the mating season, however, one or more unfamiliar males may join bisexual groups and mate with females. [Photographs courtesy of Kathy West (a) and Marina Cords (c).]

larger brains; and they retain some primitive traits, such as relatively unspecialized molars. In Old World monkeys the prominent anterior and posterior cusps are arranged to form two parallel ridges. In apes, the five cusps on the lower molars are arranged to form a side-turned Y-shaped pattern of ridges (Figure 5.12).

The family Hylobatidae includes lesser apes (gibbons and siamangs), and its living members are now found in Asia. The family Pongidae includes the larger-bodied great apes (orangutans, gorillas, bonobos, and chimpanzees). Humans are traditionally placed in their own family, the Hominidae, but many taxonomists believe that

FIGURE 5.12 The upper jaw (left) and lower jaw (right) are shown here for a male colobine (a) and a male gorilla (b). In Old World monkeys, the prominent anterior and posterior cusps of the lower molars form two parallel ridges. In apes, the five cusps of the lower molar form a Y-shaped pattern.

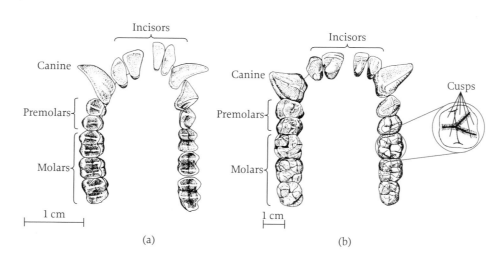
(a)   (b)

humans belong with the other great apes, in the Pongidae. Orangutans are found in Asia, while chimpanzees, bonobos, and gorillas are restricted to Africa.

The lesser apes are slightly built creatures with extremely long arms in relation to their body size (Figure 5.13). Gibbons and siamangs are strictly arboreal, and they use their long arms to perform spectacular acrobatic feats, moving through the canopy with grace, speed, and agility. Gibbons and siamangs are the only true **brachiators** among the primates, which means they propel themselves by their arms alone and are in free flight between handholds. (To picture this, think about swinging on monkey bars in your elementary school playground.) All of the lesser apes live in monogamous family groups; vigorously defend their **home ranges** (the areas they occupy); and feed on fruit, leaves, flowers, and insects. Siamang males play an active role in caring for young, frequently carrying them during the day; male gibbons are less attentive fathers. In territorial displays, mated pairs of siamangs perform coordinated vocal duets that can be heard over long distances.

Orangutans, now found only on the Southeast Asian islands of Sumatra and Borneo, are among the largest and most solitary species of primates (Figure 5.14). Orangutans have been studied extensively by Biruté Galdikas in Tanjung Puting, Borneo, for more than 20 years. Long-term studies of orangutans have also been conducted at Cabang Panti in Borneo, and at Ketambe and Suaq Balimbing in Sumatra. Orangutans feed primarily on fruit, but they also eat some leaves and bark. Adult females associate mainly with their own infants and immature offspring and do not often meet or interact with other orangutans. Adult males spend the majority of their time alone. A single adult male may defend a home range that encompasses the home ranges of several adult females; other males wander over larger areas and mate opportunistically with receptive females. When resident males encounter these nomads, fierce and noisy encounters may take place.

Gorillas, the largest of the apes, existed in splendid isolation from Western science until the middle of the nineteenth century (Figure 5.15). Today, our knowledge of the behavior and ecology of gorillas is based mainly on detailed long-term studies of one subspecies, the mountain gorilla, at the Karisoke Research Center in Rwanda, which was founded by the late Dian Fossey. Mountain gorillas live in small groups that contain one or two adult males and a number of adult females and their young. Each day, mountain gorillas ingest great quantities of various herbs, vines, shrubs, and

(a)

(b)

FIGURE 5.13 Gibbons (a) and siamangs (b) live in monogamous groups and actively defend their territories against intruders. They have extremely long arms, which they use to propel themselves from one branch to another as they swing hand over hand through the canopy, a form of locomotion called "brachiation." Siamangs and gibbons are confined to the tropical forests of Asia. Like other residents of tropical forests, their survival is threatened by the rapid destruction of tropical forests. (Photographs courtesy of John Mitani.)

(a)

(b)

FIGURE 5.14 (a) Orangutans are large, ponderous, and mostly solitary creatures. Male orangutans often descend to the ground to travel; lighter females often move through the tree canopy. (b) Today, orangutans are found only on the islands of Borneo and Sumatra, in tropical forests like this one.

FIGURE 5.15 (a) Gorillas are the largest of the primates. Mountain gorillas usually live in one-male, multifemale groups, but some groups contain more than one adult male. (b) Most behavioral information about gorillas comes from observations of mountain gorillas who live in the Virunga Mountains of central Africa, pictured here. The harsh montane habitat may influence the nature of social organization and social behavior in these animals, and the behavior of gorillas living at lower elevations may differ. (Photographs courtesy of John Mitani.)

(a)

(b)

bamboo. They eat little fruit because fruiting plants are scarce in their mountainous habitat. Adult male mountain gorillas, called **silverbacks** because the hair on their backs and shoulders turns a striking silver-gray when they mature, play a central role in the structure and cohesion of their social groups. Males sometimes remain in their natal groups to breed, but most males leave their natal groups and acquire females by drawing them away from other males during intergroup encounters. The silverback largely determines the timing of group activity and the direction of travel. As data from newly established field studies of lowland gorilla populations become available, however, we may have reason to revise some elements of this view of gorilla social organization. For example, lowland gorillas seem to eat substantial amounts of fruit, spend more of their time in trees, and form larger and less cohesive social groups than mountain gorillas do.

As humankind's closest living relatives, chimpanzees (Figure 5.16a) have played a uniquely important role in the study of human evolution. Whether reasoning by homology or by analogy, researchers have found observations about chimpanzees to be important bases for hypotheses about the behavior of early hominins.

Detailed knowledge of chimpanzee behavior and ecology comes from a number of long-term studies conducted at sites across Africa. In the 1960s, Jane Goodall began her well-known study of chimpanzees at the Gombe Stream National Park on the shores of Lake Tanganyika in Tanzania (Figure 5.16b). About the same time, a second study was initiated by Toshisada Nishida at a site in the Mahale Mountains not far from Gombe. These studies are now moving into their fourth decade. Other important study sites have been established at Boussou, Guinea; in the Taï Forest of the Ivory Coast; and at two sites in the Kibale Forest of Uganda: Kanyawara and Ngogo.

Bonobos (Figure 5.16c), another member of the genus *Pan*, live in inaccessible places and are much less well studied than common chimpanzees. Important field studies on bonobos have been conducted at two sites in the Democratic Republic of the Congo (formerly Zaire): Wamba and Lomako. Field studies of bonobos have been disrupted by civil conflicts that have ravaged central Africa over the last two decades.

Chimpanzees and bonobos form large multimale, multifemale communities. These communities differ from the social groups formed by most other species of primates in two important ways. First, female chimpanzees usually disperse from their natal groups when they reach sexual maturity, while males remain in their natal groups

(a)

(b)

(c)

FIGURE 5.16 (a) Chimpanzees live in multimale, multifemale social groups. In this species, males form the core of the social group and remain in their natal groups for life. Many researchers believe that chimpanzees are our closest living relatives. (b) Like other apes, chimpanzees are found mainly in forests like this area on the shores of Lake Tanganyika in Tanzania. However, chimpanzees sometimes range into more open areas as well. (c) Bonobos are members of the same genus as chimpanzees and are similar in many ways. Bonobos are sometimes called "pygmy chimpanzees," but this is a misnomer because bonobos and chimpanzees are about the same size. This infant bonobo is sitting in a patch of terrestial herbaceous vegetation, one of the staples of the bonobo's diet. (b, c, Photographs courtesy of John Mitani.)

throughout their lives. Second, the members of chimpanzee communities are rarely found together in a unified group. Instead, they split up into smaller parties that vary in size and composition from day to day. In chimpanzees, the strongest social bonds among adults are formed among males, while bonobo females form stronger bonds with one another and with their adult sons than males do. Chimpanzees modify natural objects for use as tools in the wild. At several sites, chimpanzees strip twigs and poke them into termite mounds and ant nests to extract insects, a much-prized delicacy. In the Taï Forest, chimpanzees crack hard-shelled nuts using one stone as a hammer and a heavy, flat stone or a protruding root as an anvil. At Gombe, chimpanzees wad leaves in their mouths and then dip these "sponges" into crevices to soak up water. New data also reveal tool use by wild orangutans, but chimpanzee tool use is more diverse and better studied.

PRIMATE CONSERVATION

Many species of primates are endangered by (1) habitat destruction, (2) hunting, or (3) live capture for trade and export.

Sadly, no introduction to the primate order would be complete without acknowledgment that the prospects for the continued survival of many primate species are grim. Today, nearly 100 primate species are considered to be endangered or critically endangered, and are in real danger of extinction. Already, primate conservation biologists believe that one subspecies, Miss Waldron's red colobus, has become extinct. The populations of some of the most gravely endangered species, such as the

FIGURE 5.17 Many populations of primates are in danger of extinction. This is a free-ranging member of a subspecies of squirrel monkeys that may number only 200 in the wild.

mountain gorilla and the golden lion tamarin, number only in the hundreds. In Africa and Asia, one-third to one-half of all primate species are endangered, and in some parts of South America an even larger fraction of species are at risk (Figure 5.17). At least two-thirds of the lemur species in Madagascar are in immediate danger of extinction.

As arboreal residents of the tropics, most primate populations are directly affected by the rapid and widespread destruction of the world's forests. Primate ecologists Colin Chapman of the University of Florida and Carlos Peres of the University of East Anglia recently reviewed the conservation status of the world's primate populations. Their analysis is quite sobering. Between 1980 and 1995, approximately 10% of the forests in Africa and Latin America were lost, and 6% of the forests of Asia disappeared (Figure 5.18). Countries that house primates are losing about 125,000 km² of forest—an area about the size of the state of Mississippi—*every year*.

The destruction of tropical forests is the product of economic and demographic pressures acting on governments and local residents. Many developing countries have huge foreign debts that must be repaid. The need to raise funds to pay off these debts generates intense pressure for timber harvesting and more intensive agricultural activity. Each year, 5 million to 6 million hectares of forest is logged, seriously disrupting the lives of the animals that live in them. (A hectare is a square measuring 100 m on a side, or about 2.5 acres.)

Forests are also cleared for agricultural activities. Rapid increases in the population of underdeveloped countries in the tropics have created intense demand for addi-

  Tropical deforestation
  Tropical forest

FIGURE 5.18 This map shows the major locations of tropical forest and the areas that have become deforested (90% of the canopy cover has been lost). Deforestation is a major threat to primates because many primate species live in tropical forests.

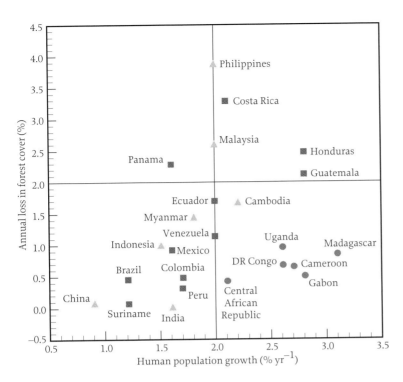

The destruction of tropical forests is often related to population pressures. Here, data on deforestation and human population growth are plotted for some of the countries that harbor free-ranging primate populations. Countries that have high rates of population growth have the highest rates of deforestation.

tional agricultural land (Figure 5.19). In West Africa, Asia, and South America, for example, vast expanses of forests have been cleared to accommodate the demands of subsistence farmers trying to feed their families, as well as the needs of large-scale agricultural projects (Figure 5.20). In Central and South America, massive areas have been cleared for large cattle ranches.

In the last two decades, a new threat to the forests of the world has emerged: wildfire. Major fires have destroyed massive tracts of forest in Southeast Asia and South America. Ecologists believe that natural fires in tropical forests are relatively rare, and that these devastating fires are the product of human activity. In Indonesia, massive

This forest, on the border of the Lomas Barbudal Biological Reserve, a national park in Costa Rica, has just been logged. In many countries, cultivation of all the land surrounding nature reserves and national parks has created forest islands. Although primates may be protected within these reserves, their isolation threatens their long-term survival. The elimination of surrounding forest corridors restricts the movement of migrating animals and limits the size and genetic diversity of local primate populations. (Photograph by Colin Chapman.)

fires in the late 1990s left thousands of orangutans dead, reducing their numbers by nearly a third.

In many areas around the world, particularly South America and Africa, primates are also hunted for meat. Although systematic information about the impact of hunting on wild primate populations is scant, some case studies reveal troubling trends. In one forest in Kenya, for example, 1200 blue monkeys and nearly 700 baboons were killed by subsistence hunters in one year. In the Brazilian Amazon, one family of rubber tappers killed 200 woolly monkeys, 100 spider monkeys, and 80 howler monkeys during an 18-month span. In addition to subsistence hunting, there is also an active market for "bushmeat" in many urban areas.

The capture and trade of live primates has been greatly reduced since the Convention on International Trade in Endangered Species of Wild Fauna and Flora (CITES) was drafted in 1973. The parties to CITES, which now number 167 countries, ban commercial trade of all endangered species and monitor the trade of those that are at risk of becoming endangered. CITES has been an effective weapon in protecting primate populations around the world. The United States imported more than 100,000 primates each year before ratifying CITES, but it had reduced this number to approximately 13,000 a decade after signing the international agreement.

Although CITES has made a major impact, some problems persist. Live capture for trade remains a major threat to certain species, particularly the great apes, whose high commercial value creates strong incentives for illegal commerce. In many communities, young primates are kept as pets (Figure 5.21). For each animal taken into captivity, many other animals are put at risk because hunters cannot obtain young primates without capturing their mothers, who are usually killed in the process. In addition, many prospective pets die after capture, from injuries suffered during capture and transport or from poor housing conditions and inappropriate diets while in captivity.

Efforts to save endangered primate populations have met with some success.

FIGURE 5.21 These two colobine monkeys were captured by poachers for the pet trade. The animals were then confiscated by the Uganda Wildlife Authority and taken to a local zoo. (Photograph by Colin Chapman.)

Although much remains to be done, conservation efforts have significantly improved the survival prospects of a number of primate species. These efforts have helped preserve muriquis and golden lion tamarins in Brazil and golden bamboo lemurs in Madagascar. But there is no room for complacency. Promising efforts to save orangutans in Indonesia and mountain gorillas in Rwanda have been seriously impeded by regional political struggles and armed conflict, putting the apes' habitats and their lives in serious jeopardy. A number of different strategies to conserve forest habitats and preserve animal populations are on the table. These include land-for-debt swaps in which foreign debts are forgiven in exchange for commitments to conserve natural habitats, to develop ecotourism projects, and to promote sustainable development of forest resources. But as conservationists study these solutions and try to implement them, the problems facing the world's primates become more pressing. More and more forests disappear each year, and many primates are lost, perhaps forever.

FURTHER READING

Chapman, C. A., and C. A. Peres. 2001. Primate conservation in the new millennium: The role of scientists. *Evolutionary Anthropology* 10: 16–33.

Cowlishaw, G., and R. I. M. Dunbar. 2000. *Primate Conservation Biology.* University of Chicago Press, Chicago.

Fleagle, J. G. 1998. *Primate Adaptation and Evolution* (2nd ed.). Academic Press, San Diego, CA, chap. 1.

Kramer, R. A., C. van Schaik, and J. Johnson, eds. 1997. *Last Stand: Protected Areas and the Defense of Tropical Biodiversity.* Oxford University Press, New York.

Martin, R. D. 1990. *Primate Origins and Evolution: A Phylogenetic Analysis.* Princeton University Press, Princeton, NJ.

STUDY QUESTIONS

1. What is the difference between homology and analogy? What evolutionary processes correspond to these terms?
2. Suppose that a group of extraterrestrial scientists lands on Earth and enlists your help in identifying animals. How do you help them recognize members of the primate order?
3. What kinds of habitats do most primates occupy? What are the features of this kind of environment?
4. Outline the taxonomy of the living primates to the superfamily level. Identify the geographic regions the animals inhabit, as well as their major features.
5. What primitive characteristics do modern prosimians retain?
6. In many ways, the superfamily Lemuroidea comprises a more diverse group than other primate superfamilies do. Why is this?
7. What genera are included in the family Pongidae? Briefly describe the social organization and geographic range of each of these genera.
8. Why are contemporary primate species threatened? In other words, what are the major hazards facing them today?
9. How can we balance the needs and rights of people living in developing nations with the needs of the primates who live around them?
10. Local peoples have been living alongside monkeys and other animals in tropical forests for thousands of years. If this is the case, then why do we face the present conservation crisis? What has changed?

KEY TERMS

viviparity
diurnal
nocturnal
conspecifics
sexual dimorphism
binocular vision
stereoscopic vision
prosimian
anthropoid

infraorder
strepsirhines
haplorhines
vertical clinging and leaping
ployandry
natal groups
brachiators
home ranges
silverbacks

CHAPTER 6

PRIMATE ECOLOGY

Much of the day-to-day life of primates is driven by two concerns: getting enough to eat and avoiding being eaten. Food is essential for growth, survival, and reproduction, and it should not be surprising that primates spend much of every day finding, processing, consuming, and digesting a wide variety of foods (Figure 6.1). At the same time, primates must always guard against predators like lions, pythons, and eagles that hunt them by day, and leopards that stalk them by night. As we will see in the chapters that follow, both the distribution of food and the threat of predation influence the extent of sociality among primates and shape the patterning of social interactions within and between primate groups.

In this chapter we describe the basic features of primate ecology. Later we will draw on this information to explore the relationships among ecological factors, social organization, and primate behavior. It is important to understand the nature of these relationships because the same ecological factors are likely to have influenced the social organization and behavior of our earliest ancestors.

THE DISTRIBUTION OF FOOD

Food provides energy that is essential for growth, survival, and reproduction.

Like all other animals, primates need energy to maintain normal metabolic processes; to regulate essential body functions; and to sustain growth, development, and reproduction. The total amount of energy that an animal requires depends on four components:

1. *Basal metabolism.* **Basal metabolic rate** is the rate at which an animal expends energy to maintain life when at rest. As Figure 6.2 shows, large animals have higher basal metabolic rates than small animals have. However, large animals require relatively fewer calories *per unit* of body weight.
2. *Active metabolism.* When animals become active, their energy needs rise above baseline levels. The number of additional calories required depends on how much energy the animal expends. The amount of energy expended, in turn, depends on the size of the animal and how fast it moves. In general, to sustain a normal range of activities, an average-sized primate like a baboon or macaque requires enough energy per day to maintain a rate about twice its basal metabolic rate.
3. *Growth rate.* Growth imposes further energetic demands on organisms. Infants and juveniles, who are gaining weight and growing in stature, require more energy than would be expected on the basis of their body weight and activity levels alone.
4. *Reproductive effort.* In addition, for female primates the energetic costs of reproduction are substantial. During the latter stages of their pregnancies, for

FIGURE 6.1   A female baboon feeds on corms in Amboseli, Kenya.

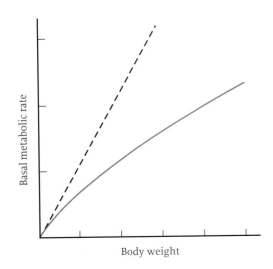

FIGURE 6.2   Average basal metabolism is affected by body size. The dashed line represents a direct linear relationship between body weight and basal metabolic rate. The solid line represents the actual relationship between body weight and basal metabolic rate. The fact that the curve "bends" means that larger animals use relatively less energy per unit of body weight.

example, primate females require about 25% more calories than usual; and during lactation, about 50% more calories than usual.

A primate's diet must satisfy the animal's energy requirements, provide specific types of nutrients, and minimize exposure to dangerous toxins.

The food that primates eat provides them with energy and essential nutrients, such as amino acids and minerals, that they cannot synthesize themselves. Proteins are essential for virtually every aspect of growth and reproduction and for the regulation of many body functions. As we saw in Chapter 2, proteins are composed of long chains of amino acids. Primates cannot synthesize amino acids from simpler molecules, so in order to build many essential proteins, they must ingest foods that contain sufficient amounts of a number of amino acids. Fats and oils are important sources of energy for animals and provide about twice as much energy as equivalent volumes of **carbohydrates**. Vitamins, minerals, and trace amounts of certain elements play an essential role in regulating many of the body's metabolic functions. Although they are needed in only small amounts, deficiencies of specific vitamins, minerals, or trace elements can cause significant impairment of normal body function. For example, trace amounts of the elements iron and copper are important in the synthesis of hemoglobin, vitamin D is essential for the metabolism of calcium and phosphorus, and sodium regulates the quantity and distribution of body fluids. Primates cannot synthesize any of these compounds and must acquire them from the foods they eat. Water is the major constituent of the bodies of all animals and most plants. For survival, most animals must balance their water intake with their water loss; moderate dehydration can be debilitating, and significant dehydration can be fatal.

At the same time that primates attempt to obtain nourishment from food, they must also take care to avoid **toxins**, substances in the environment that are harmful to them. Many plants produce toxins called **secondary compounds** to protect themselves from being eaten. Thousands of these secondary compounds have been identified: caffeine and morphine are among the secondary compounds most familiar to us. Some secondary compounds, such as **alkaloids**, are toxic to consumers because they pass through the stomach into various types of cells, where they disrupt normal metabolic functions. Common alkaloids include capsicum (the compound that brings tears to your eyes when you eat red peppers) and chocolate. Other secondary compounds, such as tannins (the bitter-tasting compound in tea), act in the consumer's gut to reduce the digestibility of plant material. Secondary compounds are particularly common among tropical plant species and are often concentrated in mature leaves and seeds. Young leaves, fruit, and flowers tend to have lower concentrations of secondary compounds, making them relatively more palatable to primates.

Primates obtain nutrients from many different sources.

Primates obtain energy and essential nutrients from a variety of sources (Table 6.1). Carbohydrates are obtained mainly from the simple sugars in fruit, but animal prey, such as insects, also provide a good source of fats and oils. **Gum**, a substance that plants produce in response to physical injury, is an important source of carbohydrates for some primates, particularly galagos, marmosets, and tamarins. Primates get most of their protein from insect prey or from young leaves. Some species have special adap-

TABLE 6.1	Sources of nutrients for primates. (×) indicates that the nutrient content is generally accessible only to animals that have specific digestive adaptations.					

Source	Protein	Carbohydrates	Fats and Oils	Vitamins	Minerals	Water
Animals	×	(×)	×	×	×	×
Fruit		×				×
Seeds	×		×	×		
Flowers		×				×
Young leaves	×			×	×	×
Mature leaves	(×)					
Woody stems	×					
Sap		×			×	×
Gum	×	(×)			×	
Underground parts	×	×				×

tations that facilitate the breakdown of cellulose, enabling them to digest more of the protein contained in the cells of mature leaves. Although seeds provide a good source of vitamins, fats, and oils, many plants package their seeds in husks or pods that shield their contents from seed predators. Many primates drink daily from streams, water holes, springs, or puddles of rainwater (Figure 6.3). Primates can also obtain water from fruit, flowers, young leaves, animal prey, and the underground storage parts (roots and tubers) of various plants. These sources of water are particularly important for arboreal animals that do not descend from the canopy of tree branches and for terrestrial animals during times of the year when surface water is scarce. Vitamins, minerals, and trace elements are obtained in small quantities from many different sources.

FIGURE 6.3 These savanna baboons are drinking from a pool of rainwater. Most primates must drink every day.

Although primates display considerable diversity in their diet, some generalizations are possible:

1. All primates rely on at least one type of food that is high in protein and another that is high in carbohydrates. Prosimians generally obtain protein from insects and carbohydrates from gum and fruit. Monkeys and apes usually obtain protein from insects or young leaves and carbohydrates from fruit.

2. Most primates rely more heavily on some types of foods than on others. Moreover, most species are fairly conservative in their dietary preferences, even though the availability of specific types of food varies over the course of the year in many habitats, and the species of food plants eaten often varies between habitats. Chimpanzees, for example, feed mainly on ripe fruit throughout their range from Tanzania to the Ivory Coast. Scientists use the terms **frugivore, folivore, insectivore,** and **gummivore** to refer to primates who rely most heavily on fruit, leaves, insects, and plant gum, respectively (Figure 6.4). Box 6.1 examines some of the adaptations in morphology among primates with different diets.

3. In general, insectivores are smaller than frugivores, and frugivores are smaller than folivores (Figure 6.5). These differences in size are related to the fact that small animals have relatively higher energy requirements than larger animals do, and they require relatively small amounts of high-quality foods that can be processed quickly. Larger animals are less constrained by the quality of their food than by the quantity because they can afford to process lower-quality foods more slowly.

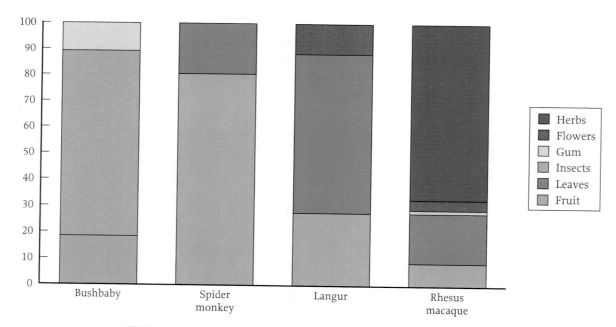

FIGURE 6.4 The diets of primates vary. The black-handed spider monkey is one of the most dedicated frugivores in the primate order; 80% of its diet consists of fruit. Langurs are among the most folivorous primate species; more than half of the purple-faced langur's diet is composed of leaves, while fruit and flowers play a much less important role. The bush baby (galago) is primarily insectivorous, relying on insects for 70% of its food supply. Macaques have more eclectic diets, feeding on herbs, fruit, gum and sap, flowers, and leaves.

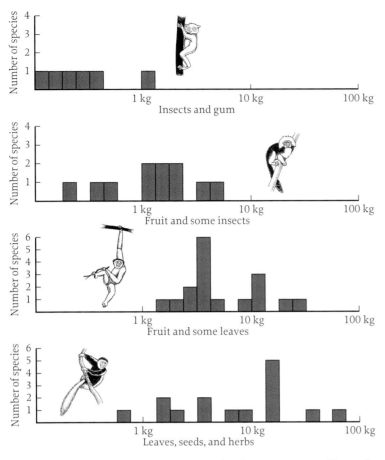

FIGURE 6.5   Body size and diet are related among primates. The smallest species eat mainly insects and gum; the largest species eat leaves, seeds, and herbs. Fruit-eating species fall in between.

The nature of dietary specializations and the challenge of foraging in tropical forests influence ranging patterns.

Nonhuman primates do not have the luxury of shopping in supermarkets where abundant supplies of food are concentrated in a single location and are constantly replenished. Instead, the availability of their preferred foods varies widely in space and time, making their food sources patchy and often unpredictable. Most primate species live in tropical forests. Although such forests, with their dense greenery, seem to provide abundant supplies of food for primates, appearances can be deceiving. Tropical forests contain a very large number of tree species, and individual trees of any particular species are few in number. Katherine Milton, a primatologist at the University of California Berkeley, has conducted detailed studies of the feeding ecology of howler monkeys in a lowland tropical forest in central Panama (Figure 6.7). In the course of her work, she carefully surveyed the composition of the forest in which the howlers live. She discovered that 65% of all tree species occur less than once per hectare. This means that potential food sources are patchily distributed in

BOX 6.1

Dietary Adaptations of Primates

Primates have evolved a number of adaptations to enhance their ability to process and digest certain types of foods. Their morphological dietary specializations include specific adaptations of the teeth and gut (Figure 6.6).

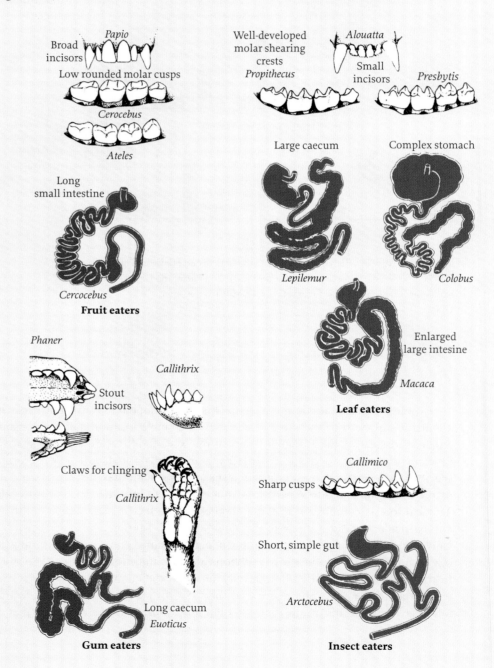

Broad incisors
Papio
Low rounded molar cusps
Cerocebus
Ateles
Long small intestine
Cercocebus
Fruit eaters

Well-developed molar shearing crests
Propithecus
Alouatta
Small incisors
Presbytis
Large caecum
Complex stomach
Lepilemur
Colobus
Enlarged large intesine
Macaca
Leaf eaters

Phaner
Stout incisors
Callithrix
Claws for clinging
Callithrix
Long caecum
Euoticus
Gum eaters

Callimico
Sharp cusps
Short, simple gut
Arctocebus
Insect eaters

TEETH

Primates who rely heavily on gum tend to have large and prominent incisors, which they use to gouge holes in the bark of trees. In some prosimian species the incisors and canines are projected forward in the jaw and are used to scrape hardened gum off the surface of branches and tree trunks. Dietary specializations are also reflected in the size and shape of the molars. Insectivores and folivores have molars with well-developed shearing crests that permit them to cut their food into small pieces when they chew. Insectivores tend to have higher and more pointed cusps on their molars, which are useful for puncturing and crushing the bodies of their prey. The molars of frugivores tend to have flatter, more rounded cusps, with broad and flat areas used to crush their food. Primates who rely on hard seeds and nuts have molars with very thick enamel that can withstand the heavy chewing forces needed to process these types of food.

GUT

Primates who feed principally on insects or animal prey have relatively simple digestive systems that are specialized for absorption. They generally have a simple small stomach, a small cecum (a pouch located at the upper end of the large intestine), and a small colon relative to the rest of the small intestine. Frugivores also tend to have simple digestive systems, but frugivorous species with large bodies have capacious stomachs to hold large quantities of the leaves they consume along with the fruit in their diet. Folivores have the most specialized digestive systems because they must deal with large quantities of cellulose and secondary plant compounds. Because primates cannot digest cellulose or other structural carbohydrates directly, folivores maintain colonies of microorganisms in their digestive systems that break down these substances. In some species these colonies of microorganisms are housed in an enlarged cecum; in other species the colon is enlarged for this purpose. Colobines, for example, have an enlarged and complex stomach divided into a number of different sections where microorganisms help process cellulose.

space. Moreover, Milton's studies showed that there is considerable variation in the production of new leaves, flowers, and fruit over the course of the year. In the forests of central Panama, young leaves are available on a single species for an average of seven months, green and ripe fruits are available for four months, and flower parts are available for three months. On individual trees, these food items may be available for even shorter periods. Ripe fruit, for example, is available on individual trees for less than one month on average. In some cases food items sustain an optimal nutritional content and edibility for only a few days. Thus, foraging in a tropical forest is not an easy task.

Primates with different dietary specializations confront different foraging challenges (Figure 6.8). Plants generally produce more leaves than flowers or fruit, and they bear leaves for a longer period during the year than they bear flowers and fruit. As a result, foliage is normally more abundant than fruit or flowers are at a given time during the year, and mature leaves are more abundant than young leaves. Insects and other suitable prey animals occur at even lower densities than plants. This means that folivores can generally find more food in a given area than frugivores or insectivores can. However, the high concentration of toxic secondary compounds in mature leaves complicates the foraging strategies of folivores. Some leaves must be avoided altogether, and others can be eaten only in

FIGURE 6.7 The forest of Barro Colorado Island, Panama, where howler monkeys have been studied for many years.

(a)

(b)

(c)

(d)

(e)

(f)

FIGURE 6.8 (a) Some primates feed mainly on leaves, though many leaves contain toxic secondary plant compounds. The monkeys shown here are red colobus monkeys in the Kibale Forest of Uganda. (b) Some primates include a variety of insects and other animal prey in their diet. This capuchin monkey in Costa Rica is feeding on a wasps' nest. (c) Mountain gorillas are mainly vegetarians. They consume vast quantities of plant material, like this fibrous stem. (d) This vervet monkey is feeding on grass stems. (e) Although many primates feed mainly on one type of food, such as leaves or fruit, no primate relies exclusively on one type of food. For example, the main bulk of the muriqui diet comes from fruit, but muriquis also eat leaves, as shown here. (f) Langurs are folivores. Here, hanuman langurs in Ramnagar, Nepal, forage for water plants. [Photographs courtesy of Lynne Isbell (a), Susan Perry (b), John Mitani (c), Carlão Limeira (e), and Carola Borries (f).]

small quantities. Nonetheless, the food supplies of folivorous species are generally more uniform and predictable in space and time than are the food supplies of frugivores or insectivores. Thus it is not surprising to find that folivores generally have smaller home ranges than frugivores or insectivores have.

ACTIVITY PATTERNS

The fact that many prosimian species are nocturnal suggests that primates evolved from a nocturnal ancestor.

Most primate species are active either during the day (diurnal) or at night (nocturnal), although a few species are active at intervals throughout the day and night (**cathemeral**). Because all monkeys and apes (except the owl monkey) are diurnal and a substantial number of prosimians are nocturnal, it seems likely that primates originally evolved from a nocturnal ancestor. Nocturnal primate species tend to be smaller, more solitary, and more exclusively arboreal than diurnal species. They also rely more heavily on olfactory signals than do diurnal species.

Primate activity patterns show regularity in seasonal and daily cycles.

Primates spend the majority of their time feeding, moving around their home ranges, and resting (Figures 6.9 and 6.10). Relatively small portions of each day are spent grooming, playing, fighting, or mating (Figure 6.11). The proportion of time devoted to various activities is influenced to some extent by ecological conditions. For primates living in seasonal habitats, for example, the dry season is often a time of scarce resources, and it is harder to find adequate amounts of appropriate types of food. In some cases this means that the proportion of time spent feeding and traveling increases during the dry season, while the proportion of time spent resting and socializing decreases.

Primate activity also shows regular patterns over the course of the day. When primates wake up, their stomachs are empty, so the first order of the day is to visit a feeding site. Much of the morning is spent eating and moving between feeding sites. As the sun moves directly overhead and the temperature rises, most species settle down in a

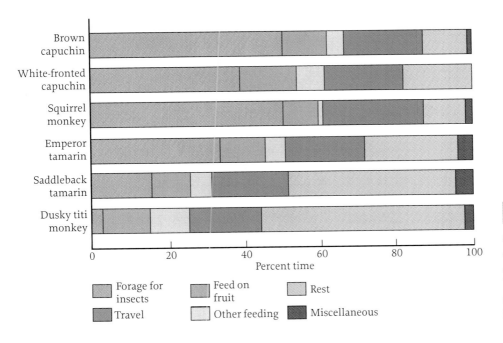

FIGURE 6.9 The amount of time that animals devote to various types of activities is called a "time budget." Time budgets of different species vary considerably. These six monkey species all live in a tropical rain forest in Manu National Park in Peru.

FIGURE 6.10   All diurnal primates, like this capuchin monkey, spend some part of each day resting. (Photograph courtesy of Susan Perry.)

FIGURE 6.11   Immature monkeys spend much of their "free" time playing. These patas monkeys are play wrestling. (Photograph courtesy of Lynne Isbell.)

FIGURE 6.12   Gorillas often rest in close proximity to other group members and socialize during a midday rest period.

FIGURE 6.13   Like most other primates, baboons sleep in trees at night to avoid predators. In Amboseli, baboons sleep in fever trees like this one. They perch on the terminal ends of the upper branches where they are safer from leopards. When this photograph was taken, most of the group was still in the sleeping tree. A number of baboons are visible in the upper right branches.

shady spot to rest, socialize, and digest their morning meals (Figure 6.12). Later in the afternoon they resume feeding. Before dusk they move to the night's sleeping site; some species sleep in the same trees every night; others have multiple sleeping sites within their ranges (Figure 6.13).

RANGING BEHAVIOR

All primates have home ranges, but only some species are territorial— defending their home range against incursions by other members of their species.

In virtually all primate species, groups range over a relatively fixed area, and members of a given group can be consistently found in a particular area over time. These areas are called home ranges, and they contain all of the resources that group members exploit in feeding, resting, and sleeping. However, the extent of overlap among adjacent home ranges and the nature of interactions with members of neighboring groups or strangers vary considerably among species. Some primate species, like gibbons, maintain exclusive access to fixed areas, called **territories** (Figure 6.14a). Territory residents regularly advertise their presence by vocalizing, and they aggressively protect the boundaries of their territories from encroachment by outsiders (Figure 6.15). Although some territorial birds defend only their nest sites, primate territories contain all of the sites at which the residents feed, rest, and sleep and the areas in which they travel. Thus, among territorial primates, the boundaries for the territory are essentially the same as for their home range, and territories do not overlap.

Nonterritorial species, like squirrel monkeys and long-tailed macaques, establish home ranges that overlap considerably with those of neighboring groups (Figure 6.14b). When members of neighboring nonterritorial groups meet, they may fight (Figure 6.16), exclude members of lower-ranking groups from resources, avoid one another, or mingle peacefully together. This last option is unusual, but in some species, adult females sexually solicit males from other groups, males attempt to mate with females from other groups, and juveniles from neighboring groups play together when their groups are in proximity.

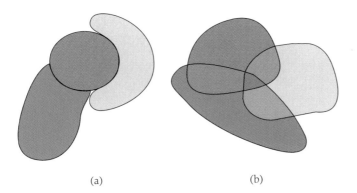

(a)   (b)

FIGURE 6.14 (a) Some primates defend the boundaries of territories (a schematic map of three territories is shown here) and do not tolerate intrusions from neighbors or strangers of the same species. (b) Other primates occupy fixed home ranges (three are shown here) but do not defend their borders against members of the same species. When two groups meet in areas of home range overlap, one may defer to the other, they may fight, or they may mingle peacefully together.

FIGURE 6.15   Gibbons perform complex vocal duets as part of territorial defense.

The two main functions suggested for territoriality are resource defense and mate defense.

To understand why some primate species defend their home ranges from intruders and others do not, we need to think about the possible costs and benefits associated with defending resources from conspecifics. Costs and benefits are measured in terms of the impact on the individual's ability to survive and reproduce successfully. Territoriality is beneficial because it prevents outsiders from exploiting the limited resources within a territory. At the same time, however, territoriality is costly because the residents must be constantly vigilant against intruders, regularly advertise their presence, and be prepared to defend their ranges against encroachment. Territoriality is expected to occur only when the benefits of maintaining exclusive access to a particular piece of land outweigh the costs of protecting these benefits.

When will the benefits of territoriality exceed the costs? The answer to this question depends in part on the kinds of resources that individuals need in order to survive and reproduce successfully, and in part on the way these resources are distributed spatially and seasonally. For reasons we will discuss more fully in Chapter 7, the reproductive strategies of mammalian males and females generally differ. In most cases, female reproductive success depends mainly on getting enough to eat for themselves and their dependent offspring, and males' reproductive success depends mainly on their ability to mate with females. As a consequence, females are more concerned about access to food, and males are more interested in access to females. Thus, territoriality has two different functions. Sometimes females defend food resources, or males defend food resources on their behalf. Other times, males defend groups of females against incursions by other males. In primates, both resource defense and mate defense seem to have influenced the evolution of territoriality.

FIGURE 6.16   Some primates defend fixed territories. Others react aggressively when they meet members of other groups. Here, a group of hanuman langurs attempts to repel an intruder. (Photograph courtesy of Carola Borries.)

(a)   (b)

FIGURE 6.17 (a) When resources are evenly distributed, like these *Ramphicarpa montana* flowers, there is little point in defending access to them. (b) When resources are clumped together, like these acacia flowers, animals may profit from driving away competitors.

Resource-defense territoriality occurs when resources are not only limited but also clumped and defendable.

Territoriality occurs when resources are economically defensible, meaning that resources are limited in abundance but occur within an area that can be defended with a reasonable amount of effort. When food resources are distributed over a wide area, it is costly to detect and evict intruders, so territoriality does not pay. Similarly, when resources are readily available, it makes little sense to defend them (Figure 6.17a). If, on the other hand, valuable resources are clumped within a small area, territoriality can be favored by natural selection (Figure 6.17b).

Mate defense also plays a role in the evolution of territoriality in some primate species.

Although territoriality seems to be linked to the defendability of resources in many species, this does not seem to be the primary factor favoring territoriality in some species. This conclusion is based on the observation that, in a number of species, males are active in intergroup encounters but females seem largely indifferent to the presence of intruders. This pattern is characteristic of gorillas, red colobus monkeys, many Southeast Asian langurs, and some species of lesser apes. Because males' reproductive success is influenced by their access to females, when males are the principal actors in territorial encounters, it seems likely that the primary function of territoriality is to defend access to mates, not food resources. By the same token, because females' ability to reproduce is related to their access to food resources, it seems likely that if females don't participate in territorial disputes, then resource defense is not the central factor determining the nature of intergroup encounters.

PREDATION

Predation is believed to be a significant source of mortality among primates, but direct evidence of predation is normally difficult to obtain.

Primates are hunted by a wide range of predators, including pythons, raptors, crocodiles, leopards, lions, tigers, and humans (Figure 6.18). In Madagascar, large lemurs

FIGURE 6.18 Primates are preyed upon by a variety of predators, including pythons (a), lions (b), leopards (c), martial eagles (d), and crocodiles (e).

are preyed upon by fossas, pumalike carnivores. Primates are also preyed on by other primates. Chimpanzees, for example, hunt red colobus monkeys, and baboons prey on vervet monkeys.

In primate populations, the estimated rates of predation vary from less than 1% of the population per year to more than 15% of the population per year. The available data suggest that small-bodied primates are more vulnerable to predation than larger ones are (Figure 6.19), and immature primates are generally more susceptible to predation than adults are. These data are not very solid, because systematic information about predation is quite hard to come by. Predation is very hard to observe directly because most predators avoid close contact with humans, and the presence of human observers probably deters attacks. Some predators, like leopards, generally

FIGURE 6.19 Small primates are more vulnerable to predators, but even large primates are vigilant. (a) This squirrel monkey is scanning the canopy for potential predators. (b) A male rhesus monkey scans the ground below for possible dangers. (a, Photograph courtesy of Sue Boinski.)

(a)   (b)   (c)

FIGURE 6.20 In some cases, researchers are able to confirm predation. Here, an adult female baboon in the Okavango Delta, Botswana, was killed by a leopard. You can see (a) the depression in the sand that was made when the leopard dragged the female's body out of the sleeping tree and across a small sandy clearing, (b) the leopard's footprints beside the drag marks, and (c) the remains of the female the following morning—her jaw, bits of her skull, and clumps of hair.

hunt at night, when most researchers are asleep. Usually predation is inferred when a healthy animal that is unlikely to have left the group abruptly vanishes without a trace (Figure 6.20). Such inferences are, of course, subject to error.

Another approach is to study the predators themselves. John Mitani of the University of Michigan and his colleagues studied the predatory behavior of crowned hawk eagles (Figure 6.21) in the Kibale Forest of Uganda by collecting the prey remains under the eagles' nests. Crowned hawk eagles are formidable predators; although they weigh only 3 to 4 kg (6.6 to 8.8 lb), they can take prey that weigh up to 30 kg (66 lb). In Kibale, crowned hawk eagles prey on all of the primates in the forest except chimpanzees, which weigh 30 to 50 kg (66 to 110 lb). Monkeys make up about 80% of the eagles' diet. Mitani and his colleagues estimated that approximately 2% of the total monkey population in Kibale is killed by hawk eagles each year.

Primates have evolved an array of defenses against predators.

Many primates give alarm calls when they sight potential predators, and some species have specific vocalizations for particular predators. Vervet monkeys, for example, give different calls when they are alerted to the presence of leopards, small carnivores, eagles, snakes, baboons, and unfamiliar humans. In many species, the most common response to predators is to flee or take cover. Small primates sometimes try to conceal themselves from predators; larger ones may confront potential predators. When slow-moving pottos encounter snakes, for example, they fall to the ground, move a short distance, and freeze. At some sites, adult red colobus monkeys aggressively attack chimpanzees who hunt their infants.

Another antipredator strategy that primates adopt is to associate with members of other primate species. In the Taï Forest of Ivory Coast, a number of monkey species share the canopy and form regular associations with one another. For example, groups of red colobus monkeys spend approximately half their time with groups of Diana monkeys. Ronald Noë and Rdeouan Bshary of the Université Louis Pasteur in Strasbourg, France, have shown that both species benefit when they are together. Red colobus monkeys sometimes detect eagles before Diana monkeys do, and Diana monkeys often detect terrestrial predators before the red colobus do. As a result, red

FIGURE 6.21 Crowned hawk eagles specialize in hunting monkeys. Primates make up 80% of the eagles' diet in the Kibale Forest of Uganda. There, crowned hawk eagles prey on all of the endemic primate species except chimpanzees.

colobus monkeys are less vigilant about dangers from terrestrial predators when they are with Diana monkeys, and Diana monkeys are less vigilant about aerial predators when they are with the red colobus.

PRIMATE SOCIALITY

Most primates, except for orangutans and some prosimians, spend most of their lives in stable groups of familiar individuals.

Although nearly all primates live in groups (Figure 6.22), not all groups are alike. The social organization among monkeys and apes encompasses great diversity, ranging from monogamous, pair-bonded groups of gibbons and owl monkeys, to polyandrous groups of tamarins; one-male groups of howlers and blue monkeys; multimale, multifemale groups of capuchin monkeys and macaques; and structured communities of gelada and hamadryas baboons (Box 6.2). The great diversity in primate social organization has prompted researchers to ask a number of related questions: Why do primates live in groups? How should groups be structured? How big should groups be?

Sociality is not unique to primates; killer whales, wolves, zebras, mongooses, and elephants (Figure 6.24) are some of the many animals that live in cohesive social groups. However, group life is not common among mammals. In many mammalian species, adult males and females occupy separate home ranges or territories and meet only for brief periods to court and mate. Females raise their young without further participation from males. Thus we need to consider why sociality has evolved among primates.

Social life has both costs and benefits.

Sociality reflects a dynamic balance between the advantages and disadvantages of living in close proximity to conspecifics. Sociality is beneficial because primates who live in social groups are better able to acquire and control resources and are less vulnerable to predators. Animals that live in groups can chase lone individuals away from

(a)

(b)

FIGURE 6.22 Most primates live in social groups. (a) Baboons often live in groups of 30 to 60 individuals. All members of the group recognize each other as individuals. (b) Ring-tailed lemurs live in female-bonded groups, from which males disperse at puberty.

BOX 6.2

Forms of Social Groups among Primates

It should be obvious by now that although most primates are social, their groups vary in size and composition. One way to order this diversity is to classify primate social systems according to the mating and residence patterns of females and males. The major forms of primate social systems are described here and diagrammed in Figure 6.23:

Solitary. Females maintain separate home ranges or territories and associate mainly with their dependent offspring. Males may establish their own territories or home ranges, or they may defend the ranges of several adult females from incursions by other males. With the exception of orangutans, all of the solitary primates are prosimians.

Monogamy. One male and one female form a pair-bond and share a territory with their immature offspring. Monogamy is characteristic of gibbons, some of the small New World monkeys, and a few prosimian species.

Polyandry. One female is paired with two or more males. They share a territory or home range with their offspring. Polyandry may occur among some of the marmosets and tamarins.

Polygyny: one-male. Groups are composed of several adult females, a single resident male, and immature individuals. (We refer to them in the text as "one-male, multifemale" groups.) In species with this form of polygyny, males who do not reside in bisexual groups often form all-male groups. These all-male groups may mount vigorous assaults on resident males of bisexual groups in an attempt to oust residents. Dispersal patterns are variable in species that form one-male, multifemale groups. One-male, multifemale groups are common among howlers, langurs, and gelada baboons.

Polygyny: multimale. Groups are composed of several adult males, adult females, and immature animals.

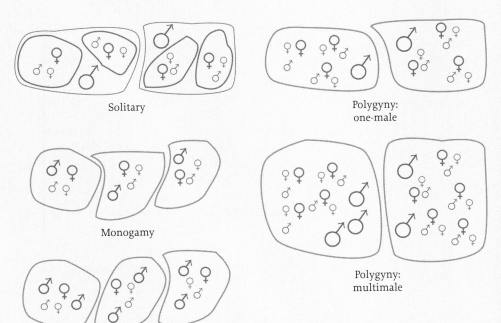

Solitary

Polygyny: one-male

Monogamy

Polygyny: multimale

Polyandry

FIGURE 6.23 The major types of social groups that primates form. When males and females share their home ranges, their home ranges are drawn here in lavender. When the ranges of the two sexes differ, male home ranges are drawn in blue and female home ranges are drawn in magenta. The sizes of the male and female symbols reflect the degree of sexual dimorphism among males and females.

(We refer to them in the text as "multimale, multifemale" groups.) Most such groups are female-bonded. Dispersing males may move directly between bisexual groups, remain solitary for short periods of time, or join all-male groups. Multimale, multifemale groups are characteristic of macaques, baboons, vervets, squirrel monkeys, capuchin monkeys, and some colobines.

This taxonomy of social groups provides one way to classify primate social organization, but readers should recognize that these categories are idealized descriptions of residence and mating patterns. The reality is inevitably more complicated. For example, in some monogamous groups, partners sometimes copulate with nonresidents or neighbors; and in species that normally form multimale groups, some groups may have only one adult male. Moreover, some species do not fit neatly within this classification system. Female chimpanzees and spider monkeys occupy individual home ranges and spend much of their time alone or with their dependent offspring. In these species, multiple males jointly defend the home ranges of multiple females. We call these fission–fusion groups because party size is variable. Hamadryas and gelada baboons form one-male, multifemale units, but several one-male, multifemale units collectively belong to larger aggregations.

feeding trees and can protect their own access to food and other resources against smaller numbers of intruders. Grouping also provides safety from predators because of the three *D*s: detection, deterrence, and dilution. Animals in groups are more likely to detect predators because there are more pairs of eyes on the lookout for predators. Animals in groups are also more effective in deterring predators by actively mobbing or chasing them away. Finally, the threat of predation to any single individual is diluted when predators strike at random. If there are two animals in a group, and a predator strikes, each has a 50% chance of being eaten. If there are 10 individuals, the individual risk is decreased to 10%.

Although there are important benefits associated with sociality, there are equally important costs. Animals that live in groups may encounter more competition over access to food and mates, become more vulnerable to disease, and face various hazards from conspecifics (such as cannibalism, cuckoldry, incest, or infanticide).

The size and composition of the groups that we see in nature are expected to reflect a compromise between the costs and benefits of sociality for individuals. The magnitude of these costs and benefits is influenced by both social and ecological factors. Thus the study of how natural selection shapes social organization in response to ecological pressures is called **socioecology**.

FIGURE 6.24 Sociality is not restricted to primates. For example, the social organization of elephants is highly structured. The basic social unit is the family group, which consists of one or more related adult females and their offspring. Certain family groups are frequently found together and form bond groups. Members of certain bond groups tend to associate with one another and are clustered into clans.

Primatologists are divided over whether feeding competition or predation is the primary factor favoring sociality among primates.

Joint defense of food resources is profitable when (1) food items are relatively valuable, (2) food resources are clumped in time and space, and (3) there is enough food within the patch to meet the needs of several individuals. Many primates feed preferentially on fruit, a food that meets these criteria (Figure 6.25). Taking note of this fact, Richard Wrangham of Harvard University suggested that primates gather into groups because they are more successful in defending access to resources when they are in groups than when they are alone.

This hypothesis is supported by several lines of evidence. Large groups generally prevail during encounters with smaller groups. Even when groups of monkeys do not meet face to face, small groups often leave feeding trees when they hear larger groups approaching. In a long-term study of vervet monkeys in Amboseli National Park in Kenya, Dorothy Cheney and Robert Seyfarth of the University of Pennsylvania found that small groups suffered more incursions into their ranges and defended their ranges more aggressively than did larger groups, which probably means that it was more costly for members of smaller groups to defend their territories. The larger groups also had bigger and more desirable territories than small groups had. These differences in territorial defense and territory quality were reflected in females' life histories: females in the small groups had lower survivorship than did females in the large groups.

FIGURE 6.25 Female baboons feed together on *Salvadora persica* fruit in Amboseli, Kenya.

Although Wrangham's idea is cogent and is supported by some evidence, University of Zürich primatologist Carel van Schaik challenged the belief that sociality is favored because it enhances resource defense. He pointed out that there is considerable competition over access to food *within* primate groups, and this may outweigh any advantages gained in competition *between* groups. For example, Charles Janson of Stony Brook University has assessed the extent of competition within and between groups of brown capuchins in Peru's Manu National Park (Figure 6.26). In this population, monkeys compete actively and often with other group members over access to food; the rate of aggression increases as group size increases (Figure 6.27). In contrast, intergroup competition was limited to relatively brief and infrequent contests over access to fruiting fig trees. Energy intake varied by 37% among individuals within the group, but by only 3% between groups that were most and least successful in intergroup competition.

The brown capuchin data suggest that if primates wanted to avoid competition with conspecifics, they might be better off avoiding them altogether. However, this option is dangerous for primates because predators present a serious threat to lone individuals. Thus, van Schaik suggested that primates live in groups because groups provide greater safety from predators. The predation model is difficult to test because

FIGURE 6.26 A brown capuchin monkey feeds in Peru's Manu National Park. (Photograph courtesy of Charles Janson.)

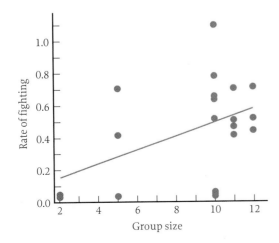

FIGURE 6.27 One possible cost of living in social groups is increased competition among group members over access to resources. For example, the rate of aggression increases as group size increases among capuchin monkeys in Peru's Manu National Park.

predation is so hard to study, as we explained earlier. However, several lines of evidence support the idea that sociality reduces predation risk. First, macaques living on islands without predators form smaller groups than do their more vulnerable conspecifics on the mainland. Second, most of the species that typically do not forage in groups (for example, spider monkeys, chimpanzees, and orangutans) are large animals and apparently face little danger from predators (Figure 6.28). Third, some of the species that regularly form mixed-species groups in the Taï Forest do not associate with each other at sites where predators are absent. Finally, in a number of species, vigilance rates decline when group members are nearby. For example, Adrian Treves and his colleagues from the University of Wisconsin found that black howler monkeys spend less time scanning for predators when companions are nearby (Figure 6.29).

Although these data suggest that grouping reduces vulnerability to predation, some data don't fit this hypothesis. Thus, in a recent review of the literature on vigilance in primates, Treves showed that vigilance does not decline smoothly as group size increases, perhaps because predation is not the only hazard that primates who live in groups face.

The jury is still out on whether resource competition or predation was the primary factor favoring the evolution of sociality in primates. However, many primatologists are convinced that the nature of resource competition affects the behavioral strategies of primates, particularly females, and influences the composition of primate groups. Females come first in this scenario because their fitness depends mainly on their nutritional status: well-nourished females grow faster, mature earlier, and have higher fertility rates than do poorly nourished females. In contrast, males' fitness depends primarily on their ability to obtain access to fertile females, not on their nutritional status. (We will discuss male and female reproductive strategies more fully in Chapter 7.)

FIGURE 6.28 Orangutans do not live in social groups. They are the only solitary anthropoid species.

The Distribution of Females

Ecologists distinguish between two kinds of resource competition: scramble competition and contest competition.

Scramble competition occurs when resources are distributed evenly across the landscape. Animals cannot effectively monopolize access to resources when they are distributed in this way, so they do not compete over resources directly. (Think about what happens when a piñata is broken open and all the candy rains down: everyone scram-

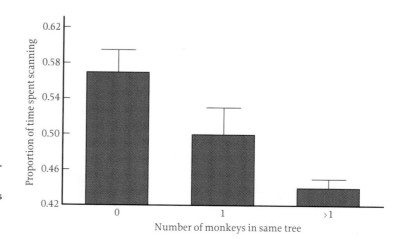

FIGURE 6.29 Primates in groups may be less vulnerable to predators and thus may be able to reduce their levels of vigilance. Black howler monkeys are less vigilant when other monkeys are in the same tree than when they are alone.

bles for the candy, but they don't fight over any particular item.) **Contest competition** occurs when resources are limited and can be monopolized profitably, generating direct confrontations over access to them. (Here, think about what happens in musical chairs: when the music stops, everyone scrambles for a chair, knowing that there are not enough to go around.) Both scramble competition and contest competition can occur within groups and between groups.

Resource competition is expected to generate dominance relationships.

In many animal species, ranging from crickets to chickens to chimpanzees, competitive encounters within pairs of individuals are common. The outcome of these contests may be related to the participants' relative size, strength, experience, or willingness to fight. In many species, for example, larger and heavier individuals regularly defeat smaller individuals. If there are real differences in power (based on size, weight, experience, or aggressiveness) between individuals, then we would expect the outcomes of dominance contests to be more or less the same from day to day. This is often the case. When dominance interactions between two individuals have predictable outcomes, we say that a **dominance** relationship has been established. We can plot the outcome of pairwise dominance contests in a matrix and construct **dominance hierarchies** to generate dominance rankings for individuals (Box 6.3).

When contest competition within groups is stronger than contest competition between groups, females will remain in their natal groups and cooperate with kin in contests with unrelated females in their group over resources.

Socioecological models consider the effects of within- and between-group competition on primate females. The model that we outline here was developed by primatologist Elisabeth Sterck of Utrecht University, David Watts of Yale University, and Carel van Schaik. Don't be put off by its complexity; many of the ideas that we introduce here will come up again in Chapters 7 and 8.

Sterck and her colleagues predict that when within-group contest competition is the primary form of competition, females' ability to control access to resources will be a function of their dominance rank. All other things being equal, high-ranking females are routinely able to exclude lower-ranking females from resources. This simple fact is expected to have a number of important consequences for primate females.

If dominance rank influences females' access to resources, and this in turn affects their fitness, females should strive for high rank. Moreover, females may benefit from helping their relatives, who share some of their genes, to gain high rank. (We will explain how kinship facilitates cooperation more fully in Chapter 8.) Because there is power in numbers, females can help their relatives win dominance contests by coming to their aid when they are challenged and supporting them when they initiate attacks. If females help their relatives in dominance encounters, then females will come to be able to defeat everyone that their relatives can defeat. We will see in Chapter 8 that cooperation among relatives leads to the formation of linear dominance hierarchies in which maternal kin occupy adjacent ranks.

If kin play an important role in the acquisition and maintenance of rank, then there are strong incentives for females to remain in their natal (birth) groups with their mothers, sisters, aunts, and cousins. We call this female **philopatry**. Finally, friendly interactions, like grooming and sitting close together, may enhance the quality of social relationships and reinforce alliances. If so, females are expected to interact preferentially with their coalition partners.

BOX 6.3

Dominance Hierarchies

When dominance interactions have predictable outcomes, we can assign dominance rankings to individuals. Consider the four hypothetical females in Figure 6.30a, which we shall call Blue, Turquoise, Green, and Lavender. Blue always beats Turquoise, Green, and Lavender. Turquoise never beats Blue but always beats Green and Lavender. Green never beats Blue or Turquoise but always beats Lavender. Poor Lavender never beats anybody. We can summarize the outcome of these confrontations between pairs of females in a **dominance matrix** such as the one in Figure 6.30b, and we can use the data to assign numerical ranks to the females. In this case, Blue ranks first, Turquoise second, Green third, and Lavender fourth. When females can defeat all the females ranked below them and none of the females ranked above them, dominance relationships are said to be **transitive**. When the relationships within all sets of three individuals (trios) are transitive, the hierarchy is linear.

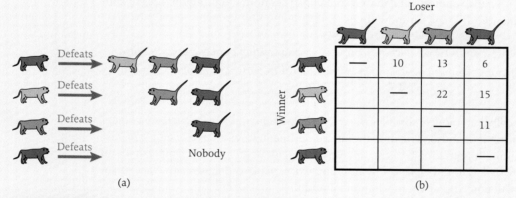

(a)
(b)

FIGURE 6.30 (a) Suppose that four hypothetical females—named Blue, Turquoise, Green, and Lavender, respectively—have the following transitive dominance relationships: Blue defeats the other three in dominance contests. Turquoise cannot defeat Blue but can defeat Green and Lavender. Green loses to Blue and Turquoise but is able to defeat Lavender. Lavender can't defeat anyone. (b) The results of data like those in part a are often tabulated in a dominance matrix, with the winners listed down the left side and the losers across the top. The value in each cell of the matrix represents the number of times one female defeated the other. Here, blue defeated Turquoise 10 times and Green defeated Lavender 11 times. There are no entries below the diagonal because females were never defeated by lower-ranking females.

When between-group contest competition is stronger than within-group contest competition, dominance will be less important, but female philopatry will still be favored.

In contrast, when between-group contest competition is strong and within-group contest competition is negligible, females' ability to gain access to resources will depend on the size of the social group that they live in, not their own dominance rank. When group size matters more than individual status, the dynamics of social interactions within groups will be affected in several ways. First, it will weaken competition for access to resources within groups, so females will not strive for high rank, and dominance relationships will tend to be more egalitarian. Second, there will be less incentive to support kin when they are involved in disputes with other group members, but females will rely on coalitionary support in intergroup encounters from all group members. Because females benefit more from supporting kin than nonkin, female philopatry will again be favored. Because females' fitness depends on collective group action, however, females are expected to cultivate more uniform social relationships with other group members.

When both within-group and between-group contest competitions are strong, the result will be intermediate between the previous scenarios.

The third possibility is that there will be strong contest competition within and between groups. In this case, females' fitness will depend on their ability to win competitive encounters with other group members, as well as their ability to compete as a group with members of neighboring groups. The result will be groups that combine the highly nepotistic structure we expect to find when within-group contest competition predominates, with the more egalitarian structure we expect to find when between-group contest competition prevails.

Finally, we might find situations in which only scramble competition prevails both within and between groups. In these cases, there will be little reason for females to establish dominance relationships, to form alliances with other females, or to participate in collective activities. Thus, ecological pressures will not guide females' behavior in any particular direction. Most researchers assume that this is a very atypical situation for primate females.

Empirical work supports the predictions of socioecological models.

In the 1990s, many researchers set to work testing the predictions derived from these socioecological models. This body of work confirmed many of the predictions. For example,

- Sue Boinski of the University of Florida and her colleagues compared two closely related species of squirrel monkeys in Costa Rica and Peru that had similar diets, formed groups of similar sizes, and experienced similarly high levels of predation. The two species differed mainly in the form of within-group competition over access to food resources. As predicted, the species that experienced high within-group contest competition formed stable, linear dominance hierarchies; developed kin-based coalitions; and exhibited strict female philopatry. And in the species that experienced low levels of within-group contest competition, a dominance hierarchy was not detected among females, female coalitions were not observed, and females sometimes left their natal groups.
- Lynn Isbell and her colleagues at the University of California Davis have compared the feeding ecology and social behavior of female monkeys in two closely related species: patas and vervets (Figure 6.31). Patas rely more heavily than vervets on arthropods and other food items that require little handling time and are consumed very quickly. This is a situation in which within-group scramble competition should prevail. On the other hand, vervets rely more heavily than patas on

FIGURE 6.31 Vervets and patas are closely related but differ in key features of their feeding ecology. Socioecological models predict that female patas monkeys (a) will have more egalitarian dominance hierarchies and less well defined social bonds than will female vervets (b).

(a)   (b)

foods such as gums, seeds, and fruits that have longer processing times and take longer to eat. Within-group contest competition would be expected to be more important for vervets. Patas have less linear dominance hierarchies and weaker social bonds than vervets do, as we would expect.

- Andreas Koenig, now at Stony Brook University, and his colleagues compared two populations of hanuman langurs living at different sites (Figure 6.32). The langurs of Ramnagar in southern Nepal relied heavily on just three food plants. These plants were low in abundance and clumped in their distribution—conditions that are expected to lead to within-group contest competition. The langurs of Kanha in India relied on a larger array of plant foods, and one of their most preferred foods is superabundant—conditions that are expected to lead to within- and between-group scramble competition. The langurs of Ramnagar behave much as expected when within-group contest competition is strong: they have linear dominance hierarchies, and females are philopatric. The langurs of Kanha have poorly developed dominance hierarchies, and females sometimes emigrate from their natal groups.

Not all of the observed variations in social organization and behavior fit predictions derived from socioecological models, perhaps because phylogeny constrains social evolution.

FIGURE 6.32 Hanuman langurs in Ramnagar, Nepal, encounter strong within-group contest competition and form stable, linear dominance hierarchies. (Photograph courtesy of Carola Borries.)

Although many of the comparisons of two closely related species provide support for socioecological models, we can also find examples that don't fit the models well. For instance, baboons occupy an extremely diverse range of habitats across Africa, but the basic structure of their social organization remains remarkably constant across habitats and groups.

The role of the environment in shaping social organization has also been challenged by comparative studies that take phylogeny explicitly into account. Anthony di Fiore of New York University and Drew Rendall of the University of Lethbridge point out that most of the characteristics associated with within-group contest competition, including linear dominance hierarchies, nepotistic alliances, and female philopatry, are found in almost all of the extant cercopithecine species, even though they now occupy an extremely diverse range of habitats. These findings suggest that there is considerable inertia in social evolution. The social systems categorized by socioecologists may represent different peaks in the adaptive landscape; once a species evolves one kind of social organization, it may be extremely difficult to shift to another kind, even if ecological conditions change. If this is true, then it seriously constrains the possible paths for change in social organization.

The Distribution of Males

In socioecological models, ecological factors shape the distribution of females, and males go where females are.

As we noted earlier, male reproductive fitness mainly depends on access to females, not food. So males go where females are. When females live in groups, they become a defensible resource for males. The more females there are, the more difficult it is for a single male to monopolize them. Moreover, if females are sexually receptive at the same time, it is more difficult for a single male to monopolize them. Comparative analyses indicate that the number of males in primate groups is generally correlated with the number of females present and with the extent of synchrony in female receptivity (Figure 6.33).

The distribution of females explains some but not all of the variation in the number of males in social groups. Some researchers think that the distribution of males in social groups is also influenced by ecological factors, particularly pressure from predators. Robert Hill and Phyllis Lee of Cambridge University conducted a systematic survey of Old World monkeys in an effort to assess the effects of predation risk on the distribution of males. They found that groups facing the highest risks of attacks by predators lived in the largest groups and had the most males. Thus the distribution of males may be a joint product of the distribution of females and the intensity of predation pressures.

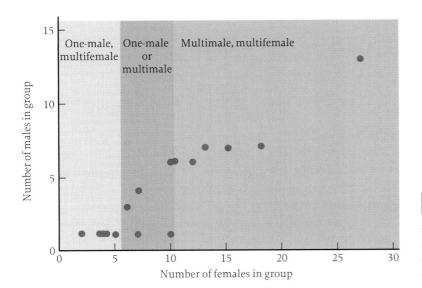

FIGURE 6.33 Among cercopithecine primate species, the number of females in social groups is positively correlated with the number of males in the same groups. Groups with only a few adult females generally contain only one adult male, and groups with many females contain several adult males.

The nature of resource competition also influences male dispersal strategies because inbreeding is disadvantageous.

Notice that female philopatry is expected to occur in three of the four competitive situations described earlier. This has important consequences for male life histories. The fitness of males depends largely on the availability of unrelated females. The presence of unrelated females is important because, in most species, mating with close kin (**inbreeding**) reduces the viability of offspring by increasing the probability that deleterious recessive traits will be expressed. Surveys of captive primate populations have shown that offspring of closely related parents are less likely to survive than are offspring of unrelated parents, and close kin generally avoid mating with one another. In all primate species studied to date, members of one or both sexes typically leave their natal groups near the time of sexual maturity. When females are philopatric, males must disperse. As you will see in Chapter 7, dispersal imposes significant costs on males.

FURTHER READING

Boinski, S., and P. A. Garber. 2000. *On the Move: How and Why Animals Travel in Groups.* University of Chicago Press, Chicago.

Janson, C. H. 2000. Primate socioecology: The end of a golden era. *Evolutionary Anthropology* 9: 73–86.

Oates, J. F. 1986. Food distribution and foraging behavior. Pp. 197–209 in *Primate Societies,* ed. by B. B. Smuts, D. L. Cheney, R. M. Seyfarth, R. W. Wrangham, and T. T. Struhsaker. University of Chicago Press, Chicago.

Richard, A. 1985. *Primates in Nature.* Freeman, New York, chaps. 4 and 5.

Sterck, E. H. M., D. P. Watts, and C. P. van Schaik. 1997. The evolution of social relationships in nonhuman primates. *Behavioral Ecology and Sociobiology* 41: 291–309.

van Schaik, C. P. 1989. The ecology of social relationships amongst female primates. Pp. 195–218 in *Comparative Socioecology: The Behavioural Ecology of Humans and Other Mammals,* ed. by V. Standon and R. A. Foley. Oxford University Press, Oxford, England.

Wrangham, R. W. 1980. An ecological model of female-bonded primate groups. *Behaviour* 75: 262–300.

STUDY QUESTIONS

1. Large primates often subsist on low-quality food such as leaves; small primates specialize in high-quality foods such as fruit and insects. Why is body size associated with dietary quality in this way?

2. For folivores, tropical forests seem to provide an abundant and constant supply of food. Why is this not an accurate assessment?

3. Territorial primates do not have to share access to food, sleeping sites, mates, and other resources with members of other groups. Given that territoriality reduces the extent of competition over resources, why aren't all primates territorial?

4. Territoriality is often linked to group size, day range, and diet. What is the nature of the association, and why does the association occur?

5. Most primates specialize in one type of food, such as fruit, leaves, or insects. What benefits might such specializations have? What costs might be associated with specialization?

6. Nocturnal primates are smaller, more solitary, and more arboreal than diurnal primates. What might be the reason(s) for this pattern?

7. Some primatologists have predicted that there should be a linear effect of group size on vigilance rates, if sociality evolved as a defense against predation. This means that every increase in the number of animals in the group should be associated with a decrease in the proportion of time that animals spend in vigilance activities. Many studies have failed to show a linear effect. How would you interpret these results? What do the results tell you about the selective forces that shape sociality in primates?

8. Sociality is a relatively uncommon feature in nature. What are the potential advantages and disadvantages of living in social groups? Why are (virtually all) primates social?

9. A number of studies have attempted to test socioecological models by comparing two closely related species. What is the logic underlying such studies? What would be another approach?

10. Males are largely left out of socioecological models. Why? What influences the distribution of males?

KEY TERMS

basal metabolic rate

carbohydrates

toxins

secondary compounds

alkaloids

gum

frugivore

folivore

insectivore

gummivore

cathemeral

territories

socioecology

scramble competition

contest competition

dominance

dominance hierarchies

philopatry

inbreeding

PRIMATE MATING SYSTEMS

Reproduction is the central act in the life of every living thing. Primates perform a dizzying variety of behaviors: gibbons fill the forest with their haunting duets, baboons threaten and posture in their struggle for dominance over other members of their group, and chimpanzees use carefully selected stone hammers to crack open tough nuts. But all of these behaviors evolved for a single ultimate purpose: reproduction. According to Darwin's theory, complex adaptations exist because they evolved step-by-step by natural selection. At each step, only those modifications that increased reproductive success were favored and retained in subsequent generations of offspring. Thus, every morphological feature and every behavior exist only because they were part of an adaptation that contributed to reproduction in ancestral populations. As a consequence, **mating systems** (the way animals find mates and care for offspring) play a crucial role in our understanding of primate societies.

Understanding the diverse reproductive strategies of nonhuman primates illuminates human evolution because we share many elements of our reproductive physiology with other species of primates.

To understand the evolution of primate mating systems, we must take into account that the reproductive strategies of living primates are influenced by their phylogenetic heritage as mammals. Mammals reproduce sexually. After conception, mammalian females carry their young internally. After they give birth, females suckle the young for an extended period of time. The mammalian male's role in the reproductive process is more variable than that of the female. In some species, males contribute little to their offspring's development besides a single sperm at the moment of conception. In other species, males defend territories; provide for their mates; and feed, carry, and protect their offspring.

Although mammalian physiology imposes some bounds on the nature of primate reproductive strategies, there is considerable room for diversity in primate mating systems and in reproductive behavior. Patterns of courtship, mate choice, and parental care vary greatly within the primate order. In some species, male reproductive success is determined mainly by success in competition with other males; in others, it is strongly influenced by female preferences. In many monogamous species, both males and females care for their offspring; in most nonmonogamous species, females care for offspring and males compete with other males to inseminate females.

What aspects of mating do humans share with other primates? Until very recent times, all pregnant women nursed their offspring for an extended period, as with other primates. In nearly all traditional human societies, fathers contribute extensively to their children's welfare, providing resources, security, and social support. An understanding of the phylogenetic and ecological factors that shape the reproductive strategies of other primates may help us understand how evolutionary forces shaped the reproductive strategies of our hominin ancestors, and the reproductive behavior of men and women in contemporary human societies.

THE LANGUAGE OF ADAPTIVE EXPLANATIONS

In evolutionary biology, the term strategy *is used to refer to behavioral mechanisms that lead to particular courses of behavior in particular functional contexts, such as foraging or reproduction.*

Biologists often use the term **strategy** when describing certain aspects of the behavior of animals. For example, folivory is characterized as a foraging strategy, and monogamy is described as a mating strategy. When evolutionary biologists use the term, they mean something very different from what we normally mean when we use *strategy* to describe, say, a general's military maneuvers or a baseball manager's tactics. In common usage, *strategy* implies a conscious plan and a goal-directed course of action. Evolutionary biologists do not think other animals consciously decide to defend their territories, wean their offspring at a particular age, monitor their ingestion of secondary plant compounds, and so on. Instead, the term is used to refer to a set of behaviors occurring in a specific functional context, such as mating, parenting, or foraging. Strategies are the product of natural selection acting on individuals to shape the motivations, reactions, preferences, capacities, and choices that influence behavior. Predispositions that produce behaviors that led to greater reproductive success in ancestral populations have been favored by natural selection and represent adaptations.

For example, howler monkeys, who are folivorous, behave as though they know that some leaves are good for them in small quantities but harmful in large quantities, and that young leaves contain more protein than older ones. Moreover, in different habitats howler monkeys adjust the mix of plants in their diet in what appears to be a deliberate attempt to balance nutrients and minimize toxins. But biologists doubt very much that howlers have any conscious knowledge of the nutritional content of the foods they eat or the components of an optimal diet. Instead, the underlying mechanisms that influence their decisions about what to eat, how much to eat, and what to avoid have been shaped by natural selection over many generations, producing the foraging behavior that we observe in nature.

The terms cost *and* benefit *refer to the effect of particular behavioral strategies on reproductive success.*

Different behaviors have different impacts on an animal's genetic fitness. Behaviors are said to be beneficial if they increase the genetic fitness of individuals, and costly if they reduce the genetic fitness of individuals. Thus, we argued in Chapter 6 that ranging behavior reflects a trade-off between the benefits of exclusive access to a particular area and the costs of territorial defense. Reproductive success is the ultimate currency in which these trade-offs are measured. Although this is a simple concept in principle, it is often very difficult to measure the costs and benefits associated with individual behavioral acts, particularly in long-lived animals like primates. Instead, researchers rely on indirect measures, such as foraging efficiency (measured as the quantity of nutrients obtained per unit time), and assume that, all other things being equal, behavioral strategies that increase foraging efficiency also enhance genetic fitness and will be favored by natural selection. We will encounter many other examples of this type of reasoning in the chapters that follow.

THE EVOLUTION OF REPRODUCTIVE STRATEGIES

Primate females always provide extensive care for their young, but males do so in only a few species.

The amount of parental care varies greatly within the animal kingdom. In most species, parents do little for their offspring. For example, most frogs lay their eggs and never see their offspring again. In such species, the nutrients that females leave in the egg are the only form of parental care. In contrast, primates—like almost all birds and mammals, and like some invertebrates and fish—provide much more than just the resources included in gametes. At least one parent, and sometimes both, shelters its young from the elements, protects them from predators, and provides them with food.

The *relative* amount of parental care provided by mothers and fathers also varies within the animal kingdom. In species without parental care, females produce large, nutrient-rich gametes, and males produce small gametes and supply only genes. Among species with parental care, however, all possible arrangements occur. Primate mothers always nurse their offspring and often provide extensive care (Figure 7.1). The behavior of fathers is much more variable. In most species, fathers give nothing to their offspring other than the genes contained in their sperm. In a minority of species, however, males are devoted fathers. In other taxa, patterns differ. For example, in most bird species, males and females form monogamous pairs and raise their young together (Figure 7.2), and there are even some bird species in which only males care for chicks.

The amount of time, energy, and resources that the males and females of a species invest in their offspring has profound consequences for the evolution of virtually every aspect of their social behavior and many aspects of their morphology. The selection pressures that affect males and females in species with equal parental investment are very different from the selection pressures that affect males and females in species in which females invest much more than males do. Thus it is important to understand why the amounts and patterns of parental investment differ among species.

FIGURE 7.1 In all primate species, females nurse their young. In baboons and many other species, females provide most of the direct care that infants receive.

Males do not care for their offspring (1) when they can easily use their resources to acquire many additional matings or (2) when caring for their offspring would not appreciably increase the offspring's fitness.

At first glance, it seems odd that most primate males fail to provide any care for their offspring. Surely if the males helped their mates, they would increase the chances that their offspring would survive to adulthood. Therefore, we might expect paternal care to be favored by natural selection (Figure 7.3).

If time, energy, and other resources were unlimited, this reasoning would be correct. In real life, though, time, energy, and material resources are always in short supply. The effort that an individual devotes to caring for offspring is effort that cannot be spent competing for prospective mates. Natural selection will favor individuals that allocate effort among these competing demands so as to maximize the number of surviving offspring that they produce.

To understand the evolution of asymmetries in parental investment, we must identify the conditions under which one sex can profitably reduce its parental effort at the expense of

FIGURE 7.2 In most species of birds, the male and female form a pair-bond and jointly raise their young. Here a bald eagle carries food to its hungry brood.

its partner. Consider a species in which most males help their mates feed and care for their offspring. Even in such a species, a few males will have a heritable tendency to invest less in their offspring. We will refer to these two types as "investing" and "non-investing" fathers. Because time, energy, and resources are always limited, males that devote more effort to caring for offspring must allocate less effort to competing for access to females. On average, the offspring of noninvesting males will receive less care than will the offspring of investing males, making them less likely to survive and to

FIGURE 7.3 In most non-monogamous species, males have relatively little contact with infants. Although males like this bonnet macaque are sometimes quite tolerant of infants, they rarely carry, groom, feed, or play with infants. (Photograph by Kathy West.)

reproduce successfully when they mature. On the other hand, because noninvesting males are not kept busy caring for their offspring, on average they will acquire more mates than will investing males. Mutations favoring the tendency to provide less parental care will increase in frequency when the benefits to males (measured in terms of the increase in fitness gained from additional matings) outweigh the costs to males (measured as the decrease in offspring fitness due to a reduction in paternal care).

This reasoning suggests that unequal parental investment will be favored when one or both of the following are true:

1. Acquiring additional mates is relatively easy, so considerable gains are achieved by allocating additional effort to attracting mates.
2. The fitness of offspring raised by only one parent is high, so the payoff for additional parental investment is relatively low.

The key factors are the costs of finding additional mates and the benefits associated with incremental increases in the amount of care that offspring receive. When females are widely separated, for example, it may be difficult for males to locate them. In these cases, males may profit more from helping their current mates and investing in their offspring than from searching for additional mates. If females are capable of rearing their offspring alone and need little help from males, then investing males may be at a reproductive disadvantage compared with males that abandon females after mating and devote their efforts to finding eligible females.

The mammalian reproductive system commits primate females to investing in their offspring.

So far, there is nothing in our reasoning to say that if only one sex invests, it should always be the female. Why aren't there primate species in which males do all the work and females compete with each other for access to males? This is not simply a theoretical possibility: female sea horses deposit their eggs in their mate's brood pouch and then swim away and look for a new mate (Figure 7.4). There are whole families of fish in which male parental care is more common than female parental care; and in several species of birds—including rheas, spotted sandpipers, and jacanas—females abandon their clutches after the eggs are laid, leaving their mates to feed and protect the young.

In primates and other mammals, the bias toward female investment evolves because females lactate and males do not. Among primates, pregnancy and lactation commit mammalian females to investing in their young, and limit the benefits of male investment in offspring. Because offspring depend on their mothers for nourishment during pregnancy and after birth, females cannot abandon their young without greatly reducing the offspring's chances of surviving. On the other hand, males are never capable of rearing their offspring without help from females. Therefore, when only one sex invests in offspring, it is invariably the female. Sometimes males can help females by defending territories or by carrying infants so that the mother can feed more efficiently, as siamangs and owl monkeys do. In most cases, however, these benefits are relatively insignificant, and selection favors males that allocate more time and energy to mating than to caring for their offspring.

You may be wondering why selection has not produced males that are able to lactate. As we noted in Chapter 3, most biologists believe the developmental changes that would enable males to lactate would also make them sterile. This is an example of a developmental constraint.

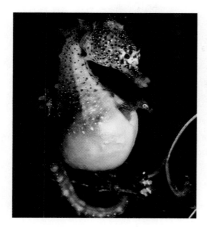

FIGURE 7.4 In sea horses, males carry fertilized eggs in a special pouch and provide care for their young as the young grow.

REPRODUCTIVE STRATEGIES OF FEMALES

So far, we have established that female primates often invest more heavily in their off-spring than do males. The next step is to consider the reproductive strategies of primate females in more detail.

Female primates invest heavily in each of their offspring.

Pregnancy and lactation are time-consuming and energetically expensive activities for female primates. The duration of pregnancy plus lactation ranges from 102 days in the tiny mouse lemur to 1839 days in the hefty gorilla. In primates, as in most other animals, larger animals tend to have longer pregnancies than do smaller animals, but primates have considerably longer pregnancies than we would expect on the basis of their body sizes alone. The extended duration of pregnancy in primates is related to the fact that brain tissue develops very slowly. Primates have very large brains in relation to their body sizes, so additional time is needed for fetal brain growth and development during pregnancy. Primates also have an extended period of dependence after birth, further increasing the amount of care mothers must provide. Throughout this period, mothers must meet not only their own nutritional requirements but also those of their growing infants. In some species, offspring may weigh as much as 30% of their mother's body weight at the time of weaning.

The energy costs of pregnancy and lactation impose important constraints on female reproductive behavior. Because it takes so much time and energy to produce an infant, each female can rear a relatively small number of surviving infants during her lifetime (Figure 7.5). For example, a female toque macaque who gives birth for the first time when she is 5 years old and survives to the age of 20 would produce 15 infants if she gave birth annually and all of her infants survived. This number undoubtedly represents an upper limit; in the wild, a substantial fraction of all toque macaque infants die before they reach reproductive age, intervals between successive live births often last two years, and some females die before they reach old age. Thus, most toque macaque females will produce a relatively small number of surviving infants over the course of a lifetime, and each infant represents a substantial proportion of a female's lifetime fitness. Therefore, we would expect mothers to be strongly committed to the welfare of their young.

A female's reproductive success depends largely on her ability to obtain enough resources to support herself and her offspring.

In most species of primates, including humans, females must achieve a minimum nutritional level in order to ovulate and to conceive. This means that females may sometimes be unable to conceive because their nutritional status is poor, they may have long periods of infertility while they recover from the rigors of their last pregnancy, and they may not always be able to nourish themselves or their newborns adequately.

For animals living in the wild, without takeout pizza or 24-hour grocery stores, getting enough to eat each day is often a serious dilemma. There is considerable evidence that female reproductive success is limited by the availability of resources within the local habitat. At a number of sites in Japan, for example, free-ranging monkeys' natural diets have been supplemented with wheat, sweet potatoes, rice, and

FIGURE 7.5 This female bonnet macaque produced twins, but only one survived beyond infancy. Twins are common among marmosets and tamarins but are otherwise uncommon among monkeys and apes. (Photograph by Kathy West.)

FIGURE 7.6 At a number of locations in Japan, indigenous monkeys are fed regularly. The size of these artificially fed groups has risen rapidly, indicating that population growth is limited by the availability of resources.

other foods by humans for many years (Figure 7.6). When their diet is supplemented in this way, females grow faster, mature earlier, survive longer, and produce infants at shorter intervals. Figure 7.7 shows changes in the size of one such group in Japan before, during, and after a period of intense supplementation. Comparisons of wild, provisioned, and captive hanuman langurs in India tell a very similar story. When females have better access to high-quality resources, they grow faster, mature earlier, and give birth at shorter intervals.

Sources of Variation in Female Reproductive Performance

High-ranking females tend to reproduce more successfully than do low-ranking females.

As we explained in Chapter 6, females often compete for access to food resources that they need in order to reproduce successfully. In some situations, females form dominance hierarchies, and high-ranking females have priority of access to valued foods (Figure 7.8). If dominance rank influences access to valuable resources, and access to resources influences female reproductive success, then we should expect to find a positive correlation between dominance rank and reproductive success. Data from a wide range of species indicate that high rank does confer reproductive advantages on females.

In multimale, multifemale groups of Old World monkeys that form quite stable matrilineal dominance hierarchies, female dominance rank is correlated with various aspects of females' reproductive performance. In Amboseli, Kenya, for example, the offspring of high-ranking female baboons grow faster and mature earlier than do the offspring of low-ranking females. In captive vervet groups, high-ranking females have shorter interbirth intervals than lower-ranking females have. In some macaque

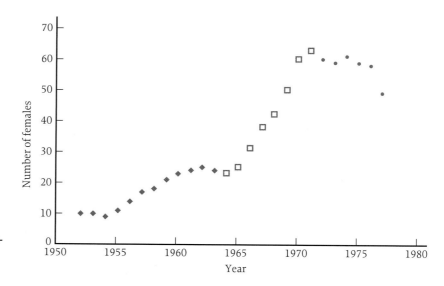

FIGURE 7.7 The size of the Koshima troop of Japanese monkeys grew rapidly when they were provisioned intensively (□), and then dropped when provisioning was restricted (•).

populations the offspring of high-ranking females are more likely to survive to reproductive age than are the offspring of lower-ranking females. Associations between dominance rank and reproductive success may produce substantial variation in lifetime fitness among females, particularly if females maintain the same rank over the course of their lives, as female macaques and baboons typically do. Thus, Maria van Noordwijk and Carel van Schaik have found substantial differences in the lifetime reproductive success of high-, middle-, and low-ranking long-tailed macaques (Figure 7.9).

Marmoset and tamarin groups in the wild often contain more than one adult female, but the dominant female is usually the only one who breeds successfully. Reproductive activity of subordinate females is suppressed in the presence of dominant females because subordinate females do not cycle normally. When subordinate females do breed, their infants may be killed by dominant females who have infants of their own.

FIGURE 7.8 Female primates sometimes fight over access to food and other resources.

Anne Pusey and her colleagues at the University of Minnesota have shown that the offspring of high-ranking female chimpanzees are more likely to survive to the age of weaning than are the offspring of low-ranking females. In addition, their daughters grow faster and mature earlier than do the daughters of low-ranking females. These differences create substantial differences in lifetime fitness for high- and low-ranking female chimpanzees at Gombe Stream National Park, Tanzania (Figure 7.10).

Red howlers live in groups that contain only two to four adult females and one or two adult males. Long-term studies of red howlers by Teresa Pope of Duke University and Carolyn Crockett of the Washington Regional Primate Research Center have shown that in sparsely populated habitats, new groups are formed when unrelated females meet, form social bonds, attract males, and establish territories. As habitats

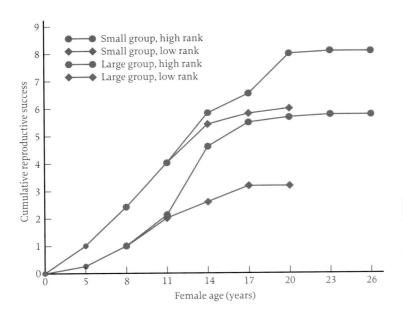

FIGURE 7.9 In free-ranging groups of long-tailed macaques, both group size and dominance rank influence females' lifetime reproductive success. In general, females living in small groups reproduce more successfully than do females living in larger groups. But in both large and small groups, high-ranking females reproduce more successfully than do low-ranking females.

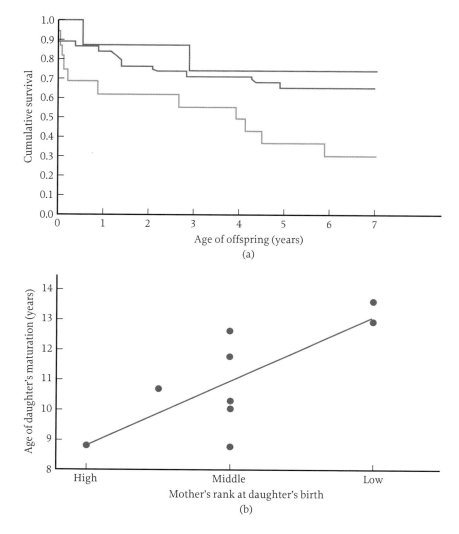

FIGURE 7.10 Among chimpanzees at Gombe Stream National Park, Tanzania, female rank influences reproductive performance. (a) The offspring of high- (red line) and middle-ranking (purple line) females are more likely to survive to weaning age than are the offspring of low-ranking females (yellow line). (b) Daughters of high-ranking females also mature at earlier ages than do the daughters of lower-ranking females.

become more crowded, it becomes harder and harder to establish new territories and female dispersal becomes more costly. Not all females can remain in their natal groups, however, because group size is confined within narrow limits. This constraint generates intense competition among females over recruitment opportunities for their daughters, and in most cases only the daughters of dominant females are able to remain in their natal groups.

In hanuman langurs, female rank also influences female reproductive performance. In this species, female rank is inversely related to age (Figure 7.11). Long-term studies of hanuman langurs near Jodhpur, India, conducted by a group of German primatologists, including Carola Borries (now at Stony Brook University) and Volker Sommer (now at University College London), have shown that young, high-ranking females reproduce more successfully than do older, lower-ranking females (Figure 7.12). Studies of hanuman langurs at Ramnagar in southern Nepal conducted by another group of German primatologists, including Andreas Koenig, Carola Borries, and Paul Winkler, have found that high-ranking females manage to commandeer higher-quality food patches and are consequently able to maintain higher levels of

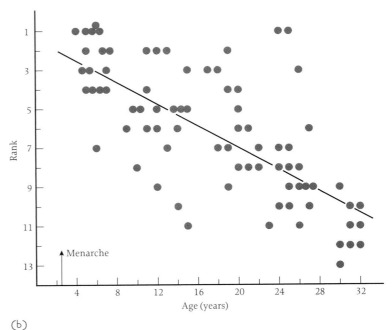

(a)

FIGURE 7.11 (a) A female hanuman langur at Jodhpur threatens another group member. (b) Among hanuman langur females, dominance rank is inversely related to age, so the youngest females hold the highest ranks and their rank declines as they age. (a, Photograph courtesy of Carola Borries.)

(b)

body fat. Females in good condition have higher fertility rates than do females in poor condition.

Some researchers question the importance of the effects of dominance rank on reproductive success because significant effects of dominance rank are not found in every study. Among wild baboons, for example, it has been very hard to find a link between dominance rank and female reproductive success. Perhaps this is because contest competition plays a less important role in regulating access to resources in

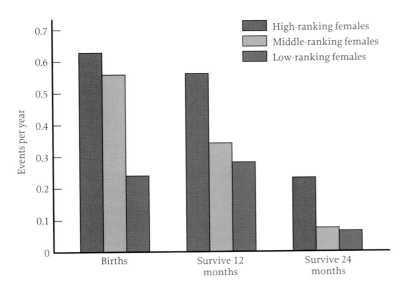

FIGURE 7.12 Female hanuman langurs reproduce more successfully when they are young and hold high ranks than when they are older and have lower ranks. The three bars on the left represent the proportion of females of each rank category who give birth each year. The other sets of bars represent the proportion of females in each rank category who give birth each year to infants who survive to 12 months and 24 months. Dominance rank influences both the likelihood of giving birth and the likelihood that infants will survive.

baboon groups than in other species or because low-ranking females find ways to compensate for the disadvantages of low rank.

Reproductive Trade-offs

Females must make a trade-off between the number of offspring they produce and the quality of care that they provide.

Just as both males and females must allocate limited effort to parental investment and mating, females must apportion resources among their offspring. All other things being equal, natural selection will favor individuals that are able to convert effort into offspring most efficiently. Because mothers have a finite amount of effort to devote to offspring, they cannot maximize both the quality and the quantity of the offspring they produce. If a mother invests great effort in one infant, she must reduce her investment in others. If a mother produces many offspring, she will be unable to invest very much in any of them.

In nature, maternal behavior reflects this trade-off when a mother modifies her investment in relation to an offspring's needs. Initially, infants spend virtually all of their time in contact with their mothers. The very young infant is entirely dependent on its mother for food and transportation and is unable to anticipate or to cope with environmental hazards. At this stage, mothers actively maintain close contact with their infants (Figure 7.13): retrieving them when they stray too far and scooping them up when danger arises.

FIGURE 7.13   A female chimpanzee sits beside her youngest infant in Gombe Stream National Park in Tanzania.

As infants grow older, however, they become progressively more independent and more competent. They venture away from their mothers to play with other infants and to explore their surroundings. They begin to sample food plants, sometimes mooching scraps of their mother's food. They become aware of the dangers around them, attending to alarm calls given by other group members and reacting to disturbances within the group. Mothers use a variety of tactics to actively encourage their infants to become more independent. Initially they may subtly resist their infants' attempts to suckle. At the same time, mothers may begin to encourage their infants to travel independently by moving off before the infants have climbed onto their backs, by shrugging their infants off their backs while moving, or by simply failing to retrieve their infants when ready to depart. Although infants may protest these developments with tantrums or whimpers, and persistently attempt to ride on their mothers and to nurse, they eventually become independent of their mothers. Nursing is gradually limited to brief and widely spaced bouts that may provide the infant with more psychological comfort than physical nourishment. At this stage, infants are carried only when they are ill, injured, or in great danger.

The changes in maternal behavior reflect the shifting balance between the requirements of the growing infant and the energy costs to the mother of catering to her infant's needs. As infants grow older, they become heavier to carry and require more food, imposing substantial burdens on mothers (Figure 7.14). As infants become more capable of feeding

FIGURE 7.14   As infants become older, they become a bigger burden on their mothers. Here a female baboon leaps across a ditch carrying her infant on her back. Later the mother will encourage her infant to leap across the ditch on its own.

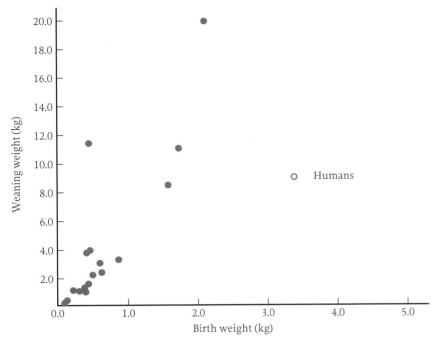

FIGURE 7.15 Primate infants are generally weaned when they reach about four times their birth weight, although the duration of lactation and infant growth rates vary considerably. The threshold weaning weight probably reflects constraints on the mother's ability to meet the energetic demands of growing infants. Each point represents the average value of birth weight and weaning weight for a given primate genus. ○ represents a human foraging group, the !Kung.

themselves and of traveling independently, mothers can gradually limit investment in their older infants without jeopardizing the infants' welfare. Mothers are thus able to conserve resources that can be allocated to subsequent infants. Moreover, because lactation inhibits ovulation in many primate species, a mother must wean her present infant before she can conceive another. Phyllis Lee of Cambridge University and her colleagues have discovered that although the duration of lactation varies within and among species, generally primate infants are weaned when they reach about four times their birth weight (Figures 7.15 and 7.16).

There is sometimes conflict between mothers and infants over the amount and extent of maternal investment.

As mothers begin to curtail investment, their infants often protest vigorously. Chimpanzee infants throw full-fledged tantrums when their mothers rebuff their efforts to nurse, and baboons whimper piteously when their mothers refuse to carry them. Weaning conflicts arise from a fundamental asymmetry in the genetic interests of mothers and their offspring. This phenomenon was labeled **parent–offspring conflict** by Rutgers University biologist Robert Trivers, who was the first to recognize the evolutionary rationale underlying the conflict between parents and their offspring.

To understand how evolution shapes parent–offspring conflict, we need to know something about how natural selection influences the evolution of social behavior. In Chapter 1 we explained that natural selection favors the evolution of behaviors that increase individual fitness. But animals sometimes perform **altruistic** acts that increase *other* individuals' fitness and reduce their own fitness. Kin selection provides one mechanism for the evolution of altruism. We will outline the basic logic of kin selection here, and provide a more complete discussion of this important evolutionary force in Chapter 8.

FIGURE 7.16 A large infant squirrel monkey is being carried by its mother. By the time they are weaned, infants in some species may weigh up to a third of their mother's weight. (Photograph by Sue Boinski.)

The theory of kin selection rests on the insight that kin are likely to share copies of the same genes because they share a common ancestor. In sexually reproducing species, mothers and fathers transmit half of their genes to their offspring, siblings share genes that they acquire from their parents, and so on. The probability that two individuals carry copies of the same gene is a function of how many links separate them from their common ancestor. For the present purposes, it's important to see that infants are related to their mothers by ½ because they receive half of their genes from their mothers (and the other half from their fathers). Infants are related to their maternal siblings by ¼ because each has a half chance of receiving a given allele from the mother; the probability that two siblings receive the *same* allele is equal to the product of these probabilities: ½ × ½ = ¼. Similarly, paternal siblings are related by ¼. Full siblings may obtain copies of the same allele from their mother or their father, so we sum values of maternal and paternal relatedness: ¼ + ¼ = ½.

The late British biologist W. D. Hamilton derived a simple rule to predict when kin selection will favor the evolution of altruism. Altruism will be favored when the benefits to the recipient (*b*) multiplied by the degree of relatedness between the actor and the recipient (*r*) is greater than the costs to the actor (*c*)—that is, when *rb* > *c*. As we explained earlier, benefits and costs are defined in terms of their effects on fitness.

To understand how evolution shapes parent–offspring conflict, imagine a mutation that increases the amount of maternal investment in the current infant a small amount and thereby reduces investment in future infants by the same amount. According to Hamilton's rule, selection will favor the expression of this gene in mothers if

$$0.5 \times (\text{increase in fitness of current infant}) >$$
$$0.5 \times (\text{decrease in fitness of future offspring})$$

Because the mother shares half of her genes with each of her offspring, 0.5 appears on both sides of the inequality. The inequality tells us that selection will increase investment in the current offspring until the benefits to the current offspring are equal to the costs to future offspring. The result is quite different if the genes expressed in the current infant control the amount of maternal investment. This time, consider a gene that is expressed in the current infant that increases the investment the infant receives by a small amount. Once again, we use Hamilton's rule, this time from the perspective of the current infant:

$$1.0 \times (\text{increase in fitness of current fetus}) >$$
$$0.5 \times (\text{decrease in fitness of future offspring})$$

In this case the infant is related to itself by 1.0 and to its full sibling by 0.5. Now selection will increase the amount of maternal investment until the incremental benefit of another unit of investment in the current infant is twice the cost to future brothers and sisters of the fetus (and four times for half-siblings). Thus, genetic asymmetries lead to a conflict of interest between mothers and their children. Selection will favor mothers that provide less investment than their infants desire, and selection will favor children that demand more investment than their mothers are willing to give. This conflict of interest plays out in weaning tantrums and sibling rivalries.

SEXUAL SELECTION AND MALE MATING STRATEGIES

Sexual selection leads to adaptations that allow males to compete more effectively with other males for access to females.

So far, we have seen that primate females invest heavily in each of their young and produce relatively few offspring over the course of their lives. Moreover, most primate females can raise their offspring without help from males. Female reproductive success is limited by access to food, not access to mates (Figure 7.17). Males can potentially produce progeny from many females, and as a result, males compete for access to females. Characteristics that increase male success in competition for mates will spread as a result of what Darwin called **sexual selection**.

It is important to understand the distinction between natural selection and sexual selection. Most kinds of natural selection favor phenotypes in both males and females that enhance their ability to survive and reproduce. Many of these traits are related to resource acquisition, predator avoidance, and offspring care. Sexual selection is a special category of natural selection that favors traits that increase success in competition for mates, and it will be expressed most strongly in the sex whose access to members of the opposite sex is most limited. Sexual selection may favor traits that increase the animal's attractiveness to potential mates, such as the peacock's tail, the red deer's antlers, and the hamadryas baboon's mane, even if those traits reduce the ability of the animal to survive or acquire resources—outcomes not usually favored by natural selection (Figure 7.18).

FIGURE 7.17 For female primates, access to resources has a bigger impact on reproduction than access to potential mates does.

Sexual selection is often much stronger than ordinary natural selection.

In mammalian males, sexual selection can have a greater effect on behavior and morphology than other forms of natural selection do. This is because male reproductive

(a)

(b)

FIGURE 7.18 Sexual selection can favor traits not favored by natural selection. (a) The peacock's tail hinders his ability to escape from predators, but it enhances his attractiveness to females. Female peahens are attracted to males that have the most eyespots in their trains. (b) Male red deer use their antlers when they fight with other males. Red deer antlers are a good example of a trait that has been favored by sexual selection.

(a)

(b)

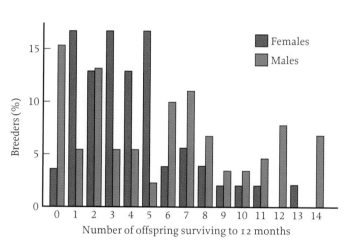

(c)

FIGURE 7.19 The reproductive success of male lions (a) is considerably more variable than the reproductive success of female lions (b). (c) In Serengeti National Park and the Ngorongoro Crater of Tanzania, few female lions fail to produce any surviving cubs, but most females produce fewer than six surviving cubs over the course of their lives. Many males fail to produce any cubs, and a few males produce many cubs.

success usually varies much more than female reproductive success. Data from long-term studies of lions conducted by Craig Packer and Anne Pusey of the University of Minnesota demonstrate that the lifetime reproductive success of the most successful males is often much greater than that of even the most successful females (Figure 7.19). The same pattern is likely to hold for nonmonogamous primates. A primate male who succeeds in competition may sire scores of offspring; a successful female might give birth to 5 or 10 offspring. Unsuccessful males and females will fail to reproduce at all. Because the strength of selection depends on how much variation in fitness there is among individuals, sexual selection acting on male primates can be much stronger than selective forces acting on female primates. (Incidentally, in species like sea horses, in which males invest in offspring and females do not, the entire pattern is reversed: sexual selection acts much more strongly on females than on males.)

There are two types of sexual selection: (1) intrasexual selection results from competition among males, and (2) intersexual selection results from female choice.

Many students of animal behavior subdivide sexual selection into two categories:

1. In species in which females cannot choose their mates, access to females will be determined by competition among males. In such species, **intrasexual selection** favors traits that enhance success in male–male competition.

2. In species in which females can choose the partner(s) with which they mate, selection favors traits that make males more attractive to females. This is **intersexual selection**.

Intrasexual Selection

Competition among males for access to females favors large body size, large canine teeth, and other weapons that enhance male competitive ability.

For primates and most other mammals, intrasexual competition is most intense among males. In the most basic form of male–male competition, males simply drive other males away from females. Males who regularly win such fights have higher reproductive success than those who lose. Thus, intrasexual selection favors features such as large body size, horns, tusks, antlers, and large canine teeth that enable males to be effective fighters. For example, male gorillas compete fiercely over access to groups of females, and males weigh twice as much as females and have longer canine teeth.

As explained in Chapter 5, when the two sexes consistently differ in size or appearance, they are said to be sexually dimorphic (Figure 7.20). The body sizes of males and females represent compromises among many competing selective pressures. Larger animals are better fighters and are less vulnerable to predation, but they also need more food and take longer to mature. Intrasexual competition favors larger body size, larger teeth, and other traits that enhance fighting ability. Males compete over females; females compete over resources but generally do not compete over mates. The effect of intrasexual competition among males, however, is quantitatively greater than the effect of competition among females because the fitness payoff to a successful male is greater than it is to a successful female. Therefore, sexual selection is much more intense than ordinary natural selection. As a result, intrasexual selection leads to the evolution of sexual dimorphism.

FIGURE 7.20 Adult male baboons are nearly twice the size of adult females. The degree of sexual dimorphism in body size is most pronounced in species with the greatest competition among males over access to females.

The fact that sexual dimorphism is greater in primate species forming one-male, multimale groups than in monogamous species indicates that intrasexual selection is the likely cause of sexual dimorphism in primates.

If sexual dimorphism among primates is the product of intrasexual competition among males over access to females, then we should expect to see the most pronounced sexual dimorphism in the species in which males compete most actively over access to females. One indirect way to assess the potential extent of competition among males is to consider the ratio of males to females in social groups. In general, male competition is expected to be most intense in social groups where males are most outnumbered by females. At first this prediction might seem paradoxical, because we might expect to have more competition when more males are present. The key to resolving this paradox is to remember that, in most natural populations, there are approximately equal numbers of males and females at birth. In species that form one-male groups there are many **bachelor males** (males who don't belong to social groups) who exert constant pressure on resident males. In species that form monogamous pair-bonds, each male is paired with a single female, reducing the intensity of competition among males over access to females.

Comparative analyses conducted by Paul Harvey of Oxford University and Tim Clutton-Brock of Cambridge University have demonstrated that the extent of sexual dimorphism in primates corresponds roughly to the form of social group in which the males live (Figure 7.21). There is little difference in body weight or canine size between males and females in species that typically form monogamous groups, such as gibbons, titi monkeys, and marmosets. At the other extreme, the most pronounced dimorphism is found in species, such as gorillas and black-and-white colobus monkeys, that live in one-male, multifemale groups. And in species that form multimale, multifemale groups, the extent of sexual dimorphism is generally intermediate between these extremes. Thus, sexual dimorphism is most pronounced in the species in which the ratio of males to females living in bisexual groups is lowest (that is, the relative number of females is the highest).

In multimale, multifemale groups, where females mate with several males during a given estrous period, sexual selection favors increased sperm production.

In most primate species, as with mammals in general, the female is receptive to mating mainly during the portion of her reproductive cycle when fertilization is possible. That period of time is called **estrus**. In primate species that live in multimale, multifemale groups, females can potentially mate with several different males during a single estrous period. In such species, sexual selection favors increased sperm production because males who deposit the largest volume of sperm in the female reproductive tract have the greatest chance of impregnating them. Competition in the quantity of sperm is likely to be relatively unimportant in monogamous species because females mate mainly with their own partners. Because sperm production involves some cost to males, monogamous males may do better by guarding their partners when they are sexually receptive than by producing large quantities of sperm. Similarly, competition in sperm quantity probably does not play an important role in species that form

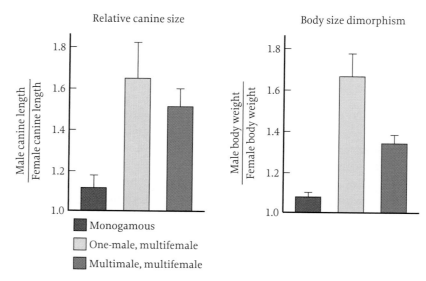

FIGURE 7.21 The degree of sexual dimorphism is a function of the ratio of males to females in social groups. Relative canine size (male canine length divided by female canine length) and body size dimorphism (male body weight divided by female body weight) are greater in species that form one-male, multifemale groups than in species that form multimale, multifemale groups or monogamous groups.

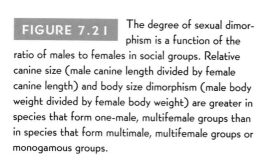

one-male, multifemale groups. In these species, competition among males is over access to groups of females, which favors traits related to fighting ability. If resident males are able to exclude other males from associating with females in their groups, there may be little need to produce additional quantities of sperm.

Social organization is associated with testis size, much as we would expect. Males with larger testes typically produce more sperm than do males with smaller testes, and males who live in multimale groups have much larger testes in relation to their body size than do males who live in either monogamous or one-male, multifemale groups (Figure 7.22). Interestingly, in polyandrous groups in which a single breeding female mates regularly with multiple males, males also have relatively large testes.

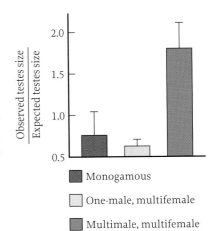

■ Monogamous

☐ One-male, multifemale

■ Multimale, multifemale

FIGURE 7.22 The average size of testes in species that typically form monogamous and one-male, multifemale groups is relatively smaller than the average size of testes in multimale, multifemale groups. Here, observed testes weight is divided by the expected testes weight to produce relative testes size. The expected testes weight is derived from analyses that correct for the effects of body size.

Intersexual Selection

Intersexual selection favors three kinds of traits in males: (1) traits that increase the fitness of their mates, (2) traits that indicate good genes and thus increase the fitness of the offspring, and (3) nonadaptive traits that make males more conspicuous to females.

Darwin also realized that females might be attracted to males that exhibit particular traits. In this case, intersexual selection favors males that are pleasing to females rather than males that can defeat other males. Modern sexual selection theory now identifies three basic modes of female choice:

1. If males with certain traits confer direct benefits on females, then selection will favor females that mate selectively with males possessing such traits. As a result, the frequency of such traits will increase among males. Thus, if females prefer males that defend superior territories, protect offspring more vigorously, provision the young better, or confer more material benefits on females than other males do, these male traits will be favored.

2. Sexual selection may also favor female preferences if females are able in some way to distinguish male genetic quality and choose mates with desirable genes. These females are at an advantage because their offspring will carry certain genes that give them a greater chance of surviving and reproducing successfully. The gaudy peacock and the noisy sage grouse may provide examples of such preferences. In both of these species, males gather together in a small area called a "lek," and each male displays his tail feathers to attract female attention (Figure 7.23). Marion Petrie of Oxford University has found that peahens never mate with the first male they encounter on the lek, and they show distinct preferences for the males with the most eyespots on their tail feathers. The offspring of the most richly ornamented males grow faster and survive better than the offspring of other males. Thus, female mate preferences enhance offspring fitness. Again, it's important to remember that we need not imagine that peahens, or any other animals, assess male genetic quality *consciously*. Instead, selection may have favored preferences for specific traits that are reliably associated with genetic quality. By choosing males with the most eyespots, hens may be inadvertently and unconsciously choosing males with desirable genotypes.

3. Females may prefer males that exhibit distinctive traits (such as conspicuous coloration, exaggerated morphological characters, or elaborate courtship behaviors) for nonadaptive reasons. Even if such characters do not increase male

FIGURE 7.23 Male sage grouse congregate on leks and display to attract females. (Photograph courtesy of Mark Chappell.)

FIGURE 7.24 Male mandrills have brightly colored faces. This trait may be favored by intersexual selection, but it is not known whether females actually prefer to mate with brightly colored males.

fitness directly, they may be favored if females can discriminate and consistently select mates that most often display preferred traits. For example, male frogs call to attract mates, and the sensory system of female frogs seems to respond more strongly to some tones in male calls than to others. Researchers have discovered that male frogs whose calls are the most readily heard attract the most mates.

Surprisingly, it turns out that female choice can also lead to the exaggeration of female preferences themselves. To see why, suppose there are two types of females: very picky females who prefer males with only the most exaggerated male trait—say, the deepest tones of a male frog's call or the longest tails of a male bird—and less picky females who will mate with a wider range of males. It is easy to see that the existence of more picky females will increase the reproductive success of males with the most exaggerated traits, and therefore increase the degree of exaggeration of the male trait in the population. However, the same process can also lead to the spread of more picky females because the offspring of more picky females will tend to carry both the genes for pickiness and the genes for the exaggerated male trait, creating a positive correlation between the female preference and the exaggerated male trait (see Chapter 3). As a result of the correlated response to selection, the female choice that increases the reproductive success of exaggerated males will also increase the pickiness of females. If conditions are right, the result might be a runaway process in which females become pickier and pickier about a male trait, and male traits become more and more exaggerated, even if they impair the male's ability to survive.

Only a few morphological traits among primates seem to have evolved because they help males attract females.

The best candidates for traits that help primate males attract females are found among mandrills, hamadryas baboons, and proboscis monkeys. Both the face and the penis of the male mandrill are brilliantly colored. The male mandrill's face is striped red, white, and blue like an exotic mask (Figure 7.24); the female mandrill's face is duller and less conspicuous. Hamadryas males have a distinctive silver mantle on their shoulders (Figure 7.25); female coats are uniformly shorter. The proboscis monkey is named for its oddly shaped nose: males have quite long, pendulous noses; females have much smaller noses that turn up at the end (Figure 7.26). There is pronounced sexual dimorphism in body size among all of these species as well. We don't really know whether females actually prefer mandrill males with brightly colored faces, hamadryas males with handsome mantles, or proboscis males with elongated noses. We also do not know which of the three modes of female choice discussed earlier may be operating. However, because these traits do not seem directly related to male–male competition and are more fully developed in males than in females, it is possible that these traits evolved in response to female preferences.

MALE REPRODUCTIVE TACTICS

Morphological evidence suggests that male–male competition is less intense in monogamous species than in nonmonogamous species. As we will see in the remainder of this chapter, sexual selection has shaped male mating strategies as well as male morphology.

FIGURE 7.25 Male hamadryas baboons have a mantle of long, gray hair on their shoulders. Females do not have this feature.

Investing Males

Monogamous pair-bonding is generally associated with high levels of paternal investment.

In species that form monogamous pair-bonds, males do not compete directly over access to females. In these species, males' reproductive success depends mainly on their ability to establish territories, find mates, and rear surviving offspring. In such pair-bonded species, mate guarding and offspring care are important components of males' reproductive tactics.

FIGURE 7.26 Male proboscis monkeys have long pendulous noses; females have much smaller upturned noses. The pendulous nose may be favored by intersexual selection, although we do not know whether females are attracted to males with long noses.

Mate guarding may be an important component of pair-bonded males' reproductive effort because monogamy doesn't necessarily imply fidelity. Numerous genetic studies of supposedly monogamous birds have demonstrated that a large fraction of the young are not sired by the female's mate. Pair-bonded titi monkeys and gibbons have also been seen copulating with members of neighboring groups. If females occasionally participate in extrapair copulations, their partners may benefit from keeping close watch on them. Ryne Palombit of Rutgers University, who has studied the dynamics of pair-bonding in gibbons, suspects that males do just that. Male gibbons are principally responsible for maintaining proximity to their female partners, and most males groom their mates more than they are groomed in return (Figure 7.27).

Pair-bonded males tend to invest heavily in their mates' offspring. In titi monkeys and owl monkeys, which have been studied in the forests of Peru by Patricia Wright of Stony Brook University, adult males play an active role in caring for infants. They carry them much of the time, share food with them, groom them, and protect them from predators. Male siamangs are also helpful fathers, carrying their infants for long periods every day.

In cooperatively breeding species, males invest heavily in offspring, but the reproductive benefits to males are not clear.

Groups of cooperatively breeding primates, which include marmosets and tamarins, typically consist of one breeding female and one or more breeding males

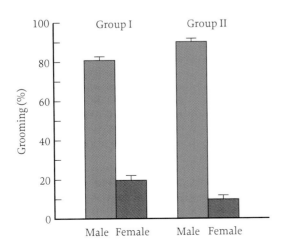

FIGURE 7.27 In two white-handed gibbon groups, males groom their mates far more than they are groomed in return. The blue bar is the proportion of grooming from the male to the female, and the red bar is the proportion of grooming from the female to the male. Males' solicitous attention to females may be a form of mate guarding.

FIGURE 7.28 Some callitrichid groups contain more than one adult male and a single breeding female. In at least some of these groups, mating activity is limited to the group's dominant male, even though all the males participate in the care of offspring.

(Figure 7.28). Behavioral and genetic data suggest that reproductive benefits are not divided equally among males in these species. In most species of marmosets and tamarins, the dominant male monopolizes matings with receptive females. These behavioral data are consistent with very limited evidence from genetic analyses. Leslie Digby of Duke University and her colleagues have analyzed the paternity of infants born in three free-ranging groups of common marmosets in Brazil. These three groups each contained two adult males and one or two breeding females. The researchers found that, in each group, the dominant male fathered nearly all of the infants. However, genetic analyses of other groups suggest that this is not always the case. Some groups include adults who are unrelated to the breeding pair, but all adults seem to lend a hand in infant care.

The presence of multiple males seems to enhance female fertility. Marmosets and tamarins are unusual among primates because they usually produce twins, and females produce litters at relatively short intervals, sometimes twice a year. Males play an active role in child care, frequently carrying infants, grooming them, and sharing food with them. Data compiled by Paul Garber of the University of Illinois at Urbana-Champaign show that groups with multiple adult males reared more surviving infants than did groups with only one male (Figure 7.29). In contrast, groups with multiple females produced slightly fewer infants than did groups with only one female resident.

Thus, females seem to benefit from having multiple males to assist them in rearing offspring, but the reproductive payoffs for males are less clear. We need more information about males' life histories to unravel this puzzle.

Male–Male Competition in Nonmonogamous Groups

In nonmonogamous groups, the reproductive success of males depends on their ability to gain access to groups of unrelated females and to obtain matings with receptive females.

As we explained in Chapter 6, males often leave their natal groups at puberty and attempt to join new groups. When females are philopatric, male dispersal is obligatory. Dispersal is often a dangerous and stressful time for males. In some species, males disperse alone and spend some time on their own before they join new groups. During this period, males are likely to become more vulnerable to predators and may have trouble gaining access to desirable feeding sites.

Males can reduce the costs of dispersal in several ways:

- Before they leave their natal groups, they can scope out neighboring groups when those groups are close by. Young males may size up the potential prospects (number of available females) and the competition (number and size of adult males) in neighboring groups.
- They can transfer to groups that contain familiar males from their natal groups. In Amboseli, Kenya, vervet males often join neighboring groups that kin or former group members have already joined.

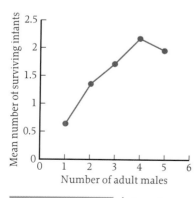

FIGURE 7.29 In tamarin groups, males clearly contribute to females' reproductive success. Tamarin groups that contain several adult males produce more surviving infants than do groups that contain only one adult male.

- Males can migrate together with peers or join all-male bands when they leave their natal groups. Peer migration is observed in a number of species, including squirrel monkeys, ring-tailed lemurs, and several species of macaques.

In species that normally form one-male groups, males compete actively to establish residence in groups of females.

In primate species that form one-male groups, resident males face persistent pressure from nonresidents. In the highlands of Ethiopia, gelada baboons challenge resident males and attempt to take over their social groups, leading to fierce confrontations that may last for several days (Figure 7.30). Robin Dunbar of Liverpool University has estimated that half of the males involved in aggressive takeover attempts are seriously wounded. Among hanuman langurs, males form all-male bands that collectively attempt to oust resident males from bisexual (coed) groups. Once they succeed in driving out the resident male, the members of the all-male band compete among themselves for sole access to the group of females. One consequence of this competition is that male tenure in one-male groups is often short.

FIGURE 7.30 Most gelada groups contain only one male. Males sometimes attempt to take over groups and oust the resident male; in other cases males join groups as followers and establish coresidence. Takeovers are risky because they do not always succeed and males are sometimes badly injured.

Residence in one-male groups does not always ensure exclusive access to females.

Even though we generally assume that male residents of one-male, multifemale groups face little competition over access to females within their groups, this is not always the case. In patas and blue monkeys, for example, researchers have discovered that the resident male is sometimes unable to prevent other males from associating with the group and mating with sexually receptive females. Such incursions are concentrated during the mating season and may last for hours, days, or weeks and involve one or several males.

Some primate species form both one-male and multimale groups, depending on the circumstances. For example, Teresa Pope and Carolyn Crockett have found that in Venezuelan forests that are relatively sparsely populated by red howlers, and where it is relatively easy to establish territories, one-male groups predominate. But when the forests become more densely populated and dispersal opportunities are more limited, males adopt a different strategy: they pair up with other males and jointly defend access to groups of females. These partnerships enable males to defend larger groups of females and to maintain residence in groups longer (Figure 7.31). Similarly, hanuman langurs live sometimes in one-male groups and sometimes in multimale groups.

FIGURE 7.31 Red howlers live in groups that contain two to four females and one or two mature males. When habitats are saturated and opportunities to form new groups are limited, males join forces to defend access to groups of females. These partnerships enable males to defend larger groups of females and to maintain residence in groups for longer periods.

For males in multimale groups, conflict arises over group membership and access to receptive females.

In multimale groups, there is more competition over gaining access to mating partners than over establishing group membership. Nonetheless, it is not necessarily easy to join a new group; no one puts out the welcome mat. In some macaque species, males

FIGURE 7.32 Male baboons compete over access to an estrous female.

hover near the periphery of social groups, avoid aggressive challenges by resident males, and attempt to ingratiate themselves with females. In chacma baboons, immigrant males sometimes move directly into the body of the group and engage high-ranking resident males in prolonged vocal duels and chases. Although there may be conflict when males attempt to join nonnatal groups, males spend most of their adult lives in groups that contain both males and females.

In multimale groups, males often compete directly over access to receptive females. Sometimes males attempt to drive other males away from females, to interrupt copulations, or to prevent other males from approaching or interacting with females. More often, however, male–male competition is mediated through dominance relationships that reflect male competitive abilities. These relationships are generally established in contests that can involve threats and stereotyped gestures but that can also lead to escalated conflicts in which males chase, wrestle, and bite one another (Figures 7.32 and 7.33). Often the outcomes of encounters between particular pairs of males are relatively predictable from day to day, and males can be ordered in a linear dominance hierarchy. Male fighting ability and dominance rank are generally closely linked to physical condition: prime-age males in good physical condition are usually able to dominate others. The correlation between physical abilities and dominance rank is imperfect, however, because old males can sometimes dominate younger and stronger males.

It seems logical that male dominance rank would correlate with male reproductive success in multimale groups, but this conclusion has been energetically debated. The issue has been difficult to resolve because it is difficult to infer paternity from behavioral data. It is hard to determine which male fathered a particular infant because females often mate with multiple males when they are receptive, clandestine matings may be overlooked, and observations typically end at sunset, whereas sexual activity may extend into the night. However, new genetic techniques make it possible to assess paternity with a much higher degree of precision. As more and more genetic information about paternity has become available, the links between male dominance rank and reproductive success have become stronger. For example, the team of primatologists studying hanuman langurs at Ramnagar in Nepal recently completed paternity analyses in several multimale groups. The number of adult males in these groups var-

FIGURE 7.33 (a) A male baboon displays his canines as he threatens a rival. (b) Male baboons' canines are formidable weapons, and males sometimes sustain serious injuries in fights with other males, as this male did.

(a)

(b)

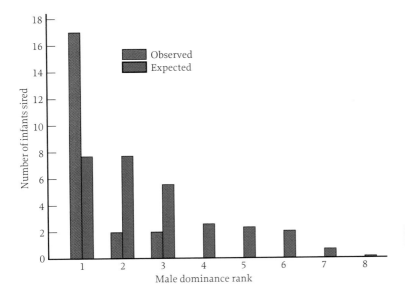

FIGURE 7.34 In hanuman langur groups at Ramnagar, high-ranking males father the majority of infants in their groups, far more than expected if all males were equally likely to sire infants.

ied from two to eight. If all males in multimale groups were equally likely to sire offspring, then the top-ranking males would have sired a total of about eight infants. However, the top-ranking males sired 17 infants—more than twice as many as expected (Figure 7.34). Genetic analyses of paternity in free-ranging baboons, long-tailed macaques, howler monkeys, and patas monkeys also show a positive relationship between male dominance rank and reproductive success.

There is growing evidence of substantial variation in the reproductive success of males over the course of their lifetimes.

As we noted earlier, male fitness is expected to vary considerably more than female fitness. Correlations between male rank and reproductive success suggest that male fitness does vary. However, we know that male dominance rank changes over time; males may be high-ranking and reproductively successful while in their prime, but low-ranking and reproductively unsuccessful the rest of their lives. Moreover, male tenure in one-male groups is often quite short, and competition for these positions is intense. If, at some point in their lives, all males are either high-ranking in multimale groups or residents in one-male groups, this might even out any variation in male fitness. However, any variation in the likelihood of attaining high rank or maintaining high rank will produce real variation in lifetime fitness among males. Thus, to evaluate fitness we need to follow individual males throughout their lives.

Infanticide

Infanticide is a sexually selected male reproductive strategy.

In the early 1970s, a young graduate student from Harvard University named Sarah Blaffer Hrdy traveled to India to study hanuman langurs. She was intrigued by what she had read about these animals. Phyllis Jay, who had observed hanuman langurs in central and northern India in the late 1950s, described them as peaceful and unaggressive animals. She wrote, "Relations among adult male langurs are relaxed. Dominance

188   CHAPTER 7 · Primate Mating Systems

(a)

(b)

FIGURE 7.35 When Sarah Hrdy began her work, it was not clear whether langurs were (a) aggressive animals who systematically killed the offspring of their rivals or (b) peaceful animals whose behavior was pathologically distorted by high levels of crowding. (Photograph Courtesy of Carola Borries.)

is relatively unimportant. . . . Aggressive threats and fighting are uncommon." This account differed sharply from the reports of Japanese primatologist Yukimaru Sugiyama and his colleagues, who had studied hanuman langurs in southern India a few years later. Sugiyama saw a band of males chase the resident male away from his troop, and then one of the members of the band took control of the troop and drove away all the other males. Shortly after he drove off his rivals, the new resident male attacked and killed the six infants in the group.

Hrdy wondered what accounted for the discrepancy in these accounts (Figure 7.35). Were langurs peaceful creatures whose behavior was pathologically distorted by high levels of crowding? Or were langurs aggressive animals who systematically killed the offspring of their rivals? Was infanticide a widespread occurrence in langur troops or an isolated aberrant incident in a few disturbed troops? To answer these questions, Hrdy began her own study of hanuman langurs near Mount Abu in northern India. During a four-year period, she recorded changes in male membership of several troops and tracked the fate of the infants in these groups.

Hrdy's observations led her to suggest that infanticide is an evolved strategy that enhances male reproductive success, not a pathological response to overcrowding. Hrdy based her hypothesis on the following reasoning: When a female langur gives birth to an infant, she nurses it for a number of months and does not become pregnant again for at least a year. After the death of an infant, however, lactation ends abruptly and females resume cycling. Thus the death of nursing infants hastens the resumption of maternal receptivity. A male who takes over a group may benefit from killing nursing infants because their deaths cause their mothers to become sexually receptive much sooner than they would otherwise. Because male tenure in langur groups is typically just over two years, and interbirth intervals last nearly three years if infants survive, infanticide may substantially increase a new resident's mating opportunities.

Hrdy's hypothesis, which has become known as the **sexual selection infanticide hypothesis,** was controversial. Some researchers were reluctant to accept the idea that lethal aggression might be adaptive. They insisted that infanticide was pathological and occurred only when langurs lived in disturbed habitats at high density. Some were skeptical of Hrdy's evidence because there were few eyewitness accounts of infanticide in langurs and she relied heavily on circumstantial evidence that linked male takeovers to infant disappearances. However, Hrdy began to comb the literature to test her ideas. Researchers had reported infant injuries, disappearances, or deaths after changes in male residence in a number of primate species, including gorillas, hamadryas baboons, and howler monkeys. Looking beyond the primates, Hrdy found evidence of similar patterns in rodents and lions.

Intrigued by these patterns, behavioral ecologists began to collect data to test Hrdy's hypothesis. Over the last 30+ years, infanticide by males has been reported in approximately 40 primate species. Researchers have now witnessed at least 60 infanticidal attacks in the wild and have documented many nonlethal attacks on infants by adult males and many more instances in which healthy infants disappeared after takeovers or changes in male rank. Although early studies suggested that sexually selected infanticide was limited to one-male groups, we now know that infanticide also occurs in multimale groups of savanna baboons, langurs, and Japanese macaques.

This body of data enables researchers to test a number of predictions derived from Hrdy's hypothesis. If infanticide is a male reproductive strategy, then we would expect that (1) infanticide would be associated with changes in male residence or status;

(2) males should kill infants whose deaths hasten their mother's resumption of cycling; (3) males should kill other males' infants, not their own; and (4) infanticidal males should achieve reproductive benefits. The data fit all four of these predictions.

Infanticide is linked to changes in male membership and status.

In a recent review of 55 infanticides in free-ranging groups that were actually seen by observers, Carel van Schaik found that 47 infanticides (85%) followed changes in male residence or dominance rank. In one-male groups, infanticide follows takeovers, as Sugiyama and his colleagues originally reported. In multimale groups, infanticide typically follows changes in male residence or dominance rank. In a savanna baboon group in Botswana, for example, observers saw five lethal attacks on infants by males. In three of these cases, the infanticidal male had recently emigrated into the study group and achieved the top-ranking position; twice infants were killed by a resident male shortly after he acquired the top-ranking position (Figure 7.36).

Infanticide shortens interbirth intervals.

Primate females generally do not conceive while they are lactating, but if their infant dies, they quickly resume cycling. Cycling typically begins again in as little as a few days or weeks. The younger an infant is when it dies, the greater is the effect of its death on the mother's interbirth interval. This means that we should expect the youngest infants to be at the greatest risk for infanticide. Van Schaik's review of observed infanticides indicates that most cases involve unweaned infants. Moreover, detailed data from studies of hanuman langurs and red howlers demonstrate that the youngest infants are at greatest risk after takeovers (Figure 7.37). On average, infanticide reduces the length of interbirth intervals by

FIGURE 7.36 A male baboon in the Moremi Game Reserve of the Okavango Delta in Botswana holds the body of an infant that he has just killed. The male had recently acquired the top-ranking position in the group. (Photograph courtesy of Ryne Palombit.)

approximately 25% in langurs and 32% in red howlers. Taken together, these data provide strong evidence that infanticide reduces the duration of the mother's interbirth interval.

Infanticidal males do not kill their own infants.

Infanticide is unlikely to be adaptive if males target their own infants for attack, or even if males are unable to differentiate between their own infants and other males' infants. We do not know whether males can recognize their own infants, but they might be able to avoid harming their own infants if they selectively attack infants who were conceived before they joined the group or before they rose to high rank. Van Schaik's analysis of 55 observed infanticides shows that 40 infants (73%) were killed by males who were not present in the group at the time the infants were conceived. In 11 (20%) of the remaining cases, the killer was present when the infant was conceived, but was not seen mating in the group. In a detailed long-term study of hanuman langurs, Volker Sommer documented 55 instances in which infanticide was

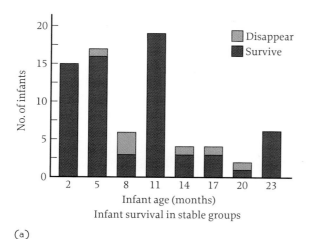

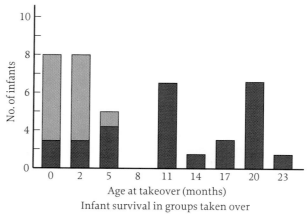

(a)

FIGURE 7.37 (a) In red howlers in Venezuela, infant mortality is much greater in groups that are taken over by new males than in groups with stable male residents. In groups that are taken over by new males, the youngest infants suffer the highest levels of mortality. (b) This infant, sitting in front of her mother, has been seriously wounded. The infant partially recovered from these wounds but was subsequently attacked by adult males a number of times. Six months later, this infant was the subject of a lethal attack by an adult male. (Photograph courtesy of Carola Borries.)

(b)

observed (12 cases) or strongly suspected (43 cases). In 52 (95%) of these cases, a genetic relationship between the killer and his victim was impossible or unlikely.

More compelling evidence comes from two studies in which researchers were able to assess the genetic relationship between males and the infants they attacked. Carola Borries and her colleagues documented 24 serious attacks on infants, including one lethal assault, by adult males among the hanuman langurs near Ramnagar. For 16 of these male–infant pairs, the team obtained DNA samples from both the male and the infant. In all 16 cases, the male was not the infant's father. Joseph Soltis of the National Institutes of Health and his colleagues from the Primate Research Institute at Kyoto University in Inuyama, Japan, observed a group of unprovisioned, free-ranging Japanese macaques following a wholesale change in the male dominance hierarchy. Soltis and his colleagues observed one infanticide directly, as well as a number of nonlethal attacks on infants by adult males. Altogether, these attacks involved 23 male–infant pairs. In 22 of these pairs, DNA analyses confirmed that the male was not the father of the infant that he attacked; in the remaining case, paternity was not resolved.

Infanticidal males gain reproductive benefits.

The final prediction is that males who commit infanticide should benefit by being able to sire the mother's next infant. Infanticidal males subsequently mated with the mother of the infant that they killed in 25 of the 55 cases that van Schaik compiled.

In 13 more cases, males may have mated with the victim's mother. Thus, infanticidal males often gained sexual access to the mothers of the infants that they killed.

Of course, mating does not guarantee paternity, but the study by Borries and her colleagues suggests that infanticidal males at Ramnagar do achieve reproductive benefits. These investigators documented five cases in which infants died after being attacked by males; in four of these cases the mother subsequently gave birth to another infant. In all of these cases, the presumed killer sired the mother's next infant.

Infanticide is sometimes a substantial source of mortality.

In some populations, infanticide is known or suspected to be a major source of mortality for infants. Among mountain gorillas in the Virunga Mountains of Rwanda, savanna baboons in the Moremi Game Reserve of Botswana, hanuman langurs in Ramnagar in Nepal, and red howlers in Venezuela, approximately one-third of all infant deaths are due to infanticide.

Females have evolved a battery of responses to infanticidal threats.

Although infanticide may enhance male reproductive success, it can have a disastrous effect on females who lose their infants. Thus, we should expect females to evolve counterstrategies to infanticidal threats.

The most obvious counterstrategy would be for females to try to prevent males from harming their infants (Figure 7.38). However, females' efforts to protect their infants are unlikely to be effective. Remember that males are larger than females in nonmonogamous species, and the extent of sexual dimorphism is the most pronounced in species that form one-male groups. Thus, in the species in which infanticide is most common, males are much larger than females and have long, dangerous canines. Females have little chance of protecting their infants from much larger males who systematically stalk the infants, waiting for an opportunity to attack. In fact, mothers are sometimes wounded when their infants are attacked.

Females sometimes enlist males' support against infanticidal males. Ryne Palombit has studied male–female relationships in the Moremi Game Reserve, where infanticide is a major source of infant mortality. He has found that mothers of young infants form close relationships with adult males. Females groom their male friends at high rates and remain close to them. In return, males are very attentive to their friends' infants and sometimes intervene when these infants are attacked by other males.

Finally, females may try to confuse males about paternity. As we have seen already, males seem to kill infants when there is no ambiguity about their paternity; if females can increase uncertainty about paternity, they may reduce the risk of infanticide. In some primate species, females mate with a number of different males while they are receptive; and in some species, females continue to mate during pregnancy. Both behaviors may reduce certainty about paternity and thereby reduce the risk of infanticide.

FIGURE 7.38 In species, like langurs, in which infanticide is common, females may lose many of their infants to infanticide. In some species, special relationships with adult males or mating with multiple partners may provide defense against infanticidal attacks.

If the data on infanticide are so consistent, why is the idea so controversial?

When Hrdy first proposed the idea that infanticide is an evolved male reproductive strategy, there was plenty of room for skepticism and dispute. The data were limited, and data needed to test the predictions derived from Hrdy's hypothesis had not yet been collected. Now, however, we have observed infanticide in 40 species distributed throughout the primate order, as well as in rodents, birds, and carnivores. We

FIGURE 7.39 In a group of brown capuchins, the dominant male allows an infant to feed beside him undisturbed. (Photograph by Charles Janson.)

have good evidence that the patterning of infanticidal attacks fits predictions derived from the sexual selection infanticide hypothesis; in fact, data from other taxa provide even stronger evidence in support of these predictions than data from primates do. But the controversy has not disappeared. In the last five years a prominent anthropological journal published two papers contending that there is no support for the hypothesis that infanticide in animal species is a sexually selected male reproductive strategy. Why does the controversy persist?

Volker Sommer, whose own work on infanticide in hanuman langurs has been attacked by critics of Hrdy's hypothesis, believes that the criticism comes in part from a tendency to commit what is called the "naturalistic fallacy"—to assume that what we see in nature is somehow right, just, and inevitable. Thus, critics are concerned that if we accept the idea that infanticide is an adaptive strategy for langurs, baboons, or other primates, it will justify similar behavior in humans. As we will discuss more fully in Part Four, however, the naturalistic fallacy is just that, a fallacy. It is misguided to try to extract moral meaning from the behavior of other animals.

Paternal Care in Nonmonogamous Groups

The extent of male care of offspring varies among nonmonogamous species.

In nonmonogamous species, males generally play little direct role in caring for infants and juveniles. In most cases, they don't carry them, provide food for them, or interact with them often. However, males sometimes contribute to the welfare of immatures in less direct ways.

Males are often quite tolerant of infants and juveniles. Silverbacks sometimes intervene in group conflicts involving infants and juveniles, and generally they support the younger of two antagonists when they do so. In brown capuchin groups, the alpha male, who monopolizes matings with receptive females, allows infants to feed near him, thus giving his offspring preferential access to resources (Figure 7.39). In both cases, male tolerance may be a low-cost form of paternal care.

In the multimale groups of langurs near Ramnagar, resident males sometimes defend infants against harassment by other males. A male protects an infant only if he was present when the infant was conceived and copulated with the infant's mother near the time of conception.

The threat of infanticide seems to influence the nature of male–female relationships in baboons.

It has been known for some time that mothers of newborn infants sometimes form close relationships ("friendships") with one or sometimes two adult males (Figure 7.40). Females are primarily responsible for maintaining proximity to their male associates and grooming them. Males defend their female associates when the females are threatened. Males also hold, carry, and groom their female associate's infants, and they sometimes intervene on behalf of immatures who become involved in aggressive encounters.

The function of these relationships has been the subject of considerable discussion. Barbara Smuts, who studied

FIGURE 7.40 After they give birth, many females begin to associate closely with one or two adult males. Here, a high-ranking female and her infant sit with the mother's male associate.

male–female relationships among baboons in Kenya, suggested that males might invest in relationships with mothers of newborn infants as a means to enhance their future mating prospects. Male care would be a form of **mating effort** if female choice played an important role in male mating success and females preferred males who had provided care for them and their infants in the past. Alternatively, male–female relationships might be a form of **parenting effort** if males selectively invested in relationships with the mothers of their own infants. Of course, in this case males would have to be able to recognize their own infants in some way.

A growing body of evidence provides strong support for the parenting effort hypothesis. In Moremi, Botswana, where male–female relationships are prominent and infanticide is common, female baboons are extremely agitated in the presence of new males. Jacinta Beehner and her colleagues from the University of Pennsylvania have found that females' cortisol levels, which provide a physiological index of stress, rise sharply when immigrant males enter the group (Figure 7.41a). The presence of a male "friend," however, reduces females' agitation (Figure 7.41b). Ryne Palombit has found that males are acutely sensitive to the distress of their female associates, but their responsiveness is directly tied to the infant's presence. If the infant dies, males stop responding. In Moremi, it is not yet clear whether males are the fathers of their female associates' infants, but behavioral data suggest that they might be. In Amboseli, Kenya, where infanticide is rare but male–female relationships are still very common, males do provide direct care to their own offspring. Adult males often intervene on behalf of infants and juveniles who are involved in agonistic disputes (Figure 7.42). Genetic analyses conducted by Susan Alberts of Duke University and her colleagues indicate that males selectively support their own offspring against unrelated individuals. Paternal kin recognition may be based on males' memories of their mating histories, females' memories of their consort partners, or some kind of phenotypic cue.

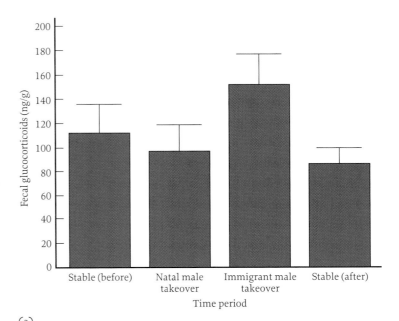

(a)

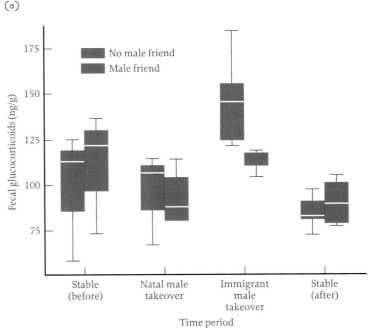

(b)

FIGURE 7.41 (a) The cortisol levels of female baboons increased when a new male immigrated into the group and took over the alpha position. (b) This effect was most pronounced for females who did not have a close male associate.

Female Mate Choice

Female preferences influence male competitive tactics.

During the last two decades there has been considerable interest in the possibility that female preferences influence mating success. Female mate choice has been demonstrated in a number of animal taxa, including cockroaches, house mice, and frogs. In some cases, females who are able to exercise mate choice have higher fitness than females whose choices are constrained.

Female primates sometimes show preferences, but there is little consensus about the kinds of traits that female primates prefer. For example, female gorillas sometimes leave their groups to join strange males, and their decisions may reflect preferences for less closely related males or for males who can provide more effective protection for their offspring. Free-ranging ring-tailed lemur females apparently prefer to mate with males from neighboring groups and rebuff mating attempts by natal males. Among provisioned groups of rhesus macaques on Cayo Santiago, a small island off the coast of Puerto Rico, females prefer unfamiliar, low-ranking males over high-ranking males. But in brown capuchin groups, females associate selectively with the highest-ranking male.

Although we don't know the basis for female preferences, there is some evidence that female preferences influence male mating success. In experimental studies of captive hamadryas baboons, Hans Kummer and his colleagues at the University of Zürich have shown that unfamiliar males are unlikely to attempt to disrupt a male–female pair if the female seems closely bonded to her mate. Joseph Soltis and his colleagues have shown that female Japanese macaques clearly prefer some males over others. Female preferences were uncorrelated with male dominance rank, but high-ranking males were still more likely to mate with females than low-ranking ones were. However, genetic analyses revealed that the males who were most attractive to females were the most successful in siring their offspring.

FIGURE 7.42 Male baboons sometimes support juveniles who become involved in agonistic disputes. Here the juvenile on the left is threatening the juvenile on the right, who huddles in his father's lap. In Amboseli, genetic analyses revealed that males selectively supported their own offspring in agonistic disputes.

FURTHER READING

Altmann, J. 1980. *Baboon Mothers and Infants*. Harvard University Press, Cambridge, MA.

Krebs, J. R., and N. B. Davies. 1993. *An Introduction to Behavioural Ecology*. Sinauer, Sunderland, MA, chaps. 8 and 9.

Nicolson, N. A. 1986. Infants, mothers, and other females. Pp. 330–342 in *Primate Societies*, ed. by B. B. Smuts, D. L. Cheney, R. M. Seyfarth, R. W. Wrangham, and T. T. Struhsaker. University of Chicago Press, Chicago.

Palombit, R. 1999. Infanticide and the evolution of pair bonds in nonhuman primates. *Evolutionary Anthropology* 7: 117–129.

van Schaik, C. P., and C. H. Janson, eds. 2000. *Infanticide by Males and Its Implications*. Cambridge University Press, Cambridge, England.

STUDY QUESTIONS

1. Explain why reproductive success is a critical element of evolution by natural selection. When biologists use the terms *cost* and *benefit*, what currency are they trying to measure?

2. What is the difference between polygyny and polyandry? It seems likely that females might prefer polyandry over polygyny, and at the same time males would favor polygyny. Explain why males and females might prefer different types of mating systems. If this conflict of interest occurs, why is polygyny more common than polyandry?

3. In many primate species, reproduction is highly seasonal. Some researchers have suggested that reproductive seasonality has evolved as a means for females to manipulate their reproductive options. How would reproductive seasonality alter females' options? Why do you think this strategy might be advantageous for females?

4. Imagine that you came upon a species in which males and females were the same size, but males had very large testes in relation to their body size. What would you infer about their social organization? Now suppose you found another species, in which males were much larger than females but had relatively small testes. What would you deduce about their social system? Why do these relationships hold?

5. Among mammalian species, male fitness is typically more variable than female fitness. Explain why this is often the case. What implications does this have for evolution acting on males and females?

6. What factors influence the reproductive success of females? How do these factors contribute to variance in female reproductive success?

7. Biologists use the term *investment* to describe parental care. What elements of the selective forces acting on parental strategies does this term capture?

8. Explain the logic underlying the sexual selection infanticide hypothesis. What predictions follow from this hypothesis? List the predictions and explain why they follow from the hypothesis.

9. In general, infanticide seems to be more common in species that form one-male groups than in species that form multimale, multifemale groups or monogamous groups. Explain why this might be the case.

10. In the text we write, "It is stupid to try to extract moral meaning from the behavior of other animals." Discuss this statement and think about whether you agree or disagree with it.

KEY TERMS

mating systems
strategy
parent-offspring conflict
sexual selection
intrasexual selection

intersexual selection
estrus
sexual selection infanticide hypothesis
mating effort
parenting effort

CHAPTER 8

THE EVOLUTION OF SOCIAL BEHAVIOR

A social primate lives in a group of known individuals. At one time or another, the other members of the group may become playmates, grooming partners, competitors for food, rivals for mates, allies in aggressive confrontations, caretakers for offspring, collaborators in intergroup encounters, and so on. Even solitary primates, like galagos and orangutans, interact regularly with their neighbors, maturing offspring, and prospective mates.

There is great diversity in the form and frequency of social interactions that occur within and among groups of primates. Just as evolutionary theory provides the framework for understanding the patterning of mating and parenting behaviors in nature, it also provides an essential foundation for understanding the form and distribution of social interactions among individuals within social groups.

KINDS OF SOCIAL INTERACTIONS

Social interactions are behaviors that affect the fitness of more than one individual.

When an animal feeds, travels, or sleeps, it has little impact on the fitness of others. But when two individuals interact—say, in competition over a prized resource or in cooperative defense of valuable food items—their activities necessarily involve each other, as collaborators, rivals, or opponents. In these situations, the behavior of one individual directly affects the fitness of the other. Interactions that involve two individuals are called **dyadic** or **pairwise** interactions. Table 8.1 classifies four kinds of pairwise interactions according to their effects on the **actor** (the individual performing the behavior) and the **recipient** (the individual affected by the behavior). An act is said to be beneficial (+) if it increases fitness, and costly or detrimental (–) if it reduces fitness. It is evident that social interactions do not necessarily have the same kinds of effects on actors and on recipients. For example, **selfish** acts are beneficial to the actor and costly to the recipient, and the costs and benefits are reversed for **altruistic** interactions. On the other hand, **mutualistic** interactions are beneficial to both actor and recipient, and **spiteful** interactions are costly to both parties.

Before we go any further, two caveats are in order. First, this classification uses ordinary English words like *altruism* and *spite* because they provide a convenient, easily remembered shorthand for describing the fitness effects of different kinds of social behaviors. However, the technical definitions sometimes differ from the meanings these words have in ordinary usage. For example, an act that is beneficial to the recipient but has no adverse impact on the fitness of the actor might be considered altruistic in common usage but is not altruistic in the biological sense of the word. So you should be careful to keep the technical definitions in mind when these terms are used.

TABLE 8.1 A classification of pairwise, or dyadic, social interactions. + indicates a positive effect on fitness, and – a negative impact on fitness.

Case	Actor	Recipient
Selfish	+	–
Mutualistic	+	+
Altruistic	–	+
Spiteful	–	–

Second, it is extremely difficult to measure the effects of particular behavioral acts on the fitness of individuals, particularly for long-lived animals like primates. A female may participate in hundreds of grooming sessions over the course of a year and in thousands during her lifetime. At the same time, she has a multitude of different experiences that may influence her reproductive career. As a result, it is virtually impossible to assess the effects of a single behavioral act, or even the effects of a class of behavioral acts, on her reproductive success. Nonetheless, we can make reasonable inferences about the immediate costs and benefits of particular acts on the basis of more general considerations, such as the energy demands or the risks associated with specific behaviors or social interactions. For example, we could assess the caloric value of a food item that a female chimpanzee shares with her daughter or measure the time required to replace the food item with another one of similar value. These kinds of assessments provide a basic estimate of the effects of particular kinds of social interactions on fitness.

ALTRUISM: A CONUNDRUM

Altruistic behavior cannot evolve by ordinary natural selection.

Notice from Table 8.1 that two of the four forms of interactions—selfish and mutualistic acts—enhance the fitness of the actor and will be favored by natural selection, all other things being equal. Thus there is no problem explaining why one female capuchin monkey supplants another from a patch of ripe fruit, or why two chimpanzees jointly corner a red colobus monkey and then share the carcass.

Altruism, however, is a puzzle because it *decreases* the fitness of the individual performing the behavior. According to Darwin's theory, complex adaptations, including behavioral adaptations, must be assembled step by small step, each change favored by natural selection. Thus, although altruistic behaviors might initially arise by accident or as side effects of other behaviors, it seems impossible for complex altruistic behaviors to be assembled by natural selection. Each small genetic change that made it more likely for an individual to perform the behavior would be selected against because of the behavior's negative effects on genetic fitness. The same argument applies to spite. Thus the existence of either spiteful or altruistic behavior would seem to contradict the fundamental logic of natural selection.

Primates perform altruistic behaviors in nature.

This theoretical conundrum would not be a problem if altruistic interactions were rare or unimportant, and this does seem to be the case for spite. However, there is a lot of evidence that primates and many other social animals regularly perform complex altruistic behaviors that play an important role in their social life. Virtually all social primates groom other group members, picking parasites, scabs, and bits of debris from their hair. When one individual in a group sights a predator, she often gives a distinctive alarm call that alerts the rest of the group. Sometimes two individuals form an alliance against a higher-ranking individual. Occasionally, one individual allows another to share its food. All of these behaviors seem to meet the biological definition of altruism. When it is grooming, the donor expends valuable time and energy that could otherwise be used to look for food, to court prospective mates, or for other vital tasks; and the recipient gets a thorough cleaning of parts of her body that she might find difficult to reach and may generally enjoy a period of pleasant relaxation (Figure 8.1). By giving an alarm call, the caller makes herself more conspicuous to predators, but the individuals hearing the warning are able to flee to safety. When animals form alliances, particularly when they come to the defense of other group members, they make themselves vulnerable to retaliation by the aggressor, but the defended individual may have less chance of being injured or defeated during the encounter (Figure 8.2). Sharing food may reduce the amount of food that the owner eats while increasing the amount of food that the recipient obtains. In each of these cases, it seems likely that the actor suffers a decrease in fitness while the recipient incurs an increase in fitness, which satisfies the criteria for altruism.

FIGURE 8.1 Gray langurs groom one another. Grooming is usually considered to be altruistic because the groomer expends time and energy when it grooms another animal, and the recipient benefits from having ticks removed from its skin, wounds cleaned, and debris removed from its hair.

Altruistic behaviors cannot be favored by selection just because they are beneficial to the group as a whole.

You might think that if the average effect of an act on all members of the group is positive, then it would be beneficial for all individuals to perform it. For example, suppose that when one monkey gives an alarm call, the other members of the group benefit, and the total benefits to all group members exceed the cost of giving the call. Then, if every individual gave the call when a predator was sighted, all members of the group would be better off than if no warning calls were ever given.

The problem with this line of reasoning is that it confuses the effect on the group with the effect on the actor. In most circumstances, the fact that alarm calls are beneficial to those hearing them doesn't affect whether the trait of alarm calling evolves; all that matters is the effect that giving the alarm call has on the caller. To see why, imagine a hypothetical monkey species in which some individuals give alarm calls when they are the first to spot a predator. Monkeys who are warned by the call have a chance to flee. Suppose that one-fourth of the population ("callers") give the call when they spot predators, and three-fourths of the individuals ("noncallers") do not give an alarm call in the same circumstance. (These proportions are arbitrary; we chose them because it's easier to follow the reasoning in examples with concrete numbers.) Let's suppose that in this species the tendency to give alarm calls is genetically inherited.

Now we compare the fitness of callers and noncallers. Because everyone in the group can hear the alarm calls and take appropriate action, alarm calls benefit everyone in the group to the same extent (Figure 8.3). Calling has no effect on the *relative* fitness of callers and noncallers because, on average, one-fourth of the beneficiaries will be callers and three-fourths of the beneficiaries will be noncallers—the same proportions we find in the population as whole.

FIGURE 8.2 Two capuchin monkeys form a coalition against an opponent who is not visible in the photograph. Coalitions are altruistic because the ally puts itself at risk and expends energy when it becomes involved in an ongoing dispute. The animal who receives support may benefit if the fight ends sooner or with less costly consequences. (Photograph by Susan Perry.)

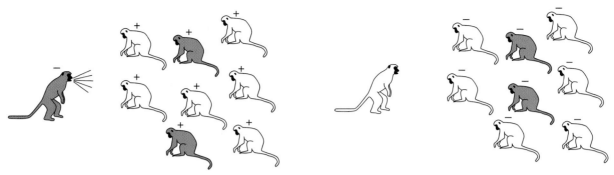

(a) Altruist gives alarm call to group.   (b) Nonaltruist doesn't give alarm call to group.

FIGURE 8.3 Two groups of monkeys are approached by a predator. (a) In one group, one individual (pink) has a gene that makes her call in this context. Giving the call lowers the caller's fitness but increases the fitness of every other individual in the group. Like the rest of the population, one out of four of these beneficiaries also carries the genes for calling. (b) In the second group, the female who detects the predator does not carry the gene for calling and remains silent. This lowers the fitness of all members of the group a certain amount because they are more likely to be caught unaware by the predator. Once again, one out of four is a caller. Although members of the caller's group are better off on average than members of the noncaller's group, the gene for calling is not favored, because callers and noncallers in the caller's group both benefit from the caller's behavior but callers incur some costs. Callers are at a disadvantage relative to noncallers. Thus, calling is not favored, even though the group as a whole benefits.

Calling reduces the risk of mortality for everyone who hears the call, but it does not change the frequency of callers and noncallers in the population, because everyone gains the same benefits. However, callers are conspicuous when they call, so they are more vulnerable to predators. Although all individuals benefit from hearing alarm calls, callers are the only ones who suffer the costs from calling. This means that, on average, noncallers will have a higher fitness than callers. Thus, genes that cause alarm calling will not be favored by selection, even if the cost of giving alarm calls is small and the benefit to the rest of the group is large. Instead, selection will favor genes that suppress alarm calling, because noncallers have higher fitness than callers. This simple example can be reformulated for grooming, food sharing, coalition formation, and so on, and in each case the conclusion is the same: individual selection is not expected to favor the evolution of these kinds of behaviors (Box 8.1).

KIN SELECTION

Natural selection can favor altruistic behavior if altruistic individuals are more likely to interact with each other than chance alone would dictate.

If altruistic behaviors can't evolve by ordinary natural selection or by group selection, then how do they evolve? A clear answer to this question did not come until 1964, when the young biologist W. D. Hamilton published a landmark paper. This paper was the first of a series of fundamental contributions that Hamilton made to our understanding of the evolution of behavior. There are several different ways to conceptualize Hamilton's basic idea, and we will adopt the approach presented in his original paper.

Group Selection

BOX 8.1

Group selection was once thought to be the mechanism for the evolution of altruistic interactions. In the early 1960s, British ornithologist V. C. Wynne-Edwards contended that altruistic behaviors like those we have been considering here evolved because they enhanced the survival of whole groups of organisms. Thus, individuals gave alarm calls, despite the costs of becoming more conspicuous to predators, because calling protected the group as a whole from attacks. Wynne-Edwards reasoned that groups that contained a higher number of altruistic individuals would be more likely to survive and prosper than groups that contained fewer altruists, and the frequency of the genes leading to altruism would increase.

Wynne-Edwards's argument is logical because Darwin's postulates logically apply to groups as well as individuals. However, group selection is not an important force in nature, because there is generally not enough genetic variation among groups for selection to act on. Group selection can occur if groups vary in their ability to survive and to reproduce, and that variation is heritable. Then group selection may increase the frequency of genes that increase group survival and reproductive success. The strength of selection among groups depends on the amount of genetic variation among groups, just as the strength of selection among individuals depends on the amount of genetic variation among individuals. However, when individual selection and group selection are opposed and group selection favors altruistic behavior while individual selection favors selfish alternatives, individual selection has a tremendous advantage. This is because the amount of variation among groups is much smaller than the amount of variation among individuals, unless groups are very small and there is very little migration among them. Thus, individual selection favoring selfish behavior will generally prevail over group selection, making group selection an unlikely source of altruism in nature.

The argument made in the previous section contains a hidden assumption: that altruists and nonaltruists are equally likely to interact with one another. We supposed that callers give alarm calls when they hear a predator, no matter who is nearby. This is why callers and noncallers were equally likely to benefit from hearing an alarm call. Hamilton's insight was to see that any process that causes altruists to be more likely to interact with other altruists than they would by chance could facilitate the evolution of altruism.

To see why this is such an important insight, let's modify the previous example by assuming that our hypothetical species lives in groups composed of full siblings, offspring of the same mother and father. The frequencies of the calling and noncalling genes don't change, but their distribution will be affected by the fact that siblings live together (Figure 8.4). If an individual is a caller, then, by the rules of Mendelian genetics, there is a 50% chance that the individual's siblings will share the genes that cause calling behavior. This means that the frequency of the genes for calling will be higher in groups that contain callers than in the population as a whole, and therefore more than one-fourth of the beneficiaries will be callers themselves. When a caller gives an alarm call, the audience will contain a higher fraction of callers than the population at large does. Thus the caller raises the average fitness of callers relative to noncallers. Similarly, because the siblings of noncallers are more likely to be noncallers than chance alone would dictate, callers are less likely to be present in such groups than in

(a) Altruist gives alarm call to siblings.   (b) Nonaltruist doesn't give alarm call to siblings.

FIGURE 8.4 Two groups of monkeys are approached by a predator. Each group is composed of nine sisters. (a) In one group there is a caller (pink), an individual with a gene that makes her call in this context. Her call lowers her own fitness but increases the fitness of her sisters. (b) In the second group, the female who detects the predator is not a caller and does not call when she spots the predator. As in Figure 8.3, calling benefits the other group members but imposes costs on the caller. However, there is an important difference between the situations portrayed here and in Figure 8.3. Here the groups are made up of sisters, so five of the eight recipients of the call also carry the calling gene. In any pair of siblings, half of the genes are identical because the siblings inherited the same gene from one of their parents. Thus, on average, half of the caller's siblings also carry the calling allele. The remaining four siblings carry genes inherited from the other parent, and like the population as a whole, one out of four of them is a caller. The same reasoning shows that in the group with the noncaller, there is only one caller among the beneficiaries of the call. Half are identical to their sister because they inherited the same noncalling gene from one of their parents; one of the remaining four is a caller. In this situation, callers are more likely to benefit from calling than noncallers, and so calling alters the relative fitness of callers and noncallers. Whether the calling behavior actually evolves depends on whether these benefits are big enough to compensate for the reduction in the caller's fitness.

the population at large. Therefore, the absence of a warning call lowers the relative fitness of noncallers relative to callers.

When individuals interact selectively with relatives, callers are more likely to benefit than noncallers and, all other things being equal, the benefits of calling will favor the genes for calling. However, we must remember that calling is costly, and this will tend to reduce the fitness of callers. Calling will be favored by natural selection only if its benefits are sufficiently greater than its costs. The exact nature of this trade-off is specified by what we call "Hamilton's rule."

Hamilton's Rule

Hamilton's theory of kin selection predicts that altruistic behaviors will be favored by selection if the costs of performing the behavior are less than the benefits discounted by the coefficient of relatedness between actor and recipient.

Hamilton's theory of **kin selection**, introduced briefly in Chapter 7, is based on the idea that selection could favor altruistic alleles if animals interacted selectively with their genetic relatives. Hamilton's theory also specifies the quantity and distribution of help among individuals. According to **Hamilton's rule**, an act will be favored by selection if

$$rb > c \qquad\qquad (8.1)$$

where

  r = the average coefficient of relatedness between the actor and the recipients
  b = the sum of the fitness benefits to all individuals affected by the behavior
  c = the fitness cost to the individual performing the behavior

The **coefficient of relatedness**, *r*, measures the genetic relationship between interacting individuals. More precisely, r is the average probability that two individuals will acquire the same allele through descent from a common ancestor. Figure 8.5 shows how these probabilities are derived in a simple genealogy. Female A obtains one allele at a given locus from her mother and one from her father. Her half-sister, female B, also obtains one allele at the same locus from each of her parents. We obtain the probability that both females receive the same allele from their mother by multiplying the probability that female A obtains the allele (0.5) by the probability that female B obtains the same allele (0.5); the result is 0.25. Thus, half-sisters have, on average, a 25% chance of obtaining the same allele from their mothers. Now consider the relatedness between female B and her brother, male C. In this case, note that female B and male C are full siblings: they have the same mother and the same father. The probability that both siblings will acquire the same allele from their mother is still 0.25, but female B and male C might also share an allele from their father. The probability of this event is also 0.25. Thus the probability that female B and male C share an allele is equal to the sum of 0.25 and 0.25, or 0.5. This basic reasoning can be extended to calculate the degrees of relatedness among various categories of kin (Table 8.2).

Hamilton's rule leads to two important insights: (1) altruism is limited to kin, and (2) closer kinship facilitates more costly altruism.

If you reflect on Hamilton's rule for a while, you will see that it produces two fundamental predictions about the conditions that favor the evolution of altruistic behaviors. First, altruism is not expected to be directed toward nonkin, because the coefficient of relatedness, *r*, between nonkin is 0. The condition for the evolution of

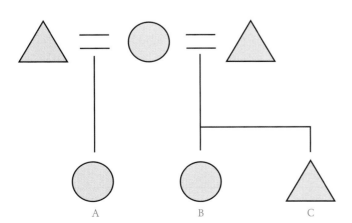

A   B   C

FIGURE 8.5 This genealogy shows how the value of r is computed. Triangles represent males, circles represent females, and the equal sign represents mating. The relationships between individuals labeled in the genealogy are described in the text.

| TABLE 8.2 | The value of r for selected categories of relatives. (When cousins are off-spring of full siblings, they are related by .125, but when they are the off-spring of half-siblings they are related by .0625.) |

Relationship	r
Parent and offspring	.5
Full siblings	.5
Half-siblings	.25
Grandparent and grandchild	.25
First cousins	.125 or .0625
Unrelated individuals	0

altruistic traits will be satisfied only for interactions between kin, when $r > 0$. Thus, altruists are expected to be nepotistic, showing favoritism toward kin.

Second, close kinship is expected to facilitate altruism. If an act is particularly costly, it is most likely to be restricted to close kin. Figure 8.6 shows how the benefit/cost ratio scales with the degree of relatedness among individuals. Compare what happens when $r = \frac{1}{16}$ (or 0.0625) and when $r = \frac{1}{2}$ (or 0.5). When $r = \frac{1}{16}$, the benefits must be more than 16 times as great as the costs for Hamilton's inequality [Equation (8.1)] to be satisfied. When $r = \frac{1}{2}$, the benefit needs to be just over twice as large as the costs. All other things being equal, altruism will be more common among close relatives than among distant ones.

It is important to understand that the fractions in Table 8.2 are not equivalent to the quantities in a recipe. If a recipe calls for two cups of flour and one cup of sugar, the dutiful cook measures out these amounts. But monkeys don't necessarily apportion altruism in precise amounts according to the coefficients of relatedness. Reality is, as always, more complicated. Sometimes dramatic asymmetries in the benefit/cost ratios within pairs of relatives influence the distribution of altruism. In many primate species, for example, it may be much less costly for a mother to defend her infant than for the infant to defend her mother. The inability to recognize certain categories

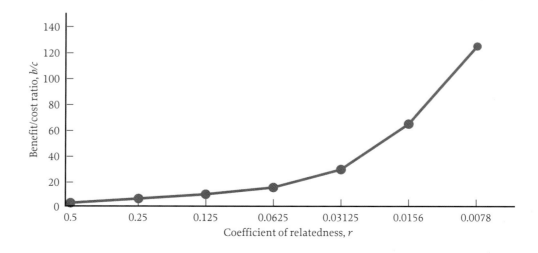

| FIGURE 8.6 | As the degree of relatedness (r) between two individuals declines, the value of the ratio of benefits to costs (b/c) that is required to satisfy Hamilton's rule for the evolution of altruism rises rapidly.

of kin, particularly distant kin or paternal kin, may also limit the distribution of altruism. Finally, it is important to remember that close kinship does not always prevent violence and aggression. In some species of birds, nestlings kill their own siblings to prevent them from competing for resources that parents bring back to the nest. Juvenile primates sometimes jostle with their newborn siblings for their mother's time and attention.

Evidence of Kin Selection in Primates

A considerable body of evidence suggests that the patterns of many forms of altruistic interactions among primates are largely consistent with predictions derived from Hamilton's rule. Later in this section we consider three examples: food sharing, grooming, and coalition formation. In each case, the behaviors are usually biased toward kin. First, however, we discuss how primates identify relatives.

Primates may use contextual cues to recognize maternal relatives.

In order for kin selection to provide an effective mechanism for the evolution of cooperative behavior, animals must be able to distinguish relatives from nonrelatives and close relatives from distant ones. Some organisms are able to recognize their kin by their smell or likeness to themselves. This is called **phenotypic matching**. Others learn to recognize relatives using contextual cues—cues such as familiarity and proximity—that predict kinship (Figure 8.7). We have generally assumed that primates rely on contextual cues to identify their relatives, but new data suggest that phenotypic matching may also play a role in primate kin recognition.

Mothers seem to make use of contextual cues to recognize their own infants. After they give birth, females repeatedly sniff and inspect their newborns. By the time their infants are a few weeks old, mothers are clearly able to recognize them. After this, females of most species nurse only their own infants and respond selectively to their own infants' distress calls. Primate mothers don't really need innate means of recognizing their young, because young infants spend virtually all of their time in physical

(a)

(b)

FIGURE 8.7 (a) Even mothers must learn to recognize their own infants. Here, a female bonnet macaque peers into her infant's face. (b) Primates apparently learn who their relatives are by observing patterns of association among group members. Here, a female inspects another female's infant. (a, Photograph courtesy of Kathy West.)

(a)

(b)

FIGURE 8.8 Monkey and ape infants grow up surrounded by various relatives. (a) These adult baboon females are mother and daughter. Both have young infants. (b) An adolescent female bonnet macaque carries her younger brother while her mother is recovering from a serious illness.

contact with their mothers. Thus, mothers are unlikely to confuse their own newborn with another.

Monkeys and apes may learn to recognize other maternal kin through contact with their mothers. Offspring continue to spend considerable amounts of time with their mothers even after their younger siblings are born. Thus they have many opportunities to watch their mothers interact with their new brothers and sisters. Similarly, the newborn infant's most common companions are its mother and siblings (Figure 8.8). Because adult females continue to associate with their mothers, infants also become familiar with their grandmothers, aunts, and cousins.

Contextual cues may play some role in paternal kin recognition as well.

Primatologists were once quite confident that primates were unable to recognize their paternal kin. This conclusion was based on the following reasoning: First, close associations between males and females are uncommon in most primate species, so proximity would not provide accurate cues of paternal kinship. Second, there may be considerable uncertainty about paternity in most primate species because females may mate with several different males. Even in pair-bonded species, like gibbons and *Callicebus* monkeys, females sometimes mate with males from outside their groups. Thus it was assumed that paternal kin recognition was unlikely. As we explained in Chapter 7, however, male baboons are able identify their own offspring. This tells us that the conventional wisdom is sometimes wrong.

New evidence from field studies on macaques and baboons suggests that monkeys use contextual cues to assess paternal kinship. Jeanne Altmann of Princeton University pointed out that age may provide a good proxy measure of paternal kinship in species in which a single male typically dominates mating activity within the group. When this happens, all infants born at about the same time are likely to have the same father. Thus, age-mates are likely to be paternal half-siblings. Recent studies suggest that Altmann's logic is correct—that is, that monkeys do use age to identify paternal kin. Anja Widdig of Humboldt University in Berlin and her colleagues studied kin recognition among rhesus macaques on Cayo Santiago, Puerto Rico. In this population, there is strong **reproductive skew** because a single male dominated matings during each mating season. Females showed strong affinities for maternal half-sisters, as expected. However, females also showed strong affinities for their paternal half-sisters, spending more time grooming them and more time in close proximity to them than to unrelated females (Figure 8.9). Their affinities for paternal kin seem to be based partly on strong preferences for interacting with age-mates. However, females also distinguished *among* their age-mates, preferring paternal half-sisters over unrelated females of the same age. Similar patterns have been observed among wild baboons in Amboseli. There, females can recognize their paternal half-sisters.

Although the mechanisms underlying paternal kin recognition are not fully understood at this time, it is clear that monkeys can sometimes identify paternal kin. They may rely on contextual cues, such as age similarity or previous mating behavior, phenotypic cues, or both.

Food sharing occurs mainly among kin.

In a small number of primate species, adults voluntarily share food items with other group members. **Food sharing,** the unforced transfer of food from one individ-

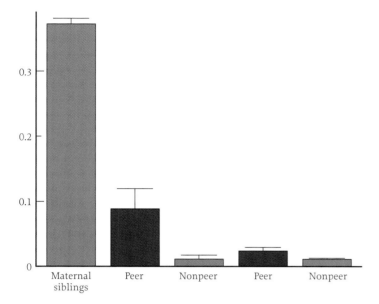

FIGURE 8.9 Female macaques are able to identify paternal siblings. Females interact far more often with maternal half-siblings (blue bar) than with paternal siblings, but they interact more often with paternal siblings than with nonkin. Age similarity seems to provide a cue for paternal kinship: females interact more often with paternal half-sibling peers (red hatched bar) than with paternal half-siblings that are not close in age (solid red bar). But note that females also distinguish among peers, preferring half-sibling peers over unrelated peers (purple hatched bar).

ual to another, generally happens between close relatives (Figure 8.10). At Gombe Stream National Park in Tanzania, researchers provided chimpanzees with bananas, a food that the chimpanzees coveted but that doesn't grow in their home range (Figure 8.11). William McGrew, now at Miami University of Ohio, found that 86% of such exchanges involved maternal kin, even though these individuals constituted only 5% of the possible pairs of individuals in the group. In most exchanges, mothers were sharing bananas with their own offspring. However, not all food sharing between chimpanzees involves kin. When male chimpanzees make a kill, they often share meat with other males and with unrelated adult females.

Food sharing also occurs in tamarins and marmosets. These tiny monkeys are omnivores, feeding on fruit, gum and sap, nectar, insects, and small vertebrates. Young marmosets and tamarins have trouble catching large and mobile insect prey and manipulating large fruits, so older group members selectively share these foods with infants. In some cases, food items are spontaneously offered to infants and juveniles; in other cases, infants use specialized begging vocalizations to solicit food from others. Infants are closely related to most other group members, so some of this food sharing might be the product of kin selection.

Grooming is also more common among kin than nonkin.

Social **grooming** plays an important role in the lives of most gregarious primates (Figure 8.12). Grooming is likely to be beneficial to the participants in at least two ways: First, grooming serves hygienic functions because bits of dead skin, debris, and parasites are removed and wounds are kept clean and open. Second, grooming may provide a means for individuals to establish relaxed, **affiliative** (friendly) contact

FIGURE 8.10 A female baboon allows her infant to share gum that she has extracted from an acacia tree.

FIGURE 8.11   A female chimpanzee allows her infant to share bananas that she has received at the feeding station in the Gombe Stream Research Center. A veteran Tanzanian field assistant, Hilali Matama, monitors the situation.

and to reinforce social relationships with other group members (Box 8.2). Grooming is costly because the actor expends both time and energy in performing these services. Moreover, Marina Cords of Columbia University has shown that blue monkeys are less vigilant when they are grooming, perhaps exposing themselves to some risk of being captured by predators.

Grooming is more common among kin, particularly mothers and their offspring, than among nonkin. For example, Ellen Kapsalis and Carol Berman of the University at Buffalo recently documented the effect of maternal relatedness among rhesus macaques on Cayo Santiago. In this population, females groom close kin at higher rates than nonkin, and close kin are groomed more often than distant kin (Figure 8.14). It is interesting to note that as relatedness declined, the differences in the proportions of time spent grooming kin and nonkin were essentially eliminated. This may mean that monkeys cannot recognize more distant kin or that the conditions of Hamilton's rule ($rb > c$) are rarely satisfied for distant kin.

Primates most often form coalitions with close kin.

Most disputes in primate groups involve two individuals. Sometimes, however, several individuals jointly attack another individual, or one individual comes to the

(a)

(b)

FIGURE 8.12   Some of the many species of primates that groom are capuchin monkeys (a), blue monkeys (b), baboons (c), and gorillas (d). [Photographs courtesy of Susan Perry (a), Marina Cords (b), and John Mitani (d).]

(c)

(d)

BOX 8.2

How Relationships Are Maintained

Conflict and competition are fundamental features of social life for many primates: females launch unprovoked attacks on unsuspecting victims, males battle over access to receptive females, subordinates are supplanted from choice feeding sites, and dominance relationships are clearly defined and frequently reinforced. Although violence and aggression are not prevalent in all primates (muriquis, for example, are so peaceful that dominance hierarchies cannot be detected), some primates can be charitably characterized as contentious. This raises an intriguing question: how is social life sustained in the face of such relentless conflict? After all, it seems inevitable that aggression and conflict will drive animals apart, disrupt social bonds, and reduce the cohesiveness of social groups. Recently, primatologists have begun to consider this issue.

Social relationships matter to primates. They spend a considerable portion of every day grooming other group members. Grooming is typically focused on a relatively small number of partners and is often reciprocated. Robin Dunbar of Liverpool University contends that in Old World monkeys, grooming has transcended its original hygienic function and now serves as a means to cultivate and maintain social bonds. Social bonds may have real adaptive value to individuals. For example, grooming is sometimes exchanged for support in coalitions, and grooming partners may be allowed to share access to scarce resources.

When tensions do erupt into violence, certain behavioral mechanisms may reduce the disruptive effects of conflict on social relationships. After conflicts end, victims often flee from their attackers—an understandable response. In some cases, however, former opponents make peaceful contact in the minutes that follow conflicts. For example, chimpanzees sometimes kiss their former opponents, female baboons grunt quietly to their former victims, and golden monkeys may embrace or groom their former adversaries. The swift transformation from aggression to affiliation prompted Frans de Waal of Emory University to suggest that these peaceful postconflict interactions are a form of reconciliation, a way to mend relationships that were damaged by conflict. Inspired by de Waal's work, a number of researchers have documented reconciliatory behavior in a wide range of primate species.

Peaceful postconflict interactions seem to have a calming effect on former opponents. When monkeys are nervous and anxious, rates of certain self-directed behaviors, such as scratching, increase. Thus, self-directed behaviors are a good behavioral index of stress. Felipo Aureli of Liverpool John Moores University and his colleagues at Utrecht University

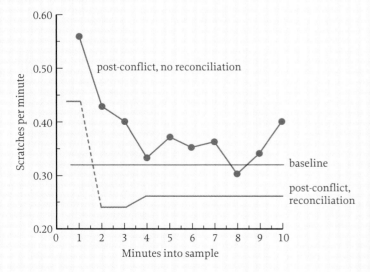

FIGURE 8.13 Rates of scratching, an observable index of stress, by victims of aggression are elevated over normal levels in the minutes that follow aggressive encounters. If some form of affiliative contact (reconciliation) between former opponents occurs during the postconflict period, however, rates of scratching drop rapidly below baseline levels. If there is no reconciliatory contact during the first few minutes of the postconflict period, rates of scratching remain elevated above baseline levels for several minutes. These data suggest that affiliative contact between former opponents has a calming effect. Similar effects on aggressors have also been detected.

and Emory University have studied the effects of fighting and reconciliation on the rate of self-directed behaviors. They have found that levels of self-directed behavior, and presumably stress, rise above baseline levels after conflicts. Both victims and aggressors seem to feel the stressful effects of conflicts. If former opponents interact peacefully in the minutes that follow conflicts, rates of self-directed behavior fall rapidly to baseline levels (Figure 8.13). If they do not reconcile, rates of self-directed behavior remain elevated above baseline levels for several minutes longer. If reconciliation provides a means to preserve social bonds, then we would expect primates to reconcile selectively with their closest associates. In a number of groups, former opponents who have strong social bonds are most likely to reconcile. Kin also reconcile at high rates in some groups, even though some researchers have argued that kin have little need to reconcile, because their relationships are unlikely to be frayed by conflict.

Reconciliation may also play a role in resolving conflicts among individuals who do not have strong social bonds. Like many other primates, female baboons are strongly attracted to newborn infants and make persistent efforts to handle them. Mothers reluctantly tolerate infant handling, but they do not welcome the attention. Female baboons reconcile at particularly high rates with the mothers of young infants, even when they do not have close relationships with them. Reconciliation greatly enhances the likelihood that aggressors will be able to handle their former victims' infants in the minutes that follow conflicts. Thus, in this case, reconciliation seems to be a means to an immediate end but not a means to preserve long-term relationships.

support of another individual involved in an ongoing dispute (Figure 8.15). We call these kinds of interactions **coalitions** or **alliances**. Support is likely to be beneficial to the individual who receives aid because support alters the balance of power among the original contestants. The beneficiary may be more likely to win the contest or less likely to be injured in the confrontation. At the same time, however, intervention may be costly to the supporter, who expends time and energy and risks defeat or injury when it becomes involved. Hamilton's rule predicts that support will be preferentially directed toward kin and that the greatest costs will be expended on behalf of close relatives.

Many studies have shown that support is selectively directed toward close kin. Female pigtail macaques defend their offspring and close kin more often than they defend distant relatives or unrelated individuals (Table 8.3). Females run some risk

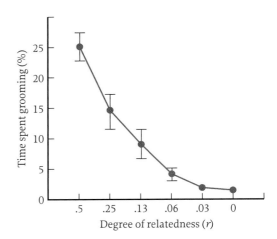

FIGURE 8.14 Rhesus monkeys on Cayo Santiago groom close relatives more often than they groom distant relatives or nonkin.

when they participate in coalitions, particularly when they are allied against higher-ranking individuals. Coalitions against high-ranking individuals are more likely to result in retaliatory attacks against the supporter than are coalitions against lower-ranking individuals. Female macaques are much more likely to intervene against higher-ranking females on behalf of their own offspring than on behalf of unrelated females or juveniles. Thus, macaque females take the greatest risks on behalf of their closest kin.

Kin-based support in conflicts has far-reaching effects on the social structure of many primate groups. Evidence of these effects comes from studies of macaque, vervet, and baboon groups.

FIGURE 8.15   Two baboons form an alliance against an adult female.

Maternal support in macaques, vervets, and baboons influences the outcome of aggressive interactions and dominance contests. Initially, an immature monkey is able to defeat older and larger juveniles only when its mother is nearby. Eventually, regardless of their age or size, juveniles are able to defeat everyone their mothers can defeat, even when the mother is some distance away. Maternal support contributes directly to several remarkable properties of dominance hierarchies within these species:

- Maternal rank is transferred with great fidelity to offspring, particularly daughters. In a group of baboons at Gilgil, Kenya, for example, maternal rank is an almost perfect predictor of the daughter's rank (Figure 8.16).
- Maternal kin occupy adjacent ranks in the dominance hierarchy, and all the members of one **matrilineage** (maternal kin group) rank above or below all members of other matrilineages.
- Ranking within matrilineages is often quite predictable. In most cases, mothers outrank their daughters, and younger sisters outrank their older sisters.

TABLE 8.3   Captive pigtail macaques support close kin more frequently than distant kin. Individuals related by r = 0.5 include parents and offspring and full siblings. Pairs of individuals related by r = 0.25 include half-siblings with each other, and grandparents and grandoffspring. Pairs of individuals related by r = 0.125 include cousins. (From Table 1 in A. Massey, 1977, Agonistic aids and kinship in a group of pigtail macaques, *Behavioral Ecology and Sociobiology* 2: 31–40.)

	DEGREES OF RELATEDNESS		
	0.5	0.25	0.125
Number of individuals	38	156	48
Number of aids	173	164	13
Number of aids per pair	4.55	1.05	0.27
Number of aids per aggressive encounter	0.15	0.04	0.01

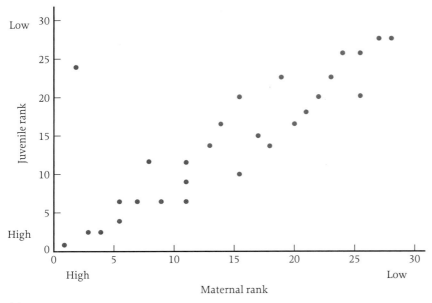

(a)

(b)

FIGURE 8.16   Juvenile female baboons acquire ranks very similar to their mother's ranks. (a) The anomalous point at maternal rank 2 belongs to a female whose mother died when she was an infant. (b) Here, the dominant female of a baboon group is flanked by her two daughters, ranked 2 and 3.

- Female dominance relationships are amazingly stable over time. In many groups, they remain the same over months and sometimes over years. The stability of dominance relationships among females may be a result of the tendency to form alliances in support of kin.

Kin-biased support also plays an important role in the reproductive strategies of red howler males.

Behavioral and genetic studies of red howlers in Venezuela conducted by Teresa Pope have shown that kinship influences howler males' behavior in important ways. Red howlers live in groups that contain two to four females and one or two males. Males sometimes join up with migrant, extragroup females and help them establish new territories. Once such groups have been established, resident males must defend their position and their progeny from infanticidal attacks by alien males. When habitats are crowded, males can gain access to breeding females only by taking over established groups and evicting male residents. This is a risky endeavor because males are often injured in takeover attempts. Moreover, as habitats become more saturated and dispersal opportunities become more limited, males tend to remain in their groups longer. Maturing males help their fathers defend their groups against takeover attempts. Collective defense is crucial to males' success because single males are unable to defend groups against incursions by rival males.

This situation leads to a kind of arms race because migrating males also form coalitions and cooperate in efforts to evict residents. After they have established residence, males collectively defend the group against incursions by extragroup males. Cooperation among males is beneficial because it helps deter rivals. But it also involves clear fitness costs because, as behavioral and genetic data have demonstrated, only one male succeeds in siring offspring within the group. Not surprisingly, kinship influences the duration and stability of male coalitions. Coalitions that are made up of related males last nearly four times as long as coalitions composed of unrelated males. Coalitions composed of related males are also less likely to experience rank reversals. In this case, the costs of cooperation may be balanced by gains in inclusive fitness.

RECIPROCAL ALTRUISM

Altruism can also evolve if altruistic acts are reciprocated.

The theory of **reciprocal altruism** relies on the basic idea that altruism among individuals can evolve if altruistic behavior is balanced between partners (pairs of interacting individuals) over time. In reciprocal relationships, individuals take turns being actor and recipient—giving and receiving the benefits of altruism (Figure 8.17). Reciprocal altruism is favored because over time the participants in reciprocal acts obtain benefits that outweigh the costs of their actions. This theory was first formulated by Robert Trivers of Rutgers University, and later amplified and formalized by others.

Three conditions occurring together favor the development of reciprocal altruism: individuals must (1) have an opportunity to interact often, (2) have the ability to keep track of support given and received, and (3) provide support only to those who help them. The first condition is necessary so that individuals will have the opportunity for their own altruism to be reciprocated. The second condition is important so that individuals can monitor and balance altruism given to and received from particular partners. The third condition produces the nonrandom interaction necessary for the evolution of altruism. If individuals are unrelated, initial interactions will be randomly distributed to altruists and nonaltruists. However, reciprocators will quickly stop helping those who do not help in return, while continuing to help those who do. Thus, as

FIGURE 8.17 Two old, male chimpanzees groom each other. Reciprocity can involve taking turns or interacting simultaneously. Male chimpanzees remain in their natal communities throughout their lives and develop close bonds with one another.

in the case of kin selection, reciprocal altruism can be favored by natural selection because altruists receive a disproportionate share of the benefits of altruistic acts. Note that altruistic acts need not be exchanged in kind; it is possible for one form of altruism (such as grooming) to be exchanged for another form of altruism (such as coalitionary support).

In primates, the conditions for the evolution of reciprocal altruism probably are satisfied often, and there is some evidence that it occurs.

Most primates live in social groups that are fairly stable, and they can recognize all of the members of their groups. We do not know whether primates have the cognitive capacity to keep track of support given and received from various partners, but we do know that they are very intelligent and can solve complex problems. Thus, primates provide a good place to look for examples of contingent forms of reciprocity.

Reciprocal grooming and coalitionary support have been observed in several species of macaques, baboons, vervet monkeys, and chimpanzees. In some cases, grooming and support are exchanged in kind; in other cases, monkeys seem to exchange grooming for support. Some monkeys switch roles during grooming bouts so that the amount of grooming given and received during each grooming bout is balanced; others balance grooming across bouts. Among male chimpanzees, social bonds seem to be based on reciprocal exchanges in many different currencies. For example, John Mitani and David Watts have found that male chimpanzees at Ngogo, a site in the Kibale Forest of Uganda, share meat selectively with males who share meat with them and with males who regularly support them in agonistic interactions. Moreover, males who hunt together also tend to groom one another selectively, support one another, and participate in border patrols together. Interestingly, close associates are not maternal kin, suggesting that males' relationships are based on reciprocity, not kinship.

These correlational findings are consistent with predictions derived from the theory of reciprocal altruism, but they do not demonstrate that altruism is contingent on reciprocation.

A few studies suggest that primates keep track of these contingencies, at least over short periods.

Robert Seyfarth and Dorothy Cheney conducted the first study to examine the contingent nature of altruistic exchanges. Like most other monkeys, vervets spend much of their free time grooming. Vervets also form coalitions and use specific vocalizations to recruit support. In this experiment, Seyfarth and Cheney played tape-recorded recruitment calls to individuals in two different situations. Vervet A's recruitment call was played to vervet B from a hidden speaker (1) after A had groomed B and (2) after a fixed period of time in which A and B had not groomed. It was hypothesized that if grooming were associated with support in the future, then B should respond most strongly to A's recruitment call after being groomed. And that's just what the vervets did (Figure 8.18).

Frans de Waal of Emory University designed another study, which focused on the relationship between grooming and food sharing in a well-established group of captive chimpanzees. At various times over a three-year period, a group of chimpanzees

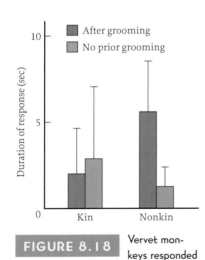

FIGURE 8.18 Vervet monkeys responded more strongly to recruitment calls played from a hidden speaker if the caller had previously groomed them than if the caller had not.

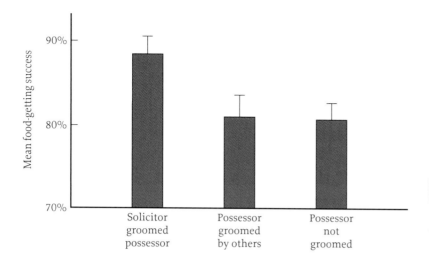

FIGURE 8.19 Chimpanzees were more successful at obtaining food from animals whom they had previously groomed (left-hand bar) than from animals who had been groomed by others or had not been groomed at all.

was provided with compact bundles of leaves. Although these bundles could be monopolized by a single individual, possessors of these bundles often allowed other individuals to share some of their booty. Before each provisioning session, de Waal monitored grooming interactions among group members. He found that the possessors were more generous to individuals who had recently groomed them than they were toward other group members (Figure 8.19). Moreover, possessors were less likely to actively resist efforts to obtain food if their solicitors had recently groomed them than if they had not groomed them.

The number of well-documented examples of reciprocal altruism in nonhuman primates is still small. It is possible that reciprocity is actually uncommon in nature. However, reciprocal altruism may occur more often than it is detected by observers. Because altruism is potentially reciprocated in different currencies (grooming for support, predator defense for food, and so on), and because the actual costs and benefits associated with specific behaviors are almost impossible to quantify, it is very difficult to establish whether altruism is actually reciprocated.

FURTHER READING

Chapais, B. 1995. Alliances as a means of competition in primates: Evolutionary, developmental, and cognitive aspects. *Yearbook of Physical Anthropology* 38: 115–136.

de Waal, F. B. M. 1997. The chimpanzees's service economy: Food for grooming. *Evolution and Human Behavior* 18: 375–386.

Dugatkin, L. A. 1997. *Cooperation among Animals.* Oxford University Press, Oxford, England.

Harcourt, A. H., and F. B. M. de Waal, eds. 1992. *Coalitions and Alliances in Humans and Other Animals.* Oxford University Press, Oxford, England.

Krebs, J. R., and N. B. Davies. 1993. *An Introduction to Behavioural Ecology.* Sinauer, Sunderland, MA, chap. 11.

Seyfarth, R. M., and D. L. Cheney. 1984. Grooming, alliances, and reciprocal altruism in vervet monkeys. *Nature* 308: 541–543.

Seyfarth, R. M., and D. L. Cheney. 1988. Empirical tests of reciprocity theory: Problems in assessment. *Ethology and Sociobiology* 9: 181–188.

Silk, J. B. 2002. Kin selection in primate groups. *International Journal of Primatology* 23: 849–875.

STUDY QUESTIONS

1. Consider the kinship diagram shown here. What is the kinship relationship (for example, mother, aunt, or cousin) and degree of relatedness (such as 0.5 or 0.25) for each pair of individuals?

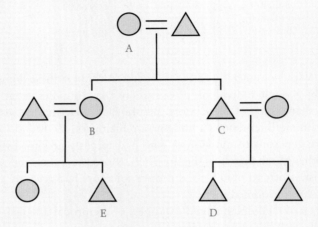

2. In biological terms, what is the difference between the following situations: (a) A male monkey sitting high in a tree gives alarm calls when he sees a lion at a distance. (b) A female monkey abandons a desirable food patch when she is approached by another female.

3. In documentaries about animal behavior, animals are often said to do things "for the good of the species." For example, when low-ranking animals do not reproduce, they are said to be forgoing reproduction in order to prevent the population from becoming too numerous and exhausting its resource base. What is wrong with this line of reasoning?

4. Why is some sort of nonrandom interaction among altruists necessary for altruism to be maintained?

5. Suppose that primates are not able to recognize paternal kin, as many primatologists have assumed. What would that tell you about how natural selection produces adaptations?

6. Data from a number of studies indicate that primates are more likely to behave altruistically toward kin than toward nonkin. However, many of the same studies show that rates of aggression toward kin and nonkin are basically the same. How does this fit with what you have learned about kin selection? Why are monkeys as likely to fight with kin as with nonkin?

7. There are relatively few good examples of reciprocal altruism in nature. Why is reciprocal altruism uncommon? Why might we expect reciprocal altruism to be more common among primates than among other kinds of animals?

8. Some of the evidence for reciprocal altruism comes from correlational studies showing that primates are most likely to groom, support, or share food with the animals who are most likely to groom, support, or share food with them. What do correlational studies like these tell you about the causal force involved in the distribution of these behaviors?

9. Seyfarth and Cheney found that vervet monkeys tended to respond more strongly to the calls of the animals who had groomed them earlier in the day than to the calls of animals who had not groomed them. However, this effect held only for unrelated animals, not for kin. The vervets responded as strongly to the calls of grooming relatives as to those of nongrooming relatives. How might we explain kinship's influence on these results?

10. In addition to kin selection and reciprocal altruism, a third mechanism leading to nonrandom interaction of altruists has been suggested. Suppose altruists had an easily detected phenotypic trait, perhaps a green beard. Then they could use the following rule: "Do altruistic acts only for individuals who have green beards." Once the allele became common, most individuals carrying green beards would not be related to one another, so this would not be a form of kin selection. However, there is a subtle flaw in this reasoning. Assuming that the genes controlling beard color are at different genetic loci than the genes controlling altruistic behavior, explain why green beards would not evolve.

KEY TERMS

dyadic	coefficient of relatedness/*r*
pairwise	phenotypic matching
actor	reproductive skew
recipient	food sharing
selfish	grooming
altruistic	affiliative
mutualistic	coalitions
spite (spiteful, adj.)	alliances
kin selection	matrilineage
Hamilton's rule	reciprocal altruism

PRIMATE LIFE HISTORIES AND THE EVOLUTION OF INTELLIGENCE

BIG BRAINS AND LONG LIVES

Large brains and long life spans are two of the features that define the primate order (Figure 9.1). Compared to many other animals, primates rely more heavily on learning to acquire the skills and knowledge that they need to survive and reproduce successfully. The complex behavioral strategies that we have explored in the last few chapters depend on primates' ability to respond flexibly in novel situations. Primates also have long periods of development and long life spans. Cognitive complexity and longevity have become even more exaggerated among modern humans, who live longer than any other primates and have relatively larger brains than any other creatures on the planet. It is not a coincidence that primates have both big brains and long lives; these traits are correlated across mammalian species. As we complete our discussion of the behavior and ecology of contemporary primates and turn our attention to the history of the human lineage, it is important to understand the evolutionary forces that have shaped the evolution of large brains and long life spans in the primate order.

Selection for larger brains generates selection for long lives.

Correlations tell us that two traits are related, but not why this relationship exists. In this case we have some reason to think that the causal arrow goes from brain size to life span, not vice versa. We come to this conclusion because brains are expensive organs to maintain. Our brains account for just 2% of our total body weight, but they consume about 20% of our metabolic energy.

Natural selection does not maintain costly features like the brain unless they confer important adaptive advantages. Moreover, the extent of investment that organisms make in a particular feature will be linked to the benefit that is derived from the investment. This is the same reason why you are usually willing to spend more for something that you will use for a long time than for something you will use only once. Animals that live for a long time will derive a greater benefit from the energy they expend on building and maintaining their brains than will animals that live for only a short time.

LIFE HISTORY THEORY

Life history theory focuses on the evolutionary forces that shape trade-offs between the quantity and quality of offspring and between current and future reproduction.

FIGURE 9.1 Primates are intelligent and long-lived. The first ape in space was a four-year-old chimpanzee named Ham, who was trained to perform a variety of tasks as he hurtled into space. In May 1961, three months after Ham's flight, Alan Shepard followed the chimp into space. Ham was one of several dozen chimpanzees that NASA used to test the safety of space travel for humans. Although Ham died at the age of 27, a number of his fellow "astrochimps" are still alive in their 40s. The survivors and some of their descendants are now living in sanctuaries in New Mexico and southern Florida.

Birth and death mark the beginning and end of every individual's life cycle. Between these two end points, individuals grow, reach sexual maturity, and begin to reproduce. Natural selection has generated considerable variation around this basic scheme. For example, Pacific salmon are hatched in freshwater but spend their adult lives in the open ocean. After years in the sea, they return to the streams where they were hatched to lay or fertilize their own eggs; they die soon after they complete this journey. Opossums, the only North American marsupial, produce their first litter at the age of one year. Females have one to two litters per year, and live less than three years (Figure 9.2). Lion females produce litters of up to six cubs at two-year intervals. Elephants conceive for the first time at 10 years of age, have 22-month pregnancies, and give birth to single infants at four- to nine-year intervals.

If natural selection favors increased reproductive success, why doesn't it extend the opossum's life span, reduce the lion's interbirth interval, or increase the elephant's litter size? The answer is that all organisms face fundamental trade-offs that constrain their reproductive options. As we explained in Chapter 7, investment in one infant limits investment in other offspring, so parents must make trade-offs between the quality and quantity of offspring they produce. Organisms also face trade-offs between current and future reproduction. All other things being equal, fast maturation and early reproduction are advantageous because they increase the length of the reproductive life span and reduce generation time. However, energy devoted to current reproduction diverts energy from growth and maintenance. If growth enhances reproductive success, then it may be advantageous to grow large before beginning to reproduce. Thus, many kinds of organisms have a juvenile phase in which they do not reproduce at all. They do not become sexually mature until they reach a size at which the payoffs of allocating energy to current reproduction exceed the payoffs of continued growth. The same kind of argument applies to maintenance. Energy that is diverted from current reproduction to maintenance enables individuals to survive and reproduce successfully in the future.

FIGURE 9.2 The Virginia opossum, *Didelphis virginiana*, is the only marsupial mammal in North America. Females produce many tiny fetuses, which make their way into the mother's pouch, attach themselves to a nipple, and nurse for two to three months. Then they emerge from the pouch and cling to their mother's back.

Aging and death result from trade-offs between reproduction at different ages and survivorship.

Like humans, other primates age; as they get older, their physical abilities deteriorate. They don't run as fast, jump as high, or react as quickly (Figure 9.3). Their teeth wear down, making it harder for them to chew their food; and their joints deteriorate. Although humans are the only primates to experience menopause, the fertility of female primates of all species declines when they reach old age. Males seem to reach peak physical condition in early adulthood and then decline.

At first glance, aging and death seem to be the inevitable effects of wear and tear on bodies. Organisms are complicated machines, like cars or computers. A machine has many components that must function together in order for it to work. It seems logical that the components in animals' bodies simply wear out and break down, like a worn clutch or faulty hard disk.

But this explanation of aging is flawed because the analogy between organisms and machines is not really apt. Every cell in an organism contains all of the genetic information necessary to build a complete, new body, and this genetic information can be used to repair damage. Wounds heal and bones mend, and some organisms, such as frogs, can regenerate entire limbs. Some organisms that reproduce asexually by budding or fission do not experience senescence at all.

If senescence is not inevitable, why doesn't natural selection do away with it? The answer has to do with the relative magnitude of the benefits that animals can derive from current reproduction or from living longer. Organisms could last longer if they were built better. A Subaru costs much less to build than a Lexus, but it is also of lower quality. As a consequence, a Lexus is not expected to break down as often as a Subaru is. The same trade-off applies to organisms. Our teeth would last longer if they were protected by a thicker covering of enamel, but building stronger teeth would require more nutrients, particularly calcium. Building higher-quality organisms consumes time and resources, thus reducing the organism's growth rate and early fertility.

The trade-off between survival and reproduction is greatly biased against characteristics that prolong life at the expense of early survival or reproduction.

Senescence is at least partly the consequence of the existence of genes with pleiotropic effects that are positive at early ages and negative at later ages. Aging is favored by selection because traits that affect fertility at young ages are favored at the expense of traits that increase longevity.

The key to understanding this idea is to realize that selective pressures are much weaker on traits that affect only the old. To see why this is the case, think about the fate of two mutant alleles. One allele kills individuals before they reach adulthood, and the other kills individuals late in their lives. Carriers of the allele that kills infants and juveniles have a fitness of zero because none survive long enough to transmit the gene to their descendants. Therefore, there will be strong selection against alleles with deleterious effects on the young. In contrast, selection will have much less impact on a mutant allele that kills animals late in their lives. Carriers of a gene that kills animals late in their life span will have already produced many offspring before the effects of the gene are felt. Thus a mutation that affects the old will have limited effects on reproductive performance, and there will be little or no selection against it. This means that genes with pleiotropic effects that enhance early fertility but reduce fitness at later ages may be favored by natural selection because they increase individual fitness.

The trade-off between current and future reproduction and between quantity and quality of offspring generates constellations of interrelated traits.

Animals that begin to reproduce early also tend to have small body sizes, small brains, short gestation times, large litters, high rates of mortality, and short life spans. Animals that begin to reproduce at later ages tend to have larger body sizes, larger brains, longer gestation times, smaller litters, lower rates of mortality, and longer life spans (Figure 9.4). Life history traits are clustered together in this way because of the inherent trade-offs between current and future reproduction and between the quantity and quality of offspring. Animals that begin to reproduce early divert energy from growth and remain small. Animals that have small litters are able to invest more in maintenance and extend their life spans. These clusters of correlated traits create a continuum of life history strategies that runs from fast to slow, or from short to long. Opossums fall somewhere along the fast/short end of the continuum; elephants lie at the slow/long end.

The benefits derived from current and future reproduction depend on a variety of ecological factors that influence the prospects for survival.

It makes little sense to divert energy to future reproduction if the prospects for surviving into the future are slim. For example, selection is likely to favor fast/short life histories in species that experience intense predation pressure. If predators are abundant and the prospects of surviving from one day to another are low, it makes little sense to postpone reproducing. In this situation, individuals that mature quickly and begin reproducing at early ages are likely to produce more surviving offspring than those that mature more slowly, so selection will favor faster/shorter life history strategies. Other kinds of ecological factors may favor slower/longer life histories. Suppose that there is severe competition for access to the resources that animals need to

FIGURE 9.3 Virtually all organisms experience senescence (aging). When this photograph was taken, this old male chimpanzee, named Hugo, was missing a lot of hair on his shoulders and back, he had lost a considerable amount of weight, and his teeth had worn to the gums. Hugo died a few weeks later.

FIGURE 9.4 Elephants are an example of a species at the slow/long end of the life history continuum. They are very large (approximately 6000 kg, or 13,200 lb) and can live up to 60 years in the wild. Females mature at about 10 years of age, have a 22-month gestation period, and give birth to single offspring at four-year intervals.

reproduce successfully and that larger animals are more successful in competitive encounters than small animals are. In this situation, small animals will be at a competitive disadvantage, and it may be profitable to invest more energy in growth, even if such an investment delays maturation. The life history strategies that characterize organisms reflect the net effects of these kinds of ecological pressures.

Natural selection not only shapes the relationship between life history traits, but it can also shift the values of these traits in response to changes in environmental conditions.

Natural selection adjusts life history traits in response to changes in prevailing conditions. Because life history traits are tightly clustered together, selection pressures acting on one trait often influence the value of other traits as well. For example, University of Idaho biologist Steven Austad compared the effect of predation on the life histories of two populations of opossums. One population lived on the mainland and was vulnerable to a variety of predators. Another population lived on an island that had very few predators for several thousand years. Opossums on the island aged more slowly, lived longer, and had smaller litters than opossums on the mainland. In this case, reduction of predation pressure favored a decelerated life history.

In some cases, organisms can adjust their life histories in relation to current ecological conditions. Recall from Chapter 7 that female monkeys mature more quickly and reproduce at shorter intervals when food is abundant. This is not simply an inevitable response to the availability of food, it is an evolved capacity to adjust development in response to local conditions.

THE EVOLUTION OF PRIMATE LIFE HISTORIES

Primates fall somewhere along the slow/long end of the life history continuum.

As a group, primates tend to delay reproduction and grow to relatively large sizes, and they also have relatively long gestation times, small litters, low rates of mortality, long life spans, and large brains in relation to their body sizes. However, there is also variation within the primate order: some primates have slower or faster life histories than others. Within the primate order, there have been several changes in the package of life history traits. The first change in this package of traits is linked to the evolution of prosimians, the second to anthropoid primates, the third to the great apes, and the fourth to humans. We will discuss the first three transitions here and come back to the human shift in Chapter 12.

The first adaptive change in the package of life history traits within the primate order involved the origins of the prosimians and their shift to an arboreal niche.

The first animals that we recognize as primates appeared about 60 million years ago. (A more detailed description of the fossil evidence that underlies our knowledge of the history of the primate order will be given in Chapter 10.) It is not entirely clear what ecological pressures favored the development of the suite of morphological traits that we see in prosimians and other primates, but there is a growing consensus that grasping hands and feet evolved before changes in the visual system. Robert Martin of the University of Chicago suggests that small-bodied prosimians, with grasping hands and feet rather than claws (Figure 9.5), were especially well adapted for clambering around in the fine, terminal branches of leafy trees and thin saplings. There they gained at least one critical advantage: they were safer from predators.

Austad and his colleagues believe that the shift to an arboreal niche led to the evolution of larger brains and longer lives among early primates. The reduction of external sources of mortality made it more profitable for prosimians to shift the balance between current and future reproduction. With lower mortality and longer life expectancy, it became profitable to invest more energy in growing larger, developing more slowly, and building bigger brains. Birds and bats, which are largely immune to terrestial predators, are similar to primates in having relatively slow life histories.

FIGURE 9.5 The ancestors of modern primates may have been adapted for maneuvering in the thin terminal branches of trees. There they were gained some measure of safety from predators. This slender loris grasps a tiny branch with his feet.

The second major adaptive shift is associated with the appearance of the anthropoid primates: the monkeys and apes.

The ancestors of contemporary monkeys and apes appeared about 35 million years ago. (We will describe the fossil evidence for these events as well in Chapter 10.) There are important differences in the brains and life histories of prosimians and anthropoid primates. First, anthropoid brains are relatively larger than prosimian brains. Second, anthropoid brains have been reconfigured. Nocturnal prosimians rely heavily on smell and sound to find food and avoid predators. Diurnal monkeys and apes rely much more heavily on vision than on olfaction and hearing, and their brains reflect this shift in processing sensory information. In anthropoid primates there has also been a substantial increase in the relative size and complexity of the forebrain, including the **neocortex**. The neocortex seems to be the part of the brain most closely associated with problem solving, learning, and planning.

Selective Pressures Favoring Large Brains

The selective pressures that favored larger brains among anthropoid primates are not fully understood.

Some researchers hypothesize that the enlargement and reorganization of the brain in anthropoids was linked to their feeding ecology. Monkeys feed mainly on plants and include many different plant species in their diets. They must evaluate the ripeness, nutritional content, and toxicity of their food items. Katherine Milton of the University of California Berkeley has argued that this task may be particularly taxing for frugivores, whose foods are patchily distributed in time and space. Successful exploitation

FIGURE 9.6 Primates sometimes exploit foods that are difficult to extract. Here, (a) a male chimpanzee pokes a long twig into a hole in a termite mound and extracts termites, and (b) a capuchin monkey punctures an eggshell and extracts the contents. (b, Photograph courtesy of Susan Perry.)

(a)

(b)

of this frugivorous feeding niche may favor better memory, greater capacities for learning, and the ability to plan a course of action.

Katherine Gibson of the University of Texas and Sue Parker of Sonoma State University offer a second ecological hypothesis. They believe that natural selection has favored enhanced cognitive capacities among primates because primates rely heavily on **extracted foods** that require complex processing techniques. For example, some primates eat hard-shelled nuts that must be cracked open with stones or smashed against a tree trunk. Others dig up roots and tubers; extract insect larvae from bark, wood pith, or dung; and crack open pods to obtain seeds (Figure 9.6). Extracted foods are valuable elements in primate diets because they tend to be rich sources of protein and energy. However, they require complicated, carefully coordinated techniques to process. For example, to feed on the pith of wild celery plants, mountain gorillas must first break the stem into manageable pieces, then peel the outer layers of the stalk away with their teeth, and finally pick out edible bits of the pith with their fingers.

Although some researchers believe that ecological factors drive the evolution of larger brains in anthropoid primates, others argue that social challenges play a more important role in this process. As we explained in Chapter 6, sociality is a response to predation pressure in diurnal species. When anthropoids began sleeping through the night, they probably began to forage and travel together during the day. This made them safer from predators, but more vulnerable to exploitation by conspecifics. In social groups, animals compete for food, mates, grooming partners, and other valuable resources. They also form social bonds that influence participation in coalitions, exchange networks, grooming partnerships, and so on. The larger groups become, the more difficult it gets to sustain social bonds and keep track of relationships within the group. The ability to operate effectively in this complicated social world may reward greater flexibility in behavior and favor expansion of the parts of the brain that are linked to learning and planning. This idea is called the **social intelligence hypothesis**.

A third model, developed by Simon Reader of Utrecht University and Kevin Laland of St. Andrews University, proposes that natural selection has favored changes in the primate brain that facilitate behavioral flexibility. In particular, Reader and Laland emphasize the ability to invent appropriate solutions to novel problems (Figure 9.7) and the ability to learn new behaviors from conspecifics. The benefits derived from innovation and social learning generated selective pressures that favored expansion and development of the parts of the brain linked to learning, planning, and behavioral flexibility. Reader and Laland emphasize that the ability to innovate and learn from others might enhance animals' ability to cope with both ecological and social challenges.

The third adaptive shift is associated with the great apes who appeared during the Miocene epoch, about 20 million years ago.

Great apes have even larger brains and slower life histories than monkeys. Ecological explanations for this adaptive shift seem more plausible than the social intelligence hypothesis because great apes generally live in smaller groups than monkeys do. Although chimpanzees and bonobos may live in communities that include as many as 50 individuals, gorillas live in much smaller groups, and orangutans are largely solitary. Thus it is hard to sustain the argument that the reason why great apes have bigger brains than monkeys is that they live in more challenging social environments.

Great apes may have relied on their cognitive skills to adapt to a changing environment. As we will see in Chapter 10, there were once many species of apes, but only a few have survived to the present. The disappearance of so many ape species may be related to global climatic changes that reduced the extent of tropical rain forests. Richard Byrne of St. Andrews University suggests that these environmental changes imposed strong selective pressures for more efficient feeding. Great apes have a particular need for efficient feeding techniques because they are large-bodied, move relatively slowly and inefficiently, and have no specialized digestive anatomy or cheek pouches to store food. Byrne hypothesizes that ecological pressures favored the evolution of certain cognitive abilities in great apes that enabled them to master technically demanding foraging techniques. In these circumstances there might have been strong selection for the ability to invent new solutions to ecological problems, to take advantage of the skills learned or invented by other group members, and to use tools to enhance foraging success.

FIGURE 9.7 Wolfgang Kohler was one of the first scientists to study systematically cognitive abilities in captive chimpanzees. He hung a bunch of bananas out of the chimpanzees' reach and put several wooden crates in the room. Eventually, one individual, named Sultan, managed to stack the crates, clamber onto the precarious tower, and grab the bananas.

Great apes rely on complex extractive foraging techniques, sometimes using tools, to a greater extent than most other primates do.

Byrne points out that great apes make use of more complicated foraging techniques than other primates do, enabling them to feed on some foods that other primates cannot process. For example, virtually all plant foods that mountain gorillas rely on are well defended by spines, hard shells, hooks, and stingers. Each of their food items requires a particular routine—a complicated sequence of steps, structured in a particular way. Many of the foods that orangutans feed upon are also difficult to process.

Great apes sometimes use tools to obtain access to certain foods that are not otherwise available to them. Chimpanzees poke twigs into holes of termite mounds and anthills, use leaves as sponges to mop up water from deep holes, and employ stones as hammers to break open hard-shelled nuts (Figure 9.8). Recently, Carel van Schaik and his colleagues observed Sumatran orangutans using sticks to probe for insects and to pry seeds out of the husks of fruit. Although bonobos and gorillas have not been observed using tools in the wild, all great apes are adept tool users in captive settings.

(a)

(b)

FIGURE 9.8 Chimpanzees use tools to help them obtain certain kinds of foods. (a) This chimpanzee at Gombe Stream National Park in Tanzania is using a long stick to dip for ants. (b) Here a chimpanzee uses a large stone to hammer open hard-shelled nuts.

Comparative analyses provide one way to examine the role of ecological and social factors in the evolution of the primate brain.

These models of the evolution of cognitive complexity in primates generate specific predictions about the pattern of variation in the brains and cognitive abilities of living primates. For example, the social intelligence hypothesis predicts a link between social complexity and cognitive complexity, and the extractive foraging hypothesis predicts that this feature of foraging behavior will be linked to cognitive ability.

In order to test these hypotheses, we need a reliable measure of cognitive ability. Although we are interested in what the brain can do, it is very difficult to assess cognitive ability in other species. Instead, most work in this area has relied on measurements of the size or organization of particular parts of the brain. Most researchers focus on the development of the forebrain, particularly the neocortex, because this is the site of the most substantial evolutionary changes in size and complexity (Figure 9.9). Moreover, as we noted already, the neocortex seems to be the part of the brain most closely associated with problem solving and behavioral flexibility. Neocortex size alone is not a very useful measure because larger animals generally have larger brains (and larger neocortexes) than smaller animals. Thus, some researchers have focused their analyses on the **neocortex ratio**, the ratio between the volume of the neocortex and the volume of the rest of the brain.

The validity of these kinds of measures has recently been confirmed by Reader and Laland, whose work on behavioral innovation was mentioned earlier. Their analyses were based on the size of the **executive brain** in relation to the **brainstem**. The executive brain is composed of the neocortex and the **striatum**, a structure in the basal ganglia that is functionally linked to the neocortex and contributes to executive function. The brainstem, which is the major route for communication between the forebrain, the spinal cord, and the peripheral nerves, has changed very little over the course of primate evolution. Reader and Laland surveyed the primate literature for information about three ecologically meaningful measures of cognitive ability: (1) reports of

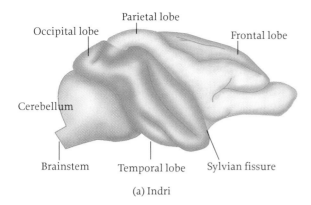

(a) Indri

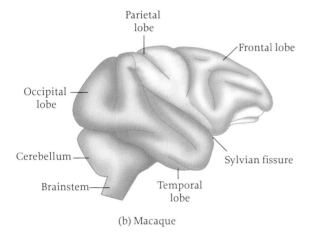

(b) Macaque

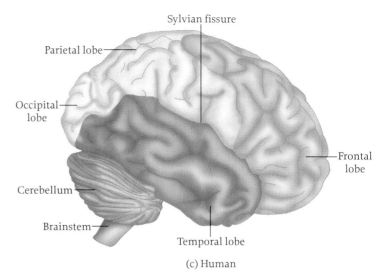

(c) Human

FIGURE 9.9 The brains of a large prosimian, the indri (a); a macaque (b); and a human (c). There are three main components of the brain: the hindbrain, which contains the cerebellum and medulla (part of the brainstem); the midbrain, which contains the optic lobe (not visible here); and the forebrain, which contains the cerebrum. The cerebrum is divided into four main lobes: occipital, parietal, frontal, and temporal. The forebrain is greatly expanded in primates and other mammals, and much of the gray matter (which is made up of cell bodies and synapses) is located on the outside of the cerebrum in a layer called the "cerebral cortex." The neocortex is a component of the cerebral cortex, and in mammals the neocortex covers the surface of virtually the entire forebrain.

behavioral innovation, which they defined as novel solutions to ecological or social problems; (2) examples of social learning, the acquisition of skills and information from others; and (3) forms of tool use. Their comparative analyses demonstrated that these measures of behavioral flexibility are closely linked to the executive brain ratio (Figure 9.10). Primates with relatively large executive brains are more likely to innovate, learn from others, and use tools than primates with relatively small executive brains are.

Ecological hypotheses for the evolution of intelligence predict that specific characteristics of the diet or the environment of particular primate species will be correlated with their cognitive abilities.

If Milton is correct, then frugivores must have greater cognitive skills than folivores because frugivores utilize a dispersed, patchy, and ephemeral (short-lived) food supply. Their reliance on fruit means that they must maintain larger home ranges and travel farther each day in search of food than folivores do. Thus, Milton's hypothesis predicts that frugivores will have larger neocortex ratios than folivores. Extractive foraging models predict that species that rely more heavily on foods that are difficult to process or extract will have larger neocortex ratios than will primates that feed on more accessible foods.

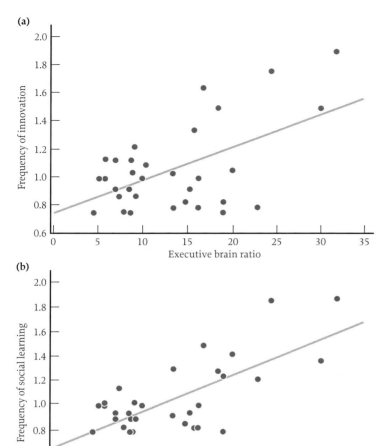

Robin Dunbar of Liverpool University suggests that group size may be taken as a rough index of social complexity because primates in social groups need to recognize, associate, and interact with all the other members of their groups. Animals must be able to keep track of their relationships with other members of their groups, particularly when they participate in social interactions that are regulated by nepotism or by reciprocity. Similarly, the decision to act aggressively or submissively, or to intervene in conflicts involving others, may depend in part on an individual's ability to remember or assess the dominance ranks of other group members. As groups grow larger, the number of pairs grows rapidly, making it considerably more difficult to keep track of relationships with all group members. Thus, Dunbar predicts there should be a positive correlation between the size of social groups and the neocortex ratio.

Comparative analyses provide support for both ecological hypotheses and the social intelligence model.

Dunbar was the one of the first investigators to assess the relationships between the neocortex ratio, diet, and group size. His analyses indicated that group size has the most consistent impact on the neocortex ratio, and he convinced many primatologists that the social intelligence hypothesis provides the most compelling explanation of the evolution of cognitive complexity in primates.

FIGURE 9.10   (a) The frequency of innovation in nonhuman primates is positively related to the executive brain ratio, which is a measure of the size of the executive brain in relation to the brainstem. (b) The frequency of social learning is also positively related to the executive brain ratio.

One problem with these analyses is that they rely on the idea that life in large groups is more cognitively challenging than life in small groups is, and they do not provide a direct measure of social knowledge or skill. Of course, social skill is not an easy parameter to measure directly. Andrew Whiten and Richard Byrne have argued that **tactical deception**, the ability to manipulate others without the use of force, is a valuable measure of social cognition. To deceive conspecifics effectively, animals rely on detailed information about the attributes of other group members and their history of interactions. In the 1980s, Byrne and Whiten compiled a list of anecdotal reports of deception in primate groups. As you might expect, great apes figure prominently in this compendium. Recently, Byrne and Nadia Corp examined the relationship between various measures of brain size and deception across the primate order. The occurrence of deception is positively related to the neocortex ratio (Figure 9.11), but not to group size.

However, there is also evidence that ecological factors are linked to cognitive abilities. Home range size, the extent of fruit in the diet, and group size are correlated with the size of the neocortex ratio. Primates who rely more heavily on fruit, have large home ranges, and live in large groups have relatively large brains in relation to body size. In addition, primates seem to be more flexible in their foraging behavior than in their social behavior. In the long list of examples of innovations and socially learned behaviors that Reader and Laland compiled, foraging innovations predominate. In these comparative analyses, there is no consistent relationship between social learning and group size.

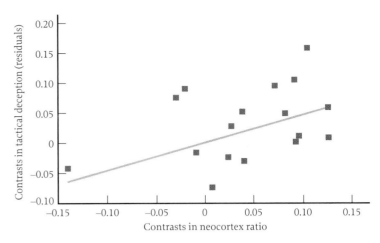

FIGURE 9.11 In primates there is a positive relationship between the size of the neocortex ratio and the incidence of tactical deception. Both measures are corrected here for phylogenetic biases.

WHAT DO MONKEYS KNOW?

Although the selective forces that favored the evolution of large brains and slow life histories among primates are not fully understood, it is clear that primates know a lot about their environments and about other members of their groups. In the remainder of this chapter we will consider what they know and how they make use of this knowledge in their daily lives. Finding out what other primates know presents major methodological obstacles, but primatologists have devised clever means of probing the minds of monkeys and apes.

Ecological Knowledge

Monkeys and apes know a lot about their environments.

Monkeys are excellent naturalists. They select appropriate food items, avoid eating foods laced with toxins, know the location of food resources within their home ranges, move efficiently from one feeding site to the next, and evade predators. Accomplishing all this can be a formidable task. Macaques living in the forests of East Kalimantan, Indonesia, where there are more than 700 species of fruiting plants, must distinguish which fruits are edible and which are not, track the status of fruit crops as they ripen, monitor the availability of fruits at different sites, make decisions about when to remain in a food patch and when to switch to a new one, and so on.

Monkeys and apes construct mental maps of their home ranges. These maps enable them to move efficiently from one food source to another.

The mental representation of the location, availability, and quality of things in the environment is called a **cognitive map**. Cognitive maps allow animals to work out efficient routes from one site to another, saving them both time and energy. To see why cognitive maps are so useful, consider the way you navigate your own environment. You begin at home, perhaps rushing to your first class of the day. Then you might head to the library to study, or to the cafeteria for lunch. Later, you might have another class

FIGURE 9.12 Tamarins feed on gum, which they harvest from many different trees in their home ranges. Most of the time, tamarins move directly from one feeding tree to the nearest tree of the same species, taking care not to return to trees that they have already visited.

at the other end of campus. At the end of the day, you might stop at the gym to work out. When you were new to campus, you probably took the same routes every day and were afraid to stray from familiar paths for fear of getting hopelessly lost in the crowded maze of buildings. As time passed, however, you worked out new routes that shortened your walk from one place to another, and you felt more confident in your ability to get around campus. You could do this because of an ability to visualize where things were and an ability to plan efficient routes. In other words, you developed a cognitive map of your home range. Many other primates apparently have the same kinds of abilities.

Paul Garber of the University of Illinois at Urbana-Champaign has studied the foraging behavior of tamarins in the forests of Peru. Some of the plant species that tamarins feed upon are highly synchronous. This means that if one individual of a particular species is fruiting or producing gum, others of the same species are doing the same thing. Thus, when tamarins have exhausted the resources of one feeding tree, they can expect to find more food at other trees of the same species. Garber found that 70% of the time, tamarin groups moved directly from one feeding tree to the nearest tree of the same species that they had not yet depleted (Figure 9.12). These tiny monkeys apparently know the location of hundreds of food trees and can remember for days or weeks when they last fed at particular sites.

Charles Janson of Stony Brook University has conducted a series of field experiments to examine capuchin monkeys' knowledge of the location and quality of their food resources. Janson constructed special feeding platforms in the forest and baited the platforms with tangerines. He conducted the experiments during the winter months, when virtually no other fruit was available to the monkeys. Janson varied the amount of food available on each platform, so some sites always contained more fruit than others, and no site was baited more than once per day. By following the monkeys' movements through the forest, he was able to examine their knowledge of the locations and quality of these artificial food sites. Janson found that monkeys moved to closer platforms more often than they moved to more distant platforms. The capuchins preferred sites where they might expect to find more fruit over sites where they might expect to find less fruit, and they avoided visiting sites that had been depleted within the previous 24 hours. These data suggest that monkeys were able to remember the location of their fruit patches, to assess the amount of fruit that would be available at each site, and to evaluate the likelihood that the crop had been renewed since their last visit.

Taken together, these data suggest that all monkeys create good cognitive maps of their home ranges and have detailed knowledge of their foods' characteristics. Advocates of the social intelligence hypothesis downplay this evidence because these skills do not necessarily distinguish primates from other kinds of animals. They point out that rodents, birds, fish, and perhaps even insects have similar skills. However, those who emphasize the complexity of ecological knowledge also contend that we don't yet understand what primates know about their environment, making it premature to conclude that other animals know as much about the nonsocial world as primates do.

Social Knowledge

One of the most striking things about primates is the interest they take in one another. Newborns are greeted and inspected with interest (Figure 9.13). Adult females are sniffed and visually inspected regularly during their estrous cycles. When a fight breaks out, other members of the group watch attentively. As we have seen in previous chapters, monkeys know a considerable amount about their own relationships to other group members. A growing body of evidence suggests that monkeys also have some knowledge of the nature of relationships among other individuals, or **third-party relationships**. Monkeys' knowledge of social relationships may enable them to form effective coalitions, to to compete effectively, and to manipulate other group members to their own advantage.

There is evidence that monkeys and apes know something about the nature of kinship relationships among other members of their groups.

FIGURE 9.13 In many primate species, all group members take an active interest in infants. Here a female baboon greets a newborn infant. Evidence from playback experiments, laboratory experiments, and naturalistic observations suggest that monkeys know something about the relationships among group members.

One of the first indications that monkeys understand the nature of other individuals' kinship relationships came from a playback experiment on vervet monkeys conducted by Dorothy Cheney and Robert Seyfarth in Kenya's Amboseli National Park. Several female vervets were played the tape-recorded scream of a juvenile vervet from a hidden speaker. When the call was played, the mother of the juvenile stared in the direction of the speaker longer than other females did. This response suggests that mothers recognized the call of their own offspring. Even before the mother reacted, however, other females in the vicinity looked directly at the juvenile's mother. This response suggests that other females understood which monkey the juvenile belonged to, and that they were aware that a special relationship existed between the mother and her offspring.

Additional evidence that monkeys may understand mother–offspring relationships comes from more controlled laboratory experiments conducted by Verena Dasser working with a captive group of long-tailed macaques in Zürich, Switzerland. In these experiments, a young female was the primary subject. In the first phase the female was shown one slide of a particular mother–offspring pair and a second slide of two monkeys who were not related as mother and infant. All of the monkeys in the slides were members of the subject's social group and were well known to her. These trials were repeated until the female was able to select the mother–offspring pair consistently. In the second phase of the experiment, the female was again presented with two slides of group members: one depicting a mother–offspring pair and one depicting another pair of animals who were not mother and infant. In this phase of the experiment, however, the subject was shown a different mother–offspring pair in each trial. If, *and only if*, she understood that there was a particular relationship between the mother and offspring that served as her model in training, she would select the new mother–offspring pair rather than the other pair of monkeys. The young female picked out the mother–offspring pair in nearly every trial. The monkey's performance in these experiments is particularly remarkable because the offspring in the mother–offspring pairs varied considerably in age, and in some cases the subjects had not seen the mother interacting with her child when it was an infant.

Monkeys may also have broader knowledge of kinship relationships (Figure 9.14). The evidence for this claim also comes from Cheney and Seyfarth's work on vervet monkeys. When monkeys are threatened or attacked, they often respond by

FIGURE 9.14 Living in complex social groups may have been the selective factor favoring large brains and intelligence among primates.

threatening or attacking a lower-ranking individual who was not involved in the original incident—a phenomenon we call **redirected aggression**. Vervets selectively redirect aggression toward the maternal kin of the original aggressor. So, if female A threatens female B, then B threatens AA, a close relative of A. If monkeys were simply blowing off steam or venting their aggression, they would choose a target at random. Thus the monkeys seem to know that certain individuals are somehow related.

Several lines of evidence suggest that monkeys understand the nature of rank relationships among other individuals.

Because kinship and dominance rank are major organizing principles in most primate groups, it makes sense to ask whether monkeys also understand third-party rank relationships. The most direct evidence that monkeys understand third-party rank relationships comes from two playback experiments conducted on a group of baboons in the Okavango Delta of Botswana who have been studied for the last 15 years by Robert Seyfarth, Dorothy Cheney, and their colleagues. In this group, dominance relationships were stable, and females never responded submissively toward lower-ranking females.

In one experiment that Seyfarth and Cheney designed, females listened to a recording of a female's grunt followed by another female's submissive fear barks. Female baboons responded more strongly when they heard a higher-ranking female responding submissively to a lower-ranking female's grunt than when they heard a lower-ranking female responding submissively to a higher-ranking female's grunt. Thus, females were more attentive when they heard a sequence of calls that did not correspond to their knowledge of dominance rank relationships among other females. Control experiments excluded the possibility that females were reacting simply to the fact that they had not heard a particular sequence of calls before. The pattern of responses suggests that females knew the relative ranks of other females in their group and were particularly interested in the anomalous sequence of calls.

In a second experiment, Thore Bergman and Jacinta Beehner collaborated with Seyfath and Cheney to probe the baboons' knowledge of the hierarchical nature of rank relationships in groups with matrilineal dominance ranks. Using the same basic experimental paradigm, researchers played sequences of vocalizations that simulated

(a)

(b)

rank reversals within lineages and rank reversals between lineages. As young female baboons mature, they often rise in rank above their older sisters and other female kin. Thus, changes in the relative rank of females in the same lineage are part of the normal course of rank acquisition. However, changes in the relative ranks of unrelated females are much less common. The baboons reacted much more strongly to simulated rank reversals between lineages than to simulated rank reversals within lineages. Again, the researchers were careful to control for confounding variables, such as rank distance and novelty. This result suggests that the females understood the relative ranks of other females and that they understood that changes in rank relationships within lineages are not the same as changes in rank relationships between lineages.

Participation in coalitions may draw on sophisticated cognitive abilities.

Even the simplest coalition is quite a complex interaction. When coalitions are formed, at least three individuals are involved, and several different kinds of interactions are going on simultaneously (Figure 9.15). Consider the case in which one monkey (the aggressor) attacks another monkey (the victim). The victim then solicits support from a third party (the ally), and the ally intervenes on behalf of the victim against the aggressor. The ally behaves altruistically toward the victim, giving support to the victim at some potential cost to itself. At the same time, however, the ally behaves aggressively toward the aggressor, imposing harm or energy costs on the aggressor. Thus the ally simultaneously has a positive effect on the victim and a negative effect on the aggressor. Under these circumstances, decisions about whether or not to intervene in a particular dispute may be quite complicated. Consider a female who witnesses a dispute between two of her offspring. Should she intervene? If so, which of her offspring should she support? When a male bonnet macaque is solicited by a higher-ranking male against a male who frequently supports him, how should he respond? In each case, the ally must balance the benefits to the victim, the costs to the opponent, and the costs to itself (Figure 9.16).

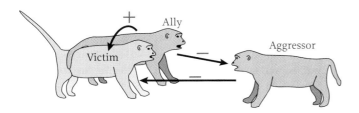

FIGURE 9.16 Complex fitness calculations may be involved in decisions about whether to join a coalition. By helping the victim against the aggressor, the ally increases the fitness of the victim and decreases the fitness of the aggressor.

Understanding of third-party relationships may be particularly useful in managing coalitions.

Given the great complexity of even the simplest coalitions, knowledge of third-party relationships may be extremely valuable because it enables individuals to predict how others will behave in certain situations. Thus, animals who understand the nature of third-party relationships may have a good idea about who will support them and who will intervene against them in confrontations with particular opponents, and they may also be able to tell which of their potential allies are likely to be most effective in coalitions against their opponents.

Monkeys seem to make use of information about third-party relationships when they recruit support in coalitions. Male bonnet macaques are most likely to intervene in ongoing conflicts if they outrank both of the participants in the original conflict. Males seem to make use of this information when they recruit support from potential allies; they selectively solicit males who outrank both themselves and their opponents. Their ability to do this consistently relies on their knowledge of the rank relationships among other males.

Susan Perry and her colleagues at the University of California Los Angeles have also studied monkeys' use of third-party information in coalitions. Capuchins follow three basic rules when they form coalitions: (1) support females against males, (2) support dominants against subordinates; and (3) support close associates against others. If capuchins "understand" these rules, particularly the last one, they are not expected to recruit support against dominant opponents or from males who have closer relationships with their opponents than with themselves. This is exactly what they do. Their ability to adhere to these rules suggests that they understand the nature of relationships among other group members.

Primates seem to form more complex coalitions than other animals do.

Many animals form alliances in defense of their territories and their young. However, Andrew Harcourt at the University of California Davis has concluded that primates use alliances in different ways from other animals. Primates seem to assess differences in competitive ability among other members of their group and cultivate relationships with powerful individuals. Primates may also attempt to manipulate the alliances among other group members. There are several accounts of chimpanzees attempting to prevent lower-ranking rivals from forming alliances with each other.

Primates may understand more about the behavior of other animals than about those other animals' feelings, thoughts, and intentions.

Monkeys and apes seem to be very good at predicting what other animals will do in particular situations. For example, we have seen that vervets groom monkeys who support them in coalitions, female langurs with newborn infants are fearful of new resident males, baboons express surprise when low-ranking animals elicit signs of submission from higher-ranking animals, and capuchins don't try to recruit support from monkeys who have closer bonds to their opponents than to themselves. These examples indicate that monkeys can predict what others will do and adjust their behavior accordingly.

The ability to predict what others will do might be based on a sophisticated ability to track contingencies between one event and another or on knowledge of other animals' mental states.

Monkeys' ability to predict what other individuals will do may not seem very remarkable. After all, we know that many animals are very good at learning to make associations between one event and another. In the laboratory, rats, pigeons, monkeys, and many other animals can learn to pull a lever, push a button, or peck a key to obtain food. These are examples of associative learning, the ability to track contingencies between one event and another. Monkeys' ability to predict what others will do in particular situations might be based on sophisticated associative learning capacities, prodigious memory of past events, and perhaps some understanding of conceptual categories like kinship and dominance. On the other hand, it is also possible that monkeys' ability to predict what others will do is based on their knowledge of the mental states of others—what psychologists call a **theory of mind**.

It may seem relatively unimportant whether monkeys and apes rely on associative learning to predict what others will do or whether they have a well-developed theory of mind. However, there may be some things that animals cannot do unless they understand what is going on in other animals' minds. For example, some researchers think that effective deception requires the ability to manipulate or take advantage of others' beliefs about the world. In the 1960s, Emil Menzel, formerly of Stony Brook University, conducted a landmark set of experiments about chimpanzees' ability to find and communicate about the location of hidden objects (Figure 9.17). In one set of experiments, Menzel showed one chimpanzee where a food item was hidden, then released the knowledgeable chimpanzee and his companions into their enclosure. The group quickly learned to follow their knowledgeable companion; but he just as quickly learned that when he led others to the hidden food, he would not get a very big share of it. Menzel noticed that the knowledgeable chimpanzee sometimes led his companions in the wrong direction and then dashed off and grabbed the hidden treasure. How did the knowledgeable chimpanzee work out this tactic? He might have understood that his knowledge differed from the knowledge of other group members and then have come up with a way to take advantage of this discrepancy effectively. If this is what he did, then we would conclude that he had a well-developed theory of mind. Although some researchers, such as Richard Byrne, suggest that deception does not rely on theory of mind, it seems clear that a theory of mind would allow for more complicated and successful deceptions. Similarly, the ability to pretend, empathize, take another's perspective, read minds, console, imitate, and teach relies on knowing what others know or how they feel. Humans do all of these things, but it is not clear whether other primates do.

FIGURE 9.17 In Menzel's experiments, a researcher showed a young chimpanzee where food was hidden in the chimp's enclosure, as pictured here. Then the chimpanzee was reunited with the other members of the group and the group was released inside the enclosure. The young chimpanzee often led the group back to the hidden food, but it also learned to divert the group so that it could get bits of food before others found the cache.

It is clearly difficult to be sure what nonhuman primates know about the minds of other individuals. However, primatologists have begun to make some progress in this area.

Again we turn to the elegant experimental work of Robert Seyfarth and Dorothy Cheney. Working with a group of captive macaques, they designed a set of experiments in which they manipulated the information available to mothers and infants. In one experiment, they asked whether mothers behaved differently when their infants were "knowledgeable" or "ignorant" about the location of desirable food items. In the "knowledgeable" condition, a mother and her infant were both able to see a human experimenter place apple slices into a bin in an adjacent enclosure. In the "ignorant" condition, the mother and infant were separated by an opaque partition so that only the mother was able to see the food being placed in the bin. After the bins were baited, the infants were allowed into the adjacent enclosure by themselves. If mothers know something about what their infants know, then mothers of "ignorant" infants would be expected to make some effort to inform their infants that apples were in the bin, and mothers of "knowledgeable" infants would not need to do this. In fact, mothers of "knowledgeable" and "ignorant" infants behaved in much the same way, suggesting that the mothers did not grasp what their infants knew.

Several lines of evidence suggest that great apes may have more knowledge of the minds of others than monkeys do.

Some researchers believe that chimpanzees display empathy and practice deception. The abilities to feel empathy and to deceive conspecifics both require knowledge of what others know and feel. In chimpanzee and bonobo groups, bystanders sometimes try to comfort or console the victims of aggression after the conflict has ended. This form of behavior is not observed among monkeys.

Chimpanzees also seem to be more devious than other primates. Andrew Whiten and Richard Byrne of St. Andrews University in Scotland compiled and catalogued examples of deception in nonhuman primate species. Chimpanzees figure most prominently in Whiten and Byrne's compendium. For example, a young chimpanzee at Gombe discovered a way to protect his access to bananas. When Jane Goodall was first starting her studies of chimpanzees, she provisioned them with bananas when they came to her camp (Figure 9.18). The chimpanzees soon began to visit her camp regularly, often in large groups. The piles of bananas were monopolized by high-ranking adults, so young and low-ranking animals rarely got many bananas. One young chimpanzee, named Figan, developed a strategy to circumvent this problem. He simply got up and walked away with a purposeful air, and the other chimpanzees followed. Figan reappeared in camp alone about 10 minutes later. This ploy was repeated on several occasions. Frans de Waal once saw one chimpanzee sitting with his back to a rival male. The rival gave an aggressive vocalization, and the listener grinned submissively. The listener used his fingers to push his retracted lips together over his teeth, altering his facial expression. He repeated this three times before the fear grin was eliminated. Then the listener turned to face his rival.

These examples might mean that chimpanzees calculate ways to take advantage of differences between what they know and what others know to deceive other chimpanzees. However, anecdotes like these are not a very satisfying form of data. It's easy to come up with alternative explanations for any given observation. Perhaps Figan simply learned that he would get many bananas when others were present and this knowledge prompted him to walk away from the feeding trench and bide his time.

This sort of ambiguity has spurred primatologists to design experi-

FIGURE 9.18   At Gombe Stream National Park in Tanzania, Jane Goodall established one of the first long-term studies of wild chimpanzees. She provisioned the chimpanzees with bananas to facilitate habituation to her presence and to make it easier to locate the animals.

ments to obtain more systematic information about what great apes know about others' minds. In one set of experiments, Daniel Povinelli and his colleagues at the University of Louisiana at Lafayette, have tried to figure out what chimpanzees understand about what others see. To understand what is involved in this process, think about gaze following. When you see someone staring up into the sky, you look up to see what she is staring at. Moreover, when you do this, you assume that the person's attention is engaged by what she is looking at. And you also know that the person who is staring upward is unlikely to notice if you wave at her or offer her your hand. Although chimpanzees follow gaze, they don't seem to understand that what other individuals are looking at is linked to what those individuals are attending to. For example, given the chance to reach out their hands and beg from two human trainers, chimpanzees do not consistently discriminate between a trainer whose face is blindfolded and a trainer whose face is uncovered, between a trainer who is turned away from them and a trainer who is looking forward, or between a trainer whose gaze is fixed

FIGURE 9.19 In one set of experiments, chimpanzees were able to beg from experimenters. One experimenter looked forward, while the other gazed up into a corner of the room. The chimpanzees begged from both experimenters, suggesting that they did not fully appreciate that only one of the two experimenters could see their gestures.

on a point above their heads and a trainer whose gaze is directed toward them (Figure 9.19). In the most striking case, the chimpanzees often followed the trainer's gaze as they entered the testing room, looking back over their own shoulders, then turned and begged from the trainer, whose gaze was still fixed on the same spot. These experiments suggest that the chimpanzees don't make the connection between what others are looking at and what others are attending to.

This body of work has been criticized because it puts the chimpanzees in a very unnatural situation. In the wild, chimpanzees don't encounter trainers or conspecifics with blindfolds over their eyes or buckets over the heads, so their performance in these tests may not reflect their true cognitive capacities. Bryan Hare, Josep Call, and Michael Tomasello of the Max Planck Institute for Evolutionary Anthropology in Leipzig, Germany, developed an alternative protocol that tests individuals' ability to take advantage of discrepancies between their own knowledge and other individuals' knowledge in a competitive situation. In their experiments, they paired subordinate and dominant chimpanzees in the configuration illustrated in Figure 9.20. These experiments take advantage of the fact that subordinates cannot obtain food when dominants are present. Here, two pieces of food are visible to the subordinate, but the dominant

Subordinate

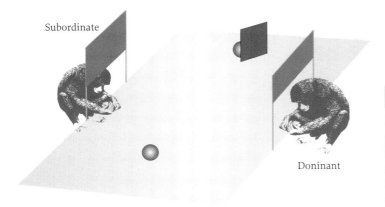

Doninant

FIGURE 9.20 In this experiment, one food item was hidden behind a barrier, so that it could be seen only by the subordinate animal, and one food item was in plain sight of the dominant. Subordinates are unlikely to obtain food rewards when dominant animals are present. So if the subordinate knew what the dominant could see, it was expected to head for the item that was hidden from the dominant. This is what the chimpanzees did most of the time.

can see only one; the other is hidden behind a barrier. Hare and his colleagues predicted that if the subordinate *knew* what the dominant could see, the subordinate would head for the piece of food that the dominant could not see, hoping to consume the hidden food item while the dominant was occupied with the other piece. This is exactly what the subordinate chimpanzees did. Interestingly, when capuchins were tested in the same task, subordinates did not succeed in obtaining hidden food items consistently. Unfortunately, this set of experiments does not fully settle the question of what chimpanzees understand about what others know. Povinelli and his colleagues have criticized some aspects of the methodology that Hare and his collaborators used, and they have not been able to replicate the results in their own laboratory. It will take more experiments on more groups of chimpanzees to settle this question.

THE VALUE OF STUDYING PRIMATE BEHAVIOR

As we come to the end of Part Two, it may be useful to remind you why information about primate behavior and ecology plays an integral role in the story of human evolution. First, humans are primates, and the first members of the human species were probably more similar to living nonhuman primates than to any other animals on Earth. Thus, by studying living primates we can learn something about the lives of our ancestors. Second, humans are closely related to primates and similar to them in many ways. If we understand how evolution has shaped the behavior of animals who are so much like ourselves, we may have greater insights about the way evolution has shaped our own behavior and the behavior of our ancestors. Both of these kinds of reasoning will be apparent in Part Three, which covers the history of our own lineage.

FURTHER READING

Byrne, R. W. 1995. *The Thinking Ape*. Oxford University Press, Oxford, England.

Byrne, R., and A. Whiten, eds. 1998. *Machiavellian Intelligence*. Oxford University Press, Oxford, England.

Cheney, D. L., and R. M. Seyfarth. 1990. *How Monkeys See the World*. University of Chicago Press, Chicago.

Kaplan, H., T. Mueller, S. Gangestad, and J. B. Lancaster. 2003. Neural capital and life span evolution among primates and humans. Pp. 69–97 in *Brain and Longevity*, ed. by C. E. Finch, J.-M. Robine, and Y. Christen. Springer, Berlin.

Reader, S. M., and K. N. Laland. 2002. Social intelligence, innovation, and enhanced brain size in primates. *Proceedings of the National Academy of Sciences, USA* 99: 4436–4441.

Whiten, A. W., and R. W. Byrne. 1997. *Machiavellian Intelligence II*. Cambridge University Press, Cambridge, England.

STUDY QUESTIONS

1. There is a positive correlation between brain size and longevity in animal species. One interpretation of this correlation is that selection for longer life spans was the

primary force driving the evolution of large brains. An alternative interpretation is that selection for larger brains was the primary force driving the evolution of longer life spans. Explain which of these interpretations is more likely to be correct, and why this is the case.

2. We have argued that natural selection is a powerful engine for generating adaptations. If that is the case, then why do organisms grow old and die? Why can't natural selection design an organism that lives forever?

3. Life history traits tend to be bundled together in particular ways. Explain how these traits are combined, and why we see these kinds of combinations in nature.

4. Primates evolved from small-bodied insectivores that were arrayed somewhere along the fast/short end of the life history continuum. What ecological factors are thought to have favored the shifts toward slower/longer life histories in early primates, monkeys, and apes?

5. Primates take a relatively long time to grow up, compared to other animals. Consider the costs and benefits of this life history pattern from the point of view of the growing primate and its mother.

6. What do comparative studies of the size and organization of primate brains tell us about the selective factors that shaped the evolution of primate brains? What are the shortcomings of these kinds of analyses?

7. Monkeys are quite skilled in navigating complicated social situations that they encounter in their everyday lives. They seem to know what others will do in particular situations and are able to respond appropriately. However, monkeys consistently fail theory-of-mind tests in the laboratory. How can we reconcile these two observations?

8. Monkeys seem to have some concept of kinship. What evidence supports this idea? What kind of variation might you expect to find in monkeys' concepts of kinship within and between species?

9. Suppose that you were studying a group of monkeys, and you discovered convincing evidence of empathy or deception. How and why would these data surprise your colleagues?

10. Detailed studies of coalitionary behavior have provided an important source of information about primate cognitive abilities. Explain why coalitions are useful sources of information about social knowledge. What does the pattern of coalitionary support tell us about what monkeys know about other group members?

KEY TERMS

neocortex

extracted foods

social intelligence hypothesis

neocortex ratio

executive brain

brainstem

striatum

tactical deception

cognitive map

third-party relationships

redirected aggression

theory of mind

THE HISTORY OF THE HUMAN LINEAGE

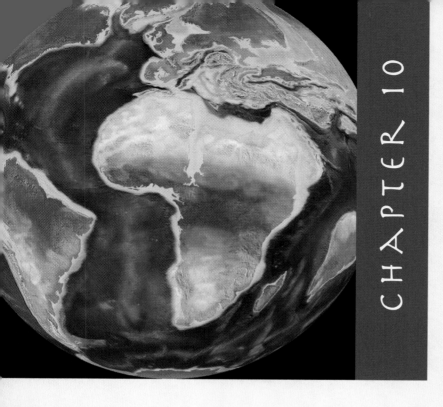

CHAPTER 10

FROM TREE SHREW TO APE

During the Permian and early Triassic periods (Table 10.1), much of the world's fauna was dominated by therapsids, a diverse group of reptiles that possessed traits, such as being warm-blooded and covered with hair (Figure 10.1), that linked them to the mammals that evolved later. At the end of the Triassic, most therapsid groups disappeared, and dinosaurs radiated to fill all of the niches for large, terrestrial animals. One therapsid lineage, however, evolved and diversified to become the first true mammals. These early mammals were probably mouse-sized, nocturnal creatures that fed mainly on seeds and insects. They had internal fertilization but still laid eggs. By the end of the Mesozoic era, 65 million years

TABLE 10.1   The geological timescale.

Era	Period	Epoch	Period Begins (mya)	Notable Events
Cenozoic	Quarternary	Recent	0.012	Origins of agriculture and complex societies
		Pleistocene	1.8	Appearance of Homo sapiens
	Tertiary	Pliocene	5	Dominance of land by angiosperms, mammals, birds, and insects
		Miocene	23	
		Oligocene	34	
		Eocene	54	
		Paleocene	65	
Mesozoic	Cretaceous		136	Rise of angiosperms, disappearance of dinosaurs, second great radiation of insects
	Jurassic		190	Abundance of dinosaurs, appearance of first birds
	Triassic		225	Appearance of first mammals and dinosaurs
Paleozoic	Permian		280	Great expansion of reptiles, decline of amphibians, last of trilobites
	Carboniferous		345	Age of Amphibians; first reptiles, first great insect radiation
	Devonian		395	Age of Fishes; first amphibians and insects
	Silurian		430	Land invaded by a few arthropods
	Ordovician		500	First vertebrates
	Cambrian		570	Abundance of marine invertebrates
Precambrian				Primitive marine life

ago (mya), placental and marsupial mammals that bore live young had evolved. With the extinction of the dinosaurs at the beginning of the next era (the Cenozoic) came the spectacular radiation of the mammals. All of the modern descendants of this radiation—including horses, bats, whales, elephants, lions, and primates—evolved from creatures something like a contemporary shrew (Figure 10.2).

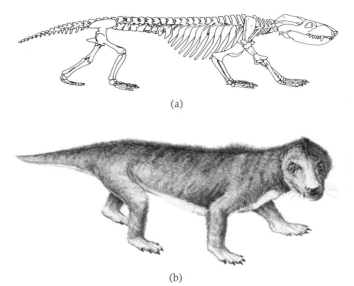

(a)

(b)

FIGURE 10.1 Therapsids dominated the Earth about 250 mya, before dinosaurs became common. The therapsids were reptiles, but they may have been warm-blooded and they had hair instead of scales. The therapsid *Thrinaxodon*, whose skeleton (a) and reconstruction (b) are shown here, was about 30 cm (12 in.) long and had teeth suited for a broad, carnivorous diet.

In order to have a complete understanding of human evolution, we need to know how the transition from a shrewlike creature to modern humans took place. Remember that, according to Darwin's theory, complex adaptations are assembled gradually, in many small steps—each step favored by natural selection. Modern humans have many complex adaptations, like grasping hands, **bipedal** locomotion (walking upright on two legs), toolmaking abilities, language, and large-scale cooperation. To understand human evolution fully, we have to consider each of the steps in the lengthy process that transformed a small, solitary, shrewlike insectivore scurrying through the leaf litter of a dark Cretaceous forest into someone more or less like you. Moreover, it is not enough to chronicle the steps in this transition. We also need to understand why each step was favored by natural selection. We want to know, for example, why claws were traded for flat nails, why quadrupedal locomotion gave way to upright bipedal locomotion, and why brains were so greatly enlarged.

In this part of the text, we trace the history of the human lineage. We begin in this chapter by describing the emergence of creatures that resemble modern lemurs and tarsiers; then we document the transformation into monkeylike creatures, and finally into

FIGURE 10.2 The first mammals probably resembled the modern-day Belanger's tree shrew.

animals something like present-day apes. In later chapters, we recount the transformation from hominoid to hominin. We will introduce the first members of the human tribe, Hominini; and then the first members of our own genus, *Homo*; and finally the first known representatives of our own species, *Homo sapiens*. We know something about each step in this process, although far more is known about recent periods than about periods in the most distant past. We will see, however, that there is still a great deal left to be discovered and understood.

CONTINENTAL DRIFT AND CLIMATE CHANGE

To understand the evolution of our species, it is important to understand the geological, climatic, and biological conditions under which these evolutionary changes occurred.

When we think about the evolution of modern humans, we usually picture early humans wandering over open grasslands dotted with acacia trees—the same breathtaking scenery that we see in wildlife documentaries. As we shift the time frame forward through millions of years, the creatures are altered but the backdrop is unchanged. Our mental image of this process is misleading, however, because the scenery has changed along with the cast of characters (Figure 10.3).

It is important to keep this fact in mind because it alters our interpretation of the fossil record. Remember that evolution produces adaptation, but what is adaptive in one environment may not be adaptive in another environment. If the environment remained the same over the course of human evolution, then the kinds of evolutionary changes observed in the hominin fossil record (such as increases in brain size, bipedalism, and prolonged juvenile dependence) would have to be seen as steady improvements in the perfection of human adaptations: evolution would progress toward a fixed goal. But if the environment varied through time, then evolution would have to track a moving target. In this scenario, new characteristics seen in fossils would not have to represent progress in a single direction. Instead, these changes might have been adaptations to changing environmental conditions. As we will see, the world has become much colder and drier in the last 20 million years and particularly variable in the last 800,000 years, and these changes likely altered the course of human evolution. If the world had become warmer rather than colder during this period, then our human ancestors would probably have remained in the safety of the trees and would not have become terrestrial or bipedal. We would probably be stellar rock climbers but poor marathon runners.

The positions of the continents have changed relative to each other and to the poles.

The world has changed a lot in the last 200 million years. One of the factors that has contributed to this change is the movement of the continents, or

FIGURE 10.3 Today, East African savannas look much like this. In the past, the scenery is likely to have been quite different.

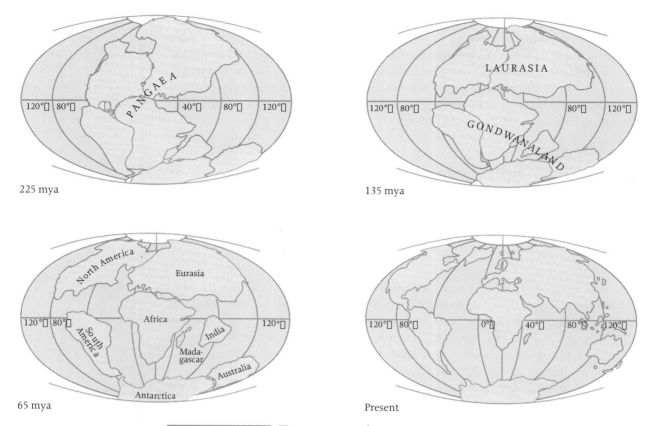

225 mya

135 mya

65 mya

Present

FIGURE 10.4 The arrangement of the continents has changed considerably over the last 180 million years.

continental drift. The continents are not fixed in place; instead, the enormous, relatively light plates of rock that make up the continents slowly wander around the globe, floating on the denser rock that forms the floor of the deep ocean. About 200 mya, all of the land making up the present-day continents was joined together in a single, huge landmass called **Pangaea**. About 125 mya, Pangaea began to break apart into separate pieces (Figure 10.4). The northern half, called **Laurasia**, included what is now North America and Eurasia minus India; the southern half, **Gondwanaland**, consisted of the rest. By the time the dinosaurs became extinct 65 mya, Gondwanaland had broken up into several smaller pieces. Africa and India separated, and India headed north, eventually crashing into Eurasia, while the remainder of Gondwanaland stayed in the south. Eventually, Gondwanaland separated into South America, Antarctica, and Australia, and these continents remained isolated from each other for many millions of years. South America did not drift north to join North America until about 5 mya.

Continental drift is important to the history of the human lineage for two reasons: First, oceans serve as barriers that isolate certain species from others, so the position of the continents plays an important role in the evolution of species. As we will see, the long isolation of South America creates one of the biggest puzzles in our knowledge of primate evolution. Second, continental drift is one of the engines of climate change, and climate change has fundamentally influenced human evolution.

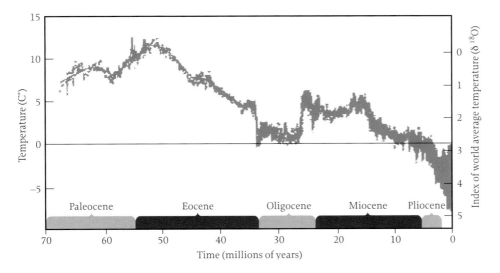

FIGURE 10.5 The gray points are estimates of average world temperature based on the ratio of ^{16}O to ^{18}O taken from deep-sea cores. The red line plots a statistically "smoothed" average value. The variability around the line represents both measurement error and rapid temperature fluctuations that last less than a million years. As we will see (Figure 13.1) most variability in the last several million years has been the result of fluctuations in world temperature, some lasting only a few centuries.

The climate has changed substantially during the last 65 million years—first becoming warmer and less variable, then cooling, and finally fluctuating widely in temperature.

The size and orientation of the continents have important effects on climate. Very large continents tend to have severe weather. This is why Chicago has much colder winters than London, even though London is much farther north. Pangaea was much larger than Asia and is likely to have had very cold weather in winter. When continents restrict the circulation of water from the tropics to the poles, world climates seem to become cooler. These changes, along with other, poorly understood factors, have led to substantial climate change. Figure 10.5 summarizes changes in global temperature during the Cenozoic era. Box 10.1 explains how climatologists reconstruct ancient climates.

To give you some idea of what these changes in temperature mean, consider that during the period of peak warmth in the early Miocene, palm trees grew as far north as what is now Alaska, rich temperate forests (like those in the eastern United States today) extended as far north as Oslo (Norway), and only the tallest peaks in Antarctica were glaciated.

THE METHODS OF PALEONTOLOGY

Much of our knowledge of the history of life comes from the study of fossils, the mineralized bones of dead organisms.

In certain kinds of geological settings, the bones of dead organisms may be preserved long enough for the organic material in the bones to be replaced by minerals (**mineralized**) from the surrounding rock. Such natural copies of bones are called **fossils**. Scientists who recover, describe, and interpret fossil remains are called **paleontologists**.

A great deal of what we know about the history of the human lineage comes from the study of fossils. Careful study of the shapes of different bones tells us about what

BOX 10.1

Using Deep-Sea Cores to Reconstruct Ancient Climates

Beginning about 35 years ago, oceanographers began a program of extracting long cores from the sediments that lie on the floor of the deep sea (about 6000 m, or 20,000 feet, below the surface). Data from these cores have allowed scientists to make much more detailed and accurate reconstructions of ancient climates. Figure 10.5 shows the ratio

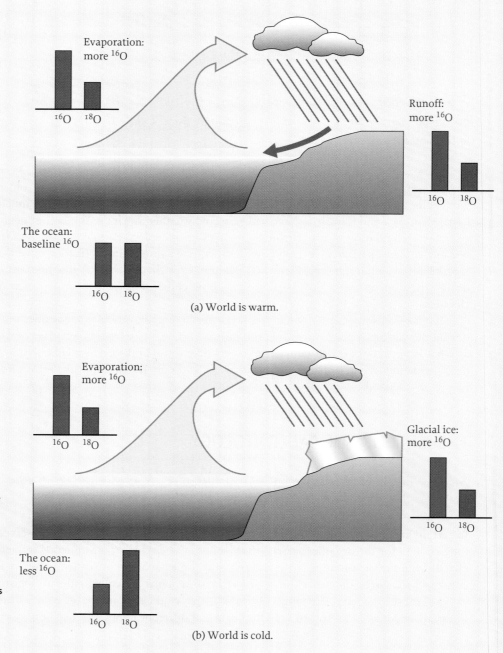

Evaporation: more ^{16}O

^{16}O ^{18}O

Runoff: more ^{16}O

^{16}O ^{18}O

The ocean: baseline ^{16}O

^{16}O ^{18}O

(a) World is warm.

Evaporation: more ^{16}O

^{16}O ^{18}O

Glacial ice: more ^{16}O

^{16}O ^{18}O

The ocean: less ^{16}O

^{16}O ^{18}O

(b) World is cold.

FIGURE 10.6 Water evaporating from the sea is enriched in ^{16}O, and therefore so is precipitation. (a) When the world is warm, this water returns rapidly to the sea, and the concentration of ^{16}O in seawater is unchanged. (b) When the world is cold, much precipitation remains on land as glacial ice, and so the sea becomes depleted in ^{16}O.

of two isotopes of oxygen, $^{16}O:^{18}O$, derived from different layers of a deep-sea cores extracted . Because different layers of the cores were deposited at different times over the last 65 million years and have remained nearly undisturbed ever since, they give us a snapshot of the relative amounts of ^{16}O and ^{18}O in the sea when the layer was deposited on the ocean floor.

These ratios allow us to estimate ocean temperatures in the past. Water molecules containing the lighter isotope of oxygen, ^{16}O, evaporate more readily than molecules containing the heavier isotope, ^{18}O, do. Snow and rain have a higher concentration of ^{16}O than the sea does because the water in clouds evaporates from the sea. When the world is warm enough that few glaciers form at high latitudes, the precipitation that falls on the land returns to the sea, and the ratio of the two isotopes of oxygen remains unchanged (Figure 10.6a). When the world is colder, however, much of the snow falling at high latitudes is stored in immense continental glaciers like those now covering Antarctica (Figure 10.6b). Because the water locked in glaciers contains more ^{16}O than the ocean does, the proportion of ^{18}O in the ocean becomes greater. Therefore, the concentration of ^{18}O in seawater increases when the world is cold and decreases when it is warm. This means that scientists can estimate the temperature of the oceans in the past by measuring the ratio of ^{16}O to ^{18}O in different layers of deep-sea cores.

early hominins were like—how big they were, what they ate, where they lived, how they moved, and even something about how they lived. When the methods of systematics described in Chapter 4 are applied to these materials, they also can tell us something about the phylogenetic history of long-extinct creatures. The kinds of plant and animal fossils found in association with the fossils of our ancestors tell us what the environment was like—whether it was forested or open, how much it rained, and whether rainfall was seasonal.

There are several radiometric methods for estimating the age of fossils.

In order to assign a fossil to a particular position in a phylogeny, we must know how old it is. As we will see in later chapters, the date that we assign to particular specimens can profoundly influence our understanding of the evolutionary history of certain lineages or traits.

Radiometric methods provide one of the most important ways to date fossils. To understand how radiometric techniques work, we need to review a little chemistry. All of the atoms of a particular element have the same number of protons in their nucleus. For example, all carbon atoms have six protons. However, different **isotopes** of a particular element have different numbers of neutrons in their nucleus. Carbon-12, the most common isotope of carbon, has six neutrons; and carbon-14 has eight. Radiometric methods are based on the fact that the isotopes of certain elements are unstable. This means that they change spontaneously from one isotope to another of the same element, or to an entirely different element. For example, carbon-14 changes to nitrogen-14, and potassium-40 changes spontaneously to argon-40. For any particular isotope, such changes (or **radioactive decay**) occur at a constant, clocklike rate that can be measured with precision in the laboratory. There are several different radiometric methods:

1. **Potassium–argon dating** is used to date the age of volcanic rocks found in association with fossil material. Molten rock emerges from a volcano at a very high temperature. As a result, all of the argon gas is boiled out of the rock. After this, any argon present in the rock must be due to the decay of potassium. Because

this occurs at a known and constant rate, the ratio of potassium to argon can be used to date volcanic rock. Then, if a fossil is discovered in a geological **stratum** ("layer"; plural *strata*) lying under the stratum that contains the volcanic rock, paleontologists can be confident that the fossil is older than the rock. A new variant of this technique, called **argon–argon dating** because the potassium in the sample is converted to an isotope of argon before it is measured, allows more accurate dating of single rock crystals.

2. **Carbon-14 dating** (or **radiocarbon dating**) is based on an unstable isotope of carbon that living animals and plants incorporate into their cells. As long as the organism is alive, the ratio of the unstable isotope (carbon-14) to the stable isotope (carbon-12) is the same as the ratio of the two isotopes in the atmosphere. Once the animal dies, carbon-14 starts to decay into carbon-12 at a constant rate. By measuring the ratio of carbon-12 to carbon-14, paleontologists can estimate the amount of time that has passed since the organism died.

3. **Thermoluminescence dating** is based on an effect of high-energy nuclear particles traveling through rock. These particles come from the decay of radioactive material in and around the rock and from cosmic rays that bombard the Earth from outer space. When they pass through rock, these particles dislodge electrons from atoms, so the electrons become trapped elsewhere in the rock's crystal lattice. Heating a rock relaxes the bonds holding the atoms in the crystal lattice together. All of the trapped electrons are then recaptured by their respective atoms—a process that gives off light. Researchers often find flints at archaeological sites that were burnt in ancient campfires. It is possible to estimate the number of trapped electrons in these flints by heating them in the laboratory and measuring the amount of light given off. If the density of high-energy particles currently flowing through the site is also known, scientists can estimate the length of time that has elapsed since the flint was burned.

4. **Electron-spin-resonance dating** is used to determine the age of **apatite crystals**, an inorganic component of tooth enamel, according to the presence of trapped electrons. Apatite crystals form as teeth grow, and initially they contain no trapped electrons. These crystals are preserved in fossil teeth and, like the burnt flints, are bombarded by a flow of high-energy particles that generate trapped electrons in the crystal lattice. Scientists estimate the number of trapped electrons by subjecting the teeth to a variable magnetic field—a technique called "electron spin resonance." To estimate the number of years since the tooth was formed, paleontologists must once again measure the flow of radiation at the site where the tooth was found.

Different radiometric techniques are used for different time periods. Methods based on isotopes that decay very slowly, such as potassium-40, work well for fossils from the distant past. However, they are not useful for more recent fossils, because their "clock" doesn't run fast enough. When slow clocks are used to date recent events, large errors can result. For this reason, potassium–argon dating usually cannot be used to date samples less than about 500,000 years old. Conversely, isotopes that decay quickly, such as carbon-14, are useful only for recent periods, because all of the unstable isotopes decay in a relatively short period of time. Thus, carbon-14 can be used only to date sites that are less than about 40,000 years old. The development of thermoluminescence dating and electron-spin-resonance dating is important because these methods allow us to date sites that are too old for carbon-14 dating, but too young for potassium–argon dating.

Absolute radiometric dating is supplemented by relative dating methods based on magnetic reversals and comparison with other fossil assemblages.

Radiometric dating methods are problematic for two reasons. First, a particular site may not always contain material that is appropriate for radiometric dating. Second, radiometric methods have relatively large margins for error. These drawbacks have led scientists to supplement these absolute methods with other, relative methods for dating fossil sites.

One such relative method is based on the remarkable fact that, every once in a while, the Earth's magnetic field reverses itself. This means, for example, that compasses now pointing north would at various times in the past have pointed south (if compasses had been around then, that is). The pattern of magnetic reversals is not the same throughout time, so for any given time period the pattern is unique. But the pattern for a given time *is* the same throughout the world. We know what the pattern is because when certain rocks are formed, they record the direction of the Earth's magnetic field at that time. Thus, by matching up the pattern of magnetic reversals at a particular site with the well-dated sequence of reversals from the rest of the world, scientists are able to date sites.

Another approach is to make use of the fact that sometimes the fossils of interest are found in association with fossils of other organisms that existed for only a limited period of time. For example, during the last 20 million years or so there has been a sequence of distinct pig species in East Africa. Each pig species lived for a known period of time (according to securely dated sites). This means that some East African materials can be accurately dated from their association with fossilized pig teeth.

THE EVOLUTION OF THE EARLY PRIMATES

The evolution of flowering plants created a new set of ecological niches. Primates were among the animals that evolved to fill these niches.

During the first two-thirds of the Mesozoic, the forests of the world were dominated by the **gymnosperms**, trees like contemporary redwood, pine, and fir. With the breakup of Pangaea during the Cretaceous, a revolution in the plant world occurred. Flowering plants, called **angiosperms**, appeared and spread. The evolution of the angiosperms created a new set of ecological niches for animals. Many angiosperms depend on animals to pollinate them, and they produce showy flowers with sugary nectar to attract pollinators. Some angiosperms also entice animals to disperse their seeds by providing nutritious and easily digestible fruits. Arboreal animals that could find, manipulate, chew, and digest these fruits could exploit these new niches. Primates were one of the taxonomic groups that evolved to take advantage of these opportunities. Tropical birds, bats, insects, and some small rodentlike animals probably competed with early primates for the bounty of the angiosperms.

The ancestors of modern primates were small-bodied nocturnal quadrupeds much like contemporary shrews.

To understand the evolutionary forces that shaped the early radiation of the primates, we need to consider two related questions. First, what kind of animal did

Plesiadapiforms   **Prosimians**

Plesiadapi-
forms were
once thought to be primates, but
closer analysis indicates they lack
many of the defining characteristics
of early modern primates. For example, they had claws instead of nails;
their eye orbits were not fully
encased in bone; their eyes were
placed on the sides of the head, so
the fields of vision of the two eyes
did not converge; and in some
species the big toe was not
opposable.

natural selection have to work with? Second, what were the selective pressures that
favored this suite of traits in ancient primates?

The **plesiadapiforms**, a group of fossil animals found in what is now Montana,
Colorado, New Mexico, and Wyoming, give us some clue about what the earliest primates were like. Plesiadapiforms are found at sites that date from the Paleocene epoch,
65 to 54 mya, a time so warm and wet that broadleaf, evergreen forests extended to
60°N (latitude 60° north, near present-day Anchorage in Alaska). The plesiadapiforms
varied from tiny, shrew-sized creatures to animals as big as a marmot (Figure 10.7).
It seems likely that they were solitary quadrupeds with a well-developed sense of smell.
The teeth of these animals are quite variable, suggesting that they had a wide range
of dietary specializations. Some members of this group were probably terrestial, some
were arboreal quadrupeds, and others may have been adaped for gliding. Most of the
plesiadapiforms had claws on their hands and feet, and they did not have binocular
vision.

The recent discovery of a 56-million-year-old plesiadapiform by Jonathan Bloch
and Doug Boyer of the University of Michigan provides important clues about primate
origins. *Carpolestes simpsoni* (Figure 10.8) had an opposable big toe with a flat nail,
but claws on its other digits. The claws on its feet and hands probably helped it climb
large-diameter tree trunks, but it was also able to grasp small supports. *C. simpsoni*
had low-crowned molars, which are suited for eating fruit. Its eyes were on the sides
of the head, and the fields of vision did not overlap. These creatures probably used

FIGURE 10.8 An artist's reconstruction of the recently discovered plesiadapiform *Carpolestes simpsoni*. This creature, which lived about 56 mya, had grasping hands and feet.

their hands and feet to grasp small branches as they climbed around in the terminal branches of fruiting trees, and used their hands to handle fruit as they were feeding.

Experts disagree about whether plesiadapiforms ought to be included within the primate order. They possess some but not all of the suite of traits that characterize modern primates, and the decision about how to classify these creatures depends on a relatively arbitrary assessment of how similar they are to other primates. The plesiadapiforms are important to know about, however, because they provide some information about the traits that characterized the common ancestor of modern primates.

The discovery of C. simpsoni *helps explain why natural selection favored the basic features of primate morphology.*

There are several different theories about why the traits that are diagnostic of primates evolved in early members of the primate order. Matt Cartmill, a Duke University anthropologist, has argued that forward-facing eyes (orbital convergence) that provide binocular stereoscopic vision, grasping hands and feet, and nails on the toes and fingers all evolved together to enhance visually directed predation on insects in the terminal branches of trees. This idea is supported by the fact that many arboreal predators, including owls and ocelots, have eyes in the front of the head. However, the discovery that grasping hands and feet evolved in a frugivorous plesiadapiform species before the eyes were shifted forward presents problems for this hypothesis.

Fred Szalay of Hunter College and Marian Dagosto of Northwestern University have suggested that grasping hands and feet and flat nails on the fingers and toes all coevolved to facilitate a form of leaping locomotion. *C. simpsoni* poses a problem for this hypothesis, too, because it had grasping hands and feet but evidently didn't leap from branch to branch.

Robert Sussman of Washington University hypothesized that the suite of traits that characterize primates may have been favored because they enhanced the ability of early primates to exploit a new array of plant resources—including fruit, nectar, flowers, and gum—as well as insects. The early primates may have foraged and handled small food items in the dimness of the forest night, and this nocturnal behavior may have

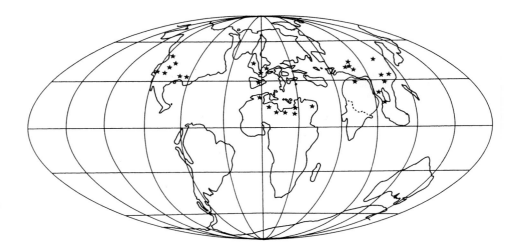

FIGURE 10.9 Sites where Eocene prosimian fossils have been found. The continents are arranged as they were during the early Eocene.

favored good vision, precise eye–hand coordination, and grasping hands and feet. However, *C. simpsoni* foraged on fruit before orbital convergence evolved.

Finally, Tab Rasmussen, also of Washington University, has proposed that grasping hands and feet allowed early primates to forage on fruit, flowers, and nectar in the terminal branches of angiosperms. Later the eyes were shifted forward to facilitate visually directed predation on insects. The idea that the evolution of grasping hands and feet preceded the movement of the eyes to the front of the face fits with the evidence from *C. simpsoni*.

Primates with modern features appeared in the Eocene epoch.

The Eocene epoch (54 to 34 mya) was even wetter and warmer than the preceding Paleocene, with great tropical forests covering much of the globe. At the beginning of the Eocene, North America and Europe were connected, but then the two continents separated and grew farther apart. The animals within these continents evolved in isolation and became progressively more different. There was some contact between Europe and Asia and between India and Asia during this period, but South America was completely isolated. Primate fossils from this period have been found in North America, Europe, Asia, and Africa (Figure 10.9). More than 200 species of fossil prosimians have now been identified. The Eocene primates were a highly successful and diverse group, occupying a range of ecological niches.

It is in these Eocene primates that we see at least the beginnings of all the features that define modern primates (see Chapter 5). They had grasping hands and feet with nails instead of claws, hindlimb-dominated posture, shorter snouts, eyes moved forward in the head and encased in a bony orbit, and relatively large brains.

The Eocene primates are classified into two families: Omomyidae and Adapidae (Figure 10.10). Although their affinities to modern primates are not known, most researchers compare the omomyids to galagos and tarsiers and the adapids to living lemurs (Figure 10.11). Some of the omomyids had huge eye orbits, like modern tarsiers. This feature suggests that they were nocturnal because nocturnal primates who do not have tapeta (singular *tapetum*), such as owl monkeys and tarsiers, have extremely large orbits. (The **tapetum** is a reflecting layer behind the retina that increases the light-capturing capacity of the eye.) Omomyid dentition was quite variable: some seem to have been adapted for frugivory, and others for more insectivorous

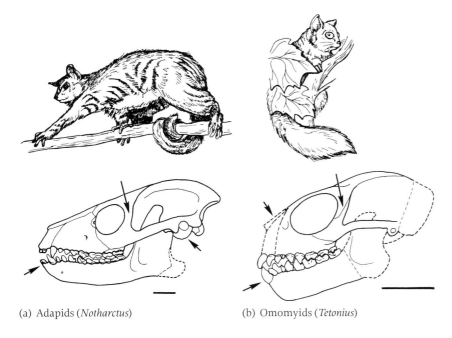

FIGURE 10.10 Adapids were larger than omomyids, and they had longer snouts and smaller orbits than the omomyids had. (a) The size of the orbits suggests that the adapids were active during the day, and the shape of their teeth suggests that they fed on fruit or leaves. (b) Omomyids were small primates that fed mainly on insects, fruit, or gum. The large orbits suggest that they may have been nocturnal. They are similar in some ways to modern tarsiers.

(a) Adapids (*Notharctus*)   (b) Omomyids (*Tetonius*)

diets. Some omomyids have elongated calcaneus bones in their feet, much like those of modern dwarf lemurs, and they may have been able to leap from branch to branch.

The adapids had smaller eye orbits and were likely diurnal. They resemble living lemurs in many aspects of their teeth, skull, nasal, and auditory regions. However, the adapids do not display some of the unique derived traits that are characteristic of modern lemurs, including the tooth comb, a specialized formation of incisors used for grooming. They had a range of dietary adaptations, including insectivorous, folivorous, and frugivorous diets. They were generally larger than the omomyids, and their **postcranial** bones (the bones that make up the skeleton below the neck) indicate that some were active arboreal quadrupeds like modern lemurs, while others were slow quadrupeds similar to contemporary lorises. At least one species showed substantial sexual dimorphism, a feature that points to life in nonmonogamous social groups.

Paleontologists are uncertain whether the modern anthropoids evolved from the omomyids or from the adapids.

The more we learn about the primate lineages of the Eocene and early Oligocene, the clearer it becomes that these early monkeys were extremely diverse and quite successful. We are not entirely sure how these fossil primates are related to modern prosimians or which of them gave rise to the first anthropoid primates.

THE FIRST ANTHROPOIDS

During the Oligocene epoch, many parts of the world became colder and drier.

By the end of the Eocene epoch, 34 mya, the continents were more or less positioned on the globe as they are today. However, South America and North America were not

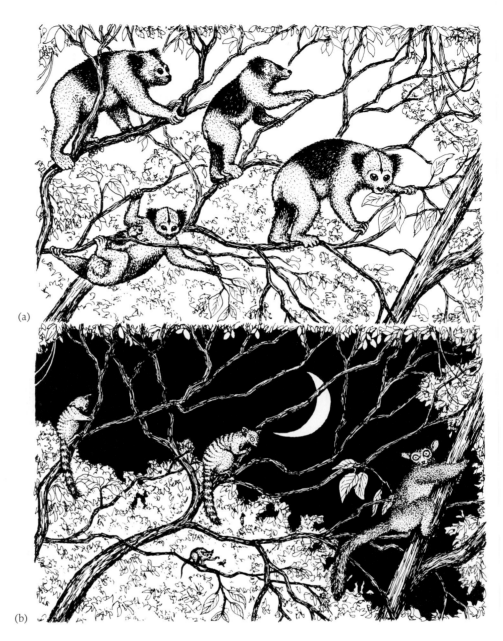

(a)

(b)

The behavior of adapids and omomyids probably differed. (a) Adapids were probably diurnal. Here a group forages for leaves. (b) Omomyids were probably nocturnal. Several species are shown here.

yet connected by Central America; Africa and Arabia were separated from Eurasia by the Tethys Sea, a body of water that connected the Mediterranean Sea and the Persian Gulf. South America and Australia had completed their separation from Antarctica, creating deep, cold currents around Antarctica (see Figure 10.4). Some climatologists believe that these cold currents reduced the transfer of heat from the equator to Antarctic regions, and may have been responsible for the major drop in global temperatures that occurred during this period. During the Oligocene epoch, 34 to 23 mya, temperatures dropped, and the range in temperature variation over the course of the year increased. Throughout North America and Europe, tropical broadleaf evergreen forests were replaced by broadleaf deciduous forests. Africa and South America remained mainly warm and tropical.

Primates similar to modern monkeys may have first evolved during the Eocene, but they radiated during the Oligocene.

The origins of the anthropoids may extend back into the Eocene epoch. *Algeripithecus minutus*, known from fragmentary fossils in North Africa, possessed some cranial features that we find in modern anthropoids, but it was otherwise quite primitive. These finds are dated to about 50 mya. *Eosimias*, a tiny primate with small incisors, large canines, and broad premolars, lived in southern China during the middle of the Eocene epoch.

The earliest unambiguous anthropoid fossils are found at a site in the Fayum (also spelled "Faiyûm") Depression of Egypt. The Fayum deposits straddle the Eocene–Oligocene boundary, 36 to 33 mya. The Fayum is now one of the driest places on Earth, but it was very different at the beginning of the Oligocene. Sediments of soil recovered from the Fayum tell us that it was a warm, wet, and somewhat seasonal habitat then. The plants were most like those now found in the tropical forests of Southeast Asia. The soil sediments contain the remnants of the roots of plants that grow in swampy areas, like mangroves, and the sediments suggest that there were periods of standing water at the site. There are also many fossils of waterbirds. All this suggests that the Fayum was a swamp during the Oligocene (Figure 10.12). Among the mammalian fauna at the Fayum are representatives of the suborder that includes porcupines and guinea pigs, opossums, insectivores, bats, primitive carnivores, and an archaic member of the hippopotamus family.

The Fayum contains one of the most diverse primate communities ever documented. This community included at least five groups of prosimians, one group of omomyids, and three groups of anthropoid primates: the oligopithecids, parapithecids, and propliopithecids (Figure 10.13). To introduce a theme that will become familiar in the chapters that follow, the more that paleontologists learn about early primates from the Fayum, the more complicated the primate family tree becomes. Instead of a neat tree with a few heavy branches that connect ancient fossils with living species, we have a messy bush with many fine branches and only the most tenuous connections between most living and exinct forms. Nevertheless, we can detect certain trends in the primate fossil record at the Fayum that provide insight about the selective pressures that shaped adaptation within the primate lineage.

The parapithecids were a very diverse group; currently they are divided into four genera and eight species. The largest parapithecids were the size of guenons (3 kg, or about 6½ lb) and the smallest were the size of marmosets (150 g,

FIGURE 10.12 Although the Fayum Depression is now a desert, it was a swampy forest during the Oligocene.

FIGURE 10.13 The Fayum was home to a diverse group of primates, including propliopithecids like *Aegyptopithecus zeuxis* (upper left) and *Propliopithecus chirobates* (upper right), and parapithecids like *Apidium phiomense* (bottom).

BOX 10.2

Facts That Teeth Can Reveal

Much of what we know about the long-dead early primates comes from their teeth. Fortunately, teeth are more durable than other bones, and they are also the most common elements in the fossil record. If paleontologists had to choose only one part of the skeleton to study, most would choose teeth. There are several reasons for this choice. First, teeth are complex structures with many independent features, which makes them very useful for phylogenetic reconstruction. Second, tooth enamel is not remodeled during an animal's life, and it carries an indelible record of an individual's life history. Third, teeth show a precise developmental sequence that allows paleontologists to make inferences about the growth and development of long-dead organisms. Finally, as we saw in Chapter 6, each of the major dietary specializations (frugivory, folivory, insectivory) is associated with characteristic dental features.

Figure 10.14 diagrams one side of the upper jaw of three modern species of primates: an insectivore, a folivore,

and a frugivore. The insectivorous tarsier (Figure 10.14a) has relatively large, sharp incisors and canines, which are used to bite through the tough external skeletons of insects. In contrast, the folivorous indri (Figure 10.14b) has relatively small incisors, and large premolars with sharp crests that allow it to shred tough leaves. Finally, the frugivorous mangabey (Figure 10.14c) has large incisors that are used to peel the rinds from fruit. Its molars are small because the soft, nutritious parts of fruit require less grinding than leaves do.

Knowing what an animal eats enables researchers to make sensible guesses about other characteristics as well. For example, there is a good correlation between diet and body size in living primates: insectivores are generally smaller than frugivores, and frugivores are generally smaller than folivores.

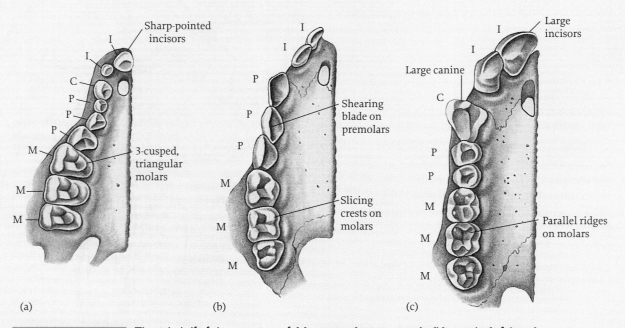

(a)　(b)　(c)

FIGURE 10.14 The right half of the upper jaws of (a) a tarsier (an insectivore), (b) an indri (a folivore), and (c) a mangabey (a frugivore). I = incisor, C = canine, P = premolar, M = molar.

or about 5 oz). These creatures have the primitive dental formula 2.1.3.3/2.1.2.3, which has been retained in New World monkeys but has been modified in Old World monkeys and apes, whose dental formula is 2.1.2.3/2.1.2.3. (Dental formulas were discussed in Chapter 5.) Many aspects of parapithecid teeth and postcranial anatomy are also primitive, suggesting that they may have been the unspecialized ancestors both of more derived Old World monkey lineages and of the New World monkeys (see Box 10.2).

The propliopithecids are represented by two genera and five species. These primates had the same dental formula as modern Old World monkeys, but they lack other derived features associated with Old World monkeys. The largest and most famous of the propliopithecids is named *Aegyptopithecus zeuxis*, who is known from several skulls and a number of postcranial bones (Figure 10.15). *A. zeuxis* was a medium-sized monkey, perhaps as big as a female howler monkey (6 kg, or 13.2 lb). It was a diurnal, arboreal quadruped with a relatively small brain. The shape and size of the teeth suggest that it ate mainly fruit. Males were much larger than females, indicating that they probably lived in nonmonogamous social groups. Other propliopithecoids were smaller than *A. zeuxis*, but their teeth suggest that they also ate fruit, as well as seeds and perhaps gum. They were probably arboreal quadrupeds with strong, grasping feet. Modern Old World monkeys and apes may have been derived from members of this family.

The third group of Fayum anthropoids, the oligopithecids, were among the earliest Fayum monkeys. The oligopithecids, who may have ranged beyond the Fayum through North Africa and the Arabian Peninsula, share many primitive features with the Eocene prosimians. However, they also share some derived features with contemporary anthropoids. For example, the orbits are fully enclosed in bone. The dental formula of some oligopithecids was the same as that of modern Old World monkeys and apes. It is not clear whether the reductions in the number of premolars in oligopithecids and propliopithecids represent independent evolutionary events or common ancestry.

Primates appear in South America for the first time during the Oligocene, but it is unclear where they came from or how they got there.

The earliest New World monkey fossils come from a late-Oligocene site in Bolivia. The monkeys at this site have three premolars like modern New World monkeys and were about the size of owl monkeys. The shape of their molars suggests they were frugivores. Sites in Argentina and Chile contain a number of monkey genera dating to the early and middle Miocene. They were part of a diverse animal population

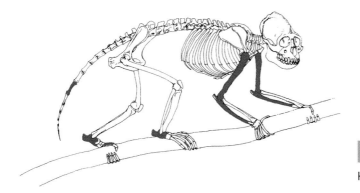

FIGURE 10.15 In this reconstruction of the skeleton of *Aegyptopithecus*, the postcranial bones that have been found are shown in red.

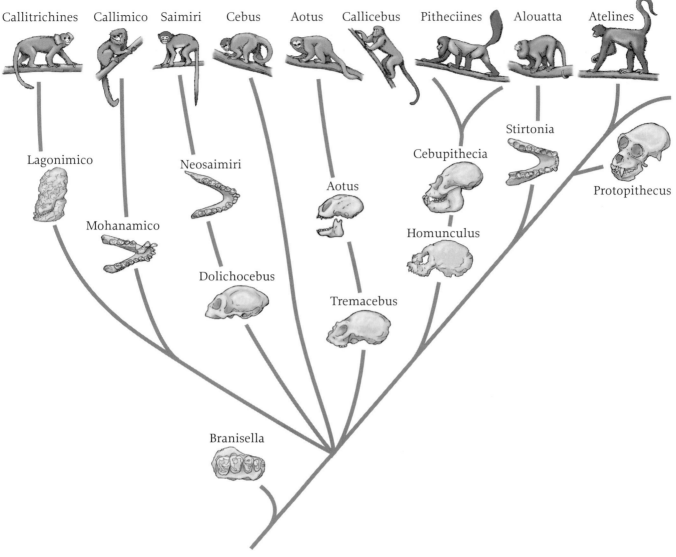

Over the last 10 years, a considerable amount of new New World fossil evidence has been discovered. A number of species bear close resemblances to modern primate species. There were once many more species of primates than we find in the same areas today.

that included rodents, ungulates, sloths, and marsupial mammals. Most of these Patagonian primates were about the size of squirrel monkeys (800 g, or 1.8 lb), though some may have been as large as sakis (3 kg, or 6.6 lb). In Colombia, Miocene sites dated to 12 to 10 mya contain nearly a dozen species of fossil primates. Many of these species closely resemble modern New World monkeys (Figure 10.16). Pleistocene sites in Brazil and the Carribean islands have yielded a mixture of extinct and extant species. Evidently, several species were considerably larger than any living New World primates. Although there are no indigenous primates in the Carribean now, these islands once housed a diverse community of primates.

The origin of New World primates is a great mystery. It is not clear how monkeys got to South America or how they found their way to the islands of the Carribean. The absence of Oligocene primate fossils in North America and the many similarities between New World monkeys and the Fayum primates suggest to many scientists that the ancestor of the New World monkeys came from Africa. The problem with this idea is that we don't know how they could have gotten from Africa to South America. By the late Oligocene, the two continents were separated by at least 3000 km (about 2000 miles) of open ocean. Some authors have suggested that primates could have rafted across the sea on islands of floating vegetation. Although there are no well-documented examples of primates rafting such distances, fossil rodents appear in South America about the same time and are so similar to those found in Africa that it seems very likely that rodents managed to raft across the Atlantic.

Other researchers suggest that New World monkeys are descended from a North American primate. But there are two problems with this hypothesis. One is that, although there is evidence of Eurasian *prosimians* reaching North America earlier (during the Eocene), there are no known *anthropoid* fossils from North America—for any time period. If we assume that an early anthropoid reached North America at about the same time, then the extensive similarities between Old World and New World monkeys could be readily explained. Otherwise we would have to reach the improbable conclusion that New World monkeys are descended from a prosimian ancestor, and that the many similarities between New World and Old World monkeys are due to convergence.

The second difficulty with this hypothesis is geographic. Because North America did not join South America until 5 mya, this scenario also requires the anthropoid ancestor to have made an ocean voyage, though it may have been possible to break up the voyage by hopping across the islands that dotted the Caribbean. Still other possibilities may exist (Figure 10.17).

The most intriguing hypothesis is that anthropoid primates actually appeared in Africa much earlier, when a transatlantic journey would have been easier to complete. Atlantic sea levels were lowest during the middle of the Oligocene. The Fayum primates with the closest affinities to New World monkeys are considerably younger than this. However, this may not be a fatal liability for the hypothesis. There is good reason to believe that the date of the earliest fossil we have discovered usually underestimates the age of a lineage. The method outlined in Box 10.3 suggests that anthropoids actually originated at least 52 mya.

FIGURE 10.17   Perhaps this is how early monkeys *really* got from Africa to South America.

BOX 10.3

Missing Links

University of Zürich anthropologist Robert Martin has pointed out that most lineages are probably older than the oldest fossils we have discovered. The extent of the discrepancy between the dates of the fossils that we have discovered and the actual origin of the lineage depends on the fraction of fossils that have been discovered. If the primate fossil record were nearly complete, then the fact that no anthropoids living more than 35 mya have been found would mean that they didn't exist that long ago. However, we have reason to believe that the primate fossil record is quite incomplete (Figure 10.18).

Just as not all fossils of a given species are ever found,

FIGURE 10.18 The sparseness of the fossil record virtually guarantees that the oldest fossil found of a particular species underestimates the age of the species. (a) The population size of a hypothetical primate species is plotted here against time. In this case, we imagine that the species becomes somewhat less populous as it approaches extinction. (b) The number of fossils left by this species is less than the population size. It is unlikely that the earliest and latest fossils date to the earliest and latest living individuals. (c) The number of fossils found by paleontologists is less than the total number of fossils.

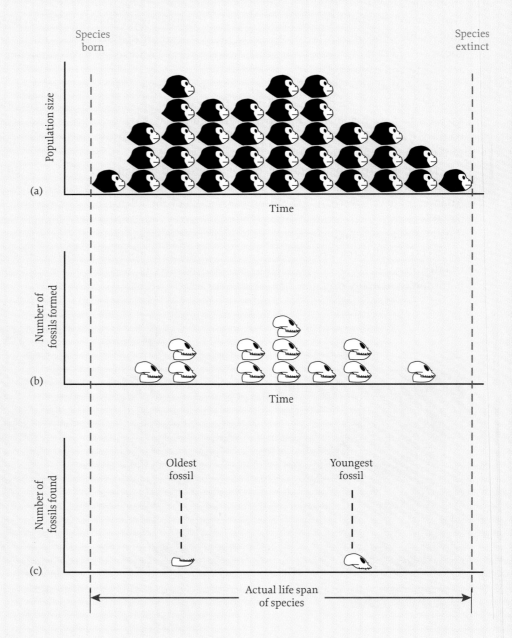

not all species are known to us either. How many species are missing from our data? Martin's method for answering this question is based on the assumption that the number of species has increased steadily from 65 mya, when the first primates appeared, to the present. This means that there were half as many species 32.5 mya as there are now, three-fourths as many species 16.25 mya as there are now, and so on. Assuming that each species has lived about 1 million years, the average life span of a mammalian species, Martin summed these figures to obtain the number of primate species that have ever lived. Then he took the number of fossil species discovered so far and divided that number by his estimate of the total number of species that have ever lived. According to these calculations, only 3% of all fossil primate species have been found so far.

Next, Martin constructed a phylogenetic tree with the same number of living species that we now find in the primate order. One such tree is shown in Figure 10.19a. He then randomly "found" 3% of the fossil species. In Figure 10.19b the gray lines give the actual pattern of descent, and the red lines show the data that would be available if we knew the characteristics of all the living species but had recovered only 3% of the fossil species. The best phylogeny possible, given these data, is shown in Figure 10.19c.

Then Martin computed the difference between the age of the lineage based on this estimated phylogeny, and the actual age of the lineage based on the original phylogeny. The discrepancy between these values represents the error in the age of the lineage that is due to the incompleteness of the fossil record. By repeating this procedure over and over on the computer, Martin was able to produce an estimate of the average magnitude of error, which turned out to be about 40%. Thus, if Martin is correct, living lineages are, on average, about 40% older than the age of the oldest fossil discovered.

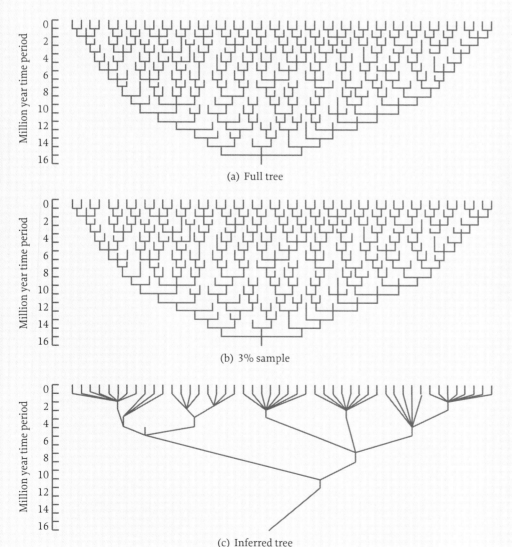

(a) Full tree

(b) 3% sample

(c) Inferred tree

FIGURE 10.19 (a) This tree shows the complete phylogeny of a hypothetical lineage. (b) This tree shows the same phylogeny when only a small percentage of the fossil species have been discovered. The red lines show the species that have been discovered, and the gray lines show the missing species. Note that all *living* species are known, but only 3% of the fossil species have been discovered. (c) The tree we would have to infer from the incomplete data at hand differs from the actual tree in two respects: (1) It links each species to its closest known ancestor, which often means assigning inappropriate ancestors to fossil and modern species, and (2) it frequently underestimates the age of the oldest member of a clade.

THE EMERGENCE OF THE HOMINOIDS

The early Miocene was warm and moist, but by the end of the epoch, the world had become much cooler and more arid.

The Miocene epoch began approximately 23 mya and ended 5 mya. In the early Miocene, the world became warmer, and once again Eurasian forests were dominated by broadleaf evergreens like those in the tropics today. At the end of the Miocene, the world became considerably colder and more arid. The tropical forests of Eurasia retreated southward, and there was more open, woodland habitat. India continued its slow slide into Asia, leading to the uplifting of the Himalayas. Some climatologists believe that the resulting change in atmospheric circulation was responsible for the late-Miocene cooling. About 18 mya, Africa joined Eurasia, splitting the Tethys Sea and creating the Mediterranean Sea. Because the Strait of Gibraltar had not yet opened, the Mediterranean Sea was isolated from the rest of the oceans. At one point, the Mediterranean Sea dried out completely, leaving a desiccated, searing hot valley thousands of feet below sea level. About the same time, the great north–south mountain ranges of the East African Rift began to appear. Because clouds decrease their moisture as they rise in elevation, there is an area of reduced rainfall, called a **rain shadow,** on the lee (downwind) side of mountain ranges. The newly elevated rift mountains caused the tropical forests of East Africa to be replaced by drier woodlands and savannas.

Apes and monkeys differ in some features of their skeletal anatomy, dentiton, brain size, and life history patterns.

Some of the anatomical features that distinguish living apes from Old World monkeys are related to their locomotor behavior. Monkeys climb along on the tops of branches, using their hands and feet to grip branches, and their tails for balance and support. They usually feed in a seated position and have fleshy sitting pads on their bottoms. Apes make much more use of suspensory postures during feeding and locomotion. Although they walk on all fours when they travel on the ground, they swing from branch to branch as they move through the trees and they often hang from branches while feeding (Figure 10.20). Their locomotor behavior is reflected in their skeletons. Compared to monkeys, apes have relatively short trunks, broad chests, long arms, and flexible shoulder joints; they have no tails. Great apes also lack fleshy sitting pads on their bottoms.

FIGURE 10.20 Apes sometimes hang below branches while they feed.

The first evidence of adaptations for suspensory locomotion comes from Miocene fossils collected nearly 40 years ago but overlooked until recently.

Fossils that are now assigned to the species *Morotopithecus bishopi* were first collected at the site of Moroto, in Uganda, in the late 1950s and early 1960s. These finds were dated to about 20 mya and classified as **hominoids,** but they were not officially named until the mid-1990s, when Daniel Gebo of Northern Illinois University, Laura MacLatchy of the University of Michigan, and their colleagues resumed work in

Moroto and collected additional material (Figure 10.21). According to MacLatchy and her colleagues, *M. bishopi* possessed a number of skeletal features that allowed it to move like an ape, not like a monkey. For example, several features of the femur suggest that it might have climbed slowly and cautiously, and the shape of the scapula indicates that it could have hung by its arms and brachiated slowly through the trees. These features are shared with modern apes, but not with the contemporary Miocene apes that we will meet next.

Other early-Miocene apes were similar to Moropithecus in their dentition, but more like monkeys in their postcranial anatomy.

Before *Moropithecus* made its debut in the pages of anthropological journals, most paleontologists believed that the oldest hominoids were members of the family Proconsulidae (Figure 10.22). This family includes 10 genera and 15 species. The smallest **proconsulids** were about the size of capuchin monkeys (3.5 kg, or about 7½ lb) and the largest were the size of female gorillas (50 kg, or a little over 100 lb). The pronconsulids seem to have occupied a range of habitats, including the tropical rain forests in which we find apes today, and open woodlands where we now find only monkeys.

Proconsul is the best-known genus of these early hominoids. The earliest members of this genus were found at a site called Losidok in northern Kenya and date to about 27 mya. The most recent fossils have been found at sites in Africa dated to 17 mya. Plant and animal fossils found with *Proconsul* indicate that they lived in rain-forest environments like those found in tropical Africa today. *Proconsul* species share several derived features with living apes and humans that we don't see in anthropoid primates. For example, *Proconsul* didn't have a tail and did not have the fleshy sitting pads that Old World monkeys and gibbons have. *Proconsul* species also had somewhat larger brains in relationship to body size than similarly sized monkeys had. Otherwise, members of the genus *Proconsul* were similar to *Aegyptopithecus* and the other Oligocene primates. Their teeth had thin enamel, which is consistent with a frugivorous diet. Their postcranial anatomy, including the relative length of their arm and leg bones and narrow and deep shape of their thorax, were much like that of quadrupedal monkeys, but certain features of their feet and lower legs were more apelike. *Proconsul* had an opposable thumb, a feature that we see in humans but not in living apes or monkeys. All of the *Proconsul* species show considerable sexual dimorphism, suggesting that they were not pair-bonded.

Besides *Moropithecus* and *Proconsul*, there were a number of other early-Miocene apes in Africa. These creatures have derived features of apes in their faces and teeth, but not in their postcranial anatomy (Figure 10.23).

The middle Miocene epoch saw a new radiation of hominoids and the expansion of hominoids throughout much of Eurasia.

Exploration of middle-Miocene (15 to 10 mya) deposits has yielded a great abundance of new hominoid species in Africa, Europe, and Asia. Well-known examples include *Kenyapithecus* from East Africa, *Oreopithecus* (Figure 10.24) and *Dryopithecus* from Europe, and *Sivapithecus* from southern Asia. The skulls and teeth of these hominoids typically differ from those of the proconsulids in a number of ways that

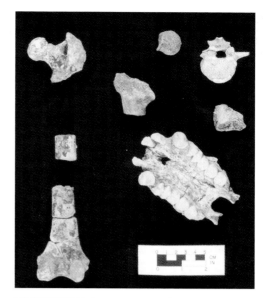

FIGURE 10.21 These fossil bones of *Morotopithecus* include parts of the right and left femurs, vertebrae, shoulder socket, and upper jaw. These remains suggest that these creatures moved like apes, not like monkeys. (Photo courtesy of Laura MacLatchy.)

FIGURE 10.22 Members of the genus *Proconsul* were relatively large (15 to 50 kg, or 33 to 110 lb), sexually dimorphic, and frugivorous. The skeleton of *Proconsul africanus*, reconstructed here, shows that it had limb proportions much like those of modern-day quadrupedal monkeys.

Turkanapithecus

Micropithecus

Afropithecus

Proconsul

FIGURE 10.23   Miocene hominoids exhibit a diverse range of cranial morphology, but they share a number derived characteristics with modern apes.

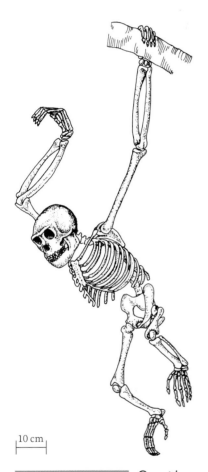

FIGURE 10.24   *Oreopithecus* had a number of traits that are associated with suspensory locomotion, including a relatively short trunk, long arms, short legs, long and slender fingers, and great mobility in all joints. The phylogenetic affinities of this late-Miocene ape from Italy are not well established.

indicate they ate harder or more fibrous foods than their predecessors. Their molars had thick enamel for longer wear, and rounded cusps, which are better suited to grinding. Their **zygomatic arches** (cheekbones) flared farther outward to make room for larger jaw muscles. And the lower jaw was more robust, to carry the forces produced by those muscles. It seems likely that these features were a response to the climatic shift from a moist, tropical environment to a drier, more seasonal environment with tougher vegetation and harder seeds. Until recently, very little was known about the postcranial skeletons of middle-Miocene apes. In 1996, however, Salvador Moyà-Solà and Meike Köhler of the Institut de Paleontología Miquel Crusafont, Spain, announced the discovery of a 9.5-million-year-old specimen of *Dryopithecus* at a site near Barcelona. This remarkable specimen includes nearly all of one hand, most of the leg and arm bones, some of the vertebrae and ribs, part of the collarbone (or clavicle), and part of a cranium (Figure 10.25). The anatomy of these bones clearly indicates that *Dryopithecus* was an arboreal creature that moved through the canopy by swinging from one branch to another, and did not walk along the top of branches as the quadrupedal proconsulids did.

There are no clear candidates for the ancestors of humans or any modern apes, except perhaps orangutans.

The evolutionary history of the apes of the Miocene is poorly understood. There were many different species, and the phylogenetic relationships among them remain largely a mystery. We have no clear candidates for the ancestors of any modern apes, except for the orangutan, who shares a number of derived skull features with *Sivapithecus* of the middle Miocene. We can establish no clear links between African apes and any of the Miocene apes. Once again, this is not too surprising, given the sparseness of the paleontological record.

We can be almost certain that the earliest hominins evolved from some type of Miocene ape, but we have no idea which one it was. Nonetheless, the study of ape and hominin origins in the Miocene has been very useful for paleontologists. John Fleagle of Stony Brook University points out that, in the process of considering and ultimately rejecting many different candidates for the "missing link" between apes and early hominins, paleoanthropologists have come to realize that the key features distinguishing the earliest hominins from ancestral hominoids were not the features that they had originally been looking for: the big brains and small teeth that characterize modern humans. Instead, it was a suite of skeletal adaptations for bipedalism that marked the primary hominin adaptation.

During the early and middle Miocene, ape species were plentiful and monkey species were not. In the late Miocene and early Pliocene, many ape species became extinct and were replaced by monkeys.

Apes flourished during the Miocene, but all but a few genera and species eventually became extinct. Today there are only gibbons, orangutans, gorillas, and chimpanzees. We don't know why so many ape species disappeared, but it seems likely that many of them were poorly suited to the drier conditions of the late Miocene and early Pliocene.

The fossil record of Old World monkeys is quite different from that of the apes. Monkeys were relatively rare and not particularly variable in the early and middle Miocene, but the number and variety of fossil monkeys increased in the late Miocene and early Pliocene. Thus, although there are many more extinct ape species than living ones, the number of living monkey species greatly exceeds the number of extinct monkey species.

Once again, the fossil record reminds us that evolution does not proceed on a steady and relentless path toward a particular goal. Evolution and progress are not synonymous. During the Miocene, there were dozens of ape species but relatively few species of Old World monkeys. Today there are many monkey species and only a handful of ape species. Despite our tendency to think of ourselves as the pinnacle of evolution, the evidence suggests that, taken as a whole, our lineage was poorly suited to the changing conditions of the Pliocene and Pleistocene.

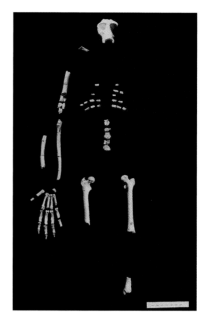

FIGURE 10.25 Much of the skeleton of *Dryopithecus* was found at a site in Spain dated to 9.5 mya. The characteristics of the skeleton indicate that it was adapted for suspensory postures, much like modern apes are.

FURTHER READING

Conroy, G. C. 1990. *Primate Evolution.* Norton, New York.

Fleagle, J. 1998. *Primate Adaptation and Evolution* (2nd ed.). Academic Press, San Diego, CA.

Klein, R. 1999. *The Human Career* (2nd ed.). University of Chicago Press, Chicago.

MacLatchy, L. 2004. The oldest ape. *Evolutionary Anthropology* 13: 30–103.

Martin, R. 1990. *Primate Origins and Evolution: A Phylogenetic Reconstruction.* Princeton University Press, Princeton, NJ.

Potts, R. 1996. *Humanity's Descent: The Consequences of Ecological Instability.* Morrow, New York.

Simons, E. L., and T. Rasmussen. 1994. A whole new world of ancestors: Eocene Anthropoideans from Africa. *Evolutionary Anthropology* 3: 128–139.

STUDY QUESTIONS

1. Briefly describe the motions of the continents over the last 180 million years. Why are these movements important to the study of human evolution?

2. What has happened to the world's climate since the end of the Age of Dinosaurs? Explain the relationship between climate change and the notion that evolution leads to steady progress.

3. What are angiosperms? What do they have to do with the evolution of the primates?

4. Why are teeth so important for reconstructing the evolution of past animals? Explain how to use teeth to distinguish among insectivores, folivores, and frugivores.

5. Which primate groups first appear during the Eocene? Give two explanations for the selective forces that shaped the morphologies of these groups.

6. Why is there a problem in explaining how primates arrived in the New World?

7. Why does the oldest fossil in a particular lineage underestimate the true age of the lineage? Explain how this problem is affected by the quality and completeness of the fossil record.

8. Explain how potassium–argon dating works. Why can it be used to date only volcanic rocks older than about 500,000 years?

KEY TERMS

bipedal	argon-argon dating
continental drift	carbon-14 dating/radiocarbon dating
Pangaea	thermoluminescence dating
Laurasia	electron-spin-resonance dating
Gondwanaland	apatite crystals
mineralized	gymnosperms
fossils	angiosperms
paleontologists	plesiadapiforms
radiometric methods	tapetum
isotopes	rain shadow
radioactive decay	hominoids
potassium-argon dating	proconsulids
stratum	zygomatic arches

CONTENT FROM *HUMAN EVOLUTION: A MULTI-MEDIA GUIDE TO THE FOSSIL RECORD*

Primate Origins
  Definition of the Primate Order
  Primate Trends
  The Earliest Primates
  Theories of Primate Origins
Prosimians and Anthropoids
  Prosimians vs. Anthropoids

The Tarsier Problem
Eocene Prosimians
Early Anthropoids
Monkeys and Apes
  Old World Monkey Origins
  Early Miocene Apes
  Hominid Connections

CHAPTER 11

FROM HOMINOID TO HOMININ

During the Miocene, the Earth's temperature began to fall. This global cooling caused two important changes in the climate of the African tropics. First, the total amount of rain that fell each year declined. Second, rainfall became more seasonal, so there were several months each year when no rain fell. As the tropical regions of Africa became drier, moist tropical forests shrank and drier woodlands and grasslands expanded. Like other animals, primates responded to these ecological changes. Some species, including many of the Miocene ape lineages, apparently failed to adapt and became extinct. The ancestors of chimpanzees and gorillas remained in the shrinking forests and carried on their lives much as before.

Changes brought about through generations of natural selection allowed a few species eventually to move down from the trees, out of the rain forests, and into the woodlands and savannas. Our ancestors, the earliest **hominins**, were among these pioneering species.

The spread of woodland and savanna led to the evolution of the first hominins about 6 mya.

Beginning about 6 mya, hominins begin to appear in the fossil record. Between 4 and 2 mya, the hominin lineage diversified, becoming community of several hominin species that ranged through eastern and southern Africa. These creatures were different from any of the Miocene apes in two ways: First, and most importantly, they walked upright. This shift to bipedal locomotion led to major morphological changes in their bodies. Second, in the new savanna and woodland habitats, new kinds of food became available. As a result, the hominin chewing apparatus—including many features of the teeth, jaws, and skull—changed. Otherwise, the behavior and life history of the earliest hominins were probably not much different from those of modern apes.

A number of shared derived characters distinguish modern humans from other living hominoids: bipedal locomotion, a larger brain, slower development, several features of dental morphology, and cultural adaptation.

When we say that these creatures were hominins, we mean that they are classified with humans in the "tribe" Hominini. The tribe is a taxonomic unit between the family and the genus. Humans belong to the family Hominidae, which includes all of the great apes and the genus *Homo*.

Five categories of derived traits distinguish modern humans from contemporary apes:

1. We habitually walk bipedally.
2. Our dentition and jaw musculature are different from those of apes in a number of ways. For example, we have a wide parabolic dental arcade, thick enamel, reduced canine teeth, and larger molars in relation to the other teeth.
3. We have much larger brains in relation to our body size.
4. We develop slowly, with a long juvenile period.
5. We depend on an elaborate, highly variable material and symbolic culture, transmitted in part through spoken language.

Because these early hominins were bipedal and shared many dental features with modern humans, they are classified as members of our subfamily. However, because their brains and life history were very similar to those of contemporary chimpanzees, early hominins are not included in the same genus (*Homo*) as modern humans.

In this chapter we describe the species that constituted this early hominin community, and we discuss the selective forces that transformed an ancestral, arboreal ape that was something like an orangutan or chimpanzee into a diverse community of bipedal apes living in the woodlands and savannas of Pliocene Africa. In subsequent chapters we will consider how one of these savanna apes became human.

Many of the morphological features that we focus on in this chapter may seem obscure, and you may wonder why we spend so much time describing them. These features help us identify hominin species and trace the origins of traits that we see in later species. We can also use some of these characteristics to reconstruct aspects of diet, social organization, and behavior.

AT THE BEGINNING

Genetic data indicate that the last common ancestor of humans and chimpanzees lived between 7 and 5 mya. Until recently, we knew very little about the hominin lineage during this period. In the last few years, however, fossil discoveries of three kinds of creatures—*Ardipithecus ramidus*, *Orrorin tugenensis*, and *Sahelanthropus tchadensis*—have begun to shed light on this important period in the history of the human lineage. So far, the materials are fragmentary, and the relationships between them and other hominins are unclear.

Ardipithecus

The genus Ardipithecus *includes two species—*A. ramidus *and* A. kadabba—*and both have similarities to both humans and chimpanzees.*

In 1992, members of an expedition led by Tim White of the University of California Berkeley found the first fossils of a primitive hominin at Aramis, a site in the Middle Awash basin of Ethiopia, just south of Hadar. These fossils, dated to 4.4 mya, include teeth and jaws, the lower part of the skull, and parts of the upper arms. White and his colleagues, Gen Suwa of Tokyo University and Berhane Asfaw of Addis Ababa, named the species *Ardipithecus ramidus*, from the words *ardi* (meaning "ground" or "floor") and *ramis* (meaning "root") in the local Afar language. These fossils date from between 5.8 and 5.2 mya.

Several features suggest that A. ramidus was a hominin. The opening on the bottom of the skull through which the spinal cord passes, called the **foramen magnum**, is located forward under the skull, as it is in humans. In apes the foramen magnum is located toward the back of the skull. This is not a definitive indication that these creatures were habitual bipeds, but the forward placement of the foramen magnum is associated with bipedal locomotion in modern humans and may mean that *A. ramidus* was also bipedal. *Ardipithecus ramidus* shares several other derived features with other hominins. These creatures have smaller, more incisorlike canine teeth that are not sharpened by the lower premolar, unlike apes, which have relatively large canines that are sharpened by the first lower premolar. *Ardipithecus ramidus* also resembles humans in several additional features of the teeth and arm bones. Other traits are much more apelike. The molars are smaller in relation to body size than in other early hominins, the enamel is thinner than in other early hominins, and the canines are smaller than in chimpanzees and gorillas but larger than in later hominins. In addition, the deciduous (baby) molars and jaw joint are very similar to those of apes, and the base of the skull is more **pneumatized** (filled with air pockets) than in later hominins.

Ardipithecus ramidus seems to have lived in a forested environment. This conclusion is based on numerous fossils of wood and seeds that were found at the site and come from plant species that grow in woodland habitats. The other animals who lived at Aramis were mainly woodland species. Thus, most of the antelope fossils are of kudus, browsers that typically live in woodlands (Figure 11.1). Moreover, more than 30% of all the mammalian fossils are of colobine monkeys, which are now found mainly in forested areas.

More recently, Yohannes Haile-Selassie, another member of White's team, found 90 more fossil specimens, including bones of the hand, foot, pelvis, arm, leg, and skull. Altogether, these fragments make up about 45% of a complete skeleton. So far,

FIGURE 11.1 *Ardipithecus ramidus* fossils are found in association with the fossils of woodland species like this kudu, suggesting that *A. ramidus* lived in a forested environment.

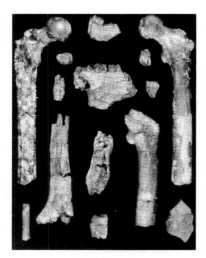

FIGURE 11.2 The fossils of *Orrorin tugenensis* include parts of the femur, lower jaw, fingers, and teeth.

Haile-Selassie has published descriptions of a partial mandible; a number of teeth; and foot, hand, and arm bones. Haile-Selassie classifies this fossil in a second species, *Ardipithecus kadabba*, because the canine teeth are considerably more chimplike than those of *A. ramidus*.

It is presently unclear whether *Ardipithecus* is related more closely to humans or to chimpanzees. White and his colleagues believe that the reduced canine size and (likely) bipedal locomotion indicate that *A. ramidus* represents the earliest stage of hominin evolution. If so, then this species would be the root of the tree linking the human lineage with that of the apes. Others have suggested that the apelike anatomical features of these two species and their apparent preference for forest environments mean that *Ardipithecus* is related more closely to chimpanzees than to modern humans. We will know much more when the remaining fossil material from Aramis has been described.

Orrorin tugenensis

Orrorin tugenensis is a second early fossil with similarities to humans.

In the spring of 2001, a French–Kenyan team led by Brigitte Senut of the National Museum of Natural History in Paris and Martin Pickford of the Kenya Palaeontology Expedition announced the discovery of 12 hominin fossils in the Tugen Hills in the highlands of Kenya. These fossils, which include parts of thigh and arm bones, a finger bone, two partial mandibles (lower jaws), and several teeth, are securely dated to 6 mya (Figure 11.2). Senut and Pickford assigned their finds to a new species, *Orrorin tugenensis*. The genus name means "original man" in the language of the local people, and it reminded Senut and Pickford of the French word for "dawn," *aurore*.

Like the fossils of *Ardipithecus*, these specimens are similar to chimpanzees in some ways and to humans in others. The incisors, canines, and one of the premolars are more like the teeth of chimpanzees than of later hominins. The molars are smaller than those of *A. ramidus* and later apelike hominins, and they have thick enamel like human molars have. As in chimpanzees and some later hominins, the arm and finger bones have features that are believed to be adaptations for climbing. The thigh bones are more similar to those of later hominins than to those of apes, which Senut and Pickford interpret as evidence that *O. tugenensis* was bipedal. Fossils of forest creatures, such as colobus monkeys, and open-country dwellers, such as impala, were found in the same strata as *O. tugenensis*, indicating that the habitat was a mix of woodland and savanna.

Sahelanthropus tchadensis

Sahelanthropus tchadensis, a third early fossil, has a surprising mix of derived and primitive features.

The first announcements of new hominin fossils sometimes read as though they were written by Hollywood publicists. But like many movies, the discoveries don't always live up to their billing. In July 2002, the discovery of a new hominin fossil was announced, and this time there is a good chance that it deserves all the fanfare.

The fossil consists of a nearly complete cranium, two fragments of the lower jaw, and several teeth found at Toros-Menalla, a site in Chad, by a team of researchers led by Michel Brunet of the University of Poitiers in France (Figure 11.3). Brunet and his colleagues have given the fossil the scientific name *Sahelanthropus tchadensis*. The word *Sahelanthropus* comes from "Sahel," the vast dry region south of the Sahara; and *tchadensis* comes from "Chad," the country where the fossil was found. This fossil has also acquired a nickname, Toumaï, which means "hope of life" in the language of the local people, and is often given to children born during the dry season.

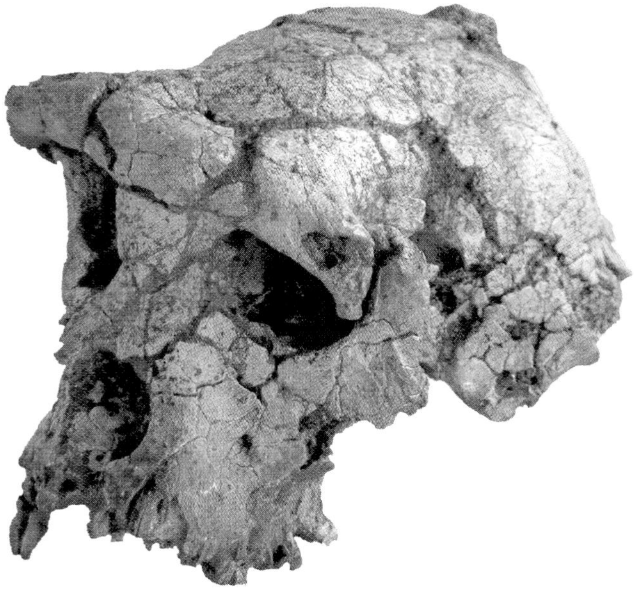

This fossil shook the paleontological community for a number of reasons. First, it comes from an unexpected place. Most work on human evolution has focused on East Africa and South Africa, but this specimen comes from the middle of the African continent. Although Toros-Menalla is now an arid desert, in the late Miocene it was a well-watered woodland. This suggests that early hominins had a much larger range than was previously believed. Second, this fossil is surprisingly old. The geology of the site does not allow radiometric or paleomagnetic dating. However, there is a close match between the other fossil animals found at the site and the fauna found at two sites in East Africa securely dated to between 7 and 6 mya. This makes Toumaï the oldest hominin cranium. *Sahelanthropus tchadensis* may give us a picture of the last common ancestor because genetic data tell us that the hominin lineage diverged from the chimpanzee lineage during this period of time. Third, and most important, is the fact that Toumaï possesses a very surprising mix of anatomical features. The face is relatively flat, and there is a massive browridge over the eyes. These are features associated with hominins who date to later than 2 mya, but not with the hominins who dominate the fossil record between 4 and 2 mya. As in *Ardipithecus*, the foramen magnum is located under the skull (rather than at the back)—a feature that is associated with bipedal locomotion. On the other hand, the braincase is very small [320 to 350 cc (cubic centimeters)], the teeth are primitive, and the back of the skull is very apelike.

Toumaï's very early date and unique mix of derived and primitive features provide yet more evidence that hominin evolution did not follow a simple progression from ape to human. Instead, it seems more and more likely that there were many different hominin species that possessed a variety of different adaptations.

FIGURE 11.3 The cranium of *Sahelanthropus tchadensis* found in Chad dates to between 7 and 6 mya. It has a flat face and large browridge. Australopithecines dated to between 4 and 2 mya have more apelike prognathic faces and lack browridges. (Photograph courtesy of Michel Brunet.)

THE HOMININ COMMUNITY DIVERSIFIES

A number of hominin species lived in Africa between 4 and 2 mya. They are divided into three genera: Australopithecus, Paranthropus, *and* Kenyanthropus.

Beginning about 4 mya the hominin lineage proliferated, and over the next 2 million years there were between four and seven species of hominins living in Africa at any given time. A number of different taxonomic schemes have been used to classify these creatures. Here we adopt a system advocated by Bernard Wood of George Washington University and Mark Collard of Washington State University. They divide these species into three genera:

1. *Australopithecus* includes six species: *A. anamensis*, *A. afarensis*, *A. africanus*, *A. garhi*, *A. habilis*, and *A. rudolfensis*. The genus name means "southern ape" and was first applied to an *A. africanus* skull found in South Africa. The **australopithecines** were small bipeds with teeth, skull, and jaws adapted to a generalized diet. Wood and Collard's scheme departs from convention; other taxonomic schemes usually place *A. habilis* and *A. rudolfensis* in the same genus as modern humans (*Homo*).

2. *Paranthropus* includes three species: *P. aethiopicus*, *P. robustus*, and *P. boisei*. The genus name means "parallel to man" and was coined by Robert Broom, who discovered the first specimens of *P. robustus*. The paranthropines were similar to the australopithecines from the neck down, but they had massive teeth and jaws adapted to heavy chewing of tough plant materials, and a skull modified to carry the enormous muscles necessary to power this chewing apparatus.

3. *Kenyanthropus* includes only one species: *K. platyops*. The genus name means "Kenyan man" and was given to a specimen recently found in northern Kenya. We do not yet know much about these creatures, but they are distinguished by a flattened face and small teeth.

The sites at which early hominin fossils have been found in Africa are identified on the map in Figure 11.4. In this section we briefly describe the history and characteristics of each of these fossil species and then turn to a discussion of their common features.

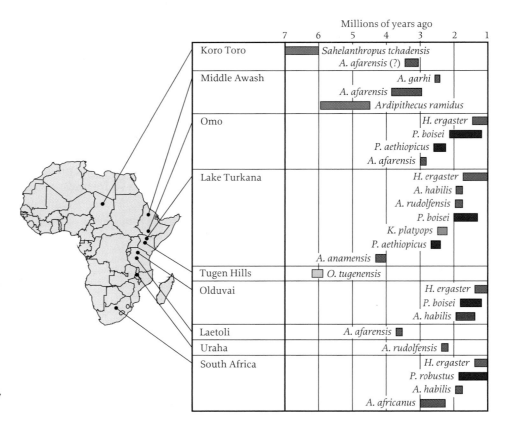

FIGURE 11.4 Hominin fossils have been discovered at a number of sites in eastern and southern Africa. Fossils belonging to each genus have been assigned a different color. Fossils discovered at Bahr el Ghazal in Chad have been provisionally assigned to the species *A. afarensis*, pending more detailed analysis.

You are excused if you feel as though you have mistakenly picked up a Russian novel with a long cast of characters filled with unpronounceable and hard-to-remember names. Keep in mind, however, that the species within each genus are quite similar: the australopithecines are small bipeds with relatively small teeth, the paranthropines are small bipeds with big teeth, and *Kenyanthropus platyops* has small teeth and a flat face.

Australopithecus

A. ANAMENSIS

Australopithecus anamensis was bipedal but had a more apelike skull than later australopithecines had.

In 1994, members of an expedition led by Meave Leakey of the National Museums of Kenya found fossils of a new hominin species at Kanapoi and Allia Bay, two sites near Lake Turkana in Kenya (Figure 11.5). Among these specimens are parts of the upper and lower jaw, part of a **tibia** (the larger of the two bones in the lower leg), and numerous teeth. In addition, parts of a **humerus** (the upper arm bone) found at Kanapoi in the early 1970s can now be associated with the new finds. Leakey assigned them to the species *Australopithecus anamensis*. The species name is derived from *anam*, the word for "lake" in the language of the people living around Lake Turkana. Subsequent work has expanded the sample to more than 50 specimens. These fossils are dated to between 4.2 and 3.8 mya.

Several features mark *A. anamensis* as a hominin. It had large molars with thick enamel and smaller canines, and the shapes of the knee and ankle joints strongly indicate that it was bipedal. The arm bones suggest that it retained adaptations for life in the trees. As we will see, these features are shared by later australopithecines as well. However, *A. anamensis* lacks several derived traits seen in later australopithecines. For example, the ear holes are small and shaped like ellipses, as they are in living apes, but the ear holes of later australopithecines are larger and more rounded; the dental arcade is more like the U shape seen in chimpanzees and gorillas, and less like the V-shaped dental arcade seen in later australopithecines; and the chin recedes more sharply than that in other australopithecines does (Figure 11.6). There are also indications that *A. anamensis* sometimes had larger canines than those seen in later hominins. This suite of more primitive characters distinguishes *A. anamensis* from later australopithecines

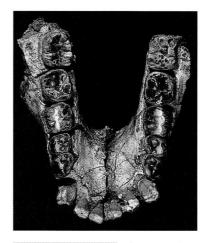

FIGURE 11.5 *A. anamensis* lived about 4 mya, half a million years before *A. afarensis*. Fragments of limb bones, the upper and lower jaws, and many teeth have been recovered.

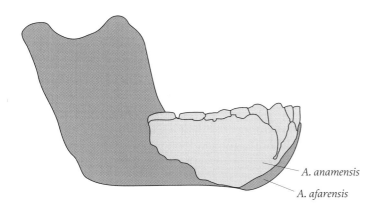

A. anamensis
A. afarensis

FIGURE 11.6 *Australopithecus anamensis* had a more receding chin than later australopithecines had, as is evident when side views of the lower jaw of *A. anamensis* (beige) and *A. afarensis* (blue) are superposed.

FIGURE 11.7 The Awash basin in Ethiopia is the site of several important paleontological discoveries, including many specimens of *A. afarensis*.

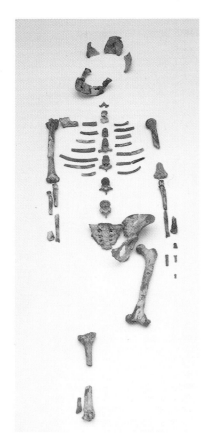

FIGURE 11.8 A sizable fraction of the skeleton of a single individual was recovered at Hadar. Because the skeleton is bilaterally symmetrical, most of the skeleton of this *A. afarensis* female, popularly known as Lucy, can be reconstructed.

and is consistent with the hypothesis that *A. anamensis* is the ancestor of these species.

Fossils of other animals found at Kanapoi and Allia Bay suggest that *A. anamensis* lived in a mixture of habitats including dry woodlands, gallery forests along rivers, and more open grasslands.

A. AFARENSIS

Hominin fossils assigned to the species Australopithecus afarensis *are found at sites in East Africa that date from 4 to 3 mya.* Australopithecus afarensis *is the best-known australopithecine species.*

Australopithecus afarensis is well known from specimens found at several sites in Africa (see Figure 11.4), but the most extensive fossil collections come from several sites in Ethiopia. In the early 1970s a French and American team headed by Maurice Taieb and Don Johanson began searching for hominid fossils at Hadar, in the Afar Depression of northeastern Ethiopia (Figure 11.7). In 1973, this team made a striking discovery: they found the bones of a 3-million-year-old knee joint that showed striking similarities to a modern human knee. The next year the team returned and found a sizable fraction of the skeleton of a single individual, an amazing discovery in a discipline in which an entire species is sometimes named and classified on the basis of a single tooth. They dubbed the skeleton "Lucy," after the Beatles song "Lucy in the Sky with Diamonds" (Figures 11.8 and 11.9). Lucy, who is 3.2 million years old, was not the only remarkable find at Hadar. During the next field season, the team found the remains of 13 more individuals. This collection of fossils is formally identified as material from Afar Locality 333 (AL 333), but it is often called simply the "First Family."

Just after these fossils were discovered, the excavations were interrupted by civil war in Ethiopia, and paleontologists were unable to resume work at Hadar for more than a decade. Then, a team led by Don Johanson returned to Hadar, while a separate team led by Tim White searched for fossils in the nearby Middle Awash basin. Both groups were successful in discovering fossils from many *A. afarensis* individuals, including a nearly complete skull (labeled AL 444-2). In 2001, Ethiopian researcher Zeresenay Alemseged announced the discovery of a very well-preserved, partial skeleton of a child in the Afar region of Ethiopia.

Australopithecus afarensis fossils have also been found at several sites elsewhere in Africa. During the 1970s, members of a team led by Mary Leakey discovered fossils of *A. afarensis* at Laetoli in Tanzania. The Laetoli specimens are 3.5 million years old, several hundred thousand years older than Lucy. In 1995, researchers announced interesting finds from Chad and South Africa. A French team published a description of a lower jaw from Bahr el Ghazal in Chad. This fossil, provisionally identified as *A. afarensis*, is between 3.1 and 3.4 million years old, judging by the age of associated fossils. In South Africa, Ronald Clarke of the Johann Wolfgang Goethe University of Frankfurt, Germany, and Phillip Tobias of the University of the Witwatersrand found several foot bones that may also belong to *A. afarensis*.

Reconstructions of the environments at these sites indicate that *A. afarensis* lived in habitats ranging from woodland to dry savanna. Between 4 and 3 mya, the environment at Hadar was a mix of woodland, scrub, and grassland; at Laetoli, it was a dry grassland sparsely dotted with trees.

The skulls of A. afarensis *share several derived features with modern humans, but the fossils also display many ancestral traits, including ape-sized brains.*

The cranium (the skull minus the lower jawbone) of *A. afarensis* is quite apelike. Its **endocranial volume** (the capacity of its brain cavity) is 404 cc—less than a pint. This is about the same size as the brain of a modern chimpanzee. The cranium of *A. afarensis* is also primitive in a number of other details. For example, the base of the cranium is flared at the bottom and the bone is pneumatized; the front of the face below the nose is pushed out—a condition known as **subnasal prognathism**; and the jaw joint is shallow (Figure 11.10). Remember that labeling a trait "primitive" does not imply that organisms possessing that trait are backward or poorly designed; it simply means that the organisms resemble earlier forms.

The teeth and jaws of *A. afarensis* are intermediate between those of apes and humans in a number of ways. Figure 11.11 shows three features that illustrate this point:

1. The dental arcade of *A. afarensis* has an intermediate V shape, the canines are medium-sized, and the **diastema** (the space between the teeth and the jaw; plural *diastemata*) is modest.
2. In its canines, *A. afarensis* displays less sexual dimorphism than chimpanzees but more than modern humans.
3. The first lower premolar of *A. afarensis* has a small inner cusp and a larger outer cusp. This is intermediate between chimpanzees, who have a single cusp on their lower premolars, and modern humans, who have two cusps of approximately equal size.

Anatomical evidence indicates that A. afarensis *was bipedal.*

The transition from a forest ape to a terrestrial biped involves many new adaptations. A number of these changes are reflected in the morphology of the skeleton.

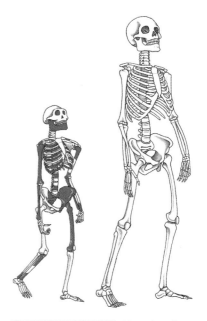

FIGURE 11.9 Here, Lucy's skeleton (left) stands beside the skeleton of a modern human female. The parts of the skeleton that have been discovered are shaded. Lucy was shorter than modern females and had relatively long arms and a relatively small brain.

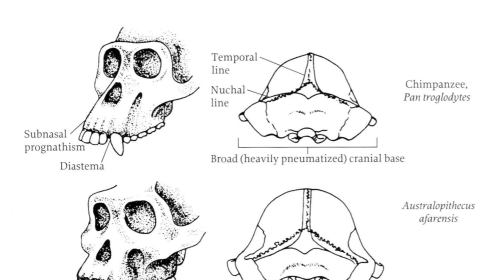

Subnasal prognathism

Diastema

Temporal line

Nuchal line

Broad (heavily pneumatized) cranial base

Chimpanzee, *Pan troglodytes*

Australopithecus afarensis

FIGURE 11.10 The cranium of *A. afarensis* possesses a number of primitive features, including a small brain, a shallow jaw joint, a pneumatized cranial base, and subnasal prognathism in the face. (Figure courtesy of Richard Klein.)

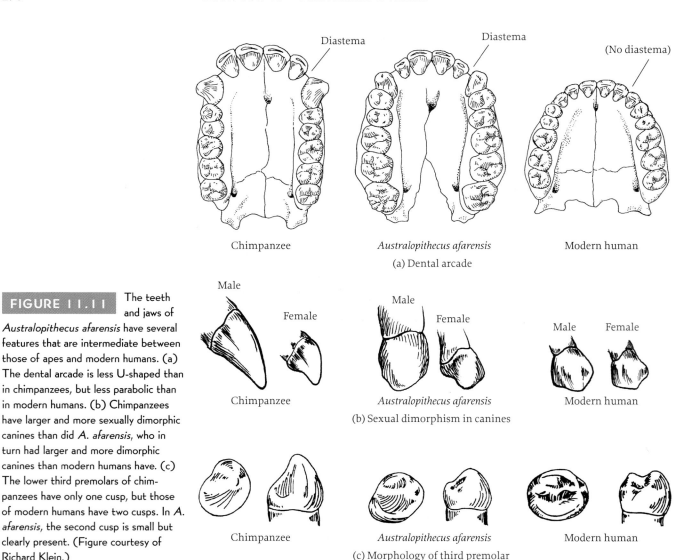

Diastema   Diastema   (No diastema)

Chimpanzee   *Australopithecus afarensis*   Modern human

(a) Dental arcade

Male
Female

Male
Female

Male   Female

Chimpanzee   *Australopithecus afarensis*   Modern human

(b) Sexual dimorphism in canines

Chimpanzee   *Australopithecus afarensis*   Modern human

(c) Morphology of third premolar

FIGURE 11.11 The teeth and jaws of *Australopithecus afarensis* have several features that are intermediate between those of apes and modern humans. (a) The dental arcade is less U-shaped than in chimpanzees, but less parabolic than in modern humans. (b) Chimpanzees have larger and more sexually dimorphic canines than did *A. afarensis*, who in turn had larger and more dimorphic canines than modern humans have. (c) The lower third premolars of chimpanzees have only one cusp, but those of modern humans have two cusps. In *A. afarensis*, the second cusp is small but clearly present. (Figure courtesy of Richard Klein.)

Thus, by studying the shape of fossils, we can make inferences about the animal's mode of locomotion. Changes in the pelvis provide a good example. In terms of shape and orientation, the human pelvis is very different from that of forest-dwelling apes like the chimpanzee (Figure 11.12). The pelvis of *A. afarensis* is much more like a human pelvis, so we infer that *A. afarensis* is likely to have been bipedal.

Other changes are more subtle but also diagnostic. When modern humans walk, a relatively large proportion of the time is spent balanced on one foot. Each time you take a step, your body swings over the foot that is on the ground, and all of your weight is balanced over that foot. At that moment, the weight of your body pulls down on the center of the pelvis, well inward from the hip joint (Figure 11.13). This weight creates a twisting force, or **torque**, that acts to rotate your torso down and away from the weighted leg. But your torso does not tip, because the torque is opposed by **abductors**, muscles that run from the outer side of the pelvis to the femur. At the appropriate moment during each stride, these muscles tighten and keep you upright. (You can demonstrate this by walking around with your open hand on the side of your hip.

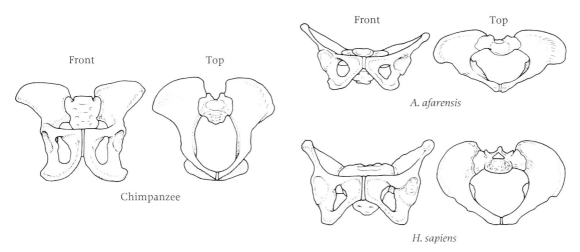

FIGURE 11.12 The pelvis of *A. afarensis* resembles the modern human pelvis more than the chimpanzee pelvis. Notice that the australopithecine pelvis is flattened and flared like the modern human pelvis. These features increase the efficiency of bipedal walking. The australopithecine pelvis is also much wider side to side and narrower front to back than that of modern humans. Some anthropologists believe these differences indicate that australopithecines did not walk the same way that modern humans do.

You'll feel your abductors tighten as you walk, and your torso tip if you relax these muscles. You might want to do this in private.) The abductors are attached to the **ilium** (plural *ilia*; a flaring blade of bone on the upper end of the pelvis). The widening and thickening of the ilium and the lengthening of the neck of the **femur** (the thigh bone) add to the leverage that the abductors can exert, and make bipedal walking more efficient. The existence of these morphological features in *A. afarensis* implies that it was bipedal.

The designs of the knee and foot of *A. afarensis* are also consistent with the idea that *A. afarensis* was bipedal. The modern human knee joint is quite different from

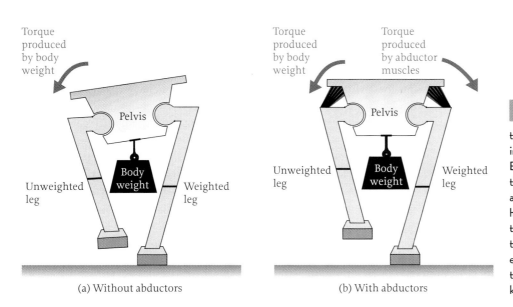

FIGURE 11.13 A schematic diagram of the lower body at the point of the stride in which all of the weight is on one leg. Body weight acts to pull down through the centerline of the pelvis. This creates a torque, or twisting force, around the hip joint of the weighted leg. (a) If this torque were unopposed, the torso would twist down and to the left. (b) During each stride the abductor muscles tighten to create a second torque that keeps the body erect.

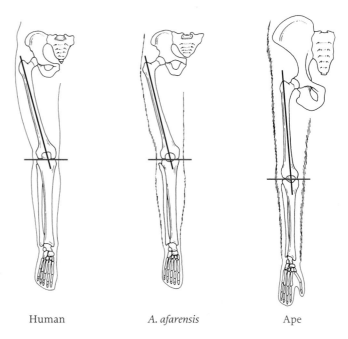

FIGURE 11.14 The knees of *A. afarensis* are more like the knees of modern humans than like the knees of chimpanzees. Consider the lower end of the femur, where it forms one side of the knee joint. In chimpanzees, this joint forms a right angle with the long axis of the femur. In humans and australopithecines, the knee joint forms an oblique angle, causing the femur to slant inward toward the centerline of the body. This slant causes the knee to be carried closer to the body's centerline, which increases the efficiency of bipedal walking.

Human   *A. afarensis*   Ape

the chimpanzee knee joint (Figure 11.14). Efficient bipedal locomotion requires that the knees lie close to the centerline of the body. As a result, the human femur slants down and inward, and its lower end is angled at the knee joint to make proper contact with the bones of the lower leg. In contrast, the chimpanzee femur descends vertically from the pelvis, and the end of the femur at the knee joint is not slanted. The femur of *A. afarensis* has the same angled shape as that of modern humans. The feet also show a number of derived features associated with bipedal locomotion, including a longitudinal arch and a humanlike ankle.

Australopithecus afarensis *may not have walked with the same efficient, striding gait that modern humans have.*

Although there is no doubt that *A. afarensis* was bipedal, it is not entirely clear whether this hominin had the same striding gait as modern humans have. Many features of the pelvis and legs of *A. afarensis* are different from the pelvis and legs in modern humans, and these differences make some researchers believe that Lucy and her colleagues walked with an inefficient, bent-legged gait. For example, the ilium is oriented more toward the back than in modern humans, and biomechanical calculations suggest that the abductors would be less efficient in this orientation. The legs of *A. afarensis* were also much shorter in relationship to body size than are the legs of modern humans, and this would reduce the efficiency and speed of their walking. However, other researchers have argued that *A. afarensis* was an efficient biped even though it did not walk in exactly the same way modern humans do. These researchers note, for example, that the pelvis of *A. afarensis* is much wider in relationship to its body size, compared with the modern human pelvis (see Figure 11.12). This extra width may have minimized vertical motion of the body during walking. The longer legs of modern humans produce the same effect.

A trail of fossil footprints proves that a striding biped lived in East Africa at the same time as A. afarensis.

The conclusion that *A. afarensis* had an efficient striding gait was dramatically strengthened by a remarkable discovery by Mary Leakey and her coworkers at Laetoli in Tanzania. They uncovered a trail of footprints 30 m (about 100 ft) long that had been made by three bipedal individuals who had crossed a thick bed of wet volcanic ash about 3.5 mya (Figure 11.15). Paleontologists estimate that the tallest of the individuals who made these prints was 1.45 m (4 ft 9 in.) and the shortest was 1.24 m (4 ft 1 in.) tall. The path of these three individuals was preserved because the wet ash solidified as it dried, leaving us a tantalizing glimpse of the past. University of Chicago anthropologist Russell Tuttle has compared the Laetoli footprints with the footprints made by members of a group of people called the Machigenga, who live in the tropical forests of Peru. Tuttle chose the Machigenga because their average height is close to that of the makers of the Laetoli footprints, and they do not wear shoes. Tuttle found that the Laetoli footprints are functionally indistinguishable from those made by the Machigenga, and he concluded that the creatures who made the footprints walked with a fully modern striding gait.

Who made these footprints? The suspects are *A. afarensis* and *Kenyanthropus platyops* (a hominin you will meet later in this chapter) because they are the only hominins known to have lived in East Africa at the time the tracks were made. *Australopithecus afarensis* is the most likely culprit because it is the only hominin whose remains have been found at Laetoli. If *A. afarensis* did make the footprints, then the researchers who doubt that they were striding bipeds are wrong. However, some anthropologists are convinced by the anatomical evidence that *A. afarensis* was not a modern biped. If they are correct, then *K. platyops* made the footprints, or there was another, as yet undiscovered, hominin species living in East Africa 3.5 mya.

FIGURE 11.15 Several bipedal creatures walked across a thick bed of wet volcanic ash about 3.5 mya. Their footprints were preserved when the ash solidified as it dried.

Australopithecus afarensis probably spent a good deal of time in trees.

All nonhuman primates, except some gorillas, spend the night perched in trees or huddled on cliffs to protect themselves from nocturnal predators like leopards. Even chimpanzees, who are about the size of Lucy, make leafy nests and sleep in trees each night (Figure 11.16). Thus it seems likely that *A. afarensis* slept in trees too, and that their skeletons might show features related to arboreality. Anatomists such as Randall Susman of Stony Brook University have argued that traits of the feet, hands, wrists, and shoulder joints suggest that Lucy and her friends were partly arboreal. For example, the bones of their fingers and toes are slender and curved like those of modern apes, and unlike modern human fingers and toes. Such hands would have been well suited to grasping branches as *A. afarensis* hung below them. Certain distinctive features of their shoulder blades are also well suited to supporting the body in a hanging position.

FIGURE 11.16 Each night, chimpanzees sleep in nests that they make by bending down branches. Sometimes chimpanzees also rest in nests during the day, as this chimpanzee is doing.

Australopithecus afarensis was sexually dimorphic in body size.

Anatomical evidence suggests that there was considerable variation in the body size of *A. afarensis* adults in the Hadar population. The bigger individuals were 1.51 m (5 ft) tall and weighed about 45 kg (100 lb). The smaller ones were about 1.05 m (3½ ft) and weighed about 30 kg (65 lb). Thus, the big ones were about 1.5 times larger than the small ones. There are two possible explanations for variation in the Hadar population. The variation could represent sexual dimorphism, with the larger adults being male and the smaller ones female. The difference between the Hadar males and females approaches the magnitude of sexual dimorphism in modern orang-utans and gorillas and is considerably greater than that in modern humans, bonobos, and chimpanzees.

On the other hand, it is possible that large and small individuals represent two different species. This interpretation was initially supported by the fact that larger individuals were much more common at Laetoli, which was then a savanna, while the smaller ones were found mainly in what were then forested environments. Some researchers also believe that the smaller individuals were less suited to bipedal loco-motion than the bigger ones were. This suggests that one species was big, fully bipedal, and adapted for savanna life. The morphology of the smaller species represented a compromise between tree climbing and walking—an adaptation for life in the forest.

The fossils recently discovered in Ethiopia by Tim White and Donald Johanson suggest that the two-species theory is wrong. The new fossils encompass the full range of variation in size at a single site and over a fairly narrow range of dates. Moreover, a large femur shows many of the same features seen in Lucy's more petite femur. The new data also suggest that both large and small forms were found in the full range of environments from forest to savanna. All of these facts suggest that this assemblage of fossils represents a single, sexually dimorphic species.

A. AFRICANUS

Australopithecus africanus is known from several sites in South Africa that date from 3 to 2.2 mya.

Australopithecus africanus was first identified in 1924 by Raymond Dart, an Aus-tralian anatomist living in South Africa. Miners brought him a piece of rock from which he painstakingly extracted the skull of an immature, small-brained creature (Figure 11.17). Dart formally named this fossil *Australopithecus africanus*, which means "the southern ape of Africa," but there is now evidence that members of the same species may have been distributed throughout southern and eastern Africa. Dart's fossil is popularly known as "the Taung child." Dart believed that the Taung child was bipedal because the position of the foramen magnum was more like that of modern humans than that of apes. At the same time, he observed many similarities between the Taung child and modern apes, including a relatively small brain. Thus, he argued that this newly discovered species was intermediate between apes and humans. His conclusions were roundly rejected by members of the scientific community because in those days most physical anthropologists believed that large brains had evolved in the human lineage before bipedal locomotion. Controversy about the taxonomic status of *A. africanus* continued for the next 30 years.

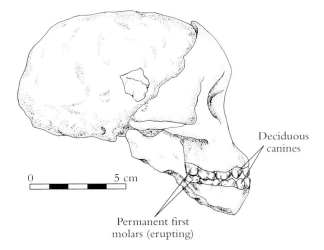

Deciduous
canines

0   5 cm

Permanent first
molars (erupting)

FIGURE 11.17 The first australopithecine specimen was identified in South Africa by Raymond Dart in 1924. Dart named it *Australopithecus africanus*, which means "the southern ape of Africa," but the specimen is often called the Taung child. Dart concluded that the Taung child was bipedal. Even though it had a very small brain, he believed it to be intermediate between humans and apes. His conclusions were not generally accepted for nearly 30 years. (Figure courtesy of Richard Klein.)

The creature who caused all the controversy was a small biped with relatively modern dentition and postcranial skeleton. As with *A. afarensis*, there is pronounced sexual dimorphism, in both canine and body size. Males stood 1.38 m (4 ft 6 in.) tall and weighed 41 kg (90 lb), and females were 1.15 m (3 ft 9 in.) tall and tipped the scales at about 30 kg (66 lb). The teeth are more modern than those of *A. afarensis* in several ways. The brain averaged 442 cc, somewhat larger than the average for *A. afarensis*. However, the difference in cranial capacity between *A. afarensis* and *A. africanus* is small compared with that among individuals and is likely due to sampling variation. Members of this species lived from 3.0 to 2.2 mya. The postcranial skeleton is virtually identical to that of *A. afarensis*. Dart's notion that the Taung child was bipedal was controversial because it relied primarily on the location of the foramen magnum, and he had no postcranial bones to corroborate his conclusions. This is partly because Dart's original find came from the rubble of mining operations where dynamite was used to excavate, and the fossils were not discovered in their original location. Subsequently, many adult skulls and postcranial bones were found at two other sites in South Africa: Makapansgat and Sterkfontein. The hip bone, pelvis, ribs, and vertebrae of *A. africanus* are much like those of Lucy and the First Family. The Taung child thus was proven to be fully bipedal, just as Dart had originally claimed.

The skull of A. africanus *has a number of derived features. Only some of these are shared with modern humans.*

Australopithecus africanus is less primitive than *A. afarensis* in a number of important details. For example, the base of the skull has fewer air pockets, the face is shorter and less protruding below the nose, the canines are less dimorphic, and the base of the cranium is bent further upward, or flexed. In all these traits *A. africanus* is more like modern humans and less like earlier primates. However, *A. africanus* exhibits a number of derived characters that it does not share with modern humans. Most of these characters seem to have to do with heavy chewing using the molars. The molars are quite big, and the lower jaw is larger and sturdier than in modern humans.

Australopithecus africanus matured rapidly like chimpanzees, not slowly like humans.

For many years it was thought that the Taung child was about six years old when it died. This estimate was based on the assumption that the australopithecines matured at the same rate that modern humans do. Recently, we have learned that this assumption is wrong; australopithecines grew up quickly. Among living primates, the age at which teeth, particularly molars, erupt is a very good predictor of the age of sexual maturity, age at first reproduction, and overall life span. Thus, if we know the rate at which teeth developed in extinct hominins, we can estimate how rapidly they developed overall. Christopher Dean and his colleagues at the University of London have found a way to tell how fast hominin teeth developed. Their method relies on the fact that enamel is secreted as teeth grow. Over the course of the day the rate of secretion varies, and this variation creates minute parallel lines that correspond to a day's growth. Dean and his colleagues used electron and polarizing light microscopes to make highly magnified images of the teeth of six australopithecines, and they carefully counted the daily layers of enamel. Then they calculated how long it took to grow to a particular stage of development. Their results indicate that australopithecines developed rapidly, somewhat faster than chimpanzees, not slowly like modern humans (Figure 11.18). So the Taung child was not six years old when it died; it was probably only about three.

These results are important because they suggest that australopithecine infants did not have as long a period of dependency as human children do. Many anthropologists believe that a number of the fundamental features of human foraging societies, such as the establishment of home bases, sexual division of labor, and extensive food sharing, were necessitated by a long period of infant dependency. If australopithecine infants matured faster than modern chimpanzees, then it seems likely that these features were not yet part of the hominin adaptation.

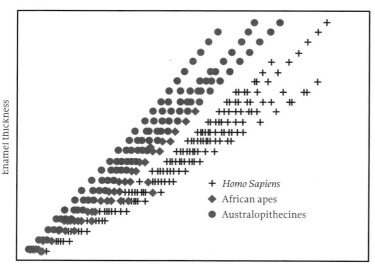

Enamel formation time

+ *Homo Sapiens*
♦ African apes
● Australopithecines

FIGURE 11.18 The rate of growth of enamel in the teeth of six australopithecines was comparable to that of African apes. The enamel growth rate of modern humans is slower.

A. GARHI

Australopithecus garhi lived about 2.5 mya in East Africa.

In 1999, a research team led by Berhane Asfaw and Tim White announced the discovery of a new species of australopithecine from the Awash valley of Ethiopia, named *Australopithecus garhi*. *Garhi* means "surprise" in the Afar language. In 1996, Asfaw, White, and their colleagues had recovered hominin remains from a site called Bouri, the ancient site of a shallow freshwater lake, not far from where *A. afarensis* was found. The findings included a number of postcranial bones and the partial skeleton of one individual. These specimens are well preserved and securely dated to 2.5

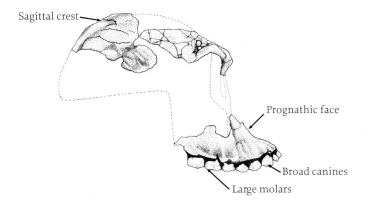

Sagittal crest

Prognathic face

Broad canines

Large molars

FIGURE 11.19 In 1999, the first descriptions of a 2.5-million-year-old hominin species from Ethiopia were published. This new species, named *A. garhi*, had a small brain and a prognathic face. It also had a sagittal crest, a fin of bone that runs along the top of the skull. Compared to *A. afarensis* and *A. africanus*, *A. garhi* had broader canines and larger molars.

mya, but they don't contain diagnostic features that can be used to assign them to a specific species. In 1997, however, the research team had discovered a number of cranial remains about 300 m (1000 ft) away from the original site. These pieces of the skull, maxilla (upper jaw), and teeth come from the same stratigraphic level as the postcranial remains, and they reveal more about the taxonomic identity of the Bouri hominins.

These creatures had small brains (approximately 450 cc), like *A. afarensis* and *A. africanus*, perched above a very prognathic face (Figure 11.19). The canines, premolars, and molars of *A. garhi* were generally larger than those of *A. afarensis* and *A. africanus*. Certain detailed features of the dentition differed from *A. afarensis* as well. This creature also had a **sagittal crest**, a fin of bone that runs along the centerline of the skull, making it look a bit like a punk rocker's Mohawk haircut. The Bouri specimens were assigned to a new species of the genus *Australopithecus* because they differ from any of the known australopithecines but lack the derived characters associated with other hominins.

So far, the postcranial remains found near the type specimen of *A. garhi* cannot be assigned to a particular species, but they do reveal some interesting developments in the hominin lineage. In chimpanzees, the humerus, the radius and ulna (the lower bones of the arm), and the femur are about the same length; in modern humans, the bones of the forearm have become shorter and the femur has become longer than the humerus. *Australopithecus afarensis* resembles chimpanzees in the proportions of these bones. However, reconstructions of the humerus, radius, ulna, and femur of the hominins at Bouri suggest that these creatures' femurs had become somewhat elongated, while the relative proportions of the humerus and forearm bones had not changed. The postcranial remains suggest that there was considerable variation in the size and robustness of the Bouri hominins. This variation may reflect the fact that males were larger than females, but there is not yet enough evidence to be certain that this was the case.

A. HABILIS/RUDOLFENSIS

Fossils with a larger brain and more humanlike teeth have been discovered at many sites in East Africa.

In 1960, while working at Olduvai Gorge with his parents, renowned paleoanthropologists Louis and Mary Leakey, Jonathan Leakey found pieces of a hominin jaw,

FIGURE 11.20 The discovery of KNM-ER 1470 at a site on Lake Turkana confirmed that there was at least one large-brained hominin in East Africa about 2 mya. (Photograph courtesy of Alan Walker.)

cranium, and hand. The Leakeys assigned this specimen, labeled Olduvai Hominin 7 (OH 7), to the genus *Homo* because they believed the cranial bones indicated that it had a much larger brain than *Australopithecus* had, and small, more humanlike teeth. Louis Leakey named the species *Homo habilis*, or "handy man," because he believed he had found the hominin responsible for the simple, flaked-stone tools discovered nearby. Later the Leakeys found several more fossils that they assigned to *H. habilis*, including pieces of a second cranium, part of a foot, and more teeth. These fossils range in age from 1.6 to 1.9 million years old.

Many paleoanthropologists doubted that the Leakeys had found the oldest member of our genus. After all, the Leakeys' estimate of OH 7's brain size was based on only part of the cranium, and even then the complete cranium was estimated to be at most 50% larger than the brains of australopithecines. There was plenty of room for skepticism. Another member of the famous Leakey family helped prove that large brains were indeed present in a hominin living 1.9 mya. Mary and Louis Leakey's son Richard set up his own research team and began work at a site called Koobi Fora on the eastern shore of Lake Turkana. In 1972, a member of Richard Leakey's team, Bernard Ngeneo, found a nearly complete skull of a hominin. This find is usually referred to by its official collection number, KNM-ER 1470 (Figure 11.20). (This identification system is not as obscure as it may seem. *KNM* stands for Kenya National Museum, and *ER* stands for East Rudolf—Lake Turkana was known as Lake Rudolf when Leakey began his work there.) The most striking thing about KNM-ER 1470 is that it had a much larger brain than any known australopithecine had. Its endocranial volume is 775 cc, 75% bigger than the brains of specimens of *A. africanus*. Compared to OH 7, however, it had a relatively large, australopithecine-like face. Since the discovery of KNM-ER 1470, similar fossils have been found at several other sites, including Olduvai Gorge, the Omo basin, Sterkfontein, and a site near Lake Malawi. The oldest of these fossils dates to 2.4 mya.

Since the discovery of KNM-ER 1470, many other fossils have been assigned to the same species as OH 7 and KNM-ER 1470. The skulls and teeth of these fossils are quite variable. Some of them, like KNM-ER 1813 (Figures 11.21 and 11.22), have more humanlike teeth and less robust skulls with smaller faces and teeth than KNM-ER 1470, but they also have smaller brains than KNM-ER 1470, often about 500 cc. Some believe that the differences among these specimens represent variation within a single species. Others have argued that there were two species of relatively large-brained hominins present in East Africa 2 mya, and that the smaller-brained, less robust individuals should be classified as *habilis*, while the more robust ones with larger brains are a second species that should be named *rudolfensis*. Until very recently, both species would have been placed in the genus

FIGURE 11.21 KNM-ER 1813 (shown here) had more modern teeth but a smaller brain than KNM-ER 1470. This variation led many investigators to argue that fossils assigned to the species *habilis* actually belonged to more than one species of hominin. (Photograph courtesy of Michael Day.)

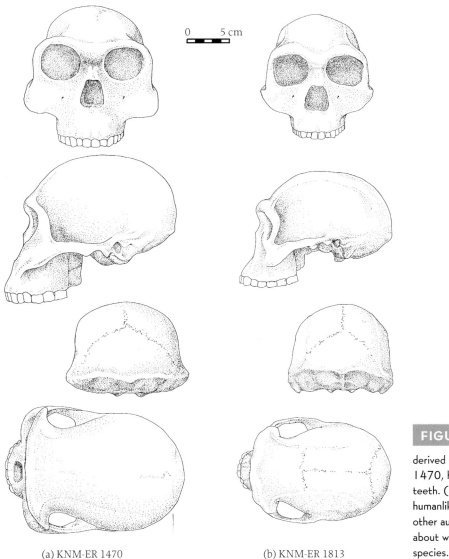

0   5 cm

(a) KNM-ER 1470   (b) KNM-ER 1813

FIGURE 11.22   Fossils assigned to the species *habilis* show a mixture of derived and ancestral traits. (a) Some, like KNM-ER 1470, have larger brains but retain robust skulls and teeth. (b) Others, like KNM-ER 1813, have more humanlike skulls and teeth but retain the small brains of other australopithecines. Anthropologists are divided about whether these fossils are members of the same species. (Figure courtesy of Richard Klein.)

Homo, as the Leakeys originally proposed; but, as you will see shortly, many now advocate their placement in the genus *Australopithecus*.

The postcrania and developmental patterns of these creatures are similar to those of australopithecines. The first fragmentary postcranial material found at Olduvai Gorge suggested that OH 7 was very similar to the australopithecines from the neck down. A new discovery from Olduvai Gorge adds credence to this picture. During an expedition led by Don Johanson, Tim White discovered a specimen including an incomplete cranium and numerous postcranial bones. This specimen, labeled OH 62, had a very short stature and surprisingly long arms, suggesting that the postcranial skeletons of the Olduvai *habilis* specimens were similar to those of Lucy 1.4 million years earlier. Analysis of the layering of dental enamel indicates that specimens assigned to the species *habilis* and *rudolfensis* had rapid apelike developmental patterns, like other australopithecines.

It has recently been proposed that these fossils should be assigned to the genus Australopithecus *because they are fundamentally more similar to other australopithecines than to other species assigned to* Homo.

The fossils assigned to *habilis* and *rudolfensis* previously were classified as members of the genus *Homo* because they are similar to us in the following two ways:

1. The brain of these fossils is bigger than that of any of the australopithecines.
2. The teeth are smaller and have thinner enamel, and the dental arcade is more parabolic, than in the australopithecines. The skulls are more rounded, there are fewer air pockets in the bottom of the skull, the face is smaller and protrudes less, and the jaw muscles are reduced in size compared with the australopithecines.

In addition, it was initially assumed that these creatures were responsible for manufacturing the flaked-stone tools that appear in the archaeological record for the first time about 2.5 mya. As we will see in the next chapter, however, we do not know which hominin is responsible for these tools; it could have been *habilis* or *rudolfensis*, but it also could have been one of the other early hominins discussed here or *Homo ergaster*, a species we will introduce in Chapter 13.

Wood and Collard, whose taxonomic scheme we have adopted here, argue that these fossils do not belong in *Homo* but should be included in the genus *Australopithecus* instead. Remember from Chapter 4 that the rules of cladistic taxonomy specify that all the species comprising a genus must be descended from a common ancestor. However, this guideline does not tell us how to group species into genera; they could all be placed in one genus or divided into several (Figure 11.23). Wood and Collard argue that species grouped in a genus should share a common **adaptive grade**, meaning that they live in basically the same way. As we will see in Chapter 13, the other members of the genus *Homo* are larger than these creatures, they are fully committed to terrestrial life, and they have reduced sexual dimorphism. These features are correlated with important behaviors, such as hunting, complex extractive foraging, food sharing, and extensive paternal investment in offspring. Thus, Wood and Collard argue that *habilis* and *rudolfensis* should be grouped with the bipedal but otherwise apelike creatures in the genus *Australopithecus*, and we follow their suggestion here.

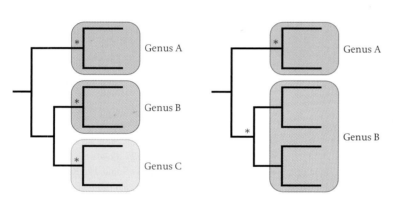

FIGURE 11.23 The same phylogeny with six species can be subdivided into different numbers of genera. In the first panel are three genera, each with two species; in the second panel are two genera, one with two species and the other with four. In each case, each genus includes all of the species that descend from the common ancestor, which is labeled with an asterisk. According to Wood and Collard, the second arrangement is better if the species in genus B share a common adaptive grade.

Paranthropus

Paranthropus aethiopicus *is a hominin who has teeth and skull structures specialized for heavy chewing.*

Alan Walker of Pennsylvania State University discovered the skull of a very robust australopithecine at a site on the west side of Lake Turkana in northern Kenya. Walker called it the "Black Skull" because of its distinctive black color, but its official name is KNM-WT 17000 (Figure 11.24) (*WT* stands for West Turkana, the site at which the fossil was found). This creature lived about 2.5 mya, and in some ways it is similar to *A. afarensis*. For example, the hinge of the jaw in KNM-WT 17000 has the same primitive structure as the jaw in *A. afarensis* has, which is very similar to the jaw joints of chimpanzees and gorillas. The later australopithecines and members of the genus *Homo* have a modified jaw hinge. In addition, both KNM-WT 17000 and *A. afarensis* were quite sexually dimorphic, they had a similar postcranial anatomy, and they were equipped with relatively small brains in relation to body size.

However, as shown in Figure 11.25, KNM-WT 17000 is very different from the skull of other australopithecines. The molars are enormous, the lower jaw is very large, and the entire skull has been reorganized to support the massive chewing apparatus. For example, it has a pronounced sagittal crest, which enlarges the surface area of bone available

FIGURE 11.24 The Black Skull (KNM-WT 17000), which was discovered on the west side of Lake Turkana in Kenya.

for attaching the **temporalis muscle**, one of the muscles that works the jaw. You can easily demonstrate for yourself the function of the sagittal crest. Put your fingertips on your temple and clench your teeth; you will feel the temporalis muscle bunch up. As you continue clenching your teeth, slowly move your fingertips upward until you can't feel the muscle any more. This is the top point of attachment for your temporalis muscle, about one inch above your temple. Australopithecines had bigger teeth than we do and so needed larger temporalis muscles, and these muscles required more space

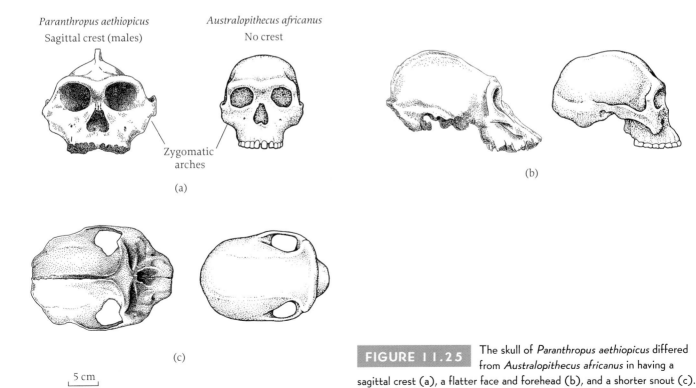

Paranthropus aethiopicus
Sagittal crest (males)

Australopithecus africanus
No crest

Zygomatic arches

(a)

(b)

(c)

5 cm

FIGURE 11.25 The skull of *Paranthropus aethiopicus* differed from *Australopithecus africanus* in having a sagittal crest (a), a flatter face and forehead (b), and a shorter snout (c).

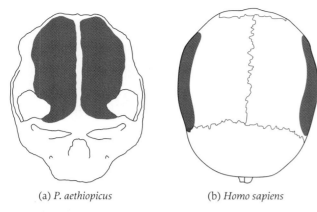

(a) *P. aethiopicus*   (b) *Homo sapiens*

FIGURE 11.26 In these two skulls—of *P. aethiopicus* (a) and a modern human (b)—the area of attachment of the temporalis muscle is shown in red.

for attachment to the skull. In *A. africanus* the muscles expanded so that they almost met at the top of the skull. The even larger teeth of *P. aethiopicus* required even more space for muscle attachment, and the sagittal crest served this function (Figure 11.26). Other distinctive features of *P. aethiopicus* are also accommodations to the enlarged teeth and jaw musculature. For example, the cheekbones (zygomatic arches) are flared outward to make room for the enlarged temporalis muscle. The flared cheekbones caused the face to be flat, or even pushed in.

Paranthropus robustus is a more recent species found in southern Africa.

At Kromdraai, a site in South Africa, retired Scottish physician and avid paleontologist Robert Broom discovered fossils that seemed very different from the *A. africanus* specimens previously found at Sterkfontein (Figure 11.27). These creatures were much more robust (with more massive skulls and larger teeth) than their neighbors, so Broom called his find *Paranthropus robustus*. Many of his colleagues questioned this choice. They thought he had discovered another species of *Australopithecus*, and so for a time most paleontologists classified Broom's fossils as *Australopithecus robustus*. Recently, however, the consensus has been that the robust early hominins are so different from the gracile (more lightly built) australopithecines that they should be placed in a separate genus, and Broom's original classification has been restored.

There is a large sample of *P. robustus* fossils from Kromdraai and also from a second site nearby, named Swartkrans. These creatures appeared about 1.8 mya and disappeared about 1 mya. Their brains averaged about 530 cc. *Paranthropus robustus* males stood about 1.3 m (4½ ft) tall and weighed about 40 kg (88 lb); females were about 1.1 m (3½ ft) tall and weighed 32 kg (70 lb). Their postcranial anatomy shows that they were undoubtedly bipedal. They share many of the same derived features with humans that *A. africanus* does. In addition, they share with *A. africanus* several specialized features for heavy chewing, but these feature are much more pronounced in *P. robustus*.

It is unclear what the paranthropines were doing with this massive chewing apparatus. Until recently, most anthropologists believed that the paranthropines relied more on plant materials that required heavy chewing than *A. africanus* did. In other animals, large grinding teeth are often associated with a diet of tough plant materials, and omnivorous animals typically have relatively large canines and incisors. Moreover, the wear patterns on *P. robustus* teeth suggest that they ate very hard foods like seeds or nuts. However, recent research on the chemistry of *P. robustus* teeth suggests that they probably also ate substantial amounts of meat (Box 11.1).

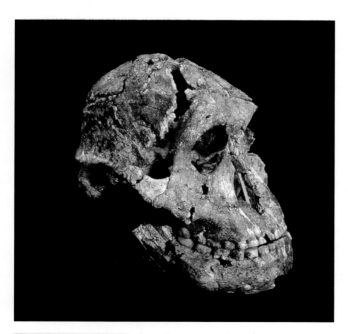

FIGURE 11.27 A nearly complete skull of *Paranthropus robustus* was found at Sterkfontein in South Africa in 1999. This species has very large jaws and molars and relatively small incisors and canines.

BOX 11.1

Chemical Clues about the Diet of Paranthropines

Two kinds of chemical analyses of the fossilized teeth of paranthropines suggest that they may have eaten substantial amounts of animal matter. The first kind of evidence comes from the obscure but useful fact that the ratio of strontium (Sr) to calcium (Ca) in the bones of living animals declines as these elements are passed up the food chain: plants have higher Sr:Ca ratios than do the grazing herbivores that eat them, and grazing herbivores that feed mainly on grasses have higher Sr:Ca ratios than do the carnivores that prey on them. Browsing herbivores constitute the major exception to this rule. Browsers, such as deer and kudu, eat mainly the leaves and shoots of woody plants, but they have low Sr:Ca ratios like carnivores.

To find out what *P. robustus* was eating, Andrew Sillen of the University of Cape Town measured the Sr:Ca ratios in the fossilized bones of *P. robustus* and in the fossilized bones of a number of animal species found at Swartkrans. He needed to survey multiple species to be sure that fossil specimens showed the same relative Sr:Ca ratios that living members of the same species show (Figure 11.28). He found that they do, so we can be confident that the Sr:Ca ratios were not changed by geochemical processes after the animals died. Sillen's measurements of the Sr:Ca ratio for *P. robustus* suggest that these creatures were either browsers or carnivores but not grazing herbivores.

To determine whether *P. robustus* was a browser or a carnivore, we turn to another line of evidence. It turns out there are two chemically distinct kinds of photosynthesis. Trees, shrubs, and other woody plants utilize one type, called "C_3 photosynthesis"; and grasses use a second type, called "C_4 photosynthesis." C_4 plants and the animals that eat them have higher concentrations of the heavy isotope of carbon, carbon-13, than do C_3 plants and the animals that eat C_3 plants. Julia Lee-Thorp of the University of Cape Town and her colleagues measured the amount of carbon-12 and carbon-13 in the fossilized teeth of a number of species at Swartkrans, including *P. robustus*. The levels of carbon-13 found in the teeth of *P. robustus* are

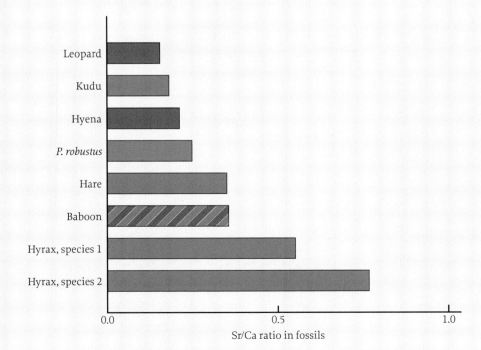

FIGURE 11.28 Measuring the ratio of strontium (Sr) to calcium (Ca) in the bones of both living and fossil species can help paleontologists deduce the kinds of diets that extinct creatures had. In general, the Sr:Ca ratio in the bones of living animals declines as you move up the food chain from pure herbivory to carnivory. An exception is browsing herbivores (like kudu), which have low Sr:Ca ratios. Here, the Sr:Ca ratios for typical herbivores (green) and carnivores (red) are plotted along with the Sr:Ca ratio for *P. robustus* (blue). Baboons, who combine herbivory and carnivory, are plotted in red and green. The low Sr:Ca ratio in *P. robustus* is consistent with carnivory or browsing herbivory, but not with general herbivory.

higher than the levels found in browsing species like kudu. This evidence indicates that *P. robustus* either ate grass and grass seeds or ate animals that ate grasses. Because the Sr:Ca ratios suggest that *P. robustus* did not eat grasses, these data indicate that *P. robustus* was at least partially carnivorous.

Taken together, these data suggest that *P. robustus* consumed some animal protein. Nonetheless, their massive teeth may have evolved for chewing tough plant material. For example, they might have specialized on tough plant material during the dry season but had a more diverse diet during the rest of the year.

FIGURE 11.29   Olduvai Gorge in Tanzania has been the site of many important paleontological and archaeological finds. Louis and Mary Leakey worked in Olduvai for more than 30 years.

FIGURE 11.30   The fossil cranium of *Kenyanthropus platyops* was found near Lake Turkana in northern Kenya. It dates to about 3.5 mya and has a unique combination of anatomical features.

Paranthropus boisei was a robust robustus.

Another paranthropine with large molars was discovered by Mary Leakey at Olduvai Gorge, Tanzania, in 1959 (Figure 11.29). This specimen, officially labeled Olduvai Hominin 5 (OH 5), was first classified as *Zinjanthropus boisei*. *Zinj* derives from an Arabic word for "East Africa," and *boisei* comes from Charles Boise, who was funding Leakey's research at the time. Leakey's find was later reclassified *Paranthropus boisei* because of its affinities to the South African forms of *P. robustus*. (Those who advocate classifying *P. robustus* as *Australopithecus robustus* would assign Leakey's find to the same genus, and call it *A. boisei*.) The discovery of OH 5 was important, partly because it ended nearly 30 frustrating years of work in Olduvai Gorge in which there had been no dramatic hominin finds. It was only the first of a very remarkable set of fossil discoveries at Olduvai.

Essentially, *P. boisei* is an even more robust *P. robustus*—that is, a hyperrobust paranthropine. Its body is somewhat larger than the body of *P. robustus*, and its molars are larger than those of *P. robustus*, even when the difference in body size is taken into account. The skulls are even more specialized for heavy chewing.

Paranthropus boisei appears in the fossil record at about 2.2 mya in eastern Africa. It became extinct about 1.3 mya, although the exact date of its disappearance is difficult to establish.

Kenyanthropus

Kenyanthropus platyops is a recently discovered species that lived in East Africa between 3.5 and 3.2 mya.

In 1999, Justus Erus, a member of a research team led by Meave Leakey, found a nearly complete hominin cranium, labeled KNM-WT 40000, on the western side of Lake Turkana in Kenya (Figure 11.30). KNM-WT 40000 has what one commentator described as "a dizzying mosaic of features." As in chimpanzees, *A. anamensis*, and *A. ramidus*, the cranium has a small ear hole. Like most of the early hominins, the specimen has a chimpanzee-sized braincase and thick enamel on its molars. However, its molars are substantially smaller than those of any other early hominin except *A. ramidus*. Its face is broad and very flat, like those of *P. boisei* and *A. rudolfensis*. Because this specimen combines features not found in other hominins, Leakey and her coauthors placed it, and a fragmentary upper jaw found nearby some years earlier, in a new genus, *Kenyanthropus*, meaning "Kenyan man." The species name, *platyops*, is from the Greek for "flat face" and refers to the most notable anatomical feature of this specimen. Argon–argon dating methods have revealed that KNM-WT 40000 is 3.5 million years old. The fossils found in association with the cranium suggest that these creatures lived in a mix of woodland and savanna environments.

HOMININ PHYLOGENIES

It is difficult to infer the phylogenetic relationships among the Plio-Pleistocene hominins.

Remember, the goal of our enterprise is to reconstruct the evolutionary history of human beings. To do so, ideally we would like to understand the phylogenetic relationships among our ancestors. Unfortunately, as the hominin fossil record has become richer, this task has become harder. The problem is illustrated by the phylogeny shown in Figure 11.31, which accompanied the first published description of *Kenyanthropus platyops*, in the journal *Nature*. Clearly, a vast number of different phylogenies could be constructed, depending on which relationships are chosen and which are rejected.

The problem arises from the extensive convergence and parallelism in hominin evolution. This means that many alternative phylogenies are equally plausible, depending on what is assumed to be homologous and what is assumed to be convergent. Randall Skelton of the University of Wyoming and Henry McHenry of the University of California Davis performed a phylogenetic analysis of 77 traits of the skulls of a number of the early hominin species discussed here. They sorted the traits into groups according to their function and their location in the body. For example, one functional group of traits included features thought to be adaptations for heavy chewing, such as large teeth and a sagittal crest, and another group included all of the features of the bottom of the cranium. Then Skelton and McHenry determined the best trees to fit different combinations of traits. They found that the branching pattern of the best tree depends on

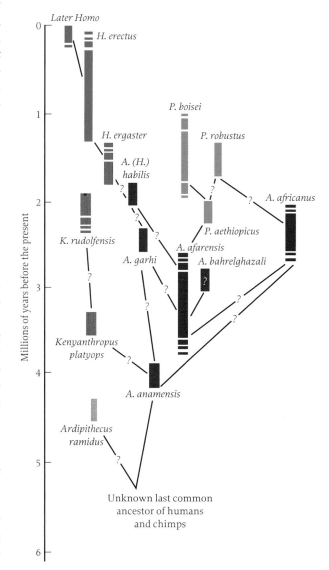

FIGURE 11.31 A phylogeny accompanying the first published description of *K. platyops* in the journal *Nature*. The question marks represent links that the article's author regards as tenuous or uncertain. Notice that nearly every link has a question mark. The author has chosen to classify *Australopithecus rudolfensis* as *Kenyanthropus rudolfensis*, and *A. bahrelghazali* is a specimen we consider to be *A. afarensis*.

which groups of traits are included. For example, the best fit based on traits in the heavy-chewing complex groups together all the paranthropines and shows *Homo* species as deriving from *A. afarensis* (Figure 11.32). In contrast, when the traits in the heavy-chewing complex are eliminated, the results are consistent with the radically different tree shown in Figure 11.33. In this case, *P. aethiopicus* is unrelated to *P. robustus* and *P. boisei*, and *Homo* species derive from *A. africanus*.

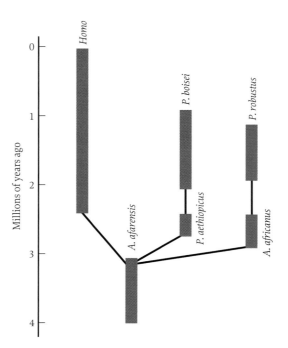

FIGURE 11.32 Before the discovery of *P. aethiopicus* (KNM-WT 17000), there was a general consensus about the hominin phylogeny: that *Homo* species derive from *A. afarensis*, and that *A. africanus* began a lineage of increasing robustness culminating in *P. boisei*.

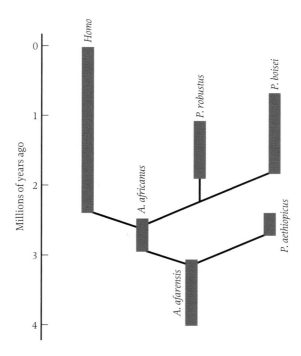

FIGURE 11.33 If we assume that the similarities in traits other than those involved in heavy chewing are homologous, then we might obtain the phylogeny shown here, in which *A. afarensis* gives rise to *A. africanus*, who in turn gives rise to *Homo* species and the later paranthropines. According to this phylogeny the similarities between *P. aethiopicus* and the other paranthropines are due to parallel evolution.

THE EVOLUTION OF EARLY HOMININ MORPHOLOGY AND BEHAVIOR

The absence of a secure phylogeny for the early hominins does not prevent us from understanding human evolution.

It is easy to be discouraged by the uncertainties in the hominin fossil record and to doubt that we know anything concrete about our earliest ancestors. Although there are many gaps in the fossil record and much controversy about the relationship among the early hominin species, we do know several important things about them. In every plausible phylogeny, the human lineage is derived from a small biped who was adept in trees. Its teeth and jaws were suited for a generalized diet. The males were considerably larger than the females, their brains were the same size as those of modern apes, and their offspring developed rapidly like modern apes. This is the kind of creature that linked the apes of the Miocene to the earliest members of our own genus, *Homo*.

In the discussion that follows, we first consider why bipedal locomotion evolved, and we then review what we can infer about the behavior and life history of these hominins.

The Evolution of Bipedalism

There are several explanations for the evolution of bipedalism.

Ever since Darwin, evolutionists have speculated about why bipedal locomotion was favored by natural selection. Here we examine the most popular hypotheses.

Walking on two legs is an efficient form of locomotion on the ground. At first glance, this hypothesis seems to be an unlikely possibility. After all, there are far more quadrupedal species than bipedal species, and studies of the energetics of locomotion show that terrestrial quadrupeds like horses move about more efficiently than humans do. If bipedalism is a superior form of locomotion, we would expect it to be more common. However, Peter Rodman and Henry McHenry at the University of California Davis, who studied the locomotion of humans and of quadrupedal primates on the ground, found that bipedal and quadrupedal locomotion are more or less equivalent in efficiency.

If bipedal and quadrupedal locomotion are equally efficient means of traveling on the ground, why did humans become bipedal when they left the trees? Rodman and McHenry suggest that the answer has to do with the kind of quadrupedalism found in hominin ancestors. They conjecture that the ancestors of terrestrial monkeys were arboreal quadrupeds who fed on the tops of branches, but the ancestors of the hominins were suspensory feeders (hanging below branches to feed, as modern orangutans do). Selection might have favored quadrupedalism among animals descended from above-the-branch feeders and favored bipedalism among animals descended from suspensory feeders—in each case because that form of ground locomotion required fewer anatomical changes.

A recent study of the wrist morphology of *A. anamensis* and *A. afarensis* casts doubt on this explanation for bipedalism. Anthropologists Brian Richmond and David Strait of George Washington University have shown that the wrist bones of chimpanzees and gorillas have a suite of distinctive features that facilitate knuckle walking. They have also shown that the wrists of *A. anamensis* and *A. afarensis* share these knuckle-walking features. However, neither of these species exhibit other knuckle-walking adaptations that are seen in the hand bones of chimpanzees and gorillas. Richmond and Strait argue that these australopithecines had already become committed bipeds, but their wrists retained some features of their knuckle-walking ancestors. If this hypothesis is correct, then human bipedalism evolved in a creature already adapted to terrestrial locomotion, not a creature exclusively adapted for suspensory locomotion. This scenario leaves us with a puzzle: why did a knuckle-walking ape shift to bipedal locomotion?

Erect posture allowed hominins to keep cool. Heat stress is a more serious problem in the open, sun-baked savanna than in the shade of the forest. Although modern apes are restricted to forested areas, hominins apparently moved into more open areas. If an animal is active in the open during the middle of the day, it must have a way to prevent its body temperature, particularly the temperature of its brain, from rising too high. Many savanna creatures have adaptations that allow their body temperatures to rise during the day and to cool off at night. They also have adaptations (such as large nasal chambers) that keep the temperature of the blood in their brain well below the temperature of the blood in the rest of their bodies. However, this option does not seem to be available to hominins. Instead, humans have evolved a system of evaporative cooling: we sweat. Peter Wheeler of Liverpool John Moores University has pointed out that the erect posture associated with bipedal locomotion helps to reduce heat stress and lessens the amount of water necessary for evaporative cooling, in three different ways (Figure 11.34):

1. The sun strikes a much smaller fraction of the body of an erect hominin than of a quadrupedal animal of the same size. An erect hominin absorbs sunlight directly only on the top of its head and shoulders, while a quadruped absorbs the energy of the sun across its whole back and head. (Think about how much harder it is to get a tan standing up than lying down.) This difference is important because near the equator the sun is overhead during much of the middle of the day.
2. The air is warmer near the ground than it is higher up. The sun heats the soil, and the soil radiates and conducts heat back to the air, so air temperatures are highest close to the ground.
3. Wind velocities are higher 2 m (about 6 ft) off the ground than they are 1 m (about 3 ft) off the ground. Because moving air increases the efficiency of evaporative cooling, an erect hominin is better able to withstand the heat of the day.

FIGURE 11.34 Bipedal locomotion helps an animal living in warm climates to keep cool by reducing the amount of sunlight that falls on the body, by increasing the animal's exposure to air movements, and by immersing the animal in lower-temperature air.

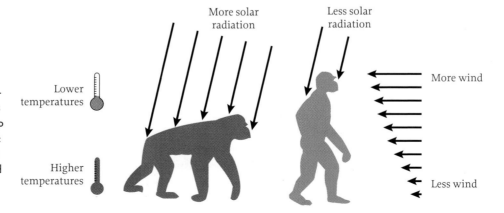

Using standard engineering methods, Wheeler calculated that these three factors would reduce heat stress and water consumption in bipedal hominins compared with quadrupeds. Interestingly, the same calculations show that loss of body hair provides a thermoregulatory advantage to a bipedal creature, but not to a quadruped of the same size.

Bipedal locomotion leaves the hands free to carry things. The ability to carry things is, in a word, handy. Quadrupeds can't carry things in their hands without interfering with their ability to walk and to climb. As a consequence, they must carry things in their mouths. Some Old World monkeys pack great quantities of food into their cheeks to be chewed and swallowed later. Other primates must eat their food where they find it—a necessity that can cause problems when food is located in a dangerous place, or when there is a lot of competition over food. Bipedal hominins can carry larger quantities of food in their hands and arms.

Bipedal posture allows efficient harvesting of fruit from small trees. University of Indiana anthropologist Kevin Hunt has argued that the anatomy of *A. afarensis* is well suited to standing erect but not well designed for efficient bipedal walking. Hunt thinks that bipedal posture was favored because it allows efficient harvesting of fruit from the small trees that predominate in African woodlands. Two kinds of data support this hypothesis. First, Hunt found that chimpanzees rarely walk bipedally, but they spend a lot of time standing bipedally as they harvest fruit from small trees (Figure 11.35). Using their hands for balance, they pick the fruit and slowly shuffle from depleted patches to fresh ones. Standing upright allows the chimpanzees to use both hands to gather fruit, and the slow bipedal shuffling allows them to move from one fruit patch to another without lowering and raising their body weight. Second, Hunt argues that many anatomical features of *A. afarensis* are consistent with standing upright but not with walking bipedally. He points out that the very broad pelvis of *A. afarensis* would make a stable platform for standing on both feet. Hunt also argues that the features of the shoulders, hands, and feet that many anthropologists believe are adaptations for climbing trees are actually adaptations that enabled *A. afarensis* to hang by one hand while standing bipedally and feeding with the other hand.

Any or all of these hypotheses may be correct. Bipedalism might have been favored by selection because it was more efficient than knuckle walking, because it allowed early hominins to keep cool, because it enabled them to carry food or tools from place to place, and/or because it enabled them to feed more efficiently. And once bipedalism had evolved, it might have facilitated other forms of behavior, such as tool use.

FIGURE 11.35 Chimpanzees sometimes stand bipedally as they harvest fruit from small trees. They use one hand for balance and feed with the other, shuffling slowly from one food patch to another.

Early Hominin Subsistence

Increased seasonality in rainfall favors a greater dependence on foods available during the dry season: corms, tubers, and meat.

Herbivorous animals that live in seasonal environments are faced with quite a variable food supply. When the rain falls, grasses sprout, flowers blossom, fruit ripens, and there are many pools of freshwater (Figure 11.36a). Food is abundant and diverse. Because food is plentiful, there is little competition among species over access to food. During the dry season, grasses go to seed, flowers die, and all but a few permanent water holes dry up (Figure 11.36b). Both the abundance and diversity of food are greatly reduced (Figure 11.37). Because there is very little to eat for most herbivorous animals, intense competition erupts among certain species. Many species respond to

FIGURE 11.36 In seasonal habitats, the landscape changes dramatically over the course of the year. Here you see the same area in Amboseli just after the seasonal rains have ended (a) and at the end of the long dry season (b).

(a)   (b)

this problem by focusing their foraging efforts on a particular type of food during the dry season. In Amboseli National Park, for example, baboons spend much of their time during the dry season digging up grass corms (the bulbous storage organs in grass stems). Another way to cope with shortages of plant foods is to become more omnivorous (Figure 11.38). Baboons at many different sites increase their meat intake during the dry season (Figure 11.39).

FIGURE 11.37 In seasonal habitats, resource availability is strongly affected by rainfall. (a) After the rains, lush growth sprouts, and baboon foods are abundant and diverse. (b) By the end of the dry season, the ground is parched, new growth has withered, and less food is available to baboons.

(a)   (b)

Cambridge University anthropologist Rob Foley has suggested that the hominin lineage may have adapted to increased seasonality in two different ways. The robust hominins in the genus *Paranthropus*, with their huge jaws and chewing muscles, may have specialized in processing tough plant materials in much the same way that modern baboons specialize on corms. The less robust species, on the other hand, may have shifted to a more omnivorous diet and relied more heavily on meat.

Chimpanzees are effective hunters who capture monkeys, bushpigs, and small antelope, but they rarely scavenge.

In the past, many anthropologists thought it was unlikely that the early hominins ate very much meat. Hunting, they argued, requires careful planning, subtle coordination, and extensive reciprocity. These interactions would require cognitive and linguistic skills that the relatively small-brained australopithecines must have lacked. Studies of chimpanzee hunting by Christophe and Hedwige Boesch in the Taï Forest of the Ivory Coast, Craig Stanford at Gombe Stream National Park in Tanzania, and John Mitani and David Watts in Ngogo, Uganda, suggest that we must revise this view.

At Taï, Gombe, and Ngogo, chimpanzees participate in hunts approximately once every two or three days. Most predatory activity focuses on monkeys, particularly red

FIGURE 11.38 A male baboon consumes a newborn bushbuck in the Okavango Delta of Botswana.

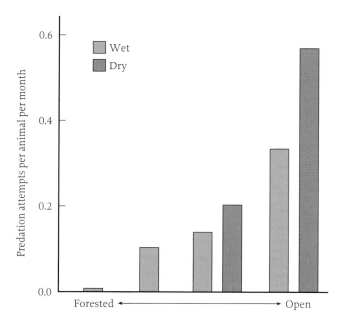

FIGURE 11.39 Observations at different sites in Africa show that baboons hunt meat more often in the dry season than in the wet season.

colobus (Figure 11.40). Gombe chimpanzees also prey on bushpigs and bushbucks (sheep-sized forest antelope) that they encounter on the forest floor (Figure 11.41). Unlike some other meat-eating species, chimpanzees never feed on the carcasses of dead animals that they encounter opportunistically, and they rarely take fresh kills from other predator species.

At Gombe, Ngogo, and Taï, many red colobus hunts begin when the chimpanzees encounter a group of monkeys in the forest or detect them at a distance. They often spend some time on the ground gazing up into the canopy before they begin to hunt. Then at least one chimpanzee climbs into the trees and gives chase. Others watch from the ground.

Although most hunts at Gombe and Ngogo seem to arise out of opportunistic encounters with groups of monkeys, about half of the hunts at Taï begin before there is any sign that monkeys are nearby. In these cases, the chimpanzees fall completely silent, stay very close to one another, frequently change direction, and stop often to search for monkeys in the canopy. The chimpanzees in the three study areas are effective hunters, making kills in approximately half of their hunts.

These observations invalidate the argument that a small-brained hominin must have lacked the necessary cognitive and linguistic skills for hunting. However, they do not tell us how likely it is that the early hominins actually hunted.

(a)

(b)

FIGURE 11.40 Chimpanzees hunt small mammals, particularly red colobus monkeys. (a) A red colobus monkey carrying an infant leaps from one tree to another at Gombe Stream National Park in Tanzania. (b) A male chimpanzee at Gombe consumes a red colobus monkey. (Photographs courtesy of Craig Stanford.)

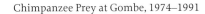
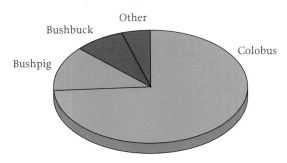

Chimpanzee Prey at Gombe, 1974–1991

Other
Bushbuck
Bushpig
Colobus

FIGURE 11.41 Chimpanzees at Gombe most commonly hunt red colobus monkeys, although they sometimes prey on bushpigs, bushbucks, and other animals.

Early hominins may have used tools like contemporary chimpanzees do.

As we will see in the next chapter, the earliest identifiable tools first appear about 2.5 mya, and it is possible that one of the early hominins discussed here manufactured these tools. However, the absence of identifiable tools does not mean that early hominins were not tool users.

Chimpanzees make and use tools to perform a variety of different tasks. They use long, thin branches, vines, stems, sticks, and twigs to poke into ant nests, termite mounds, and bees' nests to extract insects and honey. They sometimes scrape marrow and other tissue from the bones and braincases of mammalian prey with wooden twigs. They pound stone hammers against heavy flat stones, exposed rocks, and roots to crack open hard-shelled nuts (Figure 11.42). They wad up leaves, dip them into pools of rainwater that have collected in hollow tree trunks, and then suck water from the leafy "sponges."

Moreover, chimpanzees modify natural objects for specific purposes. That is, they are toolmakers, as well as tool users. For example, chimpanzees at Mount Assirik in Senegal, studied by Caroline Tutin of Stirling University and William McGrew of the Miami University of Ohio, use twigs to extract termites from their mounds. The twigs are first detached from the bush or shrub; then leaves growing from the stem are stripped off, the bark is peeled back from the stem, and the twig is clipped to an appropriate length. Chimpanzees do not modify all of the materials that they use as tools. The stone hammers and anvils that chimpanzees use to crack open hard-shelled nuts are carefully selected, used repeatedly, and sometimes moved from one site in the forest to another, but they are not deliberately altered by the chimpanzees. However, none of these tools would leave a trace in the archaeological record.

Approximately half of the tasks for which chimpanzees manufacture and use tools are related to food processing and food acquisition. In some cases, these tools allow chimpanzees to exploit food resources that other animals cannot use. For example, many animals are attracted to termite mounds when the winged insects emerge in massive swarms to form new colonies, but only chimpanzees can obtain access to the termites inside the mounds at other times of the year. Some of the shells of nuts that chimpanzees eat in the Taï Forest are so hard that they can be opened only by repeated battering of the shell with a stone hammer.

Despite the examples of tool use by chimpanzees, humans are the only primates to use tools extensively. William McGrew argues that the reason chimpanzees don't make greater use of tools is that they can't transport them easily (Figure 11.43). If a chimpanzee can't carry a tool from place to place, then it must manufacture its tools at the site where they will be used. As a consequence, it doesn't pay off for chim-

FIGURE 11.42 In West Africa, chimpanzees use stones to crack open hard-shelled nuts.

panzees to put a lot of effort into making tools, so the tools they use are crude.

Early Hominin Social Organization

Chimpanzees share food, but only transfers from mothers to offspring account for a significant fraction of total consumption.

As we will see in the next chapter, food sharing plays a crucial role in the economies of modern foraging groups. We will learn that there are reasons to believe that hunting makes sharing necessary, and sharing makes hunting feasible. Thus, if meat eating played an important role in the lives of early hominins, then reciprocal exchanges of food must have been important as well. Again we look to chimpanzees for insights about the origins of this behavior.

Unlike most other primates, chimpanzees frequently share food. Mothers regularly share with their infants, and adults share food with each other under certain circumstances. Plant foods are what mothers generally share with their infants, but on the rare occasions when adults share food, it is usually meat. The patterns of food sharing between mothers and their infants have been the most carefully analyzed at Gombe. There, mothers are most likely to share foods that are difficult for the infants to obtain or to process independently (see Chapter 8). For example, infants have a hard time opening hard-shelled fruits and extracting seeds from sticky pods. A mother will often allow an infant to take bits of these items from her own hand, or sometimes she will spontaneously offer them to her infant. When chimpanzees capture vertebrate prey, they dismember, divide, and sometimes redistribute the meat among members of the foraging party. However, the amounts are a small fraction of all food calories consumed. In the Taï Forest, small prey are generally retained by the captor, and larger prey are typically divided among several individuals (Figure 11.44). The distribution of meat spans a continuum from outright coercion to apparently voluntary donations. High-ranking males sometimes take kills away from lower-ranking

FIGURE 11.43 It is awkward for chimpanzees to carry things. Sometimes they carry small objects in their mouths. They can carry large objects only if they walk bipedally. Here a male chimpanzee drags a log as he performs an aggressive display. (Photograph courtesy of William C. McGrew.)

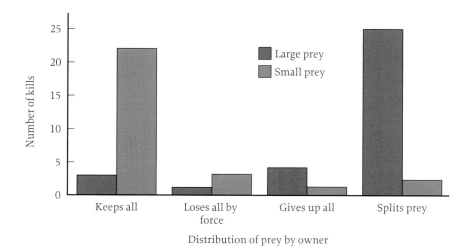

FIGURE 11.44 In the Taï Forest, chimpanzees sometimes share their kills. Infant and juvenile monkeys (small prey) are generally consumed by the captor, but adult monkey carcasses are generally divided by the captor and shared with other chimpanzees.

FIGURE 11.45 Male chimpanzees sometimes share their kills with other males, sexually receptive females, and immatures. (Photograph courtesy of Craig Stanford.)

males, and adult males sometimes take kills away from females. At Gombe, about one-third of all kills are appropriated by higher-ranking individuals. More often, however, kills are retained by the captor and shared with others who cluster closely around. Males, who are generally the ones who control the kills, share food with other males, adult females, juveniles, and infants (Figure 11.45).

It is plausible that early hominins lived in multimale, multifemale groups with little male investment in offspring.

The fossils themselves give us reason to believe that hominins lived in large groups. The australopithecines and early hominins showed pronounced dimorphism in body size: males were substantially larger than females. Sexual dimorphism in body size is consistent with the formation of multimale or one-male groups.

Comparative analyses indicate that terrestrial primates usually live in bigger groups, on average, than arboreal primates do. Baboons often live in groups of 60 individuals, but arboreal howler monkeys usually live in groups that number less than a dozen. Large group size may be a form of defense against terrestrial predators because animals living in large groups are less vulnerable to predators than animals living in smaller groups are.

Large groups tend to contain several adult males, as we saw in Chapter 7, because it is difficult for a single male to maintain exclusive access to a large number of females. In such groups there are no permanent bonds between males and females, and males invest little in offspring. In nonhuman primates, terrestrial life is generally associated with multimale, multifemale social organization.

FURTHER READING

Conroy, G. 1997. *Reconstructing Human Origins: A Modern Synthesis*. Norton, New York.

Foley, R. 1987. *Another Unique Species*. Longman Scientific and Technical, Harlow, England.

Klein, R. 1999. *The Human Career* (2nd ed.). University of Chicago Press, Chicago.

Wood, B., and M. Collard. 1999. The genus *Homo*. *Science* 284: 65–71.

STUDY QUESTIONS

1. What features distinguish modern humans from other great apes?
2. Many researchers refer to the early hominins as bipedal apes. Is this description accurate? In what ways do you think early hominins may have differed from other apes?

3. The map in Figure 11.4 indicates that early hominins were restricted to the eastern and southern parts of Africa. Why do you think this is the case?

4. What circumstances might have favored the divergence and subsequent diversification of hominin species in Africa 4 to 2 mya?

5. What do we mean when we label a trait as "primitive"?

6. Outline three reasons why natural selection may have favored bipedal locomotion in the hominin lineage. In each case, explain why hominins became bipedal but other terrestrial primates, such as baboons, did not.

7. What evidence suggests that australopithecines spent more time in trees than modern humans do?

8. What features do the *Australopithecus* species share? In what ways do they differ from *Paranthropus* and *Kenyanthropus*?

9. A number of different phylogenetic schemes have been proposed over the last few decades, including the two phylogenies illustrated in Figures 11.32 and 11.33. Note that these phylogenies do not include the newly discovered species of *Orrorin tugenensis, Kenyanthropus platyops,* and *Australopithecus garhi.* How do these new species alter these phylogenies?

10. Conventional wisdom has been that *A. habilis* was the first maker of stone tools and that this accomplishment merits its inclusion in the genus *Homo*. Does the conventional wisdom fit the current evidence? Is it likely that australopithecines were the first tool users?

11. From the comparative and morphological evidence on hand, what can we say about the behavior and social organization of the early hominins?

KEY TERMS

hominins	diastema
foramen magnum	torque
pneumatized	abductors
australopithecines	ilium
tibia	femur
humerus	sagittal crest
endocranial volume	adaptive grade
subnasal prognathism	temporalis muscle

CONTENT FROM *HUMAN EVOLUTION:* ## *A MULTI-MEDIA GUIDE TO THE FOSSIL RECORD*

Human Origins
  Overview
  Ardipithecus ramidus
  Australopithecus anamensis
  Australopithecus afarensis

The Australopithecine Adaptive Radiation
  Overview
  Robust Australopithecines
  Gracile Australopithecines
  Possible Relationships

OLDOWAN TOOLMAKERS AND THE ORIGIN OF HUMAN LIFE HISTORY

THE OLDOWAN TOOLMAKERS

The earliest identifiable stone tools are from about 2.5 mya.

Hominins have been making stone tools for at least 2.5 million years. The first identifiable stone tools were found in the Awash region of Ethiopia and are dated to about 2.5 mya. Similar tools found at a number of other sites in East Africa are between 1.7 and 2.4 million years old. In South Africa the earliest stone tools date to about 2 mya. Tools made from bone appear in the fossil record between 2 and 1 mya at Olduvai Gorge and in South Africa.

The earliest toolmakers were already quite proficient at their craft. At a West Turkana site that dates to 2.3 mya, researchers have found an extensive array of stone artifacts, including **flakes** (small, sharp chips), **cores**, hammer stones, and debris from manufacturing. Preservation conditions were so good that workers were able to fit a number of flakes back onto the cores from which they had been struck. This reconstruction revealed that early hominins removed as many as 30 flakes from a single core, maintaining precise flaking angles during the entire toolmaking sequence.

These artifacts, collectively referred to as the **Oldowan tool industry**, are very simple. They consist of rounded stones, like the cobbles once used to pave city streets, that have been flaked (chipped) a few times to produce an edge (Figure 12.1). Toolmaking (**knapping**) leaves telltale traces on the cobble cores, so it's possible to distinguish natural breakage from deliberate modification. The Oldowan artifacts are quite variable in their shape and size, but this variation does not seem to be related to differences in how the tools were used, how they were made, or how toolmakers thought their tools should look. Instead, the tools vary because they were made out of different raw materials. There is now evidence that the flakes struck from these cobbles were at least as useful as the cores themselves (Box 12.1).

The Oldowan tool industry is an example of Mode 1 technology.

Tool industries like the Oldowan refer to collections of tools that are found in a particular region and time. It is also useful to have a name for a particular method of

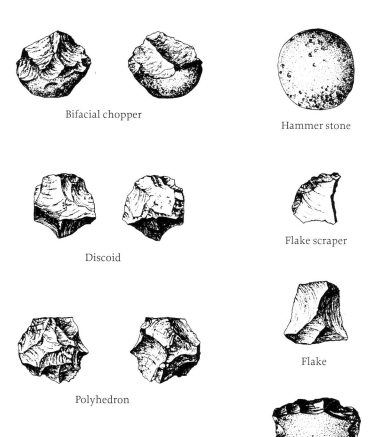

Bifacial chopper

Hammer stone

Discoid

Flake scraper

Polyhedron

Flake

Heavy-duty (core) scraper

0   5 cm

FIGURE 12.1 Stone tools like these first appear in the archaeological record about 2.5 mya. Researchers are not sure how these tools were used. Some think that the large cores shown here were used for a variety of tasks; others think that the small flakes removed from these cores were the real tools.

BOX 12.1

Ancient Toolmaking and Tool Use

Recently, Kathy Schick and Nicholas Toth of Indiana University have done many experiments with simple stone tools. They have mastered the skills needed to manufacture the kinds of artifacts found at Oldowan sites and learned how to use them effectively. Their experiments have produced several interesting results. Schick and Toth believe the Oldowan artifacts that archaeologists have painstakingly collected and described are not really tools at all. They are cores left over after striking off small, sharp flakes, and the flakes are the real tools. Schick and Toth find that the flakes can be used for an impressive variety of tasks, even for butchering large animals like elephants. In contrast, although the cores can be used for some tasks, such as chopping down a tree to make a digging stick or a spear, or cracking bones to extract marrow, they are generally much less useful. Schick and Toth's conclusion is supported by microscopic analysis of the edges of a small number of Oldowan flakes, which indicates that they were used for both woodworking and butchery.

Schick and Toth have also been able to explain the func-

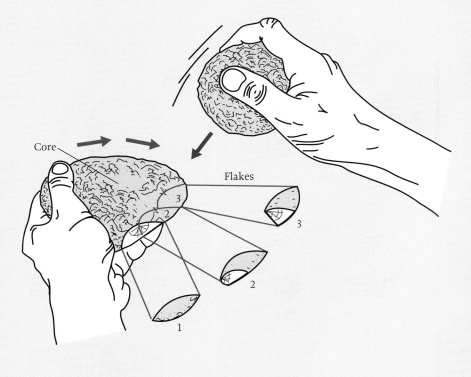

FIGURE 12.2 This diagram demonstrates why right-handed flint knappers make distinctive flakes. When a right-handed person makes a stone tool, she typically holds the hammer stone in her right hand and the core to be flaked in the left hand. When the hammer stone strikes the core, a flake spalls off, leaving a characteristic pattern of rays and ripple marks centered around the point of impact (here shown in red). The knapper then rotates the core and strikes it again. If she rotates the core clockwise as shown, the second flake has the percussion marks from the first flake on the upper left, and part of the original rough surface of the rock, or cortex (here shown in gray), on the right. If she rotates the core counterclockwise, the cortex will be on the left and the percussion marks on the upper right. Modern right-handed flint knappers make about 56% right-handed flakes and 44% left-handed flakes, and left-handed flint knappers do just the opposite. A sample of flakes from Koobi Fora dated to between 1.9 and 1.5 mya contains 57% right-handed flakes and 43% left-handed flakes, suggesting that early hominin toolmakers were right-handed.

Core

Flakes

Side view of flakes

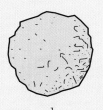

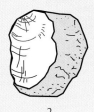

1 2 3

tion of enigmatic objects that archaeologists call spheroids. These are smooth, approximately spherical pieces of quartz about the size of a baseball. Several suggestions for their function were put forth, including processing plants and bashing bones to extract marrow. Some researchers thought the spheroids were part of a bola, a hunting tool used on the grasslands of Argentina. In a bola, three stones are connected by leather thongs and thrown so that they tangle the prey's legs. Schick and Toth have shown there is a much more plausible explanation for these stones. If a piece of quartz is used as a hammer to produce flakes, bits of the hammer stone are inadvertently knocked off. The hammer surface is no longer flat, so the hammer stone is shifted in the toolmaker's hand. Thus the hammer gradually becomes more and more spherical. After a while, a quartz hammer becomes a spheroid.

Perhaps the most remarkable conclusion that Schick and Toth drew from their experiments is that early hominin toolmakers were right-handed. Schick and Toth found that right-handers usually hold the hammer stone in their right hand and hold the stone to be flaked in the left hand. After driving off the first flake, they rotate the stone clockwise and drive off the second flake. This sequence produces flakes in which the **cortex** (the rough, unknapped surface of the stone) is typically on the right side of the flakes (Figure 12.2). On flakes made by left-handers, the cortex is typically on the left side. With this result in hand, Schick and Toth studied flakes from sites at Koobi Fora, Kenya, dated to 1.9 to 1.5 mya. Their results suggest that most of the individuals that made these flakes were right-handed.

manufacturing tools so that we can compare industries from different times and places. J. Desmond Clark of the University of California Berkeley devised a scheme for classifying modes of production that we will use here. According to this scheme, crude flaked pebble tools like those associated with the Oldowan tool industry are classified as **Mode 1** technology. The distinction between industry and mode is not very meaningful for our discussion of early hominins, who used simple techniques to create a very limited set of tools, but it will become significant later when there are major regional and temporal variations in the composition of tool kits and modes of production.

We do not know which hominin species were responsible for making the tools.

As you learned in Chapter 11, a sizable number of hominin species were running around in East Africa between 2.5 and 1.7 mya, and it is not clear which of these species made the Oldowan tools. *Australopithecus garhi* is probably the best candidate for "first engineer." At Bouri, animal bones bearing the distinctive marks made by stone tools are found in the same strata as the fossils of *A. garhi*. Although no tools were found with the fossils at Bouri, many tools have been found at the nearby site of Gona, which is dated to 2.5 mya. Thus, *A. garhi* may have been the first stone toolmaker. But remember that *A. rudolfensis*, *Kenyanthropus platyops*, and *Paranthropus aethiopicus* were probably all present in East Africa at the same time; they could have made the tools found at Gona but impolitely failed to leave their own fossils behind.

It is also possible that the earliest stone tools were made by species that do not appear until later in the fossil record. Stone tools are more durable than bones, so the archaeological record is usually more complete than the fossil record. This means that the earliest tools typically appear in the fossil record before the first fossil of the creature that made them (see Box 10.3 for more discussion of this topic). Thus the first stone toolmaker may have been *Australopithecus habilis*, *Paranthropus boisei*, or even *Homo ergaster*, the oldest member of our own genus, who will be discussed in more detail in Chapter 13. Because of this ambiguity, in the rest of this chapter we will refer to the hominins who made the tools of this period as Oldowan hominins or Oldowan toolmakers.

Because it is likely that Oldowan toolmakers were human ancestors, we can learn a lot about the processes that shaped human evolution by studying Oldowan archaeological sites.

Homo ergaster appeared in Africa about 1.8 mya. Unlike the situation with *A. habilis* or *A. rudolfensis*, there is little doubt that these creatures belong in the same genus as *Homo sapiens*. As you will learn in the next chapter, *Homo ergaster* was a fully terrestrial creature with a large body—one quite similar to our own. These creatures developed more slowly than early hominins, and males and females were about as different in size as men and women are today. Thus, about 2 mya selection must have favored these more humanlike characteristics in one of the early hominin lineages. That we don't know which early hominin is ancestral to humans is annoying but not really fatal, because it seems very likely that whoever made the Oldowan tools is the culprit. By combining knowledge of contemporary foraging peoples with careful study of Oldowan tools, the sites where these tools were found, and the marks that they make on animal bones, we can learn a lot about the transition from a bipedal ape to a creature much more like modern humans.

In the discussion that follows, we will first see how a reliance on highly productive but very hard-to-learn foraging skills distinguishes humans from other primates, and how this novel foraging niche may have led to the evolution of other novel features of the human life cycle, such as slow maturation and reduced sexual dimorphism. We will then consider the archaeological evidence from Oldowan sites indicating that these hominins had begun to shift to a subsistence economy based on more challenging foraging techniques.

COMPLEX FORAGING SHAPES HUMAN LIFE HISTORY

Anthropologists divide the foods acquired by foragers into three types according to the amount of knowledge and skill required to obtain them. These are, in order of increasing difficulty of acquisition, collected foods, extracted foods, and hunted foods.

Hillard Kaplan, Kim Hill, Madgalena Hurtado, and Jane Lancaster of the University of New Mexico recently argued that the evolution of modern human life history was driven by a shift to valuable but hard-to-acquire food resources. They rank food resources into three categories according to the difficulty of acquisition:

1. **Collected foods** can be simply collected from the environment and eaten. Examples include ripe fruit and leaves.
2. **Extracted foods** come from things that don't move but are protected in some way. These things must be processed before the food can be eaten. Examples include fruits in hard shells, tubers or termites that are buried deep underground, honey hidden in hives high in trees, and plants containing toxins that must be extracted before the plants can be eaten.

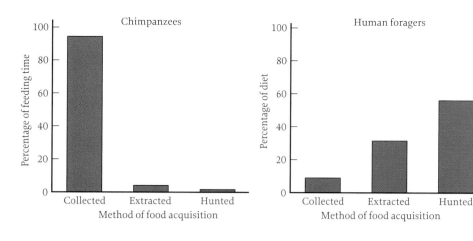

FIGURE 12.3 (a) Chimpanzees spend most of their time feeding on collected foods like fruit and leaves, which can be eaten without processing. (b) Human foragers get most of their calories from extracted foods like tubers, which must be processed before they can be eaten, and hunted foods like animal game, which must be caught or trapped.

3. **Hunted foods** come from things that run away, and thus must be caught or trapped. They may also need to be extracted and processed before consumption. Vertebrate prey are the prime example of hunted foods for both humans and chimpanzees.

Apes, especially chimpanzees, are the "brainiacs" of the primate world, at least when it comes to foraging. Gorillas and orangutans use elaborate routines to process some plant foods, but they don't hunt. Chimpanzees have a broader diet, including collected, hunted, and extracted foods, the latter two requiring considerable skill to acquire and process. Both orangutans and chimpanzees use tools to process some kinds of foods. Moreover, as we will see in Chapter 16, different ape populations use different techniques for the acquisition of extracted foods like hard-shelled nuts. However, even these clever apes do not come close to human expertise in foraging.

Humans depend on hard-to-learn skills to acquire food.

Kaplan and his colleagues note that contemporary foraging peoples depend on extracted and hunted foods to a much greater extent than chimpanzees do. Figure 12.3 compares the average dependence of chimpanzees and humans on each of the food types: collected, extracted, and hunted. The general pattern is clear: chimpanzees are overwhelmingly dependent on collected resources, but human foragers get almost all of their calories from extracted or hunted resources.

Unlike other predators, humans must learn a very diverse set of hunting skills. Most large mammalian predators capture a relatively small range of prey species using one of two methods: they wait in ambush, or they combine a stealthy approach with fast pursuit. Once the prey is captured, they process it with tooth and claw. In contrast, human hunters use a vast number of methods to capture and process a huge range of prey species. For example, the Aché, a group of foragers who live in Paraguay, have been observed to take 78 different species of mammals, 21 species of reptiles, 14 species of fish, and over 150 species of birds using a dizzying array of techniques that depend on the prey, the season, the weather, and many other factors (Figure 12.4). Some animals they track—a difficult skill that entails a great deal of ecological and environmental knowledge (see Chapter 15). Other animals they call by imitating the prey's mating or distress sounds. Still other animals they trap with snares or traps, or smoke out of burrows. They capture and kill animals using their hands,

FIGURE 12.4 Meat makes up about 70% of the diet of the Aché, a group of foragers from Paraguay. Here an Aché man takes aim at a monkey. (Photograph courtesy of Kim Hill.)

FIGURE 12.5 A !Kung San woman carries her young child on her back and digs for roots and tubers in the Kalahari Desert of Botswana. (Photograph courtesy of Nicholas Blurton Jones.)

arrows, clubs, or spears. And this is just the Aché; if we included the full range of human habitats, the list would be immeasurably longer.

It takes a long time to learn this range of skills. Among the Aché, men's hunting efficiency peaks at about age 35. Twenty-year-old men manage to capture only about a fourth of the maximum. Kaplan and Hill made strenuous efforts to become competent hunters when they were living with the Aché, but they could not come close to the production of 20-year-old Aché men.

Efficient extraction of resources also requires considerable skill. Nicholas Blurton Jones, a scientist at the University of California Los Angeles who has studied the Hadza, and the !Kung, two foraging groups, describes digging up deeply buried tubers from rocky soil as a complex mining operation involving much clever engineering of braces and levers (Figure 12.5). Among the Hiwi, a group of foragers who live in the tropical savanna of Venezuela, women do not achieve maximum efficiency in root acquisition until they are between 35 and 45 years old. Ten-year-old girls get only 10% as much as older, highly skilled women. Among the Aché, rates of starch extraction from palms and honey extraction also peak when people are in their 20s.

A reliance on hunting and extractive foraging favors food sharing and division of labor in contemporary foraging groups.

In all contemporary foraging groups, hunting and extractive foraging are associated with extensive food sharing and sexual division of labor. In nearly all foraging groups, men take primary responsibility for hunting large game, and women take primary responsibility for extractive foraging (Figure 12.6). This division of labor makes sense on two grounds: First, hard-to-learn techniques reward specialization. It takes a long time to learn how to be a good hunter, and it takes a long time to learn how to dig tubers. This means that everyone is better off if some individuals specialize in hunting and others specialize in extractive foraging. Second, because child care is more compatible with gathering than with hunting, and lactation commits women to child care for a substantial portion of their adult lives, it makes sense that men specialize in

FIGURE 12.6 Data on foraging behavior for three well-studied, modern foraging groups: the Aché, the Hadza, and the Hiwi. Men and women in these groups specialize in different foraging tasks. Men hunt, and women specialize in extractive foraging.

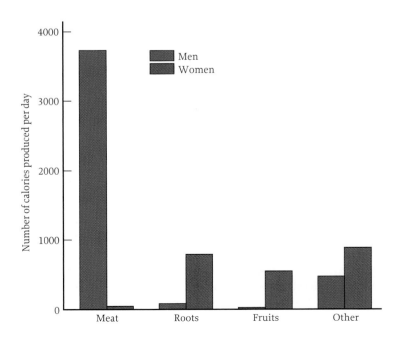

hunting and women specialize in extractive foraging. Of course, this all works only if members of the group regularly share food.

However, a reliance on meat eating favors the evolution of food sharing. Hunting is an uncertain endeavor. The best hunters can come home empty-handed, and even if hunters caught enough each day to feed themselves, a week or two of bad luck could lead to starvation. If several hunters share their catch, however, the chance of starvation is much lower (Box 12.2).

In foraging societies, food sharing and division of labor lead to extensive flows of food between people of different ages and sexes.

Remember from Chapter 11 that the total amount of food shared by chimpanzees is small: mothers share with their offspring, and males share small amounts of meat, mainly with other adult males. This means that, once they are weaned, chimpanzees obtain virtually all of their own food themselves. The story for other primates is very similar. Thus it seems likely that such self-sufficiency after weaning is the ancestral state in the hominin lineage.

The economy of human foragers is strikingly different: some people produce much more food than they consume, and others consume much more than they produce. Over the last 20 years, anthropologists have done careful quantitative studies of the subsistence economies of a number of different foraging groups. In these societies, anthropologists observed people's daily behavior, measuring how much food they produced and how much they consumed. For three of these groups—the Aché, the Hiwi, and the Hadza—researchers have meticulously computed average food production and consumption for men and women of different ages. Kaplan and his colleagues compiled these data to compare patterns of food production across societies. Their analysis reveals striking differences in the foraging economy of humans and chimpanzees, and important changes in productivity over the life course.

Figure 12.7 shows that human young continue to depend on others for food long after they are weaned. Men become self-sufficient around the age of 17, and women do not produce enough to feed themselves until they are in their late 40s. Older men

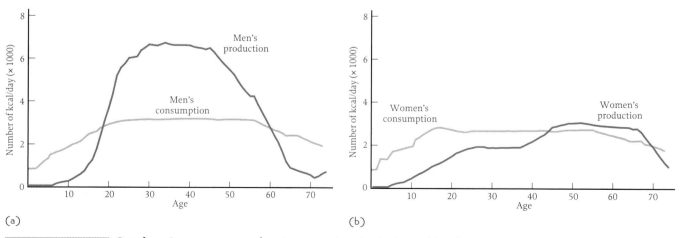

(a)   (b)

FIGURE 12.7 Data from three contemporary foraging groups show that both men (a) and women (b) do not become self-sufficient in terms of food production until they are adults. Adult men produce many more calories than they consume; adult women are approximately in balance.

BOX 12.2

Why Meat Eating Favors Food Sharing

Many anthropologists believe that a heavy dependence on meat makes food sharing necessary. Let's examine how food sharing provides insurance against the risks inherent in hunting. Hunting, especially for hunters who concentrate on large game, is a boom-or-bust activity. When a hunter makes a kill, a lot of very high quality food becomes available. Hunters are often unlucky, however, and each time one sets out to hunt there is a fairly high probability of returning empty-handed and hungry. Food sharing greatly reduces the risks associated with hunting by averaging returns over a number of hunters.

To see why this argument has such force, consider the following simple hypothetical example. Suppose there are five hunters in a group that subsists entirely on meat. Hunters are able to hunt every day, and each hunter has a 1-in-5 (0.2) chance of making a kill and a 4-in-5 (0.8) chance of bringing back nothing. Further, suppose that people starve after 10 days without food. We can calculate the probability of starvation for each hunter over any 10-day period by multiplying the probability of failing on the first day (0.8) by the probability of failing on the second day (0.8), and so on, to get

$$0.8 \times 0.8 \times 0.8 \times 0.8 \times 0.8 \times 0.8 \times$$
$$0.8 \times 0.8 \times 0.8 \times 0.8 \approx 0.1$$

Thus, there is about a 10% chance that a hunter will starve over any 10-day period. With these odds, it is impossible for people to sustain themselves by hunting alone.

A comparison with chimpanzee hunting provides good reason to think that these probabilities are realistic for early hominins. Craig Stanford of the University of Southern California and his colleagues have carefully analyzed records of hunting by chimpanzees at Gombe Stream National Park in Tanzania. In about half of the hunts, the chimpanzees succeeded in killing at least one monkey, and sometimes they made more than one kill. The average number of monkeys killed per hunt was 0.84. However, the hunting groups contained 7 males on average. Dividing 0.84 by 7 gives the average number of monkeys killed per male per hunt, which comes out to 0.12. Thus, on any given day, each male chimpanzee had only a 12% chance of making a kill, which is less than the 20% chance we posited for the five human hunters at the outset of our example.

Now let's consider how sharing food alters the probability of starvation for our human hunters. If each hunter has a 0.8 chance of coming back empty-handed, then the chance that all five hunters will come back on a given evening without food is

$$0.8 \times 0.8 \times 0.8 \times 0.8 \times 0.8 \approx 0.33$$

Thus, on each day there is a 1-in-3 chance that no one will make a kill. If the kill is large enough to feed all members of the group, then no one will go hungry as long as someone succeeds. The chance that all five hunters will face starvation during any 10-day period is

$$0.33 \times 0.33 \times 0.33 \times 0.33 \times 0.33 \times 0.33 \times$$
$$0.33 \times 0.33 \times 0.33 \times 0.33 \approx 0.000015$$

Sharing reduces the chance of starvation from one chance in 10 to roughly one chance in 60,000. Clearly, the risks associated with hunting could be reduced even further if there were alternative sources of food that unsuccessful hunters might share. For example, suppose one member of the group hunted while the others foraged, and they all contributed food to a communal pot.

The fact that food sharing is mutually beneficial is not enough to make it happen. As we pointed out in Chapter 8, food sharing is an altruistic act. Each individual will be better off if he or she gets meat but does not share it. For sharing to occur among unrelated individuals, as it often does in contemporary foraging societies, those who do not share must be punished in some way, such as by being excluded from future sharing or by being forced to leave the group.

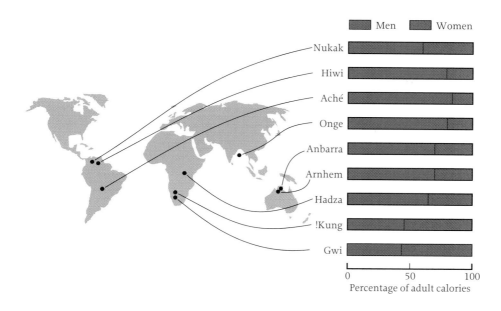

FIGURE 12.8 Males contribute significantly to subsistence in nine contemporary foraging groups from different parts of the world.

also depend on others for their daily needs. These deficits are made up by the production of young and middle-aged men and to a lesser extent, postmenopausal women. In contrast, chimpanzees obtain very little of their food from others after they are weaned.

Less detailed data from other foraging groups are consistent with this pattern. Figure 12.8 shows that men contribute more than half of the total calories consumed in seven of the nine foraging groups for which the necessary data are available. Notice that all of these groups live in tropical habitats. It seems likely from historical and ethnographic accounts that temperate and arctic foragers depended even more on meat than do tropical foragers, and thus in these societies men may make an even bigger caloric contribution.

Selection may have favored larger brains, a prolonged juvenile period, and a longer life span because these traits make it easier to learn complex foraging methods.

Complex, learned foraging techniques allow humans to acquire highly valuable or otherwise inaccessible food resources. Meat is a much better source of most nutrients that animals need than is the usual primate fare of leaves and ripe fruit. Meat is rich in energy, essential lipids, and protein. It is also dense enough to be economically transported from the kill site to home base. Some extracted resources like honey, insect larvae, and termites are also concentrated sources of important nutrients. Other kinds of extractive foraging unlock vast new food supplies. Tubers are a prime example. A number of tropical savanna plants store their reserve supplies of energy underground as various kinds of tubers, protected from the teeming herds of grazers and browsers by as much as a meter of rocky soil. By learning to recognize which plant species have tubers, how to use tools to dig them up, and when such work is likely to be profitable, humans were able to access a large supply of food for which there was relatively little competition from other organisms.

If learning is valuable, natural selection will favor adaptations that make a better learner. Thus a shift to hunting and extractive foraging would favor larger brains and greater intelligence. Reliance on complex learned foraging skills would also favor the evolution of a prolonged juvenile period. As we all know, learning takes time. You can't become a proficient skier, baker, or computer programmer in a day; practice and experience are needed. Similarly, learning the habits of animals and the lore of plants, acquiring the knowledge for tracking animals and the skills for shooting a bow or blowgun, and becoming adept at extracting starch from baobab pulp require years of practice. Thus it is plausible that selection favored a longer juvenile period to allow human children the time to acquire the skills they needed.

A prolonged juvenile period generates selection for a long life span. It is often said that time is money. In evolution, "time is fitness." To see why, suppose that two genotypes, A and B, have the same number of children on average, but type A completes reproduction in 30 years and type B in 60 years. If you do a bit of math, you will see that type A will have twice the population growth rate as type B, and it will quickly replace type B in the population. This means that a prolonged juvenile period is costly and will not be favored by natural selection unless it causes people to have sufficiently more children over their life span. Human childhood is like a costly investment; it costs time but the added time allows learning that produces more capable adults. Like any expensive investment, it will pay off more if it is amortized over a longer period. (The same logic explains why you are willing to spend more on something that you will use for a long time than on something you will use for a short time and then discard.) Selection favors a longer life because it allows people to get more benefit from the productive foraging techniques they learned during the necessary, but costly, extended juvenile period.

Food sharing and division of labor lead to reduced competition between males and reduced sexual dimorphism.

We saw in Chapter 7 that the intensity of competition between males depends on the amount of male investment in offspring. In most primate species, males do very little for their offspring, and selection consequently favors male traits that enhance their ability to compete with other males for matings. This increased competition leads to the pronounced sexual dimorphism seen in most primate species. When males do invest in offspring, there is less male–male competition and reduced sexual dimorphism.

The pattern of food sharing seen in contemporary foraging societies means that males are making substantial investments in offspring. This is clear from the data shown in Figure 12.7: males produce the bulk of the surplus calories that sustain children and teenagers. Thus we would expect selection to favor behavioral and morphological traits that make men good providers, and we would also expect selection favoring traits that enhance male–male competitive ability to be reduced. The reduction in male–male competition, in turn, should lead to reduced sexual dimorphism.

EVIDENCE FOR COMPLEX FORAGING BY OLDOWAN TOOLMAKERS

Let's stop and review for a second. So far, we have made two points. First, the Oldowan toolmakers (whoever they were) are plausible candidates for the species that links early apelike hominins to later hominins, who have more humanlike life history patterns. Second, contemporary foragers rely on complex, hard-to-learn foraging techniques to a much greater extent than other primates do, and this shift can explain the evolution of the main features of human life history. To link these points, we need to consider the evidence that Oldowan hominins had begun to rely on hunting and extractive foraging to make a living.

Contemporary experiments suggest that Oldowan tools could be used for a variety of tasks, including the butchery of large animals.

As described in Box 12.1, Kathy Schick and Nicholas Toth of Indiana University have mastered the skills needed to manufacture the kinds of artifacts found at Oldowan sites, and they have attempted to use them for different kinds of foraging activities. Schick and Toth found that stone flakes struck from cobble cores can be used for an impressive variety of tasks, even for butchering large animals like elephants. The cores can be used for a more limited number of jobs, such as chopping down a tree to make a digging stick or a spear. Microscopic analysis of the edges of a small number of Oldowan tools indicates that they were used for both woodworking and butchery.

Wear patterns on bone tools from South Africa suggest that they were used to excavate termite mounds.

Extractive foraging, which is commonly done with wooden digging sticks by modern peoples, is likely to leave few traces in the archaeological record. However, one interesting piece of evidence suggests that hominins from this period were extractive foragers. During their excavations at Swartkrans, Robert Brain and his coworkers identified a sizable number of broken bones that had wear patterns suggesting that they had been used as tools. Recently, Lucinda Backwell of the University of the Witwatersrand and Francesco d'Errico of the Institut de Préhistoire et de Géologie du Quaternaire in France carefully analyzed these bones to find out how they were used. First the researchers used freshly broken bones to do a number of foraging tasks, including digging in hard soil for tubers and digging in a termite mound. Each activity creates a distinctive wear pattern that can be detected under microscopic analysis. Then they compared these wear patterns with those on the fossil bones found at Swartkrans.

This analysis indicated that the fossil tools were used for digging in termite mounds. Figure 12.9 shows the wear patterns on the fossil tool and two experimental tools, one used for digging in soil and the other used for excavating a termite

Experimental tools used to dig

| Swartkrans fossil | Tubers | Termites |

 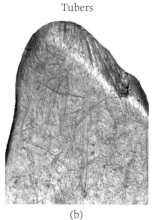 

(a)   (b)   (c)

FIGURE 12.9 Experiments indicate that bone tools found at Swartkrans were used to excavate termite mounds. The tool shown in (a) is a cast of the original tool found at Swartkrans. The experimental tools in (b) and (c) were used for subsistence tasks, digging for tubers and digging for termites. The wear pattern on the Swartkrans fossils was most similar to that on the experimental tool used to dig for termites.

mound. All of the tools have a smooth, rounded point. However, the tool experimentally used for digging in soil has deep marks of different depths going in all directions. In contrast, the tool used to dig in termite mounds has fine, parallel grooves. The fossil bone tools closely resemble the experimental tool used for digging in termite mounds, so it seems likely that this is what these tools were used for. If this is correct, then Oldowan hominins were using tools to do extractive foraging.

Archaeological Evidence for Meat Eating

At several archaeological sites in East Africa, Oldowan tools have been found along with dense concentrations of animal bones.

Archaeological sites with early stone tools occur at the Olduvai Gorge of Tanzania, Koobi Fora in Kenya, and a number of sites in Ethiopia. The sites in Bed I of the Olduvai Gorge excavated by Mary Leakey have been analyzed the most extensively. These sites, which are dated to 2 to 1.5 mya, measure only 10 to 20 m (33 to 66 ft) in diameter, but they are littered with fossilized animal bones (Figure 12.10). The densities of animal bones in the archeological sites are hundreds of times higher than those in the surrounding areas or in modern savannas. The bones belong to a wide range of animal species, including bovids (such as present-day antelope and wildebeest), pigs, equids (horses), elephants, hippopotamuses, rhinoceroses, and a variety of carnivores (Figure 12.11).

Along with these bones, Mary Leakey found several kinds of stone artifacts: cores, flakes, battered rocks that may have been used as hammers or anvils, and some stones that show no signs of human modification or use. The artifacts were manufactured

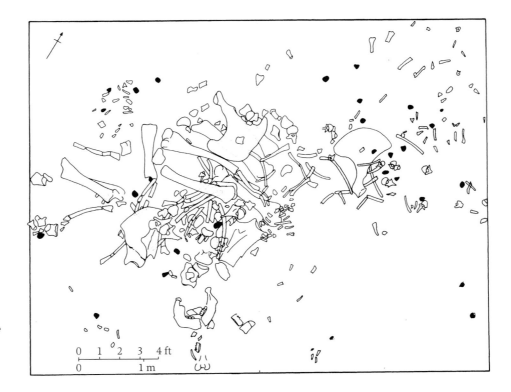

FIGURE 12.10 In this site map for one level of Bed I at Olduvai Gorge, most of the bones are from an elephant and tools are shown in black.

0 1 2 3 4 ft
0   1 m

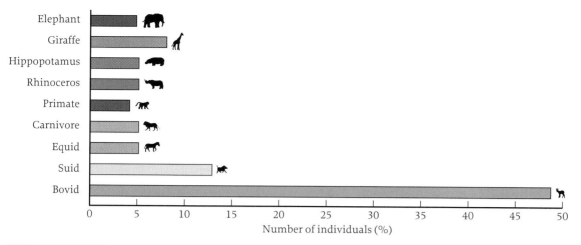

FIGURE 12.11 The bones of many different mammals were found at one archaeological site at Olduvai. Bovids (which include antelope, gazelles, sheep, goats, and cattle) clearly outnumber all other taxa.

from rocks that came from a number of different spots in the local area; some were made from rocks found several kilometers away (Figure 12.12).

The association of hominin tools and animal bones does not necessarily mean that early hominins were responsible for these bone accumulations.

It is easy to jump to the conclusion that the association of hominin tool and animal bones means that the Oldowan toolmakers hunted and processed the prey whose bones we find at these sites. If that is our assumption, these may have been sites where Oldowan hominins lived, like modern foragers' camps, or they may have been butchery sites where Oldowan hominins processed carcasses but did not live. However, there are also other possibilities. Bones may have accumulated at these sites without any help from early hominins. The bones may have been deposited there by moving water (which has now disappeared) or by other carnivores, such as hyenas. These sites also might be where many animals died of natural causes. Hominins might have visited the sites after the bones accumulated, perhaps hundreds of years later, and left their tools

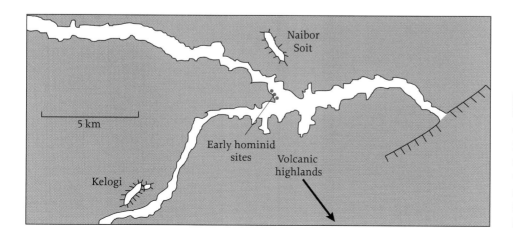

FIGURE 12.12 This map of Olduvai Gorge shows the major Bed I sites. Some of the tools were made of quartzite from Naibor Soit, others were made from gneiss from Kelogi, and some used lava cobbles that came from streams in the volcanic highlands to the south.

behind. Thus the association of stones and bones is not enough to prove that Oldowan hominins hunted or ate meat.

Archaeologists have resolved some of the uncertainty about these sites by studying how contemporary kill sites are formed.

The study of **taphonomy** provides one means to resolve questions about what happened at these sites. Taphonomy is the study of the processes that produce archaeological sites. Taphonomists examine the characteristics of contemporary kill sites—spots where animals have been killed, processed, and eaten by various predators, including contemporary human hunters. They monitor how each type of predator consumes its prey, noting whether limb bones are cracked open for marrow, which bones are carried away from the site, how human hunters use tools to process carcasses, and how bones are distributed by predators at the kill site. These data enable archaeologists to develop a profile of the characteristics of kill sites created by different types of predators. Taphonomists can also assess many of the same characteristics in archaeological sites. By comparing the features of archaeological and contemporary sites, they are sometimes able to determine what happened at an archaeological site in the past.

Taphonomic analyses at Olduvai Gorge suggest that the bones at most of these sites were not accumulated by natural processes.

Taphonomic studies of the Olduvai sites tell us, first of all, that the bones were not deposited by moving water. Animals sometimes drown as they try to cross a swollen river or when they are swept away in flash floods (Figure 12.13). The bodies are carried downstream and accumulate in a sinkhole or on a sandbar. As the bodies decompose, the bones are exposed to the surrounding elements. The study of modern sites shows that sediments deposited by rapidly moving water have a number of distinctive characteristics. For example, such sediments tend to be graded by size because particles of different size and weight sink at different spots. Sediments surrounding Olduvai sites do not show any of the features characteristic of sediments deposited by rapidly moving water.

Taphonomic analyses also tell us that the dense concentrations of bones were not due to the deaths of a large number of animals at one spot. Sometimes many animals die in the same place. In severe droughts, for example, large numbers of animals may die near water holes. Mass deaths usually involve members of a single species, and there is typically little mixing of bones from different carcasses. By contrast, the bones at the Olduvai sites come from a number of different species, and bones from different carcasses are jumbled together.

At some sites, however, bone accumulations do seem to be the product of natural processes, not hominin activity. There is one site where the pattern of bone accumulation is very similar to the pattern of bone accumulations near modern hyena dens. Hyenas often carry and drag carcasses from

FIGURE 12.13 Sometimes large numbers of wildebeest drown when trying to cross swollen rivers.

kill sites to their dens so that they can feed their young and avoid competition with other carnivores. At this site, there is no evidence of hominin activity, such as stone tools or tool marks on bones.

Taphonomic analyses suggest that hominins were active at a number of the Olduvai sites and used tools at these sites to process carcasses.

The bones at these sites provide direct evidence of hominin activity. When carnivores gnaw on meat, their teeth leave distinctive marks on the bones. Similarly, when humans use stone tools to butcher prey, their tools leave characteristic marks. Flaked-stone tools have microscopic serrations on the edges and make very fine parallel grooves when they are used to scrape meat away from bones (Figure 12.14). Many of the animal bones at Olduvai show signs of stone tool use.

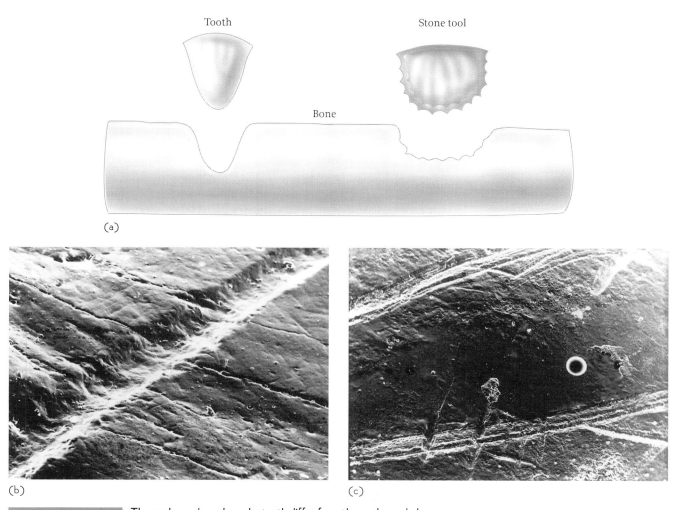

FIGURE 12.14 The marks made on bone by teeth differ from the marks made by stone tools. (a) The smooth surfaces of teeth leave broad, smooth grooves on bones; the edges of stone tools, on the other hand, have many tiny, sharp points that leave fine parallel grooves. Cut marks made by carnivore teeth (b) and stone tools (c) can be distinguished when they are examined with a scanning electron microscope. These are scanning electron micrographs of 1.8-million-year-old fossil bones from Olduvai Gorge.

Hunters or Scavengers?

There has been controversy about whether the Oldowan hominins were hunters or scavengers.

The archaeological evidence indicates that Oldowan hominins processed the carcasses of large animals, and we assume that they ate the meat they cut from the bones. But eating meat does not necessarily imply hunting. Some carnivores acquire meat by hunting, but many carnivores rely at least partly on scavenging. Scavengers steal kills from other predators or rely on opportunistic discoveries of carcasses.

There has been considerable dispute about how early hominins acquired the meat they ate. Some have argued that Oldowan hominins killed the prey found at the archaeological sites; others have argued that Oldowan hominins could not have captured large mammals, because they were too small, too poorly armed, and too poorly encephalized. They contend that the Oldowan hominins were scavengers who occasionally appropriated kills from other predators or collected carcasses they found.

For most contemporary carnivores, scavenging is as difficult and dangerous as hunting.

To resolve the debate about whether early hominins were hunters or scavengers, we must first rethink popular conceptions of scavengers. Although scavengers have an unsavory reputation, scavenging is not an occupation for the cowardly or the lazy. Scavengers must be brave enough to snatch kills from the jaws of hungry competitors, shrewd enough to hang back in the shadows until the kill is momentarily left unguarded, or patient enough to follow herds and take advantage of natural mortality. Studies of contemporary carnivores show that the great majority of scavenged meat is acquired by taking a kill away from another predator. Most predators respond aggressively to competition from scavengers. For example, lions jealously guard their prey from persistent scavengers that try to steal bits of meat or to drag away parts of the carcass. These contests can be quite dangerous (Figure 12.15).

(a)

(b)

FIGURE 12.15   Competition among predators at kill sites is frequently intense. Here lions have scavenged prey from a pack of hyenas (a), and the hyenas fight to get it back (b).

Most large mammalian carnivores practice both hunting and scavenging.

We also tend to think that some carnivores, like lions and leopards, only hunt, and that others, like hyenas and jackals, only scavenge. But the simple dichotomy between scavengers and hunters collapses when we review the data on the behavior of the five largest African mammalian carnivores: lion, hyena, cheetah, leopard, and wild dog. The fractions of meat obtained by scavenging vary from none for the cheetah, to 33% for hyenas, with the others ranging somewhere in between. Contrary to the usual stereotypes, the noble lion is not above taking prey from smaller competitors, including female members of his own pride, and hyenas are accomplished hunters (Figure 12.16). For most large carnivores in eastern Africa, hunting and scavenging are complementary activities.

No mammalian carnivores subsist entirely by scavenging. It would be difficult for any large mammal to do so. For one thing, many prey species create movable feasts, migrating in large herds over long distances. Although natural mortality in these herds might make scavenging feasible, their migratory habits eliminate this option. Mammalian carnivores cannot follow these migrating herds very far, because they have dependent young that cannot travel long distances (Figure 12.17). Only avian scavengers that can soar over great distances, like the griffon vulture, rely entirely on scavenging (Figure 12.18). When migratory herds are absent, mammalian carnivores rely on other prey, such as waterbuck and impala. Natural mortality among resident species is not high enough to satisfy the caloric demands of carnivores, so they must hunt and kill much of their own prey, though they still scavenge when the opportunity arises.

FIGURE 12.16   Male lions sometimes take kills from smaller carnivores and from female lions.

Scavenging might be more practical if carnivores switched from big game to other forms of food when migratory herds were not present. This is a plausible option for early hominins. Most groups of contemporary foraging people rely heavily on gathered foods—including tubers, seeds, fruit, eggs, and various invertebrates—in addition to meat. A few, such as the Hadza, obtain meat from scavenging as well as from

FIGURE 12.17   As dusk falls, three cheetah cubs wait for their mother to return from hunting.

FIGURE 12.18   Vultures, which soar on thermals and have enormous ranges, are the only carnivores that rely entirely on scavenging.

hunting. It is possible that early hominins relied mainly on gathered foods and scavenged meat opportunistically.

Taphonomic evidence suggests that early hominins acquired meat both by scavenging and by hunting.

As we saw earlier, predators often face stiff competition for their kills. An animal that tries to defend its kill risks losing it to scavengers. For this reason, leopards drag their kills into trees and eat the meat in safety. Other predators, like hyenas, sometimes rip off the meaty parts of the carcass, such as the hindquarters, and drag their booty away to eat in peace. This means that limb bones usually disappear from a kill site first, and less meaty bones, such as the vertebrae and skull, disappear later or remain at the kill site (Figure 12.19). If hominins obtained most of their meat from scavenging, we would expect to find cut marks made by tools mainly on bones typically left at kill sites by predators, such as vertebrae. If hominins obtained most of their meat from their own kills, we would expect to find tool cut marks mainly on large bones, like limb bones. At Olduvai Gorge, cut marks appear on all kinds of bones—those usually left to scavengers and those normally monopolized by hunters (Figure 12.20). Moreover, some bones show cut marks on top of carnivore tooth marks, and other bones show tooth marks on top of cut marks. Thus, humans may sometimes have stolen kills from carnivores, and vice versa. Taken together, the evidence suggests that early hominins acquired meat by a combination of hunting and scavenging.

FIGURE 12.19 After other predators have left, vultures consume what remains at the kill site.

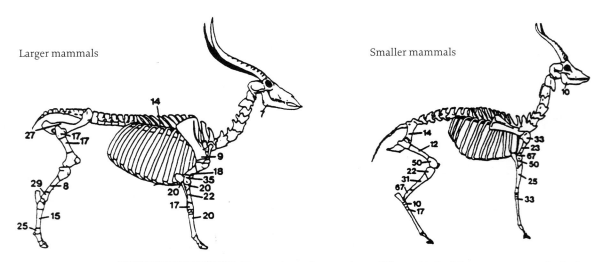

Larger mammals

Smaller mammals

FIGURE 12.20 The numbers of cut marks on different kinds of bones vary at one site in the Olduvai Gorge. There are more than twice as many pieces of bones from large mammals (more than about 115 kg, or 250 lb) as there are from small mammals (less than 115 kg). The numbers in the figure are the percentages of all bones of a particular type that bear cut marks. For example, 14% of all rib bones of larger mammals showed cut marks. In general, cut marks are concentrated on bones that have the most meat.

Domestic Lives of Oldowan Toolmakers

We have reason to believe that Oldowan toolmakers used their tools for extractive foraging and to process prey carcasses. Oldowan hominins probably obtained these carcasses through a mix of hunting and scavenging. Thus they had probably come to rely on complex foraging skills that were difficult to master. In modern foraging societies, reliance on complex foraging skills is also linked to food sharing, sexual division of labor, and the establishment of home bases. Nearly all contemporary foraging peoples establish a temporary camp (**home base**), where food is shared, processed, cooked, and eaten. The camp is also the place where people weave nets, manufacture arrows, sharpen digging sticks, string bows, make plans, resolve disputes, tell stories, and sing songs (Figure 12.21). Because foragers often move from one location to another, their camps are simple, generally consisting of modest huts or shelters built around several hearths. If Oldowan hominins established home bases, we might be able to detect traces of their occupation in the archaeological record.

FIGURE 12.21   The Efe, a foraging people of central Africa, build temporary camps in the forest and shift camps frequently. (Photograph courtesy of Robert Bailey.)

Some archaeologists have interpreted the dense accumulations of stones and bones as home bases, much like those of modern foragers, but this view is not consistent with some of the evidence.

Some archaeologists, particularly the late Glyn Isaac, have suggested that the dense accumulations of animal bones and stone tools found at some sites mark the location of hominin home bases. They have speculated that early hominins acquired meat by hunting or scavenging and then brought pieces of the carcass home, where it could be shared. The dense collections of bones and artifacts were thought to be the result of prolonged occupation of the home base (and sloppy housekeeping). At one Olduvai Bed I site, there is even a circle of stones (Figure 12.22), dated to 1.9 mya, that is similar to the circles of stones anchoring the walls of simple huts constructed by some foraging peoples in dry environments today. Many Oldowan tools and bone fragments from a variety of prey species are found at the same site.

However, a number of other observations are inconsistent with the idea that these sites were home bases:

- Both hominins and nonhominin carnivores were active at many of the Olduvai sites. Many of the bones at the Olduvai sites were gnawed by nonhominin carnivores. Sometimes the same bones show both tooth marks and cut marks, but some show only the marks of nonhominin carnivores.
- Hominins and nonhominin carnivores apparently competed over kills. The bones of nonhominin carnivores occur more often than would be expected on the basis of their occurrence in other fossil assemblages or modern carnivore densities. Perhaps the carnivores were killed (and eaten) when attempting to scavenge hominin kills or when hominins attempted to scavenge their kills. Hominins may not have

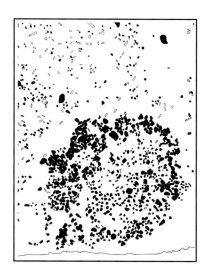

FIGURE 12.22   A circle of stones at one Olduvai Bed I site has been interpreted by some archaeologists as the remains of a simple shelter.

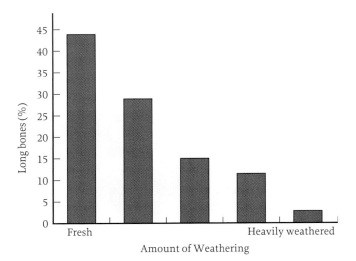

FIGURE 12.23 Many of the bones at Bed I sites at Olduvai are heavily weathered, suggesting that they were deposited and exposed to the elements over a fairly long period of time.

always won such contests; some fossilized hominin bones show the tooth marks of other carnivores.

- Modern kill sites are often the scene of violent conflict among carnivores. This conflict occurs among members of different species as well as the same species. It is especially common when a small predator, like a cheetah, makes a kill. The kill attracts many other animals, most of which are able to displace the cheetah.

- The bones accumulated at the Olduvai sites are weathered. When bones lie on the surface of the ground, they crack and peel in various ways. The longer they remain exposed on the surface, the greater the extent of weathering. Taphonomists can calibrate the weathering process and use these data to determine how long fossil bones were on the ground before being buried. Some of the bones at the Olduvai sites were exposed to the elements for at least four to six years (Figure 12.23).

- The Olduvai sites do not show evidence of intensive bone processing. The bones at these sites show cut marks and tooth marks, and many bones were apparently smashed with stone hammers to remove marrow. However, the bones were not processed intensively, as they are by modern foragers.

These observations are difficult to reconcile with the idea that the Olduvai sites were home bases—places where people eat, sleep, tell stories, and care for their children. First, foragers today do everything they can to prevent carnivores from entering their camps. They often pile up thorny branches to fence their camps and keep dogs that are meant to chase predators away. It is hard to imagine that early hominins could have occupied these sites if lions, hyenas, and saber-toothed cats were regular visitors. Second, bones at the Olduvai sites appear to have accumulated over a period of years. Contemporary foragers usually abandon their home bases permanently after a few months because the accumulating garbage attracts insects and other vermin. Even though they revisit the same areas regularly, they don't often reoccupy their old sites. Finally, the fossilized bones found at Olduvai were not processed as thoroughly as modern foragers process their kills.

Hominins may have brought carcasses to these sites and processed the carcasses with flakes made from previously cached stones.

Richard Potts, an anthropologist at the Smithsonian Institution, suggests that these sites were not home bases but butchery sites—places where hominins worked but did not live. He believes that hominins brought their kills to these sites and dismembered their carcasses there. Some of the carcasses were scavenged by hominins from other carnivores, and some of the hominins' kills were lost to scavengers. Hominins may have carried bones and meat away to other sites for more intensive processing. This would explain why bones accumulated over such a long time, why bones of nonhominin carnivores were present, and why bones were not completely processed.

At first glance, it might seem inconvenient for hominins to schlepp their kills to butchery sites. Why not process the carcass at the kill site? We have seen that hominins used tools to process the meat. However, they couldn't be sure of finding appropriate rocks for toolmaking at the sites of their kills. And they couldn't leave their kills unguarded while they went off to fetch their tools, lest a hungry scavenger steal their supper. So they must have had to carry the meat to where their tools were kept or to keep their tools with them all the time. Remember that these tools were fairly heavy, and early hominins had no pockets or backpacks. Potts suggests that the best strategy would have been to cache tools at certain places, and then to carry carcasses that they had acquired to the nearest cache.

BACK TO THE FUTURE: THE TRANSITION TO MODERN HUMAN LIFE HISTORIES

We have argued that complex foraging techniques favor food sharing, male parental investment, and sexual division of labor. These behavioral practices, in turn, create selective pressures for reduced sexual dimorphism and delayed maturation of off-spring. The Oldowan toolmakers were extractive foragers, hunters, and scavengers of meat. As we will see in the next chapter, reduced sexual dimorphism and somewhat delayed maturation characterize hominins who appear in the fossil record about 2 mya. It seems likely that these morphological and developmental features, which also characterize modern humans, represent adaptations to this new way of life.

FURTHER READING

Conroy, G. 1997. *Reconstructing Human Origins: A Modern Synthesis*. Norton, New York.

Foley, R. 1987. *Another Unique Species*. Longman Scientific and Technical, Harlow, England.

Kaplan, H., K. Hill, J. Lancaster, and A. Hurtado. 2000. A theory of human life history evolution: Diet, intelligence, and longevity. *Evolutionary Anthropology* 9: 156–185.

McGrew, W. C. 1992. *Chimpanzee Material Culture*. Cambridge University Press, Cambridge, England.

Potts, R. 1984. Home bases and early hominins. *American Scientist* 72: 338–347.

Schick, K. D., and N. Toth. 1993. *Making Silent Stones Speak: Human Evolution and the Dawn of Technology*. Simon & Schuster, New York.

STUDY QUESTIONS

1. Who made the first stone tools? What kinds of evidence complicate this question?
2. What are the differences between collected, extracted, and hunted foods? How do comparative data help us understand the unique foraging adaptations of modern humans?

3. Why do complex foraging techniques favor slow development and long childhoods? What kinds of data enable us to draw inferences about developmental patterns in early hominins?

4. Suppose we discovered that Oldowan foraging technology was very easy to master and required little skill. How would that change your ideas about the selective pressures acting on hominins? How would that change your ideas about who made Oldowan tools?

5. Why do we associate food sharing with meat eating, rather than with vegetarianism?

6. What were Oldowan tools like, and what were they used for?

7. Researchers argue about whether Oldowan toolmakers were hunters or scavengers. What are the main arguments on each side of this debate?

8. How would it change our views of human evolution if we found convincing evidence that the Oldowan hominins scavenged but did not hunt?

9. Some researchers have argued that Olduvai Bed I sites are the remains of hominin home bases; others think that they are workplaces where hominins processed carcasses. If you were an archaeologist, how would you go about testing this idea? Think about the kinds of data you would need to collect to examine the merits of each hypothesis.

10. We have discussed the Oldowan toolmakers at length, even though we don't really know who they were. Why is this a profitable exercise?

KEY TERMS

flakes
cores
Oldowan tool industry
knapping
Mode 1

collected foods
extracted foods
hunted foods
taphonomy
home base

CONTENT FROM *HUMAN EVOLUTION: A MULTI-MEDIA GUIDE TO THE FOSSIL RECORD*

Early Homo
  Overview
  Large v. Small H. habilis
  Brain Reorganization
  Early Tools

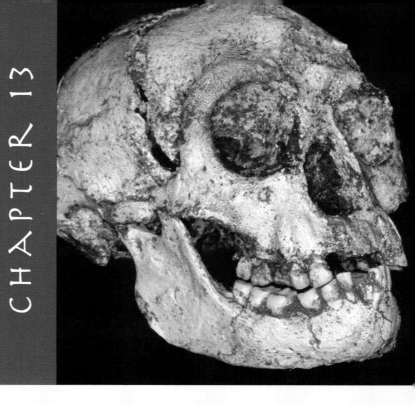

CHAPTER 13

FROM HOMININ TO *HOMO*

About 1.8 mya, a new kind of hominin appeared in Africa. These creatures, which we will call *Homo ergaster*, were much more like modern humans than the apelike hominins who preceded them were. Like modern humans, they developed slowly, and their infants must have been largely helpless at birth. They had large, robust bodies with long legs and short arms. These adaptations tell us that they were fully committed to life on the ground, as we are. In contrast to the apelike hominins, the males were only slightly larger than the females. *Homo ergaster* invented a new kind of tool technology and probably learned to master fire and to hunt large game. Although we can see much of ourselves in these creatures, some important differences remain. They had smaller brains than we have, and their subsistence technology seems to have been much less flexible than ours.

We begin this chapter by describing what is known about *H. ergaster*. Then we trace the evolution of hominins as they migrated out of Africa and spread throughout temperate Asia and into Europe. We will see that hominins gradually evolved to have even larger brains and more sophisticated subsistence technology, and these incremental changes eventually led to the emergence of our own species, *Homo sapiens*, about 100 kya (thousand years ago). We will draw on evidence from the fossil and archaeological records to reconstruct the transformation of *H. ergaster* into *H. sapiens*.

HOMININS OF THE LOWER PLEISTOCENE: *HOMO ERGASTER*

The Pleistocene epoch began 1.8 mya and saw a cooling of the world's climate.

Geologists subdivide the Pleistocene into three parts: the Lower, Middle, and Upper Pleistocene. The Lower Pleistocene began about 1.8 mya, a date that coincides with a sharp cooling of the world's climate. The beginning of the Middle Pleistocene is marked by sharply increased fluctuations in temperature and the first appearance of immense continental glaciers that covered northern Europe 900 kya, and its end is defined by the termination of the penultimate glacial period about 130 kya. The Upper Pleistocene ended about 12 kya when a warm, interglacial phase of the world climate began; this warm period has persisted into the present (Figure 13.1).

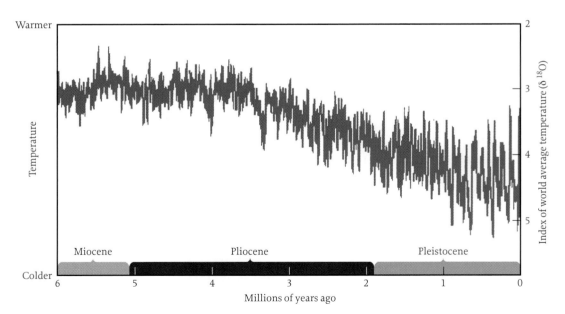

FIGURE 13.1 The pattern of world temperature over the last 6 million years. These estimates are based on the ratio of oxygen-16 to oxygen-18 in cores extracted from deep-sea sediments. During the Pliocene world temperatures declined, and in the Pleistocene climate fluctuations increased, especially during the Middle and Upper Pleistocene.

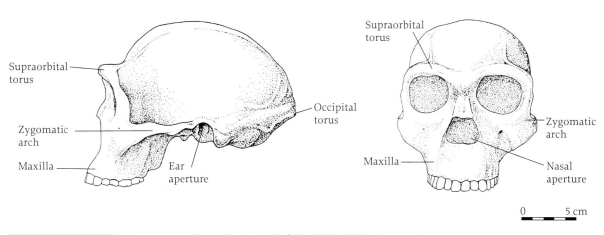

FIGURE 13.2 *Homo ergaster* skulls, like the skull of KNM-ER 3733 illustrated here, show a mix of primitive and derived features. (Figure courtesy of Richard Klein.)

Homo ergaster appears in the African fossil record about 1.8 mya and disappears about 0.6 mya.

Fossils of *H. ergaster* have been found at Konso-Gardula (Ethiopia), Daka (Ethiopia), Olduvai Gorge (Tanzania), Lake Turkana (Kenya), Olorgesailie (Kenya), Swartkrans (South Africa), and a number of other sites in Africa. Figure 13.2 shows a very well preserved skull (labeled KNM-ER 3733) from Lake Turkana that was found in 1976 by a team led by Richard Leakey. When *H. ergaster* fossils first appear on the scene, they are associated with Oldowan tools.

Until recently, most paleontologists assigned these East African fossils to the species *Homo erectus*, on the basis of their similarities to fossils that had been discovered in Indonesia many years before. However, a growing number of paleontologists have come to believe that the African specimens are distinctive enough to be assigned to a different species. According to the rules of zoological nomenclature, the Indonesian specimens retain the original name *H. erectus* because they were described first, and the African specimens must be given a new name. Advocates of the two-species view have given the African fossils the name *Homo ergaster*, or "work man." We will use *H. ergaster* to refer to African specimens of this group, and *H. erectus* for the Asian fossils. Keep in mind that it is not entirely clear whether the creatures leaving these fossils belonged to one or two species.

Homo ergaster appeared in Eurasia about the same time it was first known from Africa.

As you will recall from Chapters 11 and 12, the apelike hominins were restricted to Africa. *Homo ergaster* was the first hominin to make its way out of Africa. We know this because of a startling discovery in the Caucasus Mountains of the Republic of Georgia. In 1991, archaeologists working at a site called Dmanisi (see Figure 13.15) discovered a lower jaw and Oldowan tools in sediments dated to 1.1 or 1.8 mya. Many researchers were skeptical that hominins were present in Eurasia so early, but their doubts were soon put to rest. In 1999, a team of archaeologists from Georgia and the United States uncovered two nearly complete crania at Dmanisi that are

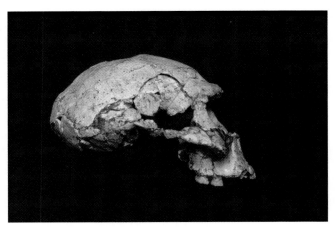

very similar to African specimens of *H. ergaster* and are securely dated to about 1.7 mya (Figure 13.3). These crania are associated with more than a thousand Oldowan tools. Thus, *H. ergaster* migrated out of Africa and into Eurasia about 1.7 mya, taking their toolmaking skills along with them.

FIGURE 13.3 One of the crania found at Dmanisi, a site in the foothills of the Caucasus Mountains in the Republic of Georgia. The Dmanisi fossils date to 1.7 mya and are very similar to *H. ergaster* fossils of the same age from Africa. The Dmanisi site is at latitude 41° north, well out of the tropics, indicating that *H. ergaster* was able to adapt to a wider range of habitats than previous hominins were.

Morphology

Skulls of H. ergaster *differ from those of both earlier hominins and modern humans.*

Skulls of *H. ergaster* retain many of the characteristics of earlier hominins (Figure 13.4), including a marked narrowing behind the eyes, a receding forehead, and no chin. *Homo ergaster* also shows many derived features. Some of these are shared by modern humans, including a smaller, less prognathic face, a higher skull, and smaller jaws and teeth. However, *H. ergaster* also has some derived features that we don't see in earlier hominins or modern humans. For example, *H. ergaster* has a horizontal ridge at the back of the skull (**occipital torus**), which gives it a pointed appearance from the side. It also had quite large browridges.

Many of the derived features of the skull of *H. ergaster* are likely to be related to diet. These hominins were probably better adapted for tearing and biting with their canines and incisors, and less suited to heavy chewing with their molars. Thus, all their teeth are smaller than the teeth of the australopithecines and paranthropines, but their molars are reduced relatively more in comparison with their incisors. The large browridges and the point at the back of the skull are needed to buttress the skull against novel stresses created by an increased emphasis on tearing and biting.

Homo ergaster had a substantially larger brain than earlier hominins. The average brain volume for these creatures was about 800 cc (for reference, 1000 cc is 1 liter, just a little more than a quart)—larger than the brains of earlier apelike hominins (500 to 700 cc) but still considerably smaller than modern human brains, which average about 1400 cc. As we will see, however, *H. ergaster* had a much larger body than the earlier hominins. Brain size is proportional to body size in mammals, so it seems likely that in comparison with earlier hominin species, *H. ergaster* did not have a substantially larger brain in relation to body size.

The postcranial skeleton of H. ergaster *is much more similar to the skeleton of modern humans than to that of earlier hominins, but it still differs from ours in interesting ways.*

Kimoya Kimeu (Figure 13.5), the leader of the Koobi Fora paleontological team, found a fossilized skeleton of a young *H. ergaster* male in 1984. The find was made on the west side of Lake Turkana, the same region where fossils of *Australopithecus anamensis, Paranthropus aethiopicus,* and *Kenyanthropus platyops* were found. The skeleton, which is formally known as KNM-WT 15000, belonged to a boy who was about 12 years old when he died. The skeleton provides us with a remarkably com-

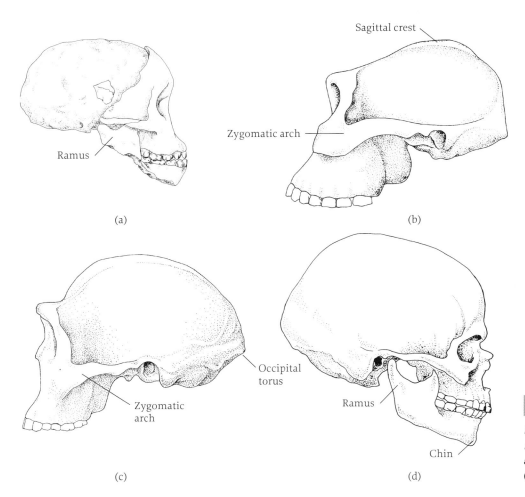

Sagittal crest

Zygomatic arch

(a)

(b)

Ramus

Occipital torus

Zygomatic arch

Ramus

Chin

(c)

(d)

FIGURE 13.4 Skulls of *Australopithecus africanus* (a), *Paranthropus robustus* (b), *Homo ergaster* (c), and modern *Homo sapiens* (d). (Figures courtesy of Richard Klein.)

plete picture of the *H. ergaster* body; even the delicate ribs and vertebrae are preserved (Figure 13.6). An extensive analysis of this skeleton, coordinated by Alan Walker, tells us a lot about this youngster's body and his way of life.

Remember that earlier hominins were bipeds but still had long arms, short legs, and other features suggesting that they spent a considerable amount of time in trees. In contrast, KNM-WT 15000 had the same body proportions as people who live in tropical savannas today: long legs, narrow hips, and narrow shoulders. KNM-WT 15000 also had short arms, compared with earlier hominins. Taken together, these features suggest that *H. ergaster* was fully committed to terrestrial life.

This skeleton, and less complete bits of the skeletons of other *H. ergaster* individuals, tells us several other interesting things about them:

- They were quite tall. KNM-WT 15000 stood about 1.625 m (5 ft 4 in.) in height. If *H. ergaster* growth patterns were comparable to those of modern humans, this boy would have been about 1.9 m (6 ft) tall when

FIGURE 13.5 Kimoya Kimeu is an accomplished field researcher. He has made many important finds, including KNM-WT 15000 (see Figure 13.6).

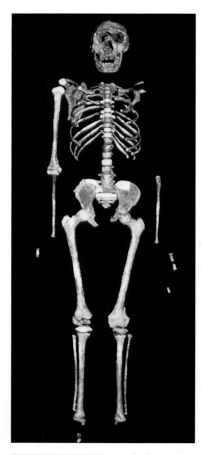

FIGURE 13.6 The fossil of the *H. ergaster* boy, KNM-WT 15000, found on the west side of Lake Turkana, is amazingly complete. (Photograph courtesy of Alan Walker.)

he was fully grown. However, he was also robust and heavily muscled. Think of him as a young shooting guard or a small forward.

- Their infants probably matured slowly and were dependent on their mothers for an extended period of time. Human infants are born well before brain growth is completed, because if they remained in the uterus long enough for the brain to finish developing, the head would be too large to pass through the birth canal. The same seems to have been true for *H. ergaster*, because the ratio of the size of the mother's birth canal to the size of the newborn's head was the same as it is in modern humans. Thus, *H. ergaster* infants were probably more like human infants in their growth and development. As we have already learned, prolonged dependence of infants may be functionally linked to the reduction of sexual dimorphism, paternal investment, and monogamous mating systems.

- Sexual dimorphism was reduced. *Homo ergaster* males were about 20% to 30% larger than females, making *H. ergaster* much less dimorphic than the apelike hominins and only slightly more dimorphic than modern humans.

- They may not have had spoken language. The vertebral canal in the thoracic (middle) region of the back is much larger in modern humans than it is in apes, and it contains a proportionally thicker spinal cord. Detailed studies of the neuroanatomy in the thoracic region suggest that all of the extra nerves that enlarge the spinal cord innervate the muscles of the rib cage and diaphragm. The thoracic vertebrae of KNM-WT 15000 are comparable to those of other primates, suggesting that this young male had less precise control over the muscles of his rib cage and diaphragm than modern humans do. Anatomist Ann McLarnon of Whitelands College in London argues that the enlargement of the thoracic vertebral canal is an adaptation for speech. When people talk, they subconsciously time their breathing to match the patterning of sentences. Sentences begin with a short, rapid intake of air, followed by a much slower outflow, and this breathing pattern allows production of the words of the sentence. Because sentences vary in length, the breathing pattern during speech is much more complex than normal rhythmic breathing is. McLarnon argues that the increased motor control of the diaphragm and thoracic muscles is an adaptation that allows people to manage this complex motor task.

- *H. ergaster* was the first hominin that could run for long distances. Compared to most other mammals, modern humans are not good sprinters. However, we are able to outrun all but a few species over distances of several kilometers. Biologist Dennis Bramble of the University of Utah and anthropologist Daniel Lieberman of Harvard University have recently argued that features of the *H. ergaster* boy like long legs and a long narrow waist are evidence that the capacity for long-distance running first appeared in this species. They believe that this talent may have been useful in long-distance scavenging and hunting in open country.

H. ergaster may have developed more rapidly than modern humans.

Remember from Chapter 11 that Christopher Dean and his colleagues used estimates of enamel growth rate to show that australopithecines developed relatively quickly, somewhat faster than chimpanzees. Using the same methods, these researchers estimate that *H. ergaster* developed more slowly than australopithecines but still faster than modern humans. If this estimate is correct, then *H. ergaster* did not have the same long childhood period that modern humans have, suggesting that learning did not play such an important role in the lives of these creatures (Figure 13.7). It is not clear

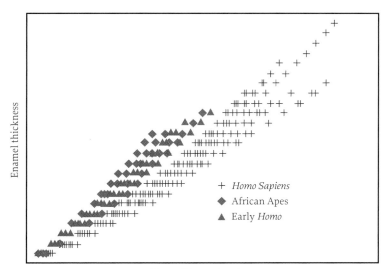

Enamel thickness

+ *Homo Sapiens*
◆ African Apes
▲ Early *Homo*

Enamel formation time

FIGURE 13.7 Rates of enamel growth estimated by counting daily growth marks for modern *Homo sapiens*, modern African apes, and early *Homo*. The latter category includes four individuals usually classified as *H. ergaster*, including KNM-WT 15000, one *Australopithecus habilis*, and one *A. rudolfensis*. Like australopithecines and modern apes, early *Homo* apparently grew rapidly.

whether this fact can be reconciled with the evidence that *H. ergaster* infants were highly dependent.

Tools and Subsistence

Homo ergaster made fancier tools than earlier hominins had made.

The earliest fossils of *H. ergaster* both in Africa and in Eurasia are associated with Oldowan tools, the same tools that we described in Chapter 12. However, sometime between 1.6 and 1.4 mya in Africa, *H. ergaster* added a new and more sophisticated tool to its kit. This totally new kind of stone tool is called a **biface**. To make a biface, the toolmaker strikes a large piece of rock from a boulder to make a core, and then flakes this core on all sides to create a flattened form with a sharp edge along its entire circumference. The most common type of biface, called a **hand ax**, is shaped like a teardrop and has a sharp point at the narrow end (Figure 13.8). A **cleaver** is a lozenge-shaped biface with a flat, sharp edge on one end; a **pick** is a thicker, more triangular biface. Bifaces are larger than Oldowan tools, averaging about 15 cm (6 in.) in length and sometimes reaching 30 cm (12 in.). Bifaces are categorized as **Mode 2** technology. Paleoanthropologists call the Mode 2 industries of Africa and western Eurasia (that is, Europe and the Middle East) the **Acheulean industry** after the French town of Saint-Acheul, where hand axes were first discovered. The oldest Acheulean tools found at West Turkana date to about 1.6 mya, just after *H. ergaster* first appeared in Africa. It is important to understand that Oldowan tools do not disappear when Acheulean tools make their debut. *Homo ergaster* continued making simple Mode 1 tools, perhaps when they needed a serviceable tool in a hurry—as foraging people do today.

The standardized form of hand axes and other Mode 2 tools in the Acheulean industry suggests that toolmakers had a specific design in mind when they made each tool. The Mode 1 tools of the Oldowan industry have a haphazard appearance; no two are alike. This lack of standardization suggests that makers of Oldowan tools simply picked up a core and struck off flakes; they didn't try to create a tool with a

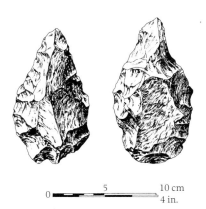

0   5   10 cm
                4 in.

FIGURE 13.8 Acheulean hand axes, like the two shown here, were teardrop-shaped tools created by removing flakes from a core. The smallest ones would fit in the palm of your hand, and the largest ones are more than 0.6 m (2 ft) long.

particular shape that they had in mind beforehand. They may have done this because the flakes were the actual tools.

It is easy to see how a biface might have evolved from an Oldowan chopper by extending the flaking around the periphery of the tool. However, Acheulean tools are not just Oldowan tools with longer edges; they are designed according to a uniform plan. Hand axes have regular proportions: the ratio of height to width to thickness is remarkably constant from one ax to another. *Homo ergaster* must have started with an irregularly shaped piece of rock and whittled it down by striking flakes from both sides until it had the desired shape. Clearly, all the hand axes would not have come out the same if their makers hadn't shared an idea for the design.

Hand axes were probably used to butcher large animals.

If hand axes were designed, what were they designed for? The answer to this question is not obvious, because hand axes are not much like the tools made by later peoples. A number of ideas about what hand axes were used for have been proposed:

1. *Butchering large animals. Homo ergaster* acquired the carcasses of animals like zebra or buffalo either by hunting or by scavenging, and then used a hand ax as a modern butcher would use a cleaver, to dismember the carcass and cut it into useful pieces.
2. *Digging up tubers, burrowing animals, water.* Contemporary foragers in savanna environments spend a lot of their time digging up edible tubers. Although modern peoples generally use sharpened sticks for digging, some archaeologists have suggested that *H. ergaster* may have used hand axes for this purpose. Digging tools would also have been useful for capturing burrowing animals like warthogs and porcupines, and for making wells to acquire water.
3. *Stripping bark from trees to get at the nutritious cambium layer underneath.*
4. *Hurling at prey animals.*
5. *Dispensing flake tools.* Hand axes weren't tools at all. Instead they were "flake dispensers" from which *H. ergaster* struck flakes to be used for many everyday purposes.

Although we are not certain how hand axes were used, two kinds of evidence support the first hypothesis that they were heavy-duty butchery tools. Kathy Schick and Nicholas Toth, whose investigations into the function of Oldowan tools we discussed in Chapter 12, have also done experiments using Acheulean hand axes for each of the tasks just listed. From these experiments, Schick and Toth conclude that hand axes are best suited to butchery. The sharp end of the hand ax easily cuts through meat and separates joints; the rounded end provides a secure handle. The large size is useful because it provides a long cutting edge, as well as the cutting weight necessary to be effective for a tool without a long handle. Schick and Toth's results are supported by the work of University of Illinois paleoanthropologist Lawrence Keely, who performed microscopic analysis of wear patterns on a small number of hand axes. He concluded that the pattern of wear is consistent with animal butchery.

Evidence from Olorgesailie, a site in Kenya at which enormous numbers of hand axes have been found, indicates that these axes may also have served as flake dispensers. A team led by Richard Potts discovered most of the fossilized skeleton of an elephant, along with numerous small stone flakes dated to about 1 mya. Chips on the edges of the flakes suggest that they were used to butcher the elephant, and the elephant bones show cut marks made by stone tools. Careful examination of the flakes reveals that they were struck from an already flaked core, such as a hand ax, not from an unflaked cobble. Moreover, when the flakes were removed from the hand axes, they did not leave a sharper hand-ax edge. Taken together, these data suggest that *H. ergaster* struck flakes from hand axes and used both hand axes and flakes as butchery tools.

The Acheulean industry remained remarkably unchanged for almost 1 million years.

There is relatively little change over space and time in the Acheulean tool kit from its first appearance about 1.6 mya until it was replaced around 300 kya. Amazingly enough, Acheulean tools that were made half a million years apart are just as similar as tools made at about the same time at sites located thousands of miles apart. Essentially the same tools were made for more than 1 million years. In fact, as we will see shortly, the Acheulean industry was longer-lived than *H. ergaster* itself, which disappeared from the fossil record in Africa about 1 mya. Most anthropologists assume that the knowledge necessary to make a proper hand ax was passed from one generation to the next by teaching and imitation. It is a remarkable notion that this form of knowledge might have been faithfully transmitted and preserved for so long in a small population of hominins spread from Africa to eastern Eurasia.

Homo ergaster clearly ate meat, but there is controversy about whether they hunted or scavenged their kills.

There is fierce controversy among paleoanthropologists about whether *H. ergaster* was a big-game hunter. The issues and the evidence are similar to the controversy surrounding hunting by Oldowan hominins. There is good evidence that *H. ergaster* ate meat, but less certainty about whether *H. ergaster* hunted or scavenged. One of the most compelling pieces of evidence that these hominins ate meat comes from the skeleton of an *H. ergaster* woman (KNM-ER 1808) that was discovered at Koobi Fora near Lake Turkana in northern Kenya by Alan Walker and his colleagues (Figure 13.9). This woman died about 1.6 mya. Her long bones are covered with a thick layer of abnormal bone tissue (Figure 13.10). This amorphous bone growth puzzled Walker, so he

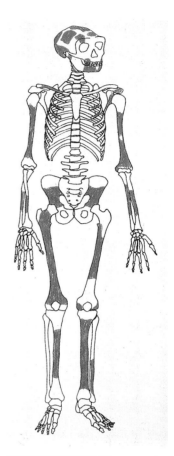

FIGURE 13.9 Much of the skeleton of an *H. ergaster* female (KNM-ER 1808) was discovered at Koobi Fora. The bones marked in red are those that were recovered. (Diagram by Alan Walker.)

FIGURE 13.10 The long bones of KNM-ER 1808 are covered in a thick, amorphous layer of abnormal bone tissue. A fragment of the tibia of KNM-ER 1808 (bottom) is compared with the tibia of a normal *H. ergaster* (top). (Photograph courtesy of Alan Walker.)

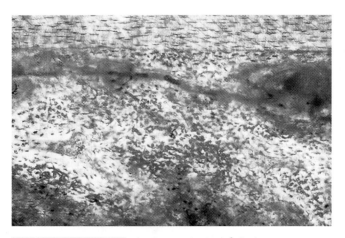

FIGURE 13.11 A microscopic view of the bone structure of KNM-ER 1808. Note that the small band of normal bone at the top looks very different from the puffy, irregular, diseased bone in the rest of the picture. (Photograph courtesy of Alan Walker.)

consulted with some colleagues at a medical school. They told him that this kind of bone growth is symptomatic of vitamin A poisoning (Figure 13.11). This, of course, posed another puzzle: how could a hunter-gatherer get enough vitamin A to poison herself? The most likely way would be to eat the liver of a large predator, like a lion or a leopard. The same symptoms have been reported for Arctic explorers who ate the livers of polar bears and seals. If this woman's bones were deformed because she ate a large predator's liver, then we can assume that *H. ergaster* ate meat. Of course, we don't know how this woman obtained the liver that poisoned her. She might have scavenged the liver from a predator's carcass, or she might have killed the predator in a contest over a kill.

As we noted earlier, the bones of the elephant found at Olorgesailie show signs of cut marks from stone tools. It is not clear how hominins obtained the elephant carcass. The archaeological evidence indicates that the site was not repeatedly used by hominins or other animals. Perhaps the hominins killed the elephant and butchered it at the kill site. Perhaps they came upon the body of the elephant after it died, and butchered it there. It does not seem likely that they stole the kill from another predator, because there are no tooth marks on the bones.

Several lines of circumstantial evidence also suggest that *H. ergaster* relied on meat, at least during part of the year. First, hand axes seem to be well suited to the butchery of large animals. Second, the teeth of *H. ergaster* are well suited for biting and tearing and less suited for chewing tough plant foods. As we pointed out in Chapter 12, meat, seeds, and fibrous plant materials are the main alternative food sources for primates during the dry season. If *H. ergaster* was not eating tough plant foods during the dry season, then it might have relied on meat. Third, Dmanisi is at about the same latitude as Chicago, so *H. ergaster* was able to live outside of the tropics. Other primates rely heavily on fruit and other tender parts of plants for survival. To survive a temperate winter when such things are not available, *H. ergaster* might have relied on meat. Very few primates except humans live in places with severe winters, and these are only small, peripheral populations of species that are usually more tropical.

Homo ergaster *may have controlled fire.*

The earliest evidence that *H. ergaster* controlled fire comes from Koobi Fora, where archaeologists found baked earth beside stone tools at a site dated to about 1.6 mya. This association doesn't prove that *H. ergaster* controlled fire, because charred bones and ash could have been produced by natural fires that swept through sites where hominins had lived, perhaps many years after hominins occupied them. Taphonomic experiments conducted by Randy Bellomo of the University of South Florida help resolve this issue. Bellomo made campfires and set fire to grass and tree stumps in several African habitats and carefully examined the remains. He found that the soil under campfires reaches much higher temperatures than the soil under grass fires or the soil surrounding smoldering tree stumps. The higher temperatures of campfires produce two distinctive characteristics in the soil. First, a bowl-shaped layer of highly oxidized soil is formed directly under the fire. Second, the soil under the fire becomes highly magnetized compared with the surrounding soil, and this magnetization is quite stable. Soil beneath stump fires and grass fires does not have either of these features. Stump fires leave a distinctive hole filled with unconsolidated sediments.

With these experimental results in hand, Bellomo examined the fire sites at Koobi Fora. One of them contains a bowl of oxidized soil that has been magnetized to the same degree as his experimental samples. Moreover, the magnetization is in the same direction as the Earth's magnetic field would have been 1.6 mya, and not in the present direction. This is just what we would expect to find if these were campfire sites.

Excavations at Swartkrans Cave in South Africa provide more evidence that *H. ergaster* used fire. There, anthropologists found numerous fragmented fossil bones of antelope, zebra, warthog, baboon, and *Paranthropus robustus* that show evidence of having been burned. Such burnt fossils were found at 20 different levels dated from 1.5 to 1 mya. Oldowan tools and *H. ergaster* fossils were also found on some of these levels. To determine whether these bones had been burnt in campfires, C. K. Brain of the Transvaal Museum in Pretoria and Andrew Sillen of the University of Cape Town compared the fossils to modern antelope bones burned at a range of temperatures. They found that the very high temperatures characteristic of long-burning campfires produce clear changes in the microscopic structure of bone—changes that can also be seen in the burnt fossils at Swartkrans. Because Bellomo's experiments suggest that wildfires rarely produce high temperatures, the presence of bones burnt at high temperature provides good evidence that these bones were burned in campfires. (Unfortunately, marshmallows leave no trace in the archaeological record.)

Homo ergaster is classified in the genus Homo because it shares the same adaptive grade as later members of the genus.

Taken together, the archaeological and paleontological data suggest that *H. ergaster* represented a marked departure from its predecessors. This hominin shared important adaptive traits with modern humans: terrestrial life, complex foraging technology, slow development, reduced sexual dimorphism, and (probably) extensive paternal investment in offspring. Because we share these features, *H. ergaster* is classified in the same genus as we are. With its smaller brain and relatively static and inflexible technology, however, *H. ergaster* was still quite different from modern humans. In the next section we explain how *H. ergaster* was transformed into a larger-brained and highly flexible creature more like us.

HOMININS OF THE EARLY MIDDLE PLEISTOCENE (900 TO 300 KYA)

The world's climate became colder and much more variable during the Middle Pleistocene (900 to 130 kya).

During the Middle and Upper Pleistocene there were many long, cold glacial periods punctuated by short, warmer interglacial periods. Figure 13.12 shows estimates of global temperatures for this time period. Notice that the world climate has fluctuated wildly in the last 700,000 years. From geological evidence, we know that during the cold periods, glaciers covered North America and Europe and arctic conditions

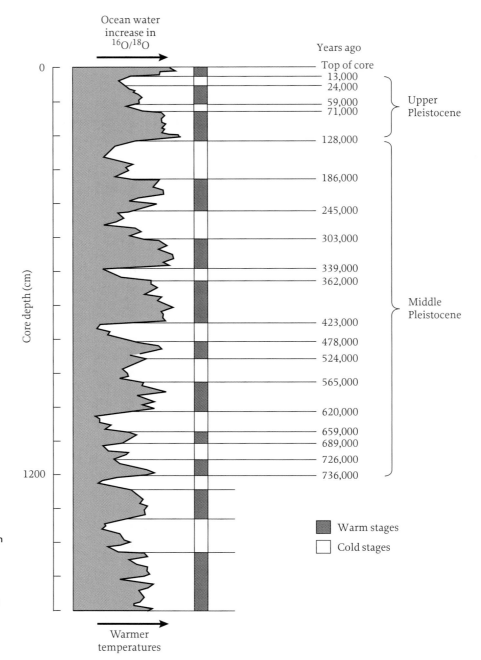

Ocean water
increase in
$^{16}O/^{18}O$

Years ago

Top of core

Core depth (cm)

Warmer
temperatures

■ Warm stages
□ Cold stages

FIGURE 13.12 The ratio of ^{16}O to ^{18}O in seawater over the last 700,000 years indicates that there have been wide swings in world temperature, but on average, the world has been colder than it is now. Larger ocean values of $^{16}O{:}^{18}O$ signify higher global temperatures.

prevailed. These cold periods were intermittently interrupted by shorter warm periods during which the glaciers receded and the forests returned.

During glacial periods the world was dry, and Africa and Eurasia were isolated from each other by a massive desert. During interglacial periods the world was much wetter, and animals moved from Africa to Eurasia.

These temperature fluctuations had massive effects on the world's biological habitats, as Figure 13.13 shows. The map in Figure 13.13a shows the distribution of habitats about 7 kya during the warmest and wettest period of the current interglacial period. At this time, much of Eurasia and Africa was covered by forests, East and North Africa and Arabia were grasslands, and deserts were limited to small bits of Southwest Africa and central Asia. The map in Figure 13.13b shows what the world was like at the depth of the last glacial period, about 20 kya. The middle latitudes were dominated by vast expanses of extreme desert, as dry as the Sahara Desert today. North and south of this desert, grassland, scrub, and open woodland predominated, and forests were restricted to small regions of central Africa and Southeast Asia. Since the beginning of the Middle Pleistocene, the planet has oscillated between these two extremes, sometimes shifting from one extreme to the other in just a few hundred years.

These fluctuations had important effects on the dispersal of animal species, including hominins, in the Middle and Upper Pleistocene. During glacial periods,

(a) Interglacial conditions (~7000 years BP)

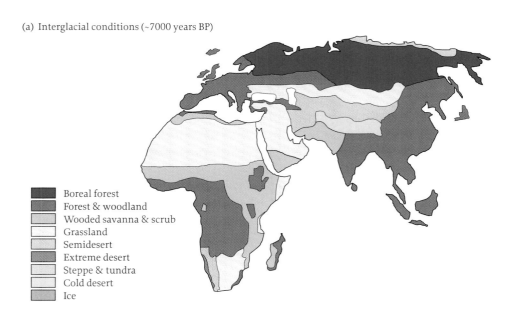

- Boreal forest
- Forest & woodland
- Wooded savanna & scrub
- Grassland
- Semidesert
- Extreme desert
- Steppe & tundra
- Cold desert
- Ice

(b) Glacial conditions (~20,000 years BP)

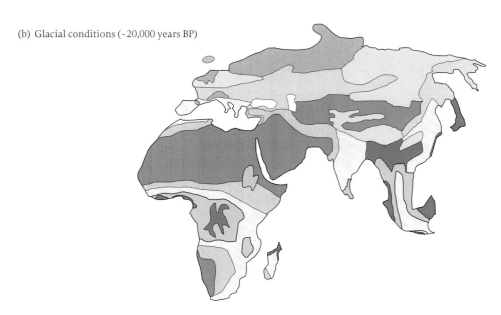

FIGURE 13.13 (a) A reconstruction of biological habitats about 7 kya during the warmest, wettest part of the present interglacial period. Much of Eurasia and Africa were covered with forest and were connected by a broad swath of grassland across northern Africa and southwestern Eurasia. (b) The habitats during the coldest, driest part of the last glacial period, about 20 kya. There was very little forest cover. Grassland and scrub predominated in central Africa and Southeast Asia. Northern Eurasia was covered with cold dry steppe and desert. Central and southern Africa were separated from Eurasia by a band of extreme desert across northern Africa, the Arabian Peninsula, and central Asia.

deserts spread across the northern part of Africa, making this region uninhabitable for most animal species. This corresponds to what we know from the fossil record: There was little movement of animal species between Africa and Asia during glacial periods. Instead, animal species moved mainly east and west across Eurasia. When the world was warmer, grasslands and savannas replaced most deserts, and animals were able to move between Africa and Eurasia much more easily. The fossil record indicates that animal species moved mainly from Africa to Eurasia, not vice versa. As we will see later, this fact has important implications for understanding human evolutionary history.

The evolutionary transition from H. ergaster *to modern humans occurred during the Middle Pleistocene.*

It was in this chaotic, rapidly changing world that natural selection reshaped the hominin lineage once again. *Homo ergaster* was transformed into a smarter and more versatile creature, which we call *Homo sapiens*. We will see that hominins were present in Africa and Eurasia for most of the Middle Pleistocene, and that hominins living in different parts of the world were morphologically distinct. A small-brained form, *H. erectus*, lived in eastern Asia until perhaps 30 kya; and a variety of larger-brained forms ranged through Africa and western Eurasia. About 300 kya, hominins in Africa and western Eurasia began to develop more sophisticated technology and behavior. This process of change continued, particularly in Africa, until the behavior and technology of hominins became indistinguishable from those of modern humans.

Although many of these events can be relatively well established in the fossil and archaeological record, there is considerable controversy about the phylogeny of the human lineage during this period. There are fierce debates about how many species of hominins there were during this period, and about how different species or forms were related. We will turn to these controversies at the end of the chapter. But first you need to learn something about the hominins of the Middle Pleistocene.

FIGURE 13.14 Eugene Dubois was a Dutch anatomist who discovered the first *Homo erectus* fossil at a site called Trinil, in what is now Indonesia.

Eastern Asia: *Homo erectus*

Homo erectus, a hominin similar to H. ergaster, *lived in eastern Asia during most of the Middle Pleistocene.*

Fossils quite similar to what we now call *H. ergaster* were unearthed by Eugene Dubois near the Solo River in Java during the nineteenth century (Figures 13.14 and 13.15). Dubois named the species *Homo erectus*, or "erect man." Dubois's fossils were very poorly dated, and for many years they were believed to be about 500,000 years old. In the 1990s, however, Carl Swisher and Garness Curtis, both from the Berkeley Geochronology Center, used newly developed argon–argon dating techniques to determine the age of small crystals of rock from the sites at which Dubois's fossils had originally been found. Their analyses indicate that the two *H. erectus* sites in Java were actually 1.6 to 1.8 million years old, a million years older than previous estimates had suggested.

The main morphological differences between *H. erectus* and *H. ergaster* lie in the skull. In *H. erectus*, the skull is thicker, the browridges are more pronounced, the sides of the skull slope more steeply, the occipital torus is more pronounced, and there is a

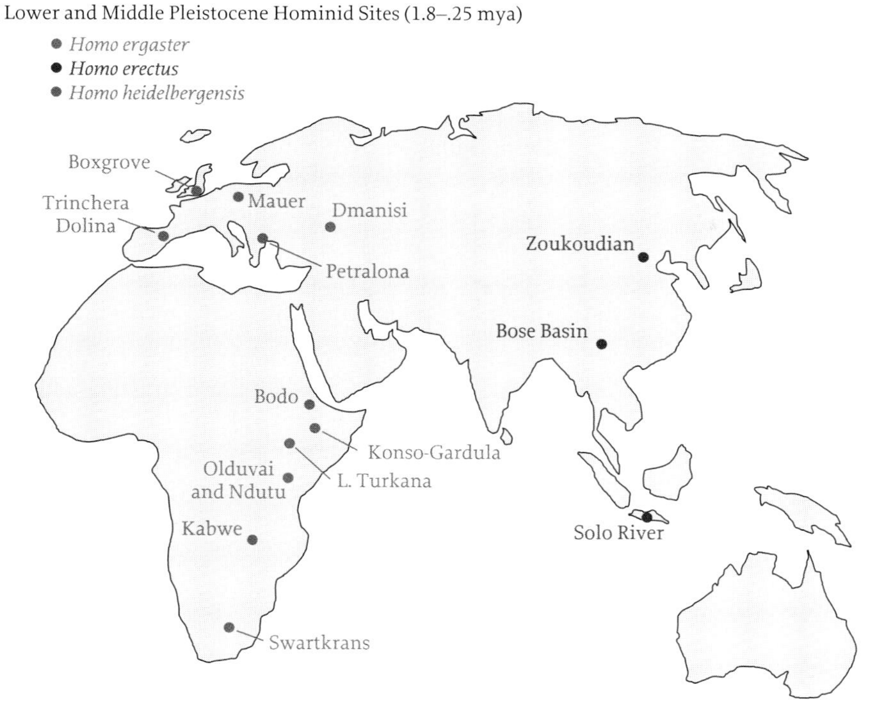

Lower and Middle Pleistocene Hominid Sites (1.8–.25 mya)

- *Homo ergaster*
- *Homo erectus*
- *Homo heidelbergensis*

Boxgrove
Trinchera
Dolina
Mauer
Dmanisi
Petralona
Zoukoudian
Bose Basin
Bodo
Konso-Gardula
Olduvai
and Ndutu
L. Turkana
Kabwe
Solo River
Swartkrans

FIGURE 13.15 The locations of fossil and archaeological sites mentioned in the text. *Homo ergaster* is found in Africa and Europe from 1.8 mya until about 1 mya. *Homo heidelbergensis* is found at sites in Africa and Europe that date to between 800 and 300 kya. Only *H. erectus* is found in East Asia during this period. The Bose Basin has yielded Mode 2 tools but no hominin fossils, so the identity of the toolmakers is unknown.

sagittal keel (a longitudinal V-shaped ridge along the top of the skull) (Figure 13.16). As you may recall from Chapter 11, the sagittal crest expands the area of attachment of the temporalis muscles in the paranthropines. The sagittal keel does not serve this function in *H. erectus*. In fact, we don't know what its function was.

Homo erectus is known from sites in Java and China and persisted until about 30 kya. Over this period, *H. erectus* changed very little. This species does not show the same increase in cranial capacity, technological sophistication, and behavioral flexibility that characterizes its contemporary hominin populations in Africa and western Eurasia. Even

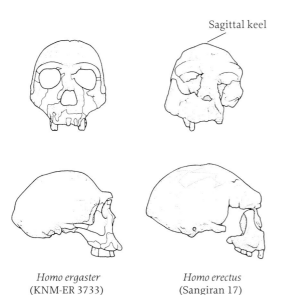

Sagittal keel

Homo ergaster
(KNM-ER 3733)

Homo erectus
(Sangiran 17)

FIGURE 13.16 *Homo ergaster* and *Homo erectus* differed in a number of ways. Most notably, *H. erectus* had a sagittal keel and a more pronounced occipital torus, and the sides of the skull sloped more than in *H. ergaster*.

the most recent *H. erectus* fossils from eastern Asian are less like modern humans than are the African specimens of *H. ergaster*, which are 1 million years older.

Nonetheless, the presence of *H. erectus* fossils in eastern Asia suggests that this species mastered difficult environmental challenges. Somewhere between 500 and 250 kya, *H. erectus* occupied a cave at Zhoukoudian (or Choukoutien) near Beijing. Fossil pollen samples indicate that *H. erectus* occupied the cave only during the warmer interglacial periods. However, *warm* is a relative term; the climate of Zhoukoudian during the interglacial periods was probably something like that of Chicago today. Imagine living through a winter in Chicago without a roof over your head or clothes to keep you warm. To survive the rigors of cold temperate winters, *H. erectus* may have relied on fire and taken shelter in caves like the one at Zhoukoudian. Even this, however, might have been a challenge. The cave at Zhoukoudian is littered with the bones of large game, particularly two species of deer. It is clear that they were brought to the cave by a predator, perhaps *H. erectus*. However, there are also fossilized hyena bones and coprolites (fossilized feces) in these caves. Many of the animal bones show evidence of hyena tooth marks, and most of the faces of the hominins are absent. Thus, it is possible that *H. erectus* was sometimes the prey at Zhoukoudian rather than the predator.

Homo erectus *is associated with Mode 1 tools.*

Acheulean tools appear by 1.6 mya in Africa, but they are rarely found in eastern Asia. Instead, *H. erectus* is usually associated with the simpler Mode 1 tools, similar to those of the Oldowan industry. Some paleoanthropologists think that the differences in tool technology between eastern Asia and Africa provide evidence for cognitive differences between the two species; others think that environmental differences may be responsible for the tool differences. Remember that *H. ergaster* was present in eastern Asia by 1.7 mya. This means that *H. ergaster* left Africa before Mode 2 tools first appeared in Africa. It is possible that a cognitive change arose in the African *H. ergaster* population after the first members of the genus had migrated north into eastern Asia, and that this cognitive adaptation enabled later members of *H. ergaster* to manufacture bifaces. The absence of more symmetrical Mode 2 tools in eastern Asia may mean that they lacked this cognitive ability.

Other researchers argue that differences in tool technology between *H. erectus* and *H. ergaster* tell us more about their habitats than about their cognitive abilities. They point out that, during the Middle Pleistocene, *H. erectus* lived in regions that were covered with dense bamboo forests. Bamboo is unique among woods because it can be used to make sharp, hard tools suitable for butchering game. *Homo erectus* may not have made hand axes because they didn't need them. The recent discovery of a large number of hand axes in the Bose Basin in southern China may support this view (Figure 13.17). About 800 kya a large meteor struck this area and set off fires that destroyed a wide area of the forest. Grasslands replaced the forest. Although grasslands predominated, the residents of this area made Mode 2 tools like those seen in Africa around the same time. Before the forest was destroyed and after it regenerated, the inhabitants of this area made Oldowanlike Mode 1 tools. This evidence suggests that hand axes were an adaptation to open-country life.

FIGURE 13.17 Hand axes found in the Bose Basin in southern China are the only Mode 2 tools discovered in East Asia. These tools date to about 800 kya. Mode 1 tools are found in the same area both before and after this date.

However, there are no hominin fossils associated with these tools, so we don't know whether *H. erectus* actually made them. It is possible that the Bose hand axes were made by members of more technologically advanced immigrant populations from farther west.

Africa and Western Eurasia: *Homo heidelbergensis*

Sometime during the first half of the Middle Pleistocene (900 to 130 kya), hominins with larger brains and more modern skulls appeared.

Hominins with substantially larger brains and more modern skulls appeared in Africa and western Eurasia during the first half of the Middle Pleistocene. Figure 13.18 shows two nearly complete crania from this period: the Petralona cranium found in Greece and the Kabwe (or Broken Hill) cranium from Zambia. These individuals had substantially

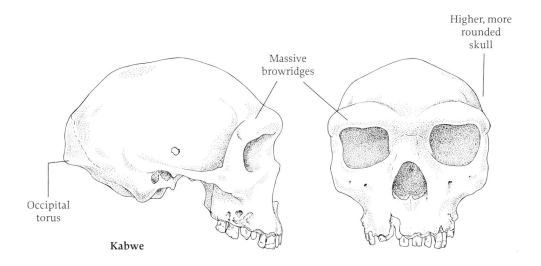

Kabwe

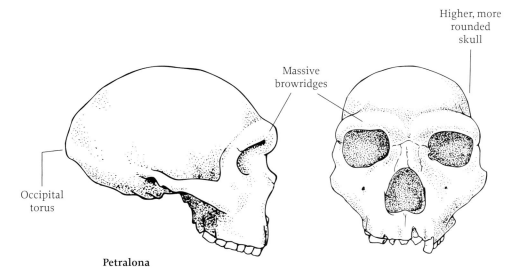

Petralona

FIGURE 13.18 Sometime between 800 and 500 kya, hominins with higher, more rounded crania and larger brains first appear in the fossil record; and by 400 kya, *Homo heidelbergensis* was common in Africa and western Eurasia. Shown here are two *H. heidelbergensis* fossils: one from Kabwe (sometimes called Broken Hill) in Zambia, and the second from Petralona in Greece. Both are approximately 400,000 years old. (Figure courtesy of Richard Klein.)

larger brains—between 1200 and 1300 cc—in relation to their body size than *H. ergaster* had. The skulls also share a number of derived features with modern humans, including more vertical sides, higher foreheads, and a more rounded back than *H. ergaster* had. However, they also retained many primitive features, such as a long, low skull; very thick cranial bones; a large prognathic face; no chin; and very large browridges. Their bodies remained much more robust than modern human bodies are. Fossils with similar characteristics have been found at other sites in Africa (for example, Ndutu in Tanzania, Bodo in Ethiopia) and western Eurasia (for example, Mauer in Germany, Boxgrove in England) that date to the same period (see Figure 13.15). There has been no sign of these kinds of fossils in eastern Asia during this period, although they do appear later.

Traditionally, paleoanthropologists have referred to these larger-brained, more modern-looking hominins as **archaic *Homo sapiens***. As we will explain shortly, however, nowadays only a minority of paleoanthropologists believe that these creatures should be classified as members of *H. sapiens*. There is considerable controversy about whether these hominins belonged to one species or several, and among the anthropologists who believe that there were several species, there is considerable disagreement over which specimens belong to which species. This dispute, which is far from resolved, creates a quandary about what we should call this group of hominins. Here we adopt what is perhaps the most common opinion and use the name ***Homo heidelbergensis*** to refer to all of the Middle Pleistocene hominins of Africa and western Eurasia. (The name *heidelbergensis* is taken from a specimen that was found near the German city of Heidelberg.) Keep in mind, though, that we are not at all sure that this category actually represents a single biological species.

Scientists are uncertain about when or where *H. heidelbergensis* first appeared. One candidate for the oldest *H. heidelbergensis* specimen is the cranium recently found at Buia in Eritrea. This skull has several derived features that are associated with *H. heilelbergensis*, but a small, *H. ergaster*–sized, braincase. A second possibility is the Kabwe fossil found in Zambia. Although it was originally dated to about 130 kya, more recent work has pushed the age of this fossil back to about 800 kya. A third candidate comes from Trinchera Dolina (also called Gran Dolina) in the Sierra de Atapuerca in northern Spain. This site has yielded a number of hominin fossils, including part of an adolescent's lower jaw and most of an adult's face. Although the Trinchera Dolina fossils are too fragmentary to provide an estimate of endocranial volume, they exhibit a number of facial features that are seen in more modern hominins. These fossils have been dated to about 800 kya by means of paleomagnetic methods, but rodent fossils found at the same site indicate a more recent date, perhaps 500 kya. Similar uncertainty afflicts our estimates of the ages of the other early *H. heidelbergensis* fossils, so the best that we can do is to bracket the first appearance of these creatures between 800 and 500 kya.

The tools used by early *H. heidelbergensis* are similar to those used by *H. ergaster*. Tool kits are dominated by Acheulean hand axes and other core tools at most sites, but in some cases the hand axes are more finely worked.

There is good evidence that H. heidelbergensis *hunted big game.*

The first solid evidence for hunting big game comes from this period. On the island of Jersey, off the coast of France, the remains of a large number of fossilized bones from mammoths and woolly rhinoceroses have been found at the base of a cliff. Some of the bones come from adults, which were too big to be vulnerable to most predators. The carcasses clearly have been butchered with stone tools. In some instances, the skull cavity has been opened, presumably to extract the brain tissue. At some places on this site, animal bones have been sorted by body parts (heads here, limbs there, and so on). All this suggests that *H. heidelbergensis* drove the animals over the headland, butchered the carcasses, and ate the meat.

More evidence for hunting comes from three wooden spears that were found in an open-pit coal mine in Schöningen, Germany. This site is dated to about 400 kya. Anthropologists are confident that these were throwing spears because they closely resemble modern javelins. They are about 2 m (6 ft) long and, like modern throwing spears, are thickest and heaviest near the pointed end, gradually tapering to the other end. Some modern people use similar spears for hunting, and it seems plausible that *H. heidelbergensis* used them in the same way. This hypothesis is strengthened by the fact that anthropologists found the bones of hundreds of horses along with the spears, and many of the bones show signs of having been processed with stone tools.

HOMININS OF THE LATER PLEISTOCENE (300 TO 50 KYA)

About 300 kya, and then slightly later in western Eurasia, hominins in Africa shifted to a new stone tool kit.

Beginning about 300 kya, hand axes became much less common and were replaced by tools that were manufactured by the production of sizable flakes, which were then further shaped or retouched. Unlike the small, irregular Oldowan flakes that had been struck from cobble cores, these tools were made from large, symmetrical, regular flakes via complicated techniques. One method, called the **Levallois technique** (after the Parisian suburb where such tools were first identified), involves three steps. First the knapper prepares a core with one precisely shaped convex surface. Then the knapper makes a striking platform at one end of the core. Finally, the knapper hits the striking platform, knocking off a flake, the shape of which is determined by the original shape of the core (Figure 13.19). A skilled knapper can produce a variety of different kinds of tools by modifying the shape of the original core. Such prepared core tools are classified as **Mode 3** technology.

Microscopic analyses of the wear patterns on Mode 3 tools made during this period suggest that some tools were **hafted** (attached to a handle). Hafting is an extremely important innovation because it greatly increases the efficiency with which humans can apply force to stone tools. (Try using a hammer without a handle.) *Homo heidelbergensis* probably hafted pointed flakes onto wooden handles to make stone-tipped spears, a major innovation for a big-game hunter.

Flake the margin of the core:

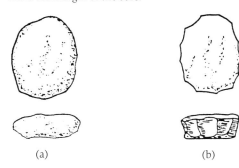

(a)   (b)

Prepare the surface of the core:

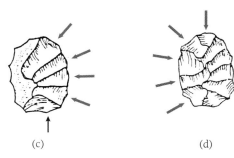

(c)   (d)

Remove Levallois flake:

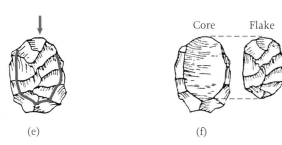

Core   Flake

(e)   (f)

FIGURE 13.19 The process of making a Levallois tool. (a) The knapper chooses an appropriate stone to use as a core. The side and top views of the unflaked core are shown. (b) Flakes are removed from the periphery of the core. (c) Flakes are removed radially from the surface of the core, with the flake scars on the periphery being used as striking platforms. Each of the red arrows represents one blow of the hammer stone. (d) The knapper continues to remove radial flakes until the entire surface of the core has been flaked. (e) Finally, a blow is struck (large red arrow) to free one large flake (outlined in red). This flake will be used as a tool. (f) At the end, the knapper is left with the remains of the core (left) and the tool (right). (Figure courtesy of Richard Klein.)

Eastern Eurasia: *Homo erectus* and *Homo heidelbergensis*

During the second half of the Middle Pleistocene, H. heidelbergensis *appeared in eastern Asia, where it may have coexisted with* H. erectus.

Hominins with larger brains and more rounded skulls have been found at several sites in China that date to approximately 200 kya. The most complete and most securely dated fossil is from Yingkou (also called Jinniushan) in northern China (Figure 13.20). This specimen (Figure 13.21), which consists of a cranium and associated postcranial bones, is similar to early *H. heidelbergensis* fossils from Africa and Europe. Like the crania found at Kabwe and Petralona, it shares a larger braincase (about 1300 cc) and more rounded skull with modern humans, but it also has massive browridges and other primitive features. Other *H. heidelbergensis* fossils have been found in Dali in northern China and Maba in southern China, and they are probably somewhat younger than the Yingkou specimen. These fossils are associated with Oldowan-type tools. It is unclear whether these hominins were immigrants from the west or the result of convergent evolution in eastern Asia.

H. heidelbergensis may have coexisted with *H. erectus* in East Asia during this period. Fossils of *H. erectus* that are between 200,000 and 300,000 years old have been found both at Zhoukoudian (Skull V) and at Hexian, also called He Xian (in southern China). The fossils of other animals found in association with *H. erectus* fossils at two sites in Java (Ngandong and Sambungmachan) are consistent with an age of 250,000 to 300,000 years. However, the teeth of bovids (buffalo and antelope) associated with the hominin fossils at these sites have been dated to about only 27 kya by means of electron-spin-resonance techniques.

Later Pleistocene Hominid Sites (250–50 kya)
- *Homo erectus*
- *Homo heidelbergensis*
- Neanderthals
- *Homo sapiens*

La Chapelle-aux-Saints, La Ferrassie, Le Moustier, Combe-Grenal, Mauran

Sima de los Huesos

Neander Valley

Shanidar

Dar es Soltan

Zoukoudian

Yinkou

Dali

Hexian

Maba

Omo-Kibish

Laetoli

Florisbad

Ngandong and Sambungmachan

Border Cave

Klasies River Mouth

FIGURE 13.20 Locations of later-Pleistocene fossil and archaeological sites mentioned in the text. *Homo erectus* and *Homo heidelbergensis* are found in East Asia, Neanderthals in Europe, and both *H. heidelbergensis* and *H. sapiens* in Africa.

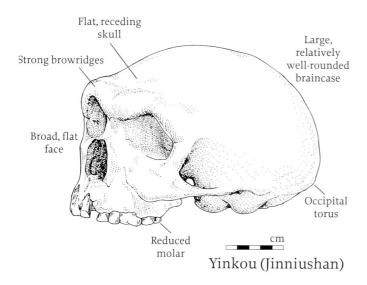

Flat, receding skull

Strong browridges

Large, relatively well-rounded braincase

Broad, flat face

Occipital torus

Reduced molar

cm

Yinkou (Jinniushan)

FIGURE 13.21 Hominins with the characteristics of *H. heidelbergensis* appeared in East Asia later than in western Eurasia. The specimen illustrated here, which is from Yingkou in northern China, is approximately 200,000 years old.

During the upper Pleistocene, H. erectus *evolved into a tiny, small-brained hominin called* Homo floresiensis *on a small Indonesian island.*

In the fall of 2004 a team of Indonesian and Australian researchers published what has been dubbed "the most surprising fossil hominin found in the last 50 years." Working at a cave site called Liang Bua on the Indonesian island of Flores (Figure 13.22), these researchers uncovered the remains of seven individuals. The fossil spec-

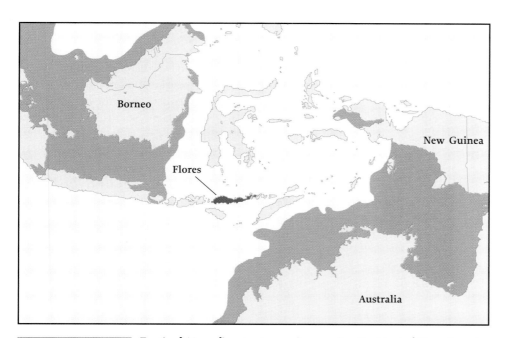

FIGURE 13.22 Fossils of *Homo floresiensis* were discovered on the island of Flores in eastern Indonesia. The shaded areas give the approximate coastlines during glacial periods showing that Flores was isolated from both Asia and Australia even when sea levels were at their lowest levels. As a result, Flores was home to an odd mix of creatures, including a dwarf elephant, huge monitor lizards, and one-meter-tall hominins.

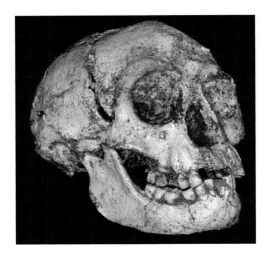

FIGURE 13.23 Three-quarter view of the skull of *Homo flore-siensis*. The skull is very small but shares a number of derived features with *H. erectus*. The brain of this creature was smaller than that of the smallest australopithecines.

imens included a complete skull (Figure 13.23), and bones from the arms, legs, and feet. The first surprise about these creatures, named *Homo flore-siensis*, was their size. They were slightly less than one meter tall (about 3 feet), much smaller than *H. erectus* elsewhere. Their brains were also very small (380 cc)—so small, in fact, that they may have been less encephalized than *H. erectus*. The second surprising thing about these creatures was their age. Using two different dating methods, the researchers estimated that *H. floresiensis* lived on Flores 35 to 14 kya.

Several lines of evidence indicate that these petite hominins are descended from *H. erectus*. First, the skulls share a number of derived characters with *H. erectus*. Second, a rich fossil record documents a long occupation of nearby Java by *H. erectus*. Third, it seems likely that *H. erectus* moved from Java to Flores because simple Mode 1 tools like those associated with *H. erectus* have been found on Flores and are dated to 800 kya.

Sophisticated flake tools were found at Liang Bua, together with the charred remains of many species of animals, including tortoises, lizards, rodents, bats, and numerous specimens of a species of dwarf elephant. Because these materials have dates that overlap with those of the hominin fossils, and there is no evidence of other hominins in the cave, the excavation team argues that *H. floresiensis* was able to hunt large game and make the kinds of sophisticated tools that are usually associated with *H. sapiens*. Other commentators have argued that these tools could not have been made by the tiny-brained *H. floresiensis*, because they are so much more sophisticated than those associated with *H. erectus*.

In addition to providing a remarkably unexpected addition to the human family tree, these fossils are important because they show us that the human lineage is subject to the same kinds of evolutionary forces as other creatures are. Flores and the islands around it have always been isolated from both Asia and Australia by wide stretches of ocean. As a result, only a few species of large animals ever reached these islands, producing a strange impoverished fauna that included Komodo dragons and an even larger species of monitor lizard, a species of dwarf elephant, and, of course, a dwarf hominin. It is not a coincidence that both the elephants and the humans were so small. Dwarfing of large mammal species on small, isolated islands has occurred a number of times. As we saw in Chapter 1, for example, red deer isolated on the island of Jersey shrank to about one-half their original size over 5000 years. Biologists think that natural selection favors smaller body size on small islands because typically such islands have less predation and more limited food supplies. If islands are isolated, there will be little gene flow from continental populations, and the animals will adapt to local conditions. On Flores this process seems to have produced pint-sized hominins and tiny elephants.

Western Eurasia: The Neanderthals

During the later Middle Pleistocene, the morphology of H. heidelbergensis *in Europe diverged from the morphology of its contemporaries in Africa and Asia.*

A large sample of fossils from a recently excavated site in Spain provides evidence that *H. heidelbergensis* in Europe had begun to diverge from other hominin populations during the Middle Pleistocene. This site, which is called Sima de los Huesos ("Pit of

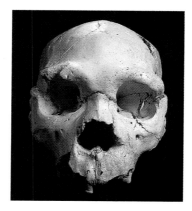
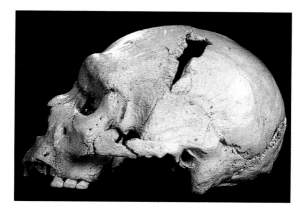

FIGURE 13.24 The many hominin fossils found at Sima de los Huesos in Spain provide evidence that hominins in Europe began to evolve a distinctive cranial morphology at least 300 kya. The features—which include rounded browridges, a large pushed-out face, a skull with a rounded back, and a large brain—are important because they are shared with the Neanderthals, the hominins who dominate the European fossil record during the Upper Pleistocene.

the Bones") is located in the Sierra de Atapuerca only a few kilometers from Trinchera Dolina. There paleoanthropologists excavated a small cave 13 m (43 ft) below the surface and found 2000 bones from at least 24 different individuals. These bones, which date to about 300 kya, include several nearly complete crania, as well as many bones from other parts of the body. Figure 13.24 shows one of the crania, labeled SH 5. Like other fossils of *H. heidelbergensis*, these skulls mix derived features of modern humans and primitive features associated with *H. ergaster*. However, the crania from Sima de los Huesos also share a number of derived characters not seen in hominins living at the same time in Africa. Their faces bulge out in the middle and have double-arched browridges, and the backs of some of the skulls are rounded. The fossils from Sima de los Huesos also have relatively large cranial capacities, averaging about 1390 cc, close to the average value for modern humans. These characters are significant because they are shared by **Neanderthals**, the hominins who dominate the European fossil record from 127 to 30 kya.

Not all of the Sima de los Huesos crania express all the Neanderthal features to the same degree, and this variability has helped to settle an important question about hominin evolution in Europe. Before the Sima de los Huesos fossils were discovered, paleoanthropologists had been perplexed by the pattern of variation in Middle Pleistocene hominins in Europe. Fossils from some European sites resemble Neanderthals; fossils from other European sites that date to the same time period do not resemble Neanderthals. It was not clear whether these differences reflected variability within a single population of hominins or whether more than one type of hominin was present in Europe during the Middle Pleistocene. The large and variable sample of individuals at Sima de los Huesos has resolved this issue. These individuals, who likely belonged to a single population, vary in their expression of Neanderthal traits, and this variation mirrors the kind of variation seen at different sites in Europe. This evidence suggests that the features defining the Neanderthals had begun to evolve in European populations by 300 kya, but that these traits evolved independently, not as a single functional complex.

The Neanderthals appeared in western Eurasia about 130 kya.

The Neanderthals were an enigmatic group of hominins who lived in Europe and western Asia from about 127 to 30 kya. These creatures were among the first fossil hominins discovered, and they are still the best known. Their distinctive morphology (large browridges, short muscular bodies, and low foreheads) has become the archetypical image of "early man" for the general public. However, there is now good evidence that the Neanderthals became isolated from the lineage that led to modern humans and eventually became extinct. Molecular analyses of Neanderthal fossils and modern peoples provide convincing evidence that all modern peoples are descended from African populations living about the same time as the Neanderthals, and that Neanderthals did not contribute significantly to the modern human gene pool.

Although the Neanderthals are a dead-end branch of the human family tree, we know more about them than we do about any other extinct hominin species. The reason for this is simple: Neanderthals lived in Europe, and paleontologists have studied Europe much more thoroughly than they have studied Africa or Asia. Sally McBrearty of the University of Connecticut and Alison Brooks of George Washington University point out that there are more than 10 times as many Middle Pleistocene archaeological sites in France as there are on the entire continent of Africa. This difference is particularly striking when you consider that Africa is about 75 times the size of France. The upshot is that we know much more about the Neanderthals than we do about their contemporaries in other parts of the world.

The last warm interglacial period lasted from about 130 kya to about 75 kya. For most of the time since then, the global climate has been colder—sometimes much colder.

The data in Figure 13.25 give a detailed picture of global temperatures over the last 123,000 years. These data are based on the $^{18}O:^{16}O$ ratios of different layers of cores taken from deep inside the Greenland ice cap. Because snow accumulates at a higher rate than sediments on the ocean bottom do, ice cores provide more detailed information on past climates than do ocean cores like those used to construct the graph in Figure 13.1. You can see from Figure 13.25 that the end of the last warm interglacial period was marked by a relatively slow decline in temperature beginning about 120 kya. During this interglacial period, the world generally was substantially warmer than it is today. During this warm interglacial period, plants and animals were distributed quite differently than they are now. Plankton species currently living in subtropical waters (like those off the coast of Florida) extended their range as far as the North Sea during the last interglacial period. Animals now restricted to the tropics had much wider ranges. Thus, the remains of a hippopotamus have been found under Trafalgar Square, which lies in the center of London. In Africa, rain forests extended far beyond their present boundaries; and in temperate areas, broadleaf deciduous forests extended farther north than they do today.

As the last glacial period began, sometime between 100 and 75 kya, the world slowly cooled. In Europe, temperate forests shrank and grasslands expanded. The glaciers grew, and the world became colder and colder—not steadily, but with wide fluctuations from cold to warm. When the glaciation was at its greatest extent (about 20 kya), huge continental glaciers covered most of Canada and much of northern Europe. Sea levels dropped so low that the outlines of the continents were altered substantially: Asia and North America were connected by a land bridge that spanned the Bering Sea; the islands of Indonesia joined Southeast Asia in a landmass called Sundaland; and Tasmania, New Guinea, and Australia formed a single continent called Sahul. Eurasia south of the glaciers was a vast, frigid grassland, punctuated by dunes of loess (fine dust produced by glaciers) and teeming with

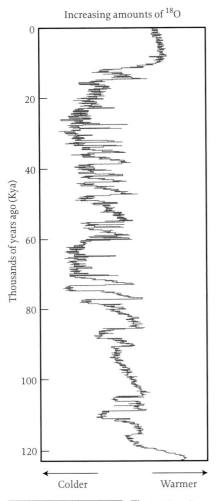

Increasing amounts of ^{18}O

Thousands of years ago (Kya)

Colder　　Warmer

FIGURE 13.25 Fluctuations in the ratios of ^{18}O to ^{16}O over the last 123,000 years taken from an ice core drilled in northern Greenland. These data indicate that, between 120 and 80 kya, the world's climate got colder and less stable. In ice cores like this one, a higher ratio of ^{18}O to ^{16}O indicates higher temperatures. (In deep-sea cores, more ^{18}O relative to ^{16}O indicates lower temperatures.)

animals—woolly mammoths, woolly rhinoceroses, reindeer, aurochs (the giant wild oxen that are ancestral to modern cattle), musk oxen, and horses.

Sometime during the last interglacial period, Neanderthals came to dominate Europe and the Near East.

In 1856, workers at a quarry in the Neander Valley in western Germany found some unusual fossil bones. The bones found their way to a noted German anatomist, Hermann Schaafhausen, who declared them to be the remains of a race of humans who had lived in Europe before the Celts. Many experts examined these curious finds and drew different conclusions. Thomas Henry Huxley, one of Darwin's staunchest supporters, suggested that they belonged to a primitive, extinct kind of human. The Prussian pathologist Rudolf Virchow, on the other hand, proclaimed them to be the bones of a modern person suffering from a serious disease that had distorted the skeleton. Initially, Virchow's view held sway, but as more fossils were discovered with the same features, researchers became convinced that the remains belonged to a distinctive, extinct kind of human. The Germans called this extinct group of people *Neanderthaler*, meaning "people of the Neander Valley" (*Thal*, now spelled *Tal*, is German for "valley"). We call them the Neanderthals, and they were characterized by several distinctive, derived features:

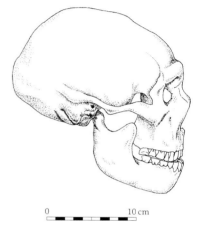

FIGURE 13.26 The skulls of Neanderthals, like this one from Shanidar Cave, Iraq, are large and long, with large browridges and a massive face. (Figure courtesy of Richard Klein.)

- *Large brains.* The Neanderthal braincase is much larger than that of *H. heidelbergensis*, ranging from 1245 to 1740 cc, with an average size of about 1520 cc. In fact, Neanderthals had larger brains than modern humans, whose brains average about 1400 cc. It is unclear why the brains of Neanderthals were so large. Some anthropologists point out that the Neanderthals' bodies were much more robust and heavily muscled than those of modern humans, and they suggest that the large brains of Neanderthals reflect the fact that larger animals usually have bigger brains than smaller animals.

- *More rounded crania than* H. erectus *or* H. heidelbergensis (Figure 13.26). The Neanderthal skull is long and low, much like the skulls of *H. heidelbergensis*, but relatively thin-walled. The back of the skull has a characteristic rounded bulge or bun and does not come to a point at the back like an *H. erectus* skull does. There are also detailed differences in the back of the cranium.

- *Big faces.* Like *H. erectus* and *H. heidelbergensis*, the skulls of Neanderthals have large browridges, but they are larger and rounder and they stick out less to the sides. Moreover, the browridges of *H. erectus* are mainly solid bone, while those of Neanderthals are lightened with many air spaces. The function of these massive browridges is not clear. The face, and particularly the nose, is enormous: every Neanderthal was a Cyrano, or perhaps a Jimmy Durante.

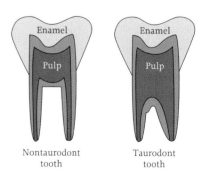

- *Small back teeth and large, heavily worn front teeth.* Neanderthal molars are smaller than those of *H. ergaster*. They had distinctive **taurodont roots** (Figure 13.27) in which the pulp cavity expanded so that the roots merged, partially or completely, to form a single broad root. Neanderthal incisors are relatively large and show very heavy wear. Careful study of these wear patterns indicates that Neanderthals may have pulled meat or hides through their clenched front teeth. There are also microscopic, unidirectional scratches on the front of the incisors, suggesting that these hominins held meat in their teeth while cutting it with a stone tool. Interestingly, the direction of the scratches suggests that most Neanderthals were right-handed, just as the Oldowan toolmakers were.

Enamel   Enamel

Pulp   Pulp

Nontaurodont tooth   Taurodont tooth

FIGURE 13.27 In Neanderthal molars, the roots often fuse together partially or completely to form a single massive taurodont root. The third root is not shown.

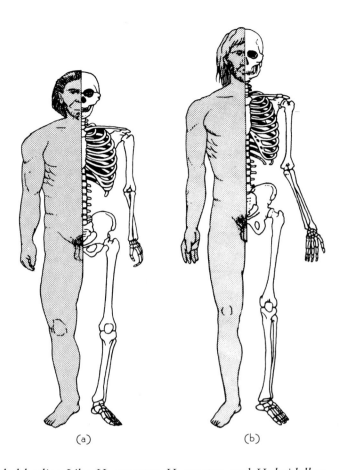

FIGURE 13.28   The Neanderthals were very robust people. Contrast the bones of a Neanderthal (a) with those of a modern human (b).

(a)   (b)

- *Robust, heavily muscled bodies.* Like *H. ergaster*, *H. erectus*, and *H. heidelbergensis*, Neanderthals were extremely robust, heavily muscled people (Figure 13.28). Their leg bones were much thicker than ours, the load-bearing joints (knees and hips) were larger, the **scapulae** (shoulder blades; singular *scapula*) had more extensive muscular attachments, and the rib cage was larger and more barrel-shaped. All these skeletal features indicate that Neanderthals were very sturdy and strong, weighing about 30% more than contemporary humans of the same height. A comparison with data on Olympic athletes suggests that Neanderthals most closely resembled the hammer, javelin, and discus throwers and shot-putters. They were a few inches shorter, on average, than modern Europeans and had larger torsos and shorter arms and legs.

This distinctive Neanderthal body shape may have been an adaptation to conserve heat in a very cold environment. In cold climates, animals tend to be larger and have shorter and thicker limbs than do members of the same species in warmer environments. This is because the rate of heat loss for any body is proportional to its surface area, so any change that reduces the amount of surface area for a given volume will conserve heat. The ratio of surface area to volume in animals can be reduced in two ways: by increasing overall body size or by reducing the size of the limbs. In contemporary human populations there is a consistent relationship between climate and body proportions. One way to compare body proportions is to calculate the **crural index**, which is the ratio of the length of the shin bone (tibia) to the length of the thigh bone (femur). As Figure 13.29 shows, people in warm climates tend to have relatively long

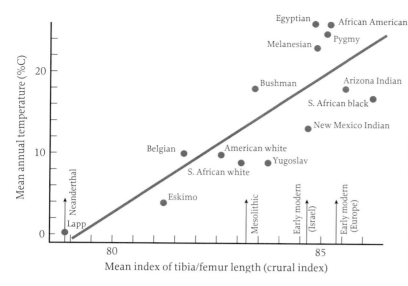

FIGURE 13.29 People have proportionally longer arms and legs in warm climates than in cold climates. Local temperature is plotted on the vertical axis; crural index is plotted on the horizontal axis. Smaller values of the crural index are associated with shorter limbs relative to body size. Populations in warm climates tend to have high crural-index values, and vice versa. Neanderthals had a crural index similar to those of present-day Lapps, who live above the Arctic Circle.

limbs in proportion to their height. Neanderthals resemble modern peoples living above the Arctic Circle.

Some authorities believe that Neanderthals lacked modern language.

The larynx (or voice box) of modern humans is much lower than in other primates—an arrangement that allows us to produce the full range of vowels used in modern human languages. On the basis of reconstructions of the Neanderthal vocal tract, anatomist Jeffrey Laitman of Mount Sinai School of Medicine argues that the Neanderthals had vocal tracts much like those of other primates, and could not have produced the full range of sounds necessary for modern speech. Laitman and his colleagues recorded the positions of several anatomical landmarks on the **basicrania** (singular *basicranium*; bottom of the skull) of humans and several other primates. In infant humans, this basicranial hump is gentle, like those of other primates; but as humans mature, the bottom of the cranium develops a pronounced upward indentation. Laitman believes that this relatively high hump helps make room for the elongated human vocal tract. Australopithecine adults have apelike basicrania, but the basicranium in *H. ergaster* is intermediate between those in apes and modern humans. Laitman argues that the initial shift in *H. ergaster* may have facilitated breathing through the mouth during heavy aerobic exercise. The basicrania in Neanderthals show a less humped profile than those of *H. ergaster*, suggesting that the Neanderthals' capacity for language was more restricted than the language capacities of even their own ancestors.

It is also possible that Laitman's reconstructions are correct, but that grammatically complex language still appeared long before modern humans evolved. It is possible for a creature with a high larynx to make all of the vowel sounds by using the nasal cavity. In modern humans, such nasalized vowels take longer to produce and are harder to understand. Thus the lowering of the larynx might have served merely to decrease the error rate of understanding a language that was otherwise very modern. Second, as Harvard University linguist Steven Pinker points out in his book *The Language Instinct*, "e lengeege weth e smell nember ef vewels cen remeen quite expresseve." So, even if Laitman is correct, early hominins could have had fancy languages.

(a)

(b)

FIGURE 13.30 Neanderthals and their contemporaries are believed to have hunted large and dangerous game, such as red deer (a) and bison (b).

There is a lot of evidence that Neanderthals made Mode 3 tools and hunted large game.

Although Neanderthals are popularly pictured as brutish dimwits, this is an unfair characterization. As we noted earlier, their brains were larger than ours. And the archaeological evidence suggests that they were skilled toolmakers and proficient big-game hunters. The Neanderthal's stone tool kit is dominated by Mode 3 tools, which are characterized by flakes struck from prepared cores. Their stone tool industry is called the **Mousterian industry** by archaeologists. Neanderthal sites are littered with stone tools and the bones of red deer (called "elk" in North America), fallow deer, bison, aurochs, wild sheep, wild goats, and horses (Figure 13.30). Archaeologists find few bones of very large animals, like hippopotamuses, rhinoceroses, and elephants, even though they were plentiful in Europe at that time.

Again, the conjunction of animal bones and stone tools does not necessarily mean that Neanderthals hunted large game, and there is a reprise of the same debate we have discussed several times before. Some archaeologists, such as Lewis Binford of Truman State University, believe that Neanderthals never hunted anything larger than small antelope, and even these prey were taken opportunistically. These researchers claim that the bones of larger animals found at these sites were acquired by scavenging. Binford believes that the hominins of this period did not have the cognitive skills necessary to plan and organize the cooperative hunts necessary to bring down large prey.

However, this position seems hard to sustain in the face of the archaeological evidence. Stanford archaeologist Richard Klein contends that Neanderthals were proficient hunters who regularly killed large animals. He points out that animal remains at sites from this period are often dominated by the bones of only one or two prey species. At Mauran, a site in the French Pyrenees, for example, over 90% of the assemblage is from bison and aurochs. The same pattern occurs at other sites scattered across Europe. It is hard to see how an opportunistic scavenger would acquire such a nonrandom sample of the local fauna. Moreover, the age distribution of prey animals does not fit the pattern for modern scavengers like hyenas, which prey mainly on the most vulnerable members of prey populations: sick or wounded animals, the old, and the very young (Figure 13.31). At these European sites, the bones of apparently healthy, prime-age adults are well represented. The distribution of animal bones is what we would expect to see at sites of catastrophic events in which whole herds of animals are killed. For example, several sites are located at the bottoms of cliffs. There are huge jumbles of bones and tools at these sites, suggesting that Neanderthals drove game over the cliffs and butchered the carcasses where they fell. At several sites, such as Combe Grenal in France, the bones from the meatiest parts of prey animals are overrepresented, and the cut marks on these bones suggest that Neanderthals stripped off the fresh flesh. (Remember that hunters often haul away the meatiest bones to eat in peace.) Taken together, these data suggest that Neanderthals were big-game hunters, not opportunistic scavengers.

There is little evidence for shelters or even organized camps at Neanderthal sites.

There are many Neanderthal sites, some very well preserved, where archaeologists have found concentrations of tools, abundant evidence of toolmaking, many animal remains, and concentrations of ash. Most of these sites are in caves or **rock shelters**, places protected by overhanging cliffs. This doesn't necessarily mean that Neanderthals preferred these kinds of sites. Cave sites are more likely to be found by researchers because they are protected from erosion, and they are relatively easy to locate because the openings to many caves from this period are still visible. Most archaeologists believe that cave sites represent home bases, semipermanent encampments from which Neanderthals sallied out to hunt and to forage.

The archaeological record suggests that Neanderthals did not build shelters. Most Neanderthal sites lack evidence of postholes and hearths, two features generally associated with simple shelters. The few exceptions to this rule occur near the end of the period. For example, hearths were built at Vilas Ruivas, a site in Portugal dated to about 60 kya.

FIGURE 13.31 Hyenas frequently scavenge kills, and they prey mainly on very young and very old individuals.

Neanderthals probably buried their dead.

The abundance of complete Neanderthal skeletons suggests that, unlike their hominin predecessors, Neanderthals frequently buried their dead. Burial protects the corpse from dismemberment by scavengers and preserves the skeleton intact. Detailed study of the geological context at sites like La Chapelle-aux-Saints, Le Moustier, and La Ferrassie in southern France also supports the conclusion that Neanderthal burials were common.

It is not clear whether these burials had a religious nature, or if Neanderthals buried their dead just to dispose of the decaying bodies. Anthropologists used to interpret some sites as ceremonial burials in which Neanderthals were interred along with symbolic materials. In recent years, however, skeptics have cast serious doubt on such interpretations. For example, anthropologists used to think that the presence of fossilized pollen in a Neanderthal grave in Shanidar Cave, Iraq, was evidence that the individual had been buried with a garland of flowers. More recent analyses, however, revealed that the grave had been disturbed by burrowing rodents, and it is quite possible that they brought the pollen into the grave.

Neanderthals seem to have lived short, difficult lives.

Careful study of Neanderthal skeletons indicates that Neanderthals didn't live very long. The human skeleton changes throughout the life cycle in characteristic ways, and these changes can be used to estimate the age at which fossil hominins died. For example, human skulls are made up of separate bones that fit together in a three-dimensional jigsaw puzzle. When children are first born, these bones are still separate, but later they fuse together, forming tight, wavy joints called **sutures**. As people age, these sutures are slowly obliterated by bone growth. By assessing the degree to which the sutures of fossil hominins have been obliterated, anthropologists can estimate how old the individual was at death. Several other skeletal features can be used in similar ways. All of these features tell the same story: Neanderthals died young. Few lived beyond the age of 40 to 45 years.

Many of the older Neanderthals suffered disabling disease or injury. For example, the skeleton of a Neanderthal man from La Chapelle-aux-Saints shows symptoms of severe arthritis that probably affected his jaw, back, and hip. By the time this fellow died, around the age of 45, he had also lost most of his teeth to gum disease. Another individual, Shanidar 1 (from the Shanidar site in Iraq), suffered a blow to his left

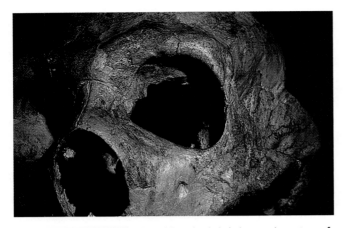

FIGURE 13.32   Many Neanderthal skeletons show signs of injury or illness. The orbit of the left eye of this individual was crushed, the right arm was withered, and the right ankle was arthritic.

temple that crushed the orbit (Figure 13.32). Anthropologists believe that his head injury probably caused partial paralysis of the right side of his body, and this in turn caused his right arm to wither and his right ankle to become arthritic. Other Neanderthal specimens display bone fractures, stab wounds, gum disease, withered limbs, lesions, and deformities.

In some cases, Neanderthals survived for extended periods after injury or sickness. For example, Shanidar 1 lived long enough for the bone surrounding his injury to heal. Some anthropologists have proposed that these Neanderthals would have been unable to survive their physical impairments—to provide themselves with food or to keep up with the group—had they not received care from others. Some researchers argue further that these fossils are evidence of the origins of caretaking and compassion in our lineage.

Katherine Dettwyler, an anthropologist at Texas A&M University, is skeptical of this claim. She points out that in some contemporary societies, disabled individuals are able to support themselves, and they do not necessarily receive compassionate treatment from others. In addition, nonhuman primates sometimes survive despite permanent disabilities. At Gombe Stream National Park, a male chimpanzee named Faben contracted polio, which left him completely paralyzed in one arm. Despite this impairment, he managed to feed himself, traverse the steep slopes of the community's home range, keep up with his companions, and even climb trees (Figure 13.33).

FIGURE 13.33   At Gombe, a male chimpanzee named Faben contracted polio, which left one arm completely paralyzed. After the paralysis, Faben received no day-to-day help from other group members, but he was able to survive for many years. Here, Faben climbs a tree one-handed.

Africa: The Road to *Homo sapiens?*

Hominins living in Africa during the later Middle Pleistocene were more similar to modern humans than were Neanderthals.

The fossil record in Africa indicates fairly clearly that African humans of this period were not like the Neanderthals, although it is not entirely clear what they were like, because the fossil record for this time in Africa is not nearly as good as it is in Europe. A number of fossils dated to later in the Middle Pleistocene (300 to 200 kya) have robust features similar to those of *H. heidelbergensis*. Examples include the Florisbad cranium found in South Africa (Figure 13.34) and the Ngaloba cranium (LH 18) from Laetoli, Tanzania. Like the Neanderthals, these African hominins show large cranial volumes, ranging from 1370 to 1510 cc. However, none of the African fossils show the complex of specialized features that are diagnostic of the European Neanderthals. Although these fossils are variable and some are quite robust, many researchers believe that they are more like modern humans than are the Neanderthals or earlier African hominins.

About 190 kya, hominins belonging to our own species began to appear in Africa, although some of the evidence is problematic. The oldest fossils classified as *Homo sapiens* were excavated in 1963 at a site in southern Ethiopia called Omo-Kibish, where paleoanthropologists found most of two fossil skulls and a number of other bone fragments. These skulls are quite robust, with prominent browridges and large faces. However, one of them has a number of modern features—most notably a high, rounded braincase (Figure 13.35). Initially these specimens could not be dated, but they were recently dated to about 190 kya by new radiometric methods. Several similar skulls were uncovered at Herto, another site in Ethiopia, and are dated to 160 kya. Other more fragmentary fossil materials from other sites in Africa also suggest that more modern hominins were present in Africa between 200 and 100 kya. The archaeological record indicates that Africans during this period developed more sophisticated technology and social behavior than did their contemporaries in Europe or Asia—a pattern that is consistent with the idea that this period saw the gradual accumulation of the cognitive and behavioral characteristics that make modern humans so different from other hominins. We turn to this evidence in the next chapter.

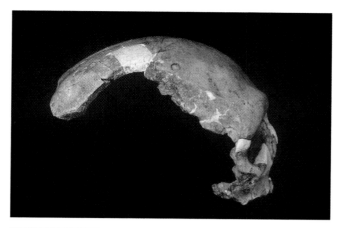

FIGURE 13.34 This cranium found at Florisbad in South Africa shows a mixture of features of *H. heidelbergensis* and modern *H. sapiens*. Although still quite robust, it has reduced browridges and a more rounded shape. It dates to between 300 and 200 kya.

THE SOURCES OF CHANGE

Europe may have been invaded repeatedly by hominins from Africa during the Middle Pleistocene.

The changes in hominin morphology and technology that we have described in this chapter may be the product of a number of different kinds of processes. It is possible that the changes we see reflect adaptive modifications in tool technology in particular regions of the world. Thus it is likely that Africa has been occupied continuously by members of the genus *Homo*, which first appeared 1.8 mya. This means that the makers of Mode 1 tools evolved slowly into the makers of Mode 2 tools, and the makers of Mode 2 tools evolved into the makers of Mode 3 tools.

However, it is also possible that some of the changes in the fossil record are the product of the replacement of one population of hominins by another. Cambridge University anthropologists Marta Lahr and Rob Foley recently suggested that Europe and western Eurasia were subjected to repeated invasions by hominins from Africa. They argue that the technological shifts seen in Eurasia were associated with the migration of hominins from Africa during interglacial periods. Remember that, during glacial periods, Africa and Eurasia were separated by a formidable desert barrier (see Figure 13.13b), and Eurasia was a cold, dry, and inhospitable habitat for primates. During these glacial periods, hominin populations in western Eurasia may have shrunk or disappeared altogether. When the glacial periods ended and the world became warmer, there was substantial movement of animal species from Africa into Eurasia. It is possible that

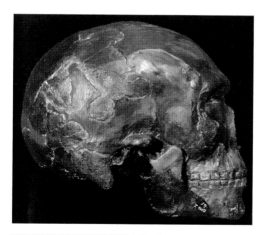

FIGURE 13.35 This fossil skull, called Omo-Kibish 1, was found in southern Ethiopia in 1963 and recently dated to 190 kya. It lacks the distinctive features of Neanderthals: the face does not protrude, and the braincase is higher and shorter than that of Neanderthals.

1000–500 kya

500–250 kya

Mode 1
Modes 1 & 2
Modes 1, 2, & 3

250–200 kya

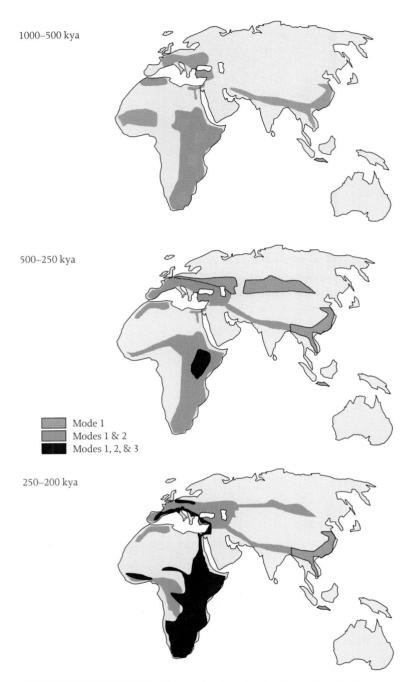

The geography of tool technologies through time suggests that Eurasia was subjected to repeated invasions of hominins from Africa. Between 1 mya and 500 kya, Mode 2 technologies were confined to Africa, and hominins in Eurasia were restricted to Mode 1 tools. Beginning about 500 kya, Mode 2 technologies appeared in Eurasia. The introduction of Mode 2 coincided with a relatively warm, moist period, which could have facilitated the movement of hominins from Africa to Eurasia. About 300 kya, elements of Mode 3 technology appeared in East Africa, and by about 250 kya, Mode 3 technology had spread throughout Africa and southern Europe. Once again, this spread coincided with a period of warmer climate.

hominins staged repeated invasions of Eurasia during each of these warm interglacial periods, bringing new technologies along with them.

Lahr and Foley's argument is buttressed by archaeological evidence. Remember that the earliest evidence for hominins in Europe dates to about 800 kya. Up until about 500 kya, only Mode 1 tools are found in Europe. Then, Mode 2 technologies appear in Europe and persist until about 250 kya, when they are replaced by Mode 3 technologies. In each case the new technology appears first in Africa and then later in Europe and (sometimes) in Asia (Figure 13.36). The appearance of new tool technologies at 500 kya and 250 kya coincides neatly with the timing of interglacial periods, consistent with the repeated replacement of Eurasian populations by African populations.

THE MUDDLE IN THE MIDDLE

Anthropologists strongly disagree about how to classify Middle Pleistocene hominins.

You will notice that we avoided assigning a species name to the Neanderthals. This is partly because there is a lot of disagreement about how to classify Middle Pleistocene hominins. The disagreement stems from different ideas about the processes that shape human evolution during this period. One school of thought, most closely associated with University of Michigan anthropologist Milford Wolpoff, contends that hominins in Africa and Eurasia formed a single interbreeding population throughout the Pleistocene. The size of Africa and Eurasia limited gene flow and allowed regional differences to evolve, but there was always enough interbreeding to guarantee that all hominins belonged to single species during this entire period. Thus, Wolpoff and others prefer to include all hominins who lived during the Middle Pleistocene in a single species, *Homo sapiens*, as shown in Figure 13.37a. As we will show in the next chapter, this point of view has become hard to sustain in the face of information that has become available from molecular genetic studies in the last few years.

Other anthropologists believe that hominins split into several new species as they migrated out of Africa and into Eurasia although there are

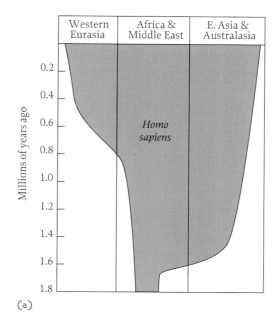

(a)

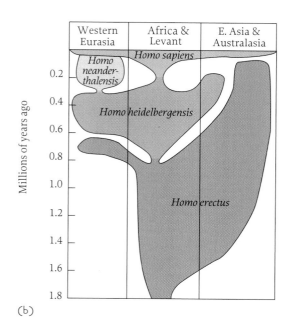

(b)

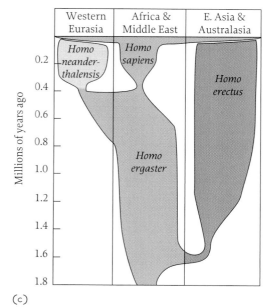

(c)

Numerous phylogenies have been proposed to account for the temporal and geographic patterns in hominin evolution during the Pleistocene. Illustrated here are three recent proposals: (a) Milford Wolpoff holds that, beginning about 1.8 mya, there was only one human species. Because there have been no speciation events since that time, all human fossils should be classified as *H. sapiens*. (b) G. Phillip Rightmire believes that both African and Asian specimens of *H. erectus* should be classified as a single species. Approximately 800 kya, a larger-brained species, *H. heidelbergensis*, evolved in Africa and eventually spread to Europe and perhaps East Asia. In Europe, *H. heidelbergensis* gave rise to the Neanderthals; in Africa, it gave rise to *H. sapiens*. (c) Richard Klein argues that *H. ergaster* evolved in Africa about 1.8 mya and soon spread to Asia, where it differentiated to become a second species, *H. erectus*. About 500 kya, *H. ergaster* spread to western Eurasia, where it evolved into *H. neanderthalensis*. In Africa, *H. ergaster* evolved into *H. sapiens* at about the same time.

disagreements about the details of phylogenetic history. Figure 13.37b and c illustrate two recent hypotheses. According to G. Phillip Rightmire of Binghamton University, the specimens that we have identified as *H. ergaster* and *H. erectus* are members of a single species (which would be called *H. erectus* because that was the name first used) (Figure 13.37b). However, he draws a distinction between early Middle Pleistocene hominins in Europe and Africa and those who came later. He groups all of the early Middle Pleistocene hominins in Europe and Africa into one species, which he labels *H. heidelbergensis*, and he lumps the hominins of the later Middle Pleistocene in Europe and the Neanderthals into one species, *H. neanderthalensis*. Although it may seem peculiar to include specimens without the distinctive Neanderthal features in the species *H. neanderthalensis*, this name is used because it has historical priority.

Richard Klein believes that hominin populations in Africa, western Eurasia, and eastern Eurasia were genetically isolated from each other during most of the Pleistocene and represent three distinct species (Figure 13.37c). In Africa, *H. ergaster* gradually evolved into *H. sapiens* about 500 kya. Isolated in Asia, *H. ergaster* evolved into *H. erectus* early in the Pleistocene. Klein considers the development of larger brains and more modern-looking skulls in the eastern Eurasia fossils, like those found at Yingkou, to be the result of convergent evolution and includes them in *H. erectus*. Once hominins reached Europe about 500 kya, they became isolated from African and East Asian populations, and diverged to become *H. neanderthalensis*.

This kind of disagreement is to be expected because the timescales of change become short as we come closer to the present. Instead of being interested in events that took place over millions of years, we are now interested in events that took place in just a hundred thousand years. But this may be roughly how long it took for two species to diverge during the process of allopatric speciation (see Chapter 4). Thus, as hominins spread out across the globe and encountered new habitats, regional populations may have become isolated and experienced selection—conditions that would eventually lead to speciation. The rapidly fluctuating climates of the Pleistocene, however, caused the ranges of hominins and other creatures to shift as well. As their ranges expanded and contracted, some populations may have become extinct, some may have become fully isolated and even more specialized, and others may have merged together. This complex set of possibilities is very difficult for paleontologists to unravel.

FURTHER READING

Conroy, G. 1997. *Reconstructing Human Origins: A Modern Synthesis*. Norton, New York.

Klein, R. 1999. *The Human Career* (2nd ed.). University of Chicago Press, Chicago.

McBrearty, S., and A. Brooks. 2000. The revolution that wasn't: A new interpretation of modern human behavior. *Journal of Human Evolution* 39: 453–563.

Rightmire, G. P. 1998. Human evolution in the Middle Pleistocene: The role of *Homo heidelbergensis*. *Evolutionary Anthropology* 6: 218–227.

Stringer, C., and C. Gamble. 1993. *In Search of the Neanderthals*. Thames and Hudson, New York.

Walker, A., and P. Shipman. 1996. *The Wisdom of the Bones: In Search of Human Origins*. Knopf, New York.

STUDY QUESTIONS

1. Which derived features are shared by modern humans and *Homo ergaster*? Which derived features are unique to *H. ergaster*?
2. What does the specimen KNM-WT 15000 tell us about *H. ergaster*?
3. What is a hand ax, and what did *H. ergaster* use hand axes for? How do we know?
4. What evidence suggests that *H. ergaster* controlled fire?

5. What evidence suggests that *H. ergaster* ate significant amounts of meat? Is the evidence for *H. ergaster* carnivory better than that for the Oldowan hominins? Explain.
6. What are the main differences between *H. ergaster* and *H. erectus*? How would you explain the evolution of these differences?
7. Using present-day examples, describe the variation in climate during the Middle Pleistocene. Why is this variation important for understanding human evolution?
8. How is *H. heidelbergensis* different from *H. erectus*?
9. What important technological transition occurred about 300 kya? Why was it important?
10. Explain why the fossils found at Sima de los Huesos are significant.
11. What is the crural index? What does it measure? How does the crural index of the Neanderthals differ from that of modern tropical peoples?
12. How did the Neanderthals differ from their contemporaries in Africa and eastern Asia?
13. What evidence suggests that western Eurasia was subjected to repeated invasions from Africa during the Middle Pleistocene?

KEY TERMS

occipital torus

biface

hand ax

cleaver

pick

Mode 2

acheulean industry

sagittal keel

archaic *Homo sapiens*

Homo heidelbergensis

Levallois technique

Mode 3

hafted

neanderthals

taurodont roots

scapulae

crural index

basicrania

Mousterian industry

rock shelters

sutures

CONTENT FROM *HUMAN EVOLUTION: A MULTI-MEDIA GUIDE TO THE FOSSIL RECORD*

Homo erectus
  Overview
  Distinguishing Features
  Homo ergaster
  Homo erectus Lifeways
Archaic Homo sapiens
  Overview
  Distinguishing Features

  Origin Theories
  Archaic H. sapiens Lifeways
The Neanderthals
  Overview
  Distinguishing Features
  Adaptive Significance
  Neanderthal Lifeways

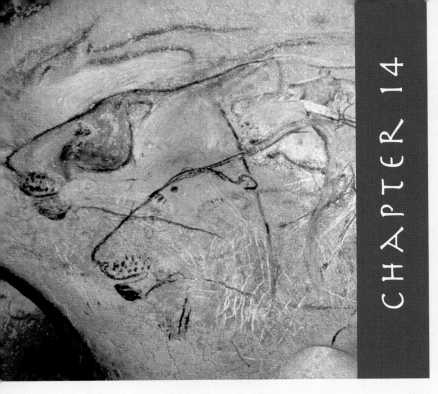

CHAPTER 14

HOMO SAPIENS AND THE EVOLUTION OF MODERN HUMAN BEHAVIOR

Between 40 and 30 kya the fossil and archaeological record in Europe undergoes a striking change: the Neanderthals disappeared and were replaced by essentially modern people. These people looked much like people in the world today, with high foreheads, sharp chins, and less robust physiques. The archaeological record suggests that they also behaved like modern people, using sophisticated tools, trading over long distances, and making jewelry and art. Anatomically and behaviorally similar people appeared in Australia around the same time. There is evidence that similar kinds of peoples also lived in Africa at this time, and most anthropologists think that modern people first appeared in Africa well before they appeared in Europe.

In this chapter we first describe the fossil and archaeological record for these early modern people in more detail. Then we consider how they evolved. You will see that evidence from the fossil record and from molecular genetic studies of DNA extracted from fossils gives us a fairly clear picture of where and when modern human morphology evolved. These data tell us that modern humans evolved in Africa between 200 and 100 kya and then migrated from Africa and spread throughout the world about 50 kya. We are less certain when and where modern human behavior evolved. According to one view, anatomically modern people inhabiting Africa 100 kya still lived much like the more robust hominins in Europe and Asia. Then about 50 kya an important cognitive innovation arose that allowed people to develop more complex technology and social lives, producing what anthropologists call the "human revolution." Adherents of a second view argue that the components of modern human behavior and technology gradually evolved in Africa along with modern human morphology over a period of about 200,000 years. There was not a human revolution in Europe or elsewhere in the world; what seems like a revolutionary change in Europe simply reflects the migration of modern people from Africa to Europe and the existence of an extremely rich archaeological record that is heavily biased in favor of European sites.

MODERN *HOMO SAPIENS*

Modern humans are characterized by several derived morphological traits, including a small face and teeth, a pointed chin, a high rounded cranium, and a less robust postcranium.

The first fossils of modern humans in Europe were found in 1868 by railroad workers in southwestern France at a site known as the Cro-Magnon rock shelter. This site eventually yielded bones from at least five different individuals who lived about 30 kya. Since 1868, many other sites in Europe have yielded fossils of modern humans or their artifacts. These creatures are classified by anthropologists as *Homo sapiens* because they shared a number of important derived features with contemporary humans that were not present in previous hominins (Figure 14.1):

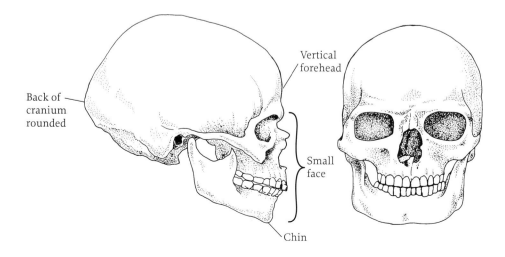

FIGURE 14.1 The skulls of modern humans have higher, rounder crania and smaller faces than earlier hominins had, as illustrated by this skull from a man who lived about 25 kya near the Don River in Russia. (Figure courtesy of Richard Klein.)

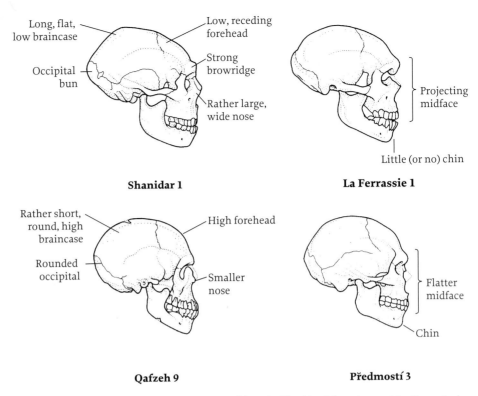

FIGURE 14.2 Neanderthals, represented here by Shanidar 1 from Iraq and La Ferrassie 1 from France, differ from modern *H. sapiens*, like Qafzeh 9 from Israel and Předmostí 3 from the Czech Republic. Modern humans had higher foreheads, smaller browridges, smaller noses, more rounded skulls, and more prominent chins than Neanderthals had.

- *Small face with protruding chin.* These people had smaller faces and smaller teeth than earlier hominins had, and the lower jaw bore a jutting chin for the first time (Figure 14.2). Some anthropologists believe that the smaller face and teeth were favored by natural selection because these people did not use their teeth as tools as much as earlier people had. There is no agreement about the functional significance of the chin.
- *Rounded skull.* Like modern people, these people had high foreheads, a distinctive rounded back of the cranium, and greatly reduced browridges (see Figure 14.2).
- *Less robust postcranial skeleton.* The skeleton of these people was much less robust than Neanderthal skeletons. These people had longer limbs with thinner-walled bones; longer, more lightly built hands; shorter, thicker pubic bones; and distinctive shoulder blades. Although these people were less robust than Neanderthals, they were still more heavily built than any contemporary human population. Eric Trinkaus of Washington University, St. Louis, argues that these people relied less on body strength and more on elaborate tools and other technological innovations to do their work.

Modern humans also differ from other living primates genetically in numerous ways. Because we lack genetic information about extinct hominins, we cannot say when these differences arose. See Box 14.1 for more details.

BOX 14.1

The Evolution of the Human Genome

The genome of an organism is all the genetic information carried on chromosomes in that species. For many years our knowledge of the genome was indirect: we could study the protein products of genes, and we could use specialized molecular techniques to track particular DNA segments called "genetic markers." In the last decade or so, technical progress has made it easier and cheaper to sequence large chunks of the genome. In 2002, a first cut at sequencing the entire human genome was announced with great fanfare. More quietly, geneticists have worked on sequencing the genomes of other organisms, including fruit flies, yeast, and mice. Of particular interest to us is the genome of one of our closest relatives, the chimpanzee. At this writing, a rough draft of the entire genome and a complete sequence of chimpanzee chromosome 22 have been produced. This genetic information gives us a new window on the kinds of evolutionary changes that have occurred in the human lineage.

Chromosome 22 contains about 33 million nucleotides and makes up about 7% of the entire chimpanzee genome. The chimpanzee chromosome 22 is homologous to the human chromosome 21. By comparing these two chromosomes, geneticists can construct a reasonable picture of the genetic similarities and differences between humans and chimpanzees.

Geneticists were able to estimate the number of positions at which one nucleotide was substituted for another by lining up the two chromosomes and comparing the sequences nucleotide by nucleotide. In about 1.4% of the nucleotides, one nucleotide has been substituted for another. There have also been many (68,000) insertions and deletions of bits of DNA in or out of the human genome or the chimpanzee genome. Most of these insertions and deletions involve a small number of nucleotides, and they add only a small percentage to the difference between the two genomes.

Many people find this fact puzzling. Humans and chimpanzees are really different, and it seems implausible that such a puny genetic difference could produce the sizable differences between the two. This would be very hard to explain if we assumed that if x percent of the DNA differed, then x percent of the genes would also differ. However, this conclusion doesn't follow. If DNA differences were distributed evenly across the genome, then even a 1% difference could cause every gene in the two species to differ.

This is where we benefit from having sequenced a sizable chunk of the genome of humans and chimpanzees. Geneticists identified 231 homologous structural genes (DNA sequences that code for proteins) on chromosome 22 in chimpanzees and chromosome 21 in humans. Only 39 of these 231 structural genes had exactly the same amino acid sequences. Among those that differed, 140 showed substitution of one or more amino acids but conserved overall length, and 47 of the genes showed significant structural changes. Thus, even though the DNA sequences of chimpanzees and humans differ by only a small percentage, 84% of the proteins produced by their genes differ.

Many of the differences in morphology and behavior between humans and chimpanzees are likely to be the result of natural selection, not just random mutations. But which ones? Another recent study provides a fascinating answer to this question. A team of geneticists determined the DNA sequence for 7645 homologous structural genes in the human, chimpanzee, and mouse genomes. Most of these genes showed some difference in DNA sequence.

To determine which changes were due to selection, the geneticists made use of the fact that the DNA code is redundant (see Chapter 2). This means that some nucleotide substitutions (**synonymous substitutions**) do not produce any change in the amino acid sequence of the protein that results from the gene, while **nonsynonymous substitutions** do produce a change in the amino acid sequence. Directional selection typically favors a particular protein that produces a particular phenotype. Thus, structural genes that have been subjected to directional selection are expected to show fewer nonsynonymous substitutions than synonymous ones show, and random change is expected to affect synonymous and nonsynonymous substitutions to the same extent.

Data on the mouse genome were included because that information allows us to determine which differences between humans and chimpanzees arose in the human lineage since our last common ancestor with chimpanzees, and which changes arose in the chimpanzee lineage after this point. When the mouse and the chimpanzee are the same and humans differ, it is likely that mice and chimpanzees have the ancestral DNA sequence, and humans have the derived one. Similarly, when mice and humans are the same and chimpanzees differ, the chimpanzee sequence is derived.

The results of these comparisons indicate that about 20% of the changes in the structural genes of humans and chimpanzees are the product of natural selection. These genes code for proteins with a number of different functions. The kinds of selective changes that have occurred in these 7645 structural genes are related mainly to functional differences in olfaction, hearing, amino acid metabolism, and development:

- *Olfaction.* Of the 7645 structural genes that code for proteins, 48 are involved in the human sense of smell, and 27 show significant change since the last common ancestor with chimpanzees. Most of this change seems to be due to a loss of function in the human lineage; apparently humans rely much less on their sense of smell than do chimpanzees.

- *Hearing.* About 15% of the proteins that affect human hearing have been subject to selection. These changes may be due to the necessity to process spoken language.
- *Amino acid metabolism.* A number of proteins that have been subjected to selection are involved in the metabolism of amino acids, including amino acids that are consumed in the form of meat. Amino acid metabolism is likely to reflect changes in the human diet and the increased importance of hunting and scavenging.
- *Development.* A number of proteins that control skeletal development and early embryonic development also show evidence of selection.

Many geneticists suspect that changes in regulatory genes are likely to be more important than changes in structural genes in the evolution of human morphology and behavior. However, it is not yet possible to identify regulatory genes from DNA sequences, so geneticists can't study the changes in regulatory genes in the same way that they study changes in structural genes.

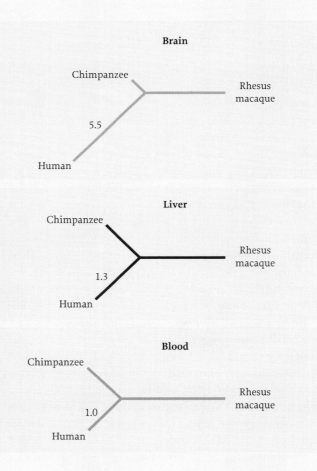

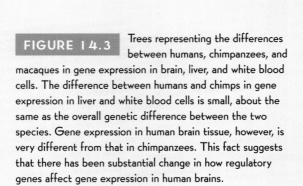

FIGURE 14.3 Trees representing the differences between humans, chimpanzees, and macaques in gene expression in brain, liver, and white blood cells. The difference between humans and chimps in gene expression in liver and white blood cells is small, about the same as the overall genetic difference between the two species. Gene expression in human brain tissue, however, is very different from that in chimpanzees. This fact suggests that there has been substantial change in how regulatory genes affect gene expression in human brains.

A team of geneticists at the Max Planck Institute for Evolutionary Anthropology, led by Svante Pääbo, has studied the sequences of messenger RNA. Their work indicates that the levels of gene expression in the human brain are very different from those of chimpanzees, and this, in turn, suggests that there have been important changes in gene regulation during human evolution. The team used an array of 21,504 DNA segments, each about 1000 bases long. These were taken from a library of coding sequences for human genes, so that each DNA segment encoded a substantial part of a structural gene. Then they extracted messenger RNA from the brain tissue, liver tissue, and white blood cells of humans, chimpanzees, and rhesus macaques. When a particular gene is expressed, messenger RNA with the coding sequence for that gene will be present, and it will bind to the complementary DNA sequence on the array. Thus, Pääbo and his colleagues could determine which of the 21,504 genes were expressed in each tissue in each species.

The results from this study are shown in Figure 14.3. The three trees depict the distances (or differences) between humans, chimpanzees, and rhesus macaques based on the genes expressed in the brain, liver, and white blood cells. For the liver and blood, chimpanzees and humans have both changed about the same amount since their last common ancestor. Morever, this is about the amount of change we would expect based on the overall genetic difference between the two species. The story for the brain is very different: there has been much more change in the patterns of gene expression in human brains than in chimpanzee brains since we diverged from our last common ancestor. These results suggest that gene expression in human brains has changed much more than gene expression in other tissues.

ARCHAEOLOGICAL EVIDENCE FOR MODERN HUMAN BEHAVIOR

Archaeological evidence indicates that early modern humans in Europe were able to accumulate complex adaptive and symbolic behavior in the same way as people living today.

Modern human behavior is extraordinarily complex and varied. Remember that *H. ergaster* and *H. heidelbergensis* used Acheulean tools throughout Africa and western Eurasia for more than a million years. Thus the same tools, and presumably much the same adaptive strategies, were used over a wide range of habitats for a very long time. In contrast, present-day foragers use a vast range of different, highly specialized tools and techniques to adapt to a very large range of environments. They also engage in elaborate and varied symbolic, artistic, and religious behavior that is unparalleled among other creatures. The extraordinary geographic range and sophistication of modern humans is due partly to our cognitive abilities. Individual people can solve problems that would completely stump other creatures (see Chapter 9). However, being smarter than the average bear (or primate) is only part of our secret. Our other trick is having the ability to accumulate and transmit complex adaptive and symbolic behavior over successive generations. People rarely solve difficult problems on their own. Even the most brilliant of us couldn't construct a seaworthy kayak or invent perspective drawing from scratch. Instead, we gain skills, knowledge, and techniques from watching and being instructed by others. The variety and sophistication of modern human behavior is a product of our ability to acquire information in this way (see Chapter 16).

The archaeological record has numerous indications that the first *H. sapiens* in Europe had achieved this kind of complex and varied behavior:

- *Ecological range.* They were able to occupy a difficult, cold, dry habitat and extended their range farther north and east than previous hominins did.
- *Technology.* They assembled more sophisticated and more highly standardized tools made from a wider variety of materials, including antler, ivory, and bone. They also constructed elaborate shelters.
- *Social organization.* They used raw materials that came from sources hundreds of kilometers away for toolmaking, suggesting that they had long-distance exchange networks.
- *Symbolic expression.* They created art and ornamentation, performed ritual burials, and practiced other forms of symbolic behavior.

As we will explain more fully later in this chapter, the first modern humans in western Eurasia created a number of different tool industries in different parts of their range. Archaeologists refer to these tool industries collectively as **Upper Paleolithic** industries to distinguish them from the earlier tool industries associated with Neanderthals and other earlier hominins. The earliest Upper Paleolithic tools appear in the Near East about 45 kya, and they disappear from the archaeological record about 10 kya. Archaeologists refer to this period in western Eurasia as the Upper Paleolithic period, and the people who made the tools as Upper Paleolithic peoples.

Modern humans first entered Australia at least 40 kya, bringing technology similar to that of the Upper Paleolithic people in Europe.

The fossil and archaeological record in Australia is also quite good. Numerous well-dated archaeological sites indicate that modern *H. sapiens* entered Australia at least 40 kya. Several sites may be older, but the dates for each are controversial. What is certain is that people occupied the entire continent by 30 kya. Although the fossil record for the earliest sites is poor, there are good data for sites dated to 30 kya. The most complete early site is in southeastern Australia at Lake Mungo. The finds here include three hominin skulls, hearths, and ovens that have been dated to 32 kya using carbon-14 dating. The skull specimens are fully modern and well within the range of variation of contemporary Australian aborigines.

The earliest Australians seem to have practiced some of the same kinds of cultural behaviors as the Upper Paleolithic Europeans did. Some of the tools found at these early sites were made from bone, there are cave paintings dated to 17 kya, and there is evidence of both ceremonial burials and cremation. About 15 kya, Australians seem to have been the first people to use polished stone tools, which are made by grinding rather than flaking. In other parts of the world, polished tools do not appear in the record until agriculture is introduced, about 7 kya.

The fact that people got to Australia at all is evidence of their technological sophistication. Much of the world's water was tied up in continental glaciers 40 kya, so New Guinea, Australia, and Tasmania formed a single continent that geologists call Sahul. However, there was still at least 100 km (62 miles) of open ocean lying between Asia and Sahul, a gap wide enough to prevent nearly all movement of terrestrial mammals across it and thus to preserve the unique, largely marsupial fauna of Australia and New Guinea. The colonization of Sahul cannot be dismissed as a single lucky event, because people also crossed another 100 km of ocean to reach the nearby islands of New Britain and New Ireland at about the same time. Thus the people who first settled these islands must have been able to build seaworthy boats, a sophisticated form of technology.

More sophisticated stone tools appeared in eastern Asia around the same time.

Modern humans must have come to Australia by way of southern Asia, but signs of modern humans in southern Asia are scarce. Only one site there contains evidence of the kind of sophisticated artifacts seen in Upper Paleolithic Europe. The Batadomba-lena Cave in Sri Lanka, is dated to about 28 kya, and it has yielded modern human remains, elaborate stone tools, and tools made from bone.

Upper Paleolithic tools have been found at a number of sites in northern China, Mongolia, and Siberia. Especially notable is Berelekh, a site 500 km (about 300 miles) north of the Arctic Circle where the Yana River empties into the Arctic Ocean. Here archaeologists have recovered sophisticated stone tools and a number of artifacts made from bone, ivory, and horn (Figure 14.4). Radiocarbon methods indicate that these tools were made about 30,000 years ago. Pollen data tell us that this area then had a cool, dry climate in which grasslands were mixed with stands of larch and birch trees. Numerous processed bones of horse, musk ox, bison, and mammoth indicate that big game was plentiful. Nonetheless, life at this site must have been challenging for a recent African emigrant. Today the winters are long and dark, and January temperatures average −50°C (about −60°F)—a bracing thought, given that the world was much colder 30,000 years ago than it is today.

There is controversy over what was happening in Africa between 60 and 30 kya.

The fossil record during this period is very sparse, and there is disagreement among paleonthropologists about how modern human behavior evolved in Africa. Some researchers believe that modern human behavior came to Africa and Europe at about the same time. Others believe that modern human behavior evolved slowly in Africa over several hundred thousand years. We will return to this question at the end of the chapter.

Upper Paleolithic Technology and Culture

Upper Paleolithic peoples manufactured blade tools, which made efficient use of stone resources.

During this period, humans shifted from the manufacture of round flakes to the manufacture of blade tools. **Blades** are stone flakes that look like modern knife blades: they are long, thin, and flat and have a sharp edge. Blades have a longer cutting edge than flakes do, so blade technology made more efficient use of raw materials than older tool technologies did. However, there was a cost: although blades made more efficient use of materials, they also took more time to manufacture, requiring more preparation and more finishing strokes. Archaeologists label tool kits that emphasize blades as **Mode 4** technologies.

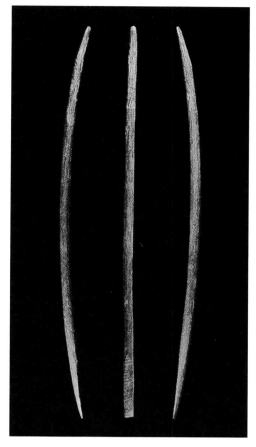

FIGURE 14.4 Spear foreshafts found at Yana, a site 500 km (about 300 miles) north of the Arctic Circle in eastern Siberia. These rhino horn foreshafts were fitted to the front end of the spear. If a hunter struck an animal without killing it, he could remove his spear, fit a second foreshaft to the spear, and reuse it. (From V. V. Pitulko et al., 2004, The Yana RHS site: Humans in the Arctic before the last glacial maximum, *Science*, 303: 52–56.)

The Upper Paleolithic tool kit includes a large number of distinctive, standardized tool types.

Upper Paleolithic peoples made many more kinds of tools than earlier hominins had made. Chisels, various types of scrapers, a number of different kinds of points, knives, burins (pointed tools used for engraving), drills, borers, and throwing sticks are just some of the items from the Upper Paleolithic tool kit.

Even more striking is the fact that the different tool types have distinctive, stereotyped shapes. It is as if the Upper Paleolithic toolmakers had a sheaf of engineering drawings on which they recorded their plans for various tools. When the toolmaker needed a new 10-cm (4-in.) burin, for example, she would consult the plan and produce one just like all the other 10-cm burins. Of course, these toolmakers didn't really use drawings as modern engineers do, but the fact that we find standardized tools suggests that they carried these plans in their minds. The final shape of Upper Paleolithic tools was not determined by the shape of the raw material; instead, the toolmakers seem to have had a mental model of what the tool was supposed to look like, and they imposed that form on the stone by careful flaking. The craft approach to toolmaking reached a peak in the Solutrean tool tradition, which predominated in southern France between 21 and 16 kya (Figure 14.5). Solutrean toolmakers crafted exquisite blades shaped like laurel leaves that were sometimes 28 cm (about 1 ft) long, 1 cm (about ½ in.) thick, and perfectly symmetrical.

FIGURE 14.5 Long, thin, and delicate blades characterized the Solutrean tool tradition.

Upper Paleolithic people were also the first to shape tools from bones, antlers, and teeth. Earlier hominins had made limited use of bone, but Upper Paleolithic people transformed bone, ivory, and antler into barbed spear points, awls, sewing needles, and beads.

Upper Paleolithic industries also varied in time and space.

The first Upper Paleolithic industry in Europe, the Aurignacian, was widespread in Europe by 35 kya (Figure 14.6). It is characterized by certain types of large blades, burins, and bone points. About 27 kya the Aurignacian was replaced in southern France by a new tool kit, the Gravettian, in which small, parallel-sided blades predominated and bone points were replaced by bone awls. Then about 21 kya, the Solutrean, with its beautiful leaf-shaped points, developed in the region. And 16 kya, the Solutrean gave way to the Magdalenian, a tool kit dominated by carved, decorated bone and antler points. Other parts of Europe are characterized by different sequences of tool complexes, so after the Aurignacian, each region typically has a distinctive material culture. Over roughly 25,000 years of the Upper Paleolithic in Europe there were dozens of distinctive tool kits emphasizing Mode 4 tools. This diversity stands in striking contrast to the Acheulean tool kit, which remained unchanged throughout more than half of the Old World for over a million years.

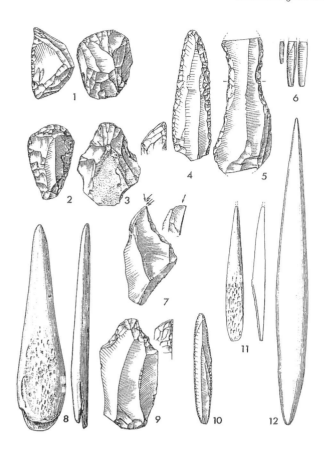

The Aurignacian tool kit, the earliest of several tool industries in the Upper Paleolithic of Europe, contains a wide variety of standardized tool types: 1, 2, 3, 9 = scrapers; 4, 5 = edged, retouched blades; 6, 10 = bladelets; 7 = a burin (a tool now used for engraving); 8, 11, 12 = bone points.

Stones and other raw materials for toolmaking were often transported hundreds of kilometers from their place of origin.

At Bacho Kiro, a 40,000-year-old Aurignacian site in Bulgaria, more than half of the flint used to make blades was brought from a source 120 km (about 75 miles) away. Distinctive, high-quality flint quarried in Poland has been found in archaeological sites more than 400 km (about 250 miles) away. Seashells, ivory, soapstone, and amber used in ornaments were especially likely to be transported long distances. By comparison, most of the stone used at a Mousterian site in France was transported less than 5 km (3 miles). The long-distance movements of these resources may mean that Upper Paleolithic peoples ranged over long distances, or that they participated in long-distance trade networks.

Modern humans exploited a wider range of prey species than did the Neanderthals, but the subsistence economies of the two populations were similar.

The Upper Paleolithic spanned the depths of the last ice age. The area in Europe that contains the richest archaeological sites was a cold, dry grassland that supported large populations of a diverse assemblage of large herbivores, including reindeer (called "caribou" in North America), horse, mammoth, bison, woolly rhinoceros, and a variety of predators, such as cave bears and wolves.

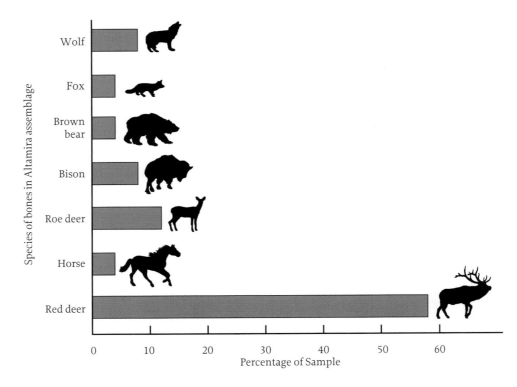

FIGURE 14.7 The bones of red deer dominate the assemblage at Altamira in northern Spain, as illustrated by these data, which come from a Magdalenian site dated to 15 to 13 kya.

Bones found at Upper Paleolithic sites indicate that large herbivores played an important role in the diets of Upper Paleolithic peoples. Everywhere, Upper Paleolithic peoples hunted herbivores living in large herds, fished, and hunted birds. In some places they concentrated on a single species—reindeer in France, red deer in Spain (Figure 14.7), bison in southern Russia, and mammoths farther north and east. In some areas, such as the southern coast of France, rich salmon runs may have been an important source of food. There are also places where these peoples harvested several different kinds of animals.

The peoples of the Upper Paleolithic developed more complex forms of shelter and clothing than the Neanderthals had.

In what is now western and central Europe, Russia, and the Ukraine, the remains of small villages have been found. Living on a frigid, treeless plain, Upper Paleolithic peoples either hunted or scavenged mammoth and used the hairy beasts for food, shelter, and warmth. At the site of Předmostí in the Czech Republic, the remains of almost 100,000 mammoths have been discovered. These peoples constructed huts by arranging mammoth bones in an interlocking pattern and then draping them with skins. [Temperatures of about –45°C (–50°F) outside provided a strong incentive for chinking the cracks!] The huge quantities of bone ash that have been found at these sites indicate that the mammoth bones were also used for fuel. A site about 470 km (about 300 miles) southeast of Moscow contains the remains of even larger shelters. They were built around a pit about a meter deep and were covered with hides supported by mammoth bones. Some of these huts had many hearths, suggesting that a number of families may have lived together (Figure 14.8).

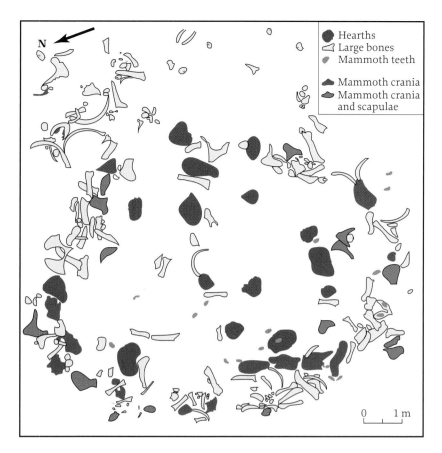

Hearths
◁ **Large bones**
🦷 **Mammoth teeth**

Mammoth crania
Mammoth crania and scapulae

0   1 m

FIGURE 14.8 On the Russian plain, early humans had to cope with harsh climatic conditions. They may have used mammoth bones to construct shelters. Mammoth bones and teeth litter this Upper Paleolithic site of Moldova. The presence of multiple hearths suggests that this site may have been occupied by several families.

Several lines of evidence indicate that Upper Paleolithic peoples living in glacial Europe manufactured fur clothing. First, when modern hunters skin a fur-bearing animal, they usually leave the feet attached to the pelt and discard the rest of the carcass. Numerous skeletons of foxes and wolves that are complete except for their feet have been found at several Upper Paleolithic sites in Russia and the Ukraine, suggesting that early modern humans kept warm in sumptuous fur coats. Second, bone awls and bone needles are common at Upper Paleolithic sites, so sewing may have been a common activity. Finally, three individuals at a burial site in Russia seem to have been buried in caps, shirts, pants, and shoes lavishly decorated with beads.

Upper Paleolithic peoples were better able to cope with their environment than the Neanderthals were.

The richness of the fossil and archaeological record in Europe provides a detailed comparison of Neanderthal and Upper Paleolithic peoples. Three kinds of data suggest that Upper Paleolithic peoples were better adapted to their environment than Neanderthals were:

1. Upper Paleolithic peoples lived at higher population densities in Europe than Neanderthals did. Archaeologists estimate the relative population size of vanished peoples by comparing the density of archaeological sites per unit of time. Thus, if one group of people occupied a particular valley for 10,000 years and

left 10 sites, and a second group occupied the same valley for 1000 years and left 5 sites, then archaeologists would estimate that the second group had five times as many people. (In using this method it is important to be sure that the sites were occupied for approximately the same lengths of time.) By these criteria, Upper Paleolithic peoples had far higher population densities in Europe than the Neanderthals did.

2. Upper Paleolithic peoples lived longer than Neanderthals. Anthropologists Rachel Caspari of the University of Michigan and Sang-Hee Lee of the University of California Riverside have estimated that Upper Paleolithic peoples lived much longer than the Neanderthals. About a third of their sample of 113 Neanderthals reached twice the age of first reproduction, which is about 30 years of age in modern hunter-gatherer populations. In contrast, two-thirds of a sample of 74 Upper Paleolithic fossil individuals reached that age. The fact that Upper Paleolithic populations included a substantial fraction of older individuals, while Neanderthals were dominated by the young, may have allowed Upper Paleolithic peoples to retain and transmit more complex cultural knowledge than Neanderthals could.

3. Upper Paleolithic peoples were less likely to suffer serious injury or disease than were Neanderthals. In sharp contrast to the Neanderthals, the skeletons of Upper Paleolithic people rarely show evidence of injury or disease. Among the few injuries that do show up in the fossil record, there is a child buried with a stone projectile point embedded in its spine, and a young man with a projectile point in his abdomen and a healed bone fracture on his right forearm. Evidence of disease is slightly more prevalent than evidence of injury among the remains of Upper Paleolithic peoples; affected specimens include a young woman who probably died as the result of an abscessed tooth, and a child whose skull seems to have been deformed by hydrocephaly (a condition in which fluid accumulates in the cranial cavity and the brain atrophies). Nonetheless, there is still less evidence for disease among these peoples than among Neanderthals.

There is good evidence for ritual burials during the Upper Paleolithic period.

Like the Neanderthals, Upper Paleolithic peoples frequently buried their dead. Upper Paleolithic sites provide the first unambiguous evidence of both multiple burials and burials outside of caves. Unlike the Neanderthals, Upper Paleolithic burials appear to have been accompanied by ritual. The people frequently buried their dead with tools, ornaments, and other objects that suggest they had some concept of life after death. Figure 14.9 diagrams the grave of a child who died about 15 kya at the Siberian site of Mal'ta. The child was buried with a number of items, including a necklace, a crown (diadem), a figurine of a bird, a bone point, and a number of stone tools.

Upper Paleolithic peoples were skilled artisans, sculpting statues of animals and humans and creating sophisticated cave paintings.

It is their art that distinguishes Upper Paleolithic peoples most dramatically from the hominins who preceded them. They engraved decorations on their bone and antler tools and weapons, and they sculpted statues of animals and female figures (Figure 14.10). The female statues are generally believed to be fertility figures because they usually emphasize female sexual characteristics. Upper Paleolithic peoples also

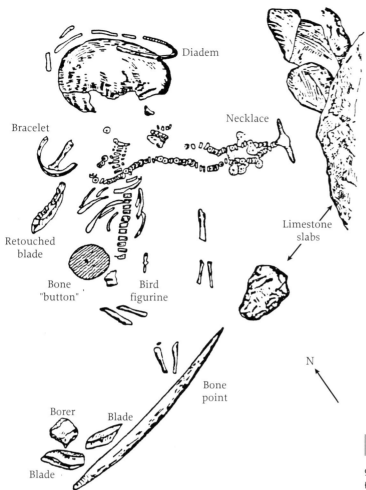

Diadem

Necklace

Bracelet

Limestone
slabs

Retouched
blade

Bone
"button"

Bird
figurine

N

Bone
point

Borer   Blade

Blade

FIGURE 14.9 Diagram of a child's grave from the Upper Paleolithic illustrating the rich collection of goods frequently included in Upper Paleolithic burials. (Figure courtesy of Richard Klein.)

adorned themselves with beads, necklaces, pendants, and bracelets and may have decorated their clothing with beads.

All of these artistic efforts are remarkable, but it is their cave art that seems most amazing now. Upper Paleolithic peoples painted, sculpted, and engraved the walls of caves with a variety of animal and human figures. They used natural substances that included red and yellow ocher, iron oxides, and manganese to create a variety of paint colors. They used horsehair, sticks, and their fingers to apply the paint. Their cave paintings frequently depict animals that they must have hunted: reindeer, mammoths, horses, and bison. Some figures are half human and half animal. Sometimes the creators incorporated the natural contours of the cave walls in their work, and sometimes they drew one set of figures on top of others. Some of their work is spectacular: the perspective is accurate, the characterization of behavior is lifelike, and the scenes are complex. We don't know precisely why or under what circumstances these cave paintings were made, but they represent a remarkable cultural achievement that sets the people of this period apart from earlier hominins.

Recently developed techniques for dating pigments have allowed scientists to determine the age of the Upper Paleolithic cave paintings. Many of the famous caves,

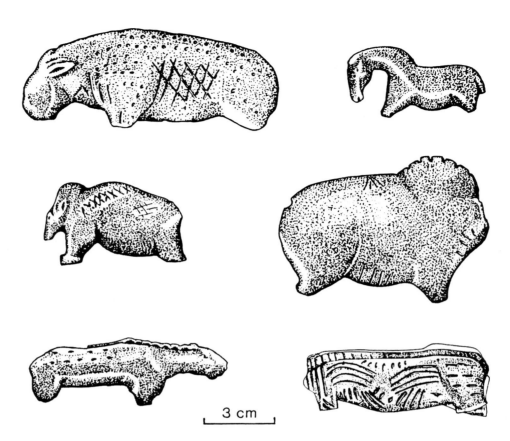

FIGURE 14.10 Small animal figures were carved from mammoth tusks during the Aurignacian period.

3 cm

such as Lascaux, are dated to the end of the Upper Paleolithic, about 17 kya, but spectacular paintings in the Le Chauvet Cave in France are about 30,000 years old (Figure 14.11). Most of the famous carved figurines are also less than 20,000 years old, but there are good examples of representational sculpture that date to the earliest Aurignacian. For example, Figure 14.12 shows an ivory figure with a human body and a lion's head that was found at a site in Germany dated to 32 kya. There is also abundant evidence of the manufacture of beads, pendants, and other body ornaments at Aurignacian sites in France. The beads are standardized and often manufactured from materials that had been transported hundreds of kilometers. Archaeologists have found a well-preserved musical instrument that looks like a flute at an Aurignacian site in southwestern France. It is 10 cm (4 in.) long and has four holes on one side and two on the other. When this instrument was played by a professional flutist, it produced musical sounds.

THE ORIGIN AND SPREAD OF MODERN HUMANS

A dramatic shift in hominin morphology occurred during the last glacial epoch. About 100 kya the world was inhabited by a morphologically heterogeneous collection of hominins: Neanderthals in Europe; other robust hominins in East Asia; and more modern humans in the Middle East and Africa. By about 30 kya, most of this morphological diversity had disappeared; modern humans occupied all of the Old World. Obviously, inquiring minds want to know, how did this transition occur?

Thirty years ago most paleoanthropologists would have given the same answer: the robust hominins of the late Middle Pleistocene formed a single morphologically variable species—"archaic *Homo sapiens*"—from which a more modern morphology gradually evolved throughout the world. New evidence, however, both from the fossil record and from molecular genetics, has cast substantial doubt on this hypothesis. The new evidence indicates that the genes that gave rise to modern human morphology evolved in an African population sometime between 200 and 100 kya. People carrying these genes subsequently spread throughout Africa and differentiated into a number of morphologically modern but genetically variable populations. Then about 50 kya, people from one of these regional populations left Africa and spread across the world, replacing other hominin populations with little gene flow between them.

In the rest of this chapter, we will describe the genetic, paleontological, and archaeological evidence pertaining to the origin and spread of modern humans. As you will see, researchers working in a diverse set of academic disciplines have brought a wide range of methods to bear on this problem.

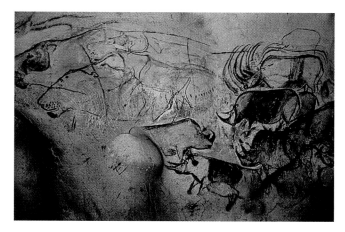

FIGURE 14.11 This image from Le Chauvet in France was created about 30 kya. At the upper left, several lions are depicted. On the other side are a number of rhinoceroses. At the top right, the multiple outlines suggest a rhinoceros in motion.

Genetic Data

Contemporary patterns of genetic variation, particularly in genes carried on mitochondria, provide information about the origin of modern humans.

The patterns of genetic variation within living people and genetic material extracted from fossils tell us a lot about the history of human populations. Much of the information used in such studies comes from genes that are carried on mitochondria. Eukaryotic cells, including those of humans, contain small cellular organelles called **mitochondria** (singular *mitochondrion*), which are responsible for the basic energy processing that goes on in the cells. Mitochondria contain small amounts of DNA (about 0.05% of the DNA contained in chromosomes), called, appropriately, **mitochondrial DNA (mtDNA)**. In humans and other primates, mtDNA codes for 13 proteins used in the mitochondria, for some of the RNA that makes up the structure of ribosomes, and for several kinds of transfer RNA. There are also two major noncoding regions on mtDNA.

Mitochondrial genes have three properties that make them especially useful for reconstructing the recent evolutionary history of human populations:

1. *Maternal inheritance.* Mitochondrial genes are inherited maternally. Both males and females get their mitochondria from their mother's egg cell. The mitochondria present in sperm are not transferred to the fertilized ovum. Therefore, there is no recombination in mtDNA, so a child has exactly the same mitochondrial genes as the mother, unless the child's mitochondria carry a novel mutation. This makes it possible to construct phylogenetic trees for mitochondria that can tell us a lot about human demographic history.

FIGURE 14.12 This ivory figure depicts a human body with a lion's head. It comes from an Aurignacian site in southern Germany and is 32,000 years old.

2. *High mutation rate.* Mitochondrial genes accumulate mutations faster than genes carried on chromosomes do. This means that the mitochondrial genetic clock provides more accurate dating of events in the last few hundred thousand years than do clocks based on nuclear genes. The mitochondrial clock is like a stopwatch that keeps track of hundredths of seconds, but the genes on chromosomes are like a clock with only a second hand. The mitochondrial genetic clock is particularly useful for dating recent events, for the same reason that a digital stopwatch is more useful than an ordinary watch for timing Justin Gatlin in the 100-m run; the mitochondrial clock changes rapidly enough to distinguish events that aren't very far apart in time.

3. *High copy number.* In each cell there are only two copies of each gene carried on chromosomes, but there are hundreds of mitochondria, each with its own copy of each mitochondrial gene. The larger number of gene copies makes it much easier to recover mitochondrial genes than nuclear genes from the tiny amounts of DNA that remain in fossils.

The human species is less genetically variable than are other species.

One measure of genetic variation is the average number of genetic differences per base pair (nucleotide) in a particular bit of DNA within a population. To compute this measure for mtDNA, geneticists collect tissue samples from members of a population and assess the same part of the DNA sequence of each individual's mitochondria. Then, for each pair of individuals in the sample, they check whether mtDNA at each position is the same or different. They repeat this for all pairs of individuals in the sample and then compute the average number of differences per nucleotide across the sample. As populations become more diverse genetically, the average number of genetic differences per nucleotide increases.

By this measure, humans are less genetically variable than chimpanzees, in two quite different ways. Figure 14.13 plots the amount of genetic variation within and between three geographically separated human populations—Africans, Asians, and Europeans —and the amount of genetic variation within and between chimpanzee populations in eastern, central, and western Africa. The bars on the diagonal (red) show that the amount of variation within each of these human populations is much lower than that within any of the chimpanzee populations. These data tell us that any two humans taken at random from a single population are much more similar to one another than are any two chimpanzees taken at random from a single population. The off-diagonal bars (blue) plot the average differences between human populations and between chimpanzee populations. The differences between chimpanzee populations are also much greater than the differences between human populations. This tells us that any two humans taken at random from *different* human populations are more similar to one another than any two chimpanzees taken from different chimpanzee populations are.

The most likely explanation for the small amount of genetic variation observed within the human species is that we have undergone a population explosion.

The data on genetic distance suggest that much of the human genome evolved mainly under the influence of drift and mutation. Remember that mutation introduces new genes at a very low rate, and that genetic drift eliminates genetic variation at a rate that depends on the size of the population. In large populations, drift removes genetic variants slowly, but in small populations, drift eliminates variation more

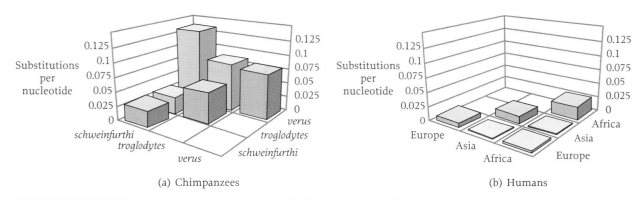

(a) Chimpanzees   (b) Humans

FIGURE 14.13 Humans are less genetically variable than chimpanzees. (a) Plotted here is the mean number of differences per nucleotide in one region of mtDNA for three geographically separated chimpanzee populations: *Pan troglodytes schweinfurthi* (from East Africa), *P. t. troglodytes* (from central Africa), and *P. t. verus* (from West Africa). (b) Here the same data are plotted for three geographically separated human populations (in Asia, Europe, and Africa). In both graphs, each red bar at the intersection of a row and a column gives data for comparing the population(s) identified at the ends of that row and column. Thus the height of each red bar represents the mean number of differences between pairs of individuals drawn from the same population. For example, each pair of chimpanzees belonging to the *verus* population differs at about 7.5% of the nucleotides sampled. Each pair of individuals drawn from the most variable human population, Africans, differs at only about 2.5% of the nucleotides assessed. The height of the blue bars represents the mean number of differences between pairs of populations. On average, an individual drawn from the *verus* population differs from an individual drawn from the *schweinfurthi* population at about 13% of the nucleotides, while Africans and Europeans, the most different pair of human populations, differ at less than 0.3% of the nucleotides. Thus the two graphs show that there is more variation within each of the chimpanzee populations than within any of the human populations, and human populations on the average are more similar to one another than chimpanzee populations are.

quickly. The effects of mutation and genetic drift will eventually balance (Figure 14.14), as the amount of variation arrives at an equilibrium. The amount of variation, which we label m, at equilibrium will depend on the size of the population. For a given mutation rate, bigger populations will have larger amounts of genetic variation at equilibrium. Population genetic theory says that $m = 2Nu$, where N is the number of females and u is the rate at which mutations occur.

We can substitute the measured values of m and u into the formula for m, to calculate N. Using an estimate for the mutation rate of mtDNA of $u \approx 0.0015$, and the average of the values for m given in Figure 14.13, we get $N \approx 5000$. This means that the existing amount of genetic variation within human populations today is consistent with the drift–mutation equilibrium value for a population of 5000 women. Obviously this estimate of N is far lower than the size of human populations today or even in the recent past. How can that be?

The simplest answer is that a small human population must have expanded very rapidly sometime in the past. When a population grows, the equilibrium amount of genetic variation also grows because genetic drift removes variation more slowly in large populations than it does in smaller ones. If the population expands rapidly, however, it may take a long time to reach a new equilibrium level because mutation adds variation very slowly. Thus the low levels of genetic variation in contemporary human populations may be the result of a rapid expansion from a small population sometime in the past.

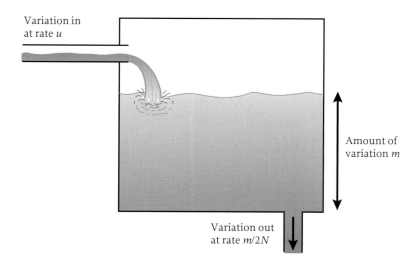

FIGURE 14.14 A physical analogy for the mutation–drift equilibrium can help explain why large populations have more genetic variation at equilibrium than small ones do. Mutation introduces new genetic variants at a constant rate u, much like a stream of water entering a tank. Genetic drift removes variants, like a drain at the bottom of the tank. The amount of variation (m) is analogous to the volume of water in the tank: as the level of water in the tank rises, the pressure increases and the water drains out more rapidly. Eventually the rate of outflow equals the rate of inflow, and the depth of the water remains constant. In the same way, as the amount of variation in a population increases, drift removes variation faster; eventually a steady state is reached at which the amount of variation is constant. To calculate the equilibrium amount of variation, set the rate of inflow (u) equal to the rate of outflow ($m/2N$), and solve for m.

This conclusion supports any hypothesis that suggests that all contemporary people are descended from a small local population living sometime in the past. A population of 5000 adult women corresponds to an overall population of roughly 20,000 men, women, and children. It is very difficult to see how much a small population could have been spread across all of Africa and Eurasia. It is much more plausible that the human population at the time was much larger, but that modern humans are descended from members of a single local population.

Natural selection provides an alternative explanation for the lack of genetic diversity in mitochondrial genes. If a strongly beneficial mutation arose in the mitochondrial genome, it would rapidly sweep through the human population. Because there is no recombination, the particular mitochondrial genotype in which the beneficial mutation occurred would also spread, and all other mitochondrial genotypes would disappear. In this way, natural selection could reduce the number of distinct mitochondrial genotypes without any reduction in the number of people. However, studies of the X and Y chromosomes are consistent with results of analyses of mtDNA and suggest that there was a bottleneck between 200 and 100 kya. Natural selection could account for these data only if a number of separate beneficial mutations occurred throughout the nuclear genome at about the same time. This seems implausible but cannot be ruled out as a possibility.

The fact that modern African populations are genetically more variable than other populations is consistent with the idea that the population explosion began in an African population.

Notice in Figure 14.13 that the amount of mitochondrial genetic variation within contemporary African populations is greater than the variation within European or Asian populations. Data from studies of ordinary nuclear genes tell the same story: Africans are more genetically variable than other peoples.

This pattern of genetic variation is consistent with the hypothesis that modern humans first evolved in Africa. Suppose that populations in different parts of Africa were isolated from each other. If that were the case, they would have diverged genetically (Figure 14.15). If emigrants to Europe and Asia were drawn from one local African population, they would carry with them only a fraction of the total genetic

100 kya: Modern humans evolve and disperse throughout Africa.

100 kya: Modern humans evolve and disperse throughout Africa.

50 kya: People from one African population migrate to Eurasia and Australasia.

50 kya: People from one African population migrate to Eurasia and Australasia.

FIGURE 14.15 A late African origin for modern humans explains why African populations today have greater genetic variation than modern human populations in other parts of the world have: (a) Modern humans evolved and dispersed throughout Africa approximately 100 kya. (b) Over the next 50,000 years, dispersed populations were isolated from each other by distance, and these populations became genetically different. (c) About 50 kya, people from one of the local African populations dispersed throughout Eurasia. Because these emigrants were from one local population, they were less variable than African populations taken as a whole. (d) Over the next 50,000 years, dispersed populations in Africa and Asia continued to diverge. Modern Eurasian populations are less variable than African populations today because Eurasian populations are descended from a single localized African population that migrated northward 50 kya. Modern African populations are more variable because they are the descendants of a number of local African populations that were spread across most of the continent.

variation present in Africa. Then the European and Asian descendants of this regional African population would be less variable than modern populations across Africa are. As we will see, phylogenetic trees for mitochondria are consistent with this scenario.

The fact that mitochondria are inherited maternally means that the mitochondria of every living person are copies of the mitochondrion of a single woman who lived in ages past.

This incredible idea is a simple consequence of asexual inheritance. The logic is based on the fact that, in each generation, some women leave no female descendants. (Because mitochondria are transmitted through the female line, sons are irrelevant.) If females leave no surviving female descendants, then no copies of their mitochondria survive into the next generation. As the generations go on, more and more mitochondrial lineages are lost, until eventually everyone carries the mitochondrion of a single woman, the most recent common ancestor—who is sometimes given the colorful but misleading name "Eve." (Box 14.2 gives a more detailed version of this argument.) Note that the most recent common ancestor is not the mother of us all, but only the mother of our mitochondria. Because genes carried on chromosomes recombine, it is likely that we have ordinary chromosomal genes from many of the women in the population containing the most recent common ancestor. And it is important to realize that this woman was not the first human woman. She may have been a

BOX 14.2

Mitochondrial Eve

To see why everyone carries the mitochondrion of a single woman, consider Figure 14.16. Each colored circle represents one woman, and each row of circles represents a population of 12 women during a given generation. The circles in row 1 represent the women in generation 1, the circles in the second row are women in generation 2—all of whom are daughters of women in generation 1—and so on. (We ignore men because their mitochondria are not transmitted to their children.) To keep things simple, we assume that population size is constant, so women must produce on average just one daughter. However, the number of daughters born to any particular woman varies; here we will assume that some have no daughters, some have one, and some have two. For the moment, we will also assume that there is no mutation.

To indicate that daughters inherit their mother's mitochondria, mothers and their daughters are assigned the same color. Thus, in any generation, all of the women with the same color carry copies of the mitochondrion of the same woman in generation 1. (This color scheme does not imply that the mitochondria carried by different women have different mtDNA sequences; they could all be genetically identical or all unique.)

In generation 1 there are three women (black, dark gray, and light gray) who have no daughters. As a result, after one generation, only 9 of the original 12 women and their mitochondria have descendants. However, three of the mitochondria in the first generation have two descendants in generation 2. During the second generation, three women have no daughters. However, because one of the women

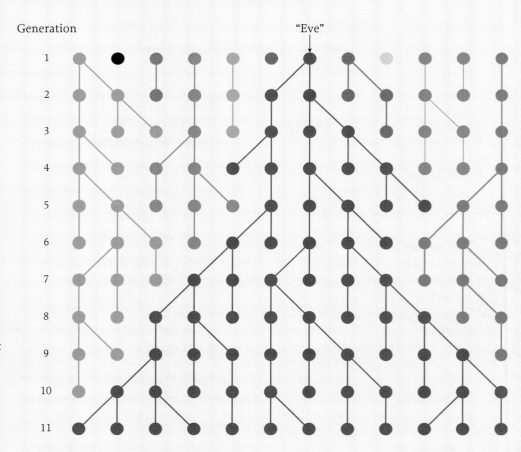

FIGURE 14.16 This mtDNA tree for a hypothetical population of 12 women shows why all the mitochondria in any population are descended from a single woman's mitochondria sometime in the past.

without a daughter has a sister who produced a daughter, copies of that woman's mitochondria are not lost. This means that, by generation 3, the number of women's mitochondria present in the population is reduced from 12 to 7. The final outcome should be evident: as long as there are copies of mitochondria from more than one of the original women present, there is a chance that one of them will be lost because all the descendants of one of the women will have no daughters. Eventually the mitochondria of all but one of the original women will disappear. At this point, all members of the population will be descendants of one of the original females. We have run this argument forward in time, but it can also be run backward. That is to say, we can pick any generation and then move backward through time to find the single mitochondrial ancestor of every individual in that generation.

In this example, there are only 12 women in the population, but exactly the same thing will happen in all populations.

modern human, or she may have been an earlier form of hominin. Eve was an otherwise unremarkable member of a (perhaps large) population whose only claim to fame is that her mitochondrion happened to survive while the mitochondria of other women did not.

The accumulation of mutations allows us to derive the phylogenetic relationships of the mtDNA of modern humans.

Even though the mitochondria of all living people are replicas of Eve's mitochondrion, they are not identical, because mutation occasionally introduces new mitochondrial variants. Since there is no recombination, once a mutation occurs in a particular woman, it is carried by all of her descendants. The existence of mutation has two important consequences. First, anthropologists can use the methods of phylogenetic reconstruction discussed in Chapter 4 to reconstruct a tree of descent among mitochondria based on derived similarities. To make such reconstructions (sometimes called **gene trees**) feasible, mutations must occur frequently to allow the branches of the gene tree to be distinguished. It turns out that mitochondria are particularly well suited to this sort of genetic reconstruction. Second, mutation makes it possible to estimate the time that has passed since our common mitochondrial ancestor lived, using genetic-distance measures (also discussed in Chapter 4).

Gene trees based on mtDNA support an African origin for modern humans.

Beginning with the pioneering work of the late Alan Wilson and Rebecca Cann, now at the University of Hawaii, numerous studies have attempted to construct gene trees from the patterning of variation in human mtDNA. These early studies were based on a relatively short noncoding section of the mtDNA, called the "D-loop," that has a highly variable mutation rate across sites. As a result, different mitochondrial lineages often experienced the same mutation, and this made it hard to construct and interpret gene trees. Newer data based on the entire mitochondrial genome yield highly consistent gene trees, like the one in Figure 14.17.

This tree supports the idea that modern humans originated in Africa and then migrated out of Africa and spread across the rest of the world. Notice that all of the deepest branches of the tree represent African populations, and all non-Africans are descendants of a single node in the tree. This indicates that humans originated in Africa and their mitochondrial lineages diversified. Later, one group left Africa and

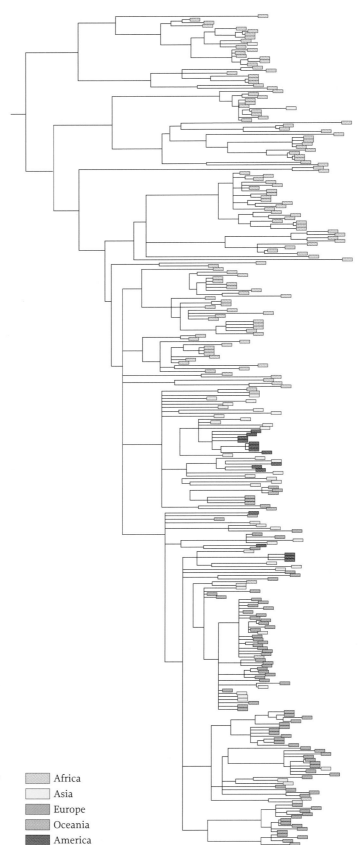

FIGURE 14.17 This gene tree for the complete mitochondrial genomes of 277 people from all over the world indicates that all modern humans are descended from an African population. Each branch represents the mtDNA sequence of a single individual, and the length of the branch represents the number of mutations along that branch. Thus the longer branches for Africans reflect the greater genetic variation on that continent. Almost all non-Africans are descended from a single node in the tree and from a more recent common ancestor than many Africans are. (Redrawn from L. Cavalli-Sforza and M. Feldman, 2003, The application of molecular genetic approaches to the study of human evolution, *Nature Genetics Supplement*, 33: 271.)

Africa
Asia
Europe
Oceania
America

gave rise to the mitochondrial lineages we observe today in the rest of the world. One group of Africans is more closely related to non-Africans than to other Africans. These could be the descendants of the African population that originally emigrated and settled the rest of the world. You can also see in Figure 14.17 that the branches that separate different African groups are much longer than the branches that separate non-African groups, indicating that non-African populations have grown rapidly. Finally, analysis of two different mtDNA gene trees tells us that the most recent common ancestor lived between 220 and 120 kya.

Trees based on the Y chromosome support the conclusion of mtDNA studies.

It is worrisome to depend completely on mtDNA data for our understanding of human demographic history, even though these data have been extremely useful. Mitochondrial genes are transmitted only by women, so mtDNA trees tell us only about the population dynamics of half of our species. Moreover, the fact that mitochondrial genes are not subject to recombination means that natural selection on even one mitochondrial gene could have strong effects on the rest of the mitochondrial genome. For example, a new beneficial mutant that arose about 150 kya and swept through the population would generate much the same pattern as a population bottleneck because only those initially rare individuals carrying the beneficial mutant would leave descendants.

Genes on the Y chromosome provide an independent source of information about human origins. Y chromosomes are carried only by men, and most of the Y chromosome does not experience any recombination. Thus, Y chromosome genes give us the same kind of information about the demographic history of men as mitochondrial genes give us for women. Because there is no linkage of genes on the Y chromosome and those on mitochondria, it is unlikely that selection would affect mtDNA and the Y chromosome in the same way. Mutation rates on the Y chromosome are lower than those on mtDNA. However, the Y chromosome is roughly 20 million DNA bases long, while mtDNA contains only about 16,000 bases. This means that there are enough detectable mutations on the Y chromosome, even though the rate of mutation is much lower, to construct accurate gene trees.

Two different gene trees for the Y chromosome have been constructed, based on two different sets of mutations. Both trees are consistent with trees based on mtDNA (Figure 14.18). The deepest branch occurs in Africa, and all but one of the non-Africans in the sample are the descendants of a more recently derived lineage. The estimated age of the most recent common ancestor is between 40,000 and 140,000 years old, and that of the most recent common ancestor of all non-Africans is between 35,000 and 89,000 years old. These ages are somewhat younger than those derived from the mtDNA, probably because the number of males who left children was smaller than the number of females who had descendants, as is typical in most mammalian populations (see Chapter 7).

Mitochondrial DNA extracted from a number of Neanderthals indicates that the last common mitochondrial ancestor of Neanderthals and modern humans lived about 500 kya.

In the film *Jurassic Park*, scientists extract dinosaur DNA from the bodies of bloodsucking insects trapped in amber and use the DNA to clone living dinosaurs. The gap between science fiction and science has narrowed in the laboratory of Svante

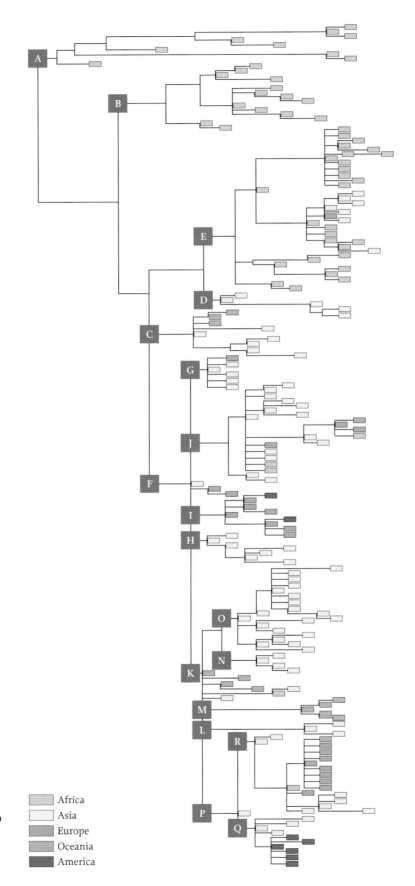

FIGURE 14.18 A gene tree for the human Y chromosome based on 167 genetic loci, and more than 1062 individual men. (Redrawn from L. Cavalli-Sforza and M. Feldman, 2003, The application of molecular genetic approaches to the study of human evolution, *Nature Genetics Supplement*, 33: 271.)

Africa
Asia
Europe
Oceania
America

Pääbo at the Max Planck Institute for Evolutionary Anthropology in Leipzig, Germany. Pääbo and his colleagues extracted and sequenced mtDNA from the original Neanderthal fossil found in the Neander Valley of Germany in 1856. More recently, Pääbo and his team sequenced mtDNA from several other Neanderthal fossils, and attempted to extract mtDNA from four early modern *Homo sapiens* from Europe.

These data allow researchers to construct gene trees and to compute the amount of genetic variation within and between populations. Figure 14.19 shows the tree for three of the Neanderthals and a large sample of modern humans from around the globe. The tree indicates that all modern humans are equally distant from Neanderthals. This is consistent with the idea that the modern human lineage diversified *after* it split from the lineage leading to the Neanderthals.

Researchers from Pääbo's laboratory have also computed the genetic distance within and between Neanderthal, modern human, and ape populations. These values indicate that the genetic distance between modern humans and Neanderthals is much greater than the distances among modern humans or among the Neanderthals. The magnitude of the differences between the Neanderthals and modern humans suggests that the last common mitochondrial ancestor of humans and Neanderthals lived between 690 and 357 kya. The data also indicate that Neanderthals, like modern humans, were much less variable genetically than other primates, such as chimpanzees. The relatively small genetic distances between the three Neanderthals in Figure 14.19 indicates that their last common ancestor lived between 352 and 151 kya.

These findings are consistent with the fossil and archaeological records of the Middle Pleistocene. Recall from Chapter 13 that the fossils from Sima de los Huesos suggest that the ancestors of Neanderthals had reached Europe and begun to evolve the distinctive Neanderthal morphology by about 300 kya. What we know about the climate and biogeography of Middle Pleistocene Eurasia and Africa suggests that these proto-Neanderthals colonized Europe during one of the interglacial periods and were subsequently isolated from African hominin populations. Morphologically and behaviorally, modern people gradually evolved in Africa between 200 and 100 kya.

These data do not exclude the possibility that there was some interbreeding between Neanderthals and modern humans. Mathematical models suggest that all

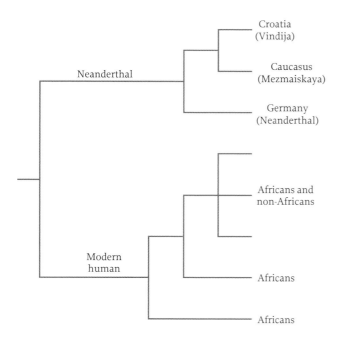

FIGURE 14.19 The phylogenetic relationships of mtDNA from three Neanderthals and modern humans are plotted. The length of each horizontal branch symbolizes the genetic distance between individuals or groups. The branches leading to modern humans represent a complex branching tree, each tip ending at a unique mtDNA genotype of at least one modern human. This tree indicates that the three Neanderthals are much more closely related to each other than any one of them is to any modern human.

modern human mitochondria are descended from about five individuals living 30,000 years ago. It is quite possible that there was a modest amount of interbreeding but the Neanderthal mitochondria were subsequently lost by chance. However, attempts to extract DNA from early modern human fossils didn't turn up any of the mtDNA sequences that characterize Neanderthals. This suggests that the Neanderthals and early humans did not interbreed.

Although we can say with some confidence that the DNA extracted from early modern human fossils does not match the DNA extracted from Neanderthals, it is harder to say how closely the DNA of early modern humans resembles the DNA of present-day humans or the DNA of Neanderthals. The problem is that the samples have apparently been contaminated. After 30,000 years in the ground, only minuscule amounts of DNA remain in fossilized bone. When archaeologists, museum curators, and molecular biologists handle bones, they inadvertently leave traces of their own DNA on the bones, and this modern DNA may swamp the DNA in the fossils. When scientists try to amplify the DNA from the fossils, they may inadvertently be amplifying modern DNA instead. This possibility is supported by the results of studies of DNA extracted from the fossilized bones of cave bears from the same period of time. These samples also yield modern human DNA, which must be the product of contamination rather than common ancestry!

Evidence from Fossils and Tool Kits

The oldest modern human fossils have been found in Africa and date to about 190 kya.

The oldest modern human fossil were unearthed in 1963 at Omo-Kibish a site in southern Ethiopia. One of the skulls found there was relatively modern, while the other was similar to *H. heidelbergensis*. Originally the dates for this site were uncertain, but new methods now indicate a date of 190 kya. More recently a number of fossils were uncovered at Herto, a site in Ethiopia. A team led by University of California anthropologist Tim White found the fossilized crania of two adults and one immature individual along with a number of other fragments. These skulls are intermediate between those of modern humans and older African hominins classified as *Homo heidelbergensis*. The skulls are longer and more robust than those of most living people. They also have prominent browridges, pointed occipital bones, and other features that link them to earlier African hominins. In addition, however, they have some very modern features—most notably a high, rounded braincase (Figure 14.20). Argon–argon dating was used to date these fossils to about 160 kya.

In those days the Herto site was on the shores of a lake, and there is ample evidence that hominins butchered hippopotamuses. The Middle Stone Age stone tools found at the site are much like those found in association with earlier African *H. heidelbergensis*. Multiple, unambiguous cut marks give evidence that the skulls were defleshed by stone tools, and polished surfaces are consistent with repeated handling by other hominins. White and his coworkers point out that the mortuary practices used by some contemporary peoples in New Guinea leave a similar combination of marks on skulls, and they suggest that the Herto fossils may provide early evidence of similar ritual behavior.

There is evidence that modern humans also lived in other parts of Africa around the same time, but the fossils are too fragmentary to be completely conclusive. The

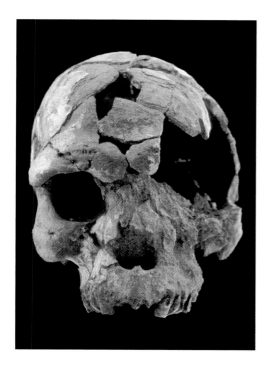

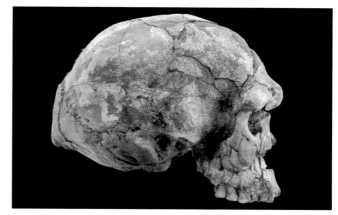

FIGURE 14.20 Side and front views of one of the hominin crania found at Herto, Ethiopia. This specimen (BOU-VP-16/1) is intermediate between *H. heidelbergensis* and modern *H. sapiens*, displaying prominent browridges but a high, rounded braincase.

most extensive finds come from the mouth of the Klasies River in South Africa. Excavations at this site have yielded five lower jaws, one upper jaw, part of a forehead, and numerous smaller skeletal fragments. Electron-spin-resonance dating indicates that the site is 74,000 to 134,000 years old. Although one of the lower jaws clearly has a modern jutting chin and the forehead has modern-looking browridges, the fragmentary nature of the fossils makes it difficult to be sure that these are modern humans.

Modern people appeared in the Middle East and probably in Africa long before the Neanderthals disappeared in Europe.

Even more modern fossils have been found at the Qafzeh and Skhul caves in Israel. Thermoluminescence and electron-spin-resonance dating techniques have shown that these fossils are 115,000 years old. This was during a relatively warm, wet period when animals would have been able to move from Africa to the Middle East. This inference is supported by the fact that other fossilized animals found at these sites are primarily African species.

Neanderthal fossils also have been found at three other sites located quite close to Qafzeh and Skhul, named Kebara, Tabun, and Amud. For many years there were no reliable absolute dates for any of these sites because they are too young for potassium–argon dating and too old for carbon-14 dating. However, most anthropologists assumed that the Neanderthals came first and were succeeded by the modern hominins at Qafzeh and Skhul. When the sites were dated by thermoluminescence and electron-spin-resonance methods, the results were a big surprise. The Neanderthals occupied Tabun about 110 kya, around the same time that modern humans lived nearby. However, the Neanderthals at Kebara and Amud lived 60 to 55 kya, which is about 30,000 years *after* anatomically modern peoples inhabited Skhul and Qafzeh.

The long period of overlap between Neanderthals and modern humans in the Middle East means that modern humans cannot be descended from Neanderthals. If anatomically modern people appeared 60,000 years before the Neanderthals disappeared, then it can hardly be true that all Neanderthals evolved into modern humans in that area. Some paleoanthropologists, most notably Milford Wolpoff of the University of Michigan, deny that the Skhul and Qafzeh fossils are really morphologically modern. Instead, they believe that they and the more robust fossils from Kebara, Amud, and Tabun are part of a variable population that is unrelated to either European Neanderthals or modern human populations. However, most anthropologists believe that Skhul and Qafzeh fossils are modern, because the skulls from these sites have relatively high, rounded crania and small faces, while a cranium found at Tabun is very similar to that of European Neanderthals. Similarly, postcranial fossils from Qafzeh are modern, but the pelvis and femur from Tabun show all the features of classic Neanderthals (see Figure 14.2).

Archaeological data suggest that Upper Paleolithic industries spread rapidly across western Eurasia.

Additional support for the idea that Neanderthals overlapped with and were later replaced by modern peoples comes from archaeological data indicating that Neanderthals and modern people overlapped in Europe, as well as in the Middle East. Recall that the earliest *Homo sapiens* fossils in Europe are associated with the Aurignacian tool kit. Moreover, no Aurignacian tools have been found with Neanderthal fossils. Aurignacian tools appear in Europe about 40 kya—4000 years before the Neanderthals disappeared. This brief overlap suggests that the modern humans using Aurignacian tools appeared and rapidly replaced Neanderthals. This transition seems too rapid to be the result of genetic change.

However, the situation is not quite so simple. In southern France, the transition between the Mousterian and the Aurignacian is marked by the presence of a third industry, called the **Châtelperronian**, which is intermediate between the Mousterian and the Aurignacian. Many anthropologists believe that the Châtelperronian is the result of Neanderthals' borrowing ideas and technology from modern humans. Cambridge University archaeologist Paul Mellars argues that three kinds of evidence support this view. First, Châtelperronian tools are associated with Neanderthal fossils at Saint Césaire and Arcy-sur-Cure in France. Second, archaeological data indicate that the Châtelperronian and Aurignacian industries coexisted in southern France for hundreds of years. Finally, other transitional tool kits have been found. As Figure 14.21 shows, each of these transitional industries is localized: the Châtelperronian in southern France, the Uluzzian in northern Italy, and the Szeletian in central Europe. Mellars points out that each of these transitional industries is quite distinctive, and he argues that it is not likely that the very widespread Aurignacian evolved from several distinctive localized transitional industries.

MODERN HUMAN BEHAVIOR: REVOLUTION OR EVOLUTION?

Anthropologists differ sharply about the causes of the striking shift that we see in the European archaeological record 40 kya. The debate turns on what was going on in Africa between 250 and 50 kya. According to one view, championed by Richard Klein, the African archaeological record is very similar to that in Europe. Between 250 and

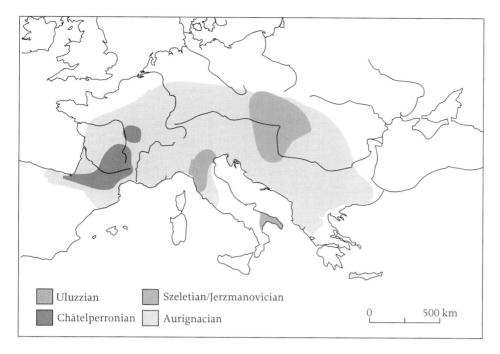

Uluzzian

Châtelperronian

Szeletian/Jerzmanovician

Aurignacian

0   500 km

FIGURE 14.21 The earliest Upper Pale-olithic industry, the Aurignacian, is found throughout Europe; three industries intermediate between the Mousterian and the Aurignacian (the Châtelperronian, the Uluzzian, and the Szeletian) are found in localized areas. Because each of these transitional industries is distinctive, some scientists believe it unlikely that the Aurignacian evolved from one of them.

50 kya, African hominins had bodies much like those of modern people, but they behaved more like their European contemporaries, the Neanderthals. Then, about 50 kya there was a rapid transition to much more modern behavior in Africa, just as in Europe. Klein believes that this behavioral transition marks the appearance of cognitively modern people. About 50 kya some key cognitive innovation evolved in at least one human population and allowed for the rapid development of modern human behavior—a transition that is commonly called the "human revolution." More recently, Sally McBrearty and Alison Brooks have challenged the idea that there was a human revolution at all. Instead, they argue that the archaeological record from Africa shows that the complex behaviors that appeared suddenly in Europe 40 kya was gradually evolving in Africa between 250 and 50 kya. What seems like a revolutionary event in Europe actually was due to the rapid replacement of Neanderthals by African immigrants. It wasn't a revolution; it was an occupation by a technologically advanced people.

The African Archaeological Record during the Later Pleistocene

Human behavior increased in complexity at about the same time in Africa and Europe.

From about 250 to 40 kya, the African archaeological record is dominated by a variety of stone tool kits that, for the most part, emphasize Mode 3 tools. These industries are collectively labeled the **Middle Stone Age (MSA)**. Until recently, most archaeologists saw the MSA in Africa as qualitatively similar to the much better known Mousterian tool industries associated with Neanderthals in Europe. Although there is some evidence for blade production in the MSA, it was thought that the other signatures of modern human behavior seen in Upper Paleolithic Europe were rare or absent in Africa. According to this view, MSA peoples in Africa produced few bone tools, developed little regional variation in tools, engaged in little long-distance trade, didn't bury their dead, and had no important symbolic behavior. As we will see, this view of the MSA has been challenged, by McBrearty and Brooks.

About 40 kya, new tool industries came to predominate in Africa. These industries, labeled the **Later Stone Age** (**LSA**), emphasize very small, carefully shaped flakes called **microliths** (literally "tiny stones") that are thought to have been mounted on wood and bone handles to form complex, composite tools—spears, axes, and so on. Industries of this kind are referred to as **Mode 5** technology. There is also ample evidence of ornaments, intentional burial, long-distance trade, and more intensive foraging activity. The Mode 5 tools; the associated evidence for hunting large, dangerous game; and the clear proof of symbolic expression have led many archaeologists to regard the LSA as qualitatively similar to the Upper Paleolithic in Europe.

Some evidence suggests that LSA peoples were more proficient foragers than MSA peoples.

MSA deposits at the mouth of the Klasies River in South Africa and LSA deposits at nearby Nelson Bay Cave provide a "natural experiment" that allows us to compare the behavioral complexity of MSA and LSA peoples. The MSA peoples occupied this area between 130 and 115 kya during a warm, moist interglacial period; and LSA peoples occupied the same area about 12 kya at the beginning of the present warm, moist interglacial period. Richard Klein argues that several features of this comparison support a qualitative difference between the behavior of MSA and LSA peoples.

- *LSA peoples hunted larger, more dangerous game.* The MSA peoples hunted large game, particularly eland, a cow-sized antelope. They also consumed other species, mainly buffalo and bushpig. LSA peoples also hunted large game, but they concentrated more on buffalo and bushpig, species that are more common but more dangerous than eland.
- *LSA peoples engaged in more sophisticated planning.* MSA peoples hunted for fur seals throughout the year, but LSA humans hunted them only seasonally. They killed mainly young seal pups, which were available only during the short period when the mothers and pups were on the beach. This behavior suggests that LSA peoples could keep track of the seasons and timed their visits to the coast to coincide with the availability of seal pups.
- *LSA peoples engaged in more intensive foraging.* The presence of artifacts that archaeologists interpret as sinkers and fish gorges suggests that LSA peoples fished. There is no evidence of fishing for MSA peoples.

Not all archaeologists accept these conclusions. It has been pointed out that MSA peoples did take dangerous creatures, both at the Klasies River mouth and elsewhere. The quantitative difference could be due to the fact that higher population densities forced LSA peoples to take less desirable prey. There is also evidence from sites in the Sudan, Congo, and Botswana that MSA peoples captured fish and shellfish.

Several of the signatures of modern human behavior appear in association with MSA archaeological sites.

Recently, Sally McBrearty and Alison Brooks argued that the MSA is *not* qualitatively similar to the Mousterian in Europe. According to their reading of the

African archaeological record, most of the signatures of modern human behavior gradually appear between 250 mya and 50 kya, when the transition is complete. Here is the kind of evidence they cite:

- *Blades appear early in the MSA.* At the Kapthurin Formation in Kenya, roughly 25% of the tools are blades. Archaeologists can show that the knappers at this site were highly skilled at blade production, wasting little raw material and making few mistakes. This site has been dated to between 280 and 240 kya by the argon–argon technique. Blades have also been found at a number of other MSA sites that date between 250 and 50 kya.

- *There is regional variation in MSA industries.* On a continental scale there is quite a bit of variation within MSA industries (Figure 14.22). In North Africa the Aterian industry is characterized by tanged projectile points; the Lupemban industry of the Congo basin is characterized by thin, long, lanceolate points; and in South Africa the Howieson's Poort industry is dominated by small points that foreshadow the LSA. We do not know whether the kind of regional variability that characterizes the Upper Paleolithic in Europe is also characteristic of Africa, because many fewer sites in Africa have been excavated.

- *Refined bone tools occur at several MSA sites.* Brooks and her coworkers have recovered a number of exquisite bone points, some of them elaborately barbed, from a site on the northeastern border of the Republic of the Congo called Katanda (Figure 14.23). The MSA layer containing the points has been dated to between 174 and 82 kya by three independent dating methods. At Blombos Cave

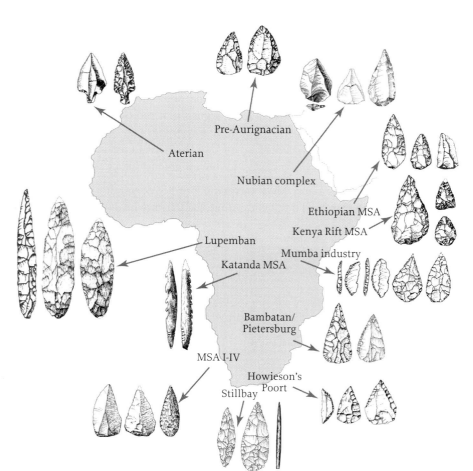

FIGURE 14.22 There was substantial variation in tool industries across Africa during the Middle Stone Age (MSA). Compare the large, leaf-shaped blades of the Lupemban industry with the tanged points of the Aterian industry or the small blades of the Howieson's Poort industry. It is not known whether similar variation existed on smaller scales.

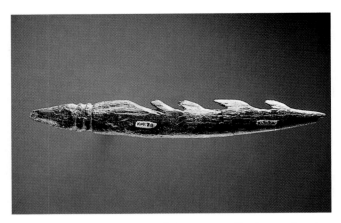

FIGURE 14.23 Beautiful bone points like this one found at Katanda in the Republic of the Congo have been dated to between 174 and 82 kya. If these dates are correct, then Middle Stone Age peoples were capable of producing bone tools that rival the best of the European Upper Paleolithic tools.

in South Africa, a number of polished bone points have been found in association with MSA tools and date to about 72 kya.

- *MSA peoples sometimes transported raw material great distances.* Most of the stone used to make tools at MSA sites comes from a short distance away. At several MSA sites, however, small amounts of raw materials were transported much farther. At several sites in East Africa, for example, tools were made from obsidian carried 140 to 240 km (about 90 to 150 miles).

- *Evidence for shelters and hearths appears in the MSA.* At several sites, MSA peoples built shelters. At the Mumbwa caves in Zambia, for example, there are three arcs of stone blocks that probably served as windbreaks. Elaborate stone hearths were constructed inside these structures. There are many more MSA sites at which people seem to have constructed hearths and built huts, but for each of these it is impossible to rule out the possibility that natural processes produced the features that archaeologists have documented.

- *There is some evidence of decorative carving, beads, and pigment use.* Recently, a team of archaeologists led by Christopher Henshilwood of the Iziko Museums of Cape Town discovered two elaborately engraved pieces of red ocher at Blombos Cave in South Africa (Figure 14.24). These artifacts were found in association with MSA tools and date to 77 kya. Ostrich shell beads that have been found at several sites dated to the very end of the MSA, 50 to 40 kya. There is good evidence for the use of red ocher at a number of sites. The earliest evidence comes from the Kapthurin Formation, which dates between 280 and 240 kya. Many present-day African peoples use red ocher to decorate themselves and for symbolic purposes, but some archaeologists have argued that red ocher might be used for utilitarian purposes, such as tanning hides. Few images are associated with MSA sites, but McBrearty and Brooks suggest that this lack of artwork may be due to the absence of deep limestone caves in which pigments are preserved.

On the basis of these data, McBrearty and Brooks conclude that there was no human revolution in Africa. They believe that the various behavioral elements that

FIGURE 14.24 One of two engraved pieces of red ocher found at Blombos Cave in South Africa. The artifact is dated to 77 kya and is associated with MSA tools. The engravings are similar to cave art found in other parts of the world.

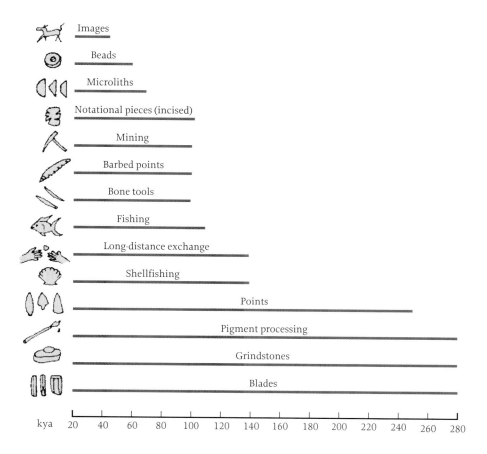

FIGURE 14.25 McBrearty and Brooks believe that many of the archaeological signatures of modern behavior in Europe appeared gradually in Africa between 280 and 40 kya. If they are correct, then the genetic and cultural changes that underlie modern human behavior accumulated gradually in Africa during this period.

suddenly appear with modern humans in Europe were assembled piecemeal over a period of 200,000 years in Africa (Figure 14.25). They argue that archaeologists have been misled about what happened because the archaeological record from Europe is much more complete than the archaeological record from Africa. Remember, there are over 1000 archaeological sites in France alone, but just 100 sites in all of Africa.

Others, like Klein, interpret the same evidence differently. He argues that for the most part the MSA *is* like the Mousterian. There is evidence of a little long-distance trade, as well as a few bone points and rare examples of shelters and art. Most MSA sites lack these features, suggesting, he argues, that the spectacular exceptions, such as the bone points and the ostrich shell beads, are likely to be intrusions from more modern sites. If they really were part of the typical behaviors of MSA peoples, he argues, they would be more common, appearing at numerous MSA sites instead of only a couple.

How Modern Human Behavior Evolved

If there was a human revolution, it may have been caused by a shift in cognitive ability that did not lead to any detectable changes in the skeleton.

Remember that, in the fossil record, the first anatomically modern humans appeared about 100 kya. This means that if there was a human revolution, it occurred long after the first appearance of people who *look* fully modern. However, we need to remember that morphology and behavior can be decoupled. People who look fully modern could have evolved new cognitive abilities that were not reflected in their skeletal anatomy. For example, Richard Klein has suggested that the human revolution may have been caused by a mutation that allowed fully modern speech. It could be that this linguistic ability evolved late in the human lineage and gave rise to the technological sophistication and symbolic behavior of the Upper Paleolithic peoples.

One problem with this explanation is that it requires very rapid genetic change. Spoken language relies on a number of quite complicated features, such as rules of grammar and specialized production and processing of sound. There is considerable evidence that the brain is specially adapted for language. The cognitive and morphological adaptations underlying human speech are likely to be every bit as complicated as specialized adaptations such as the eye. Remember from Chapter 1 that most biologists think complex, beneficial adaptations are unlikely to arise by chance. Instead, complex adaptations usually require the accumulation of many small changes. It is not so easy to see how such adaptations could have arisen in just a few thousand years.

Another problem is that it is difficult to reconcile the idea that modern humans had unique, innate abilities that gave rise to modern behavior, with recent work suggesting that Neanderthals adopted some of the technological innovations of modern humans after they came into contact with them (Box 14.3).

If there was a human revolution, it may have been caused by a technological advance that gave rise to larger human groups.

Alternatively, the human revolution may have resulted from cultural, not genetic, changes. We know that something like this happened later in human history. About 10 kya an equally profound transformation was associated with the adoption of agriculture. Agriculture led to sedentary villages, social inequality, large-scale societies, monumental architecture, writing, and many other innovations that we see in the archaeological record. Harvard archaeologist Ofer-Bar Yosef has suggested that the rapid transition we see in the archaeological record in Europe about 40 kya could have been caused by a similar kind of technological innovation that allowed for much more efficient food acquisition, which in turn led to a greater economic surplus, more economic specialization, and greater symbolic and ritual activity.

There are two related problems with this explanation. First, there is no obvious technological change, analogous to plant domestication, that could cause such a shift in a subsistence economy. Second, morphologically modern people appeared at least 100 kya. If these people were also cognitively modern, why did it take so long for the human revolution to occur? It is possible that environmental conditions were not conducive to innovation. It seems likely that the stabilization of the climate at the end of the Pleistocene gave rise to agriculture. Perhaps an analogous change created conditions that facilitated the human revolution 50 kya. The problem with this idea is that there is no obvious candidate for such a factor.

Modern human behavior could have evolved gradually in Africa through a combination of genetic and cultural changes.

The gradual evolution of human behavior in Africa is more easily reconciled with the usual picture of adaptive evolution, which is based on the gradual accumulation of small changes. Such a process requires no special macromutations or unlikely

BOX 14.3

Arcy-sur-Cure and the Causes of Modern Human Behavior

The Grotte du Renne, a site at the French town of Arcy-sur-Cure, provides important clues about whether the human revolution resulted from biological or cultural change in human populations. When the site was excavated between 1947 and 1963, archaeologists uncovered many layers that showed evidence of human activity. The lowest, and therefore oldest, levels yielded many Mousterian tools, and the highest and most recent levels contained Upper Paleolithic artifacts. In between, archaeologists found Châtelperronian tools and hominin fossils, including a small part of a cranium. More significantly, they also unearthed a large number of worked bone tools and 36 personal ornaments, artifacts typically associated with Upper Paleolithic peoples (Figure 14.26). However, it was not clear whether the hominins should be classified as Neanderthal or as anatomically modern humans. Thus it was not clear what kind of hominin made the ornaments and bone tools.

Recently, two clever bits of detective work have shown that the tools and ornaments at the Grotte du Renne were made by Neanderthals. The first piece of evidence was provided by Fred Spoor and his colleagues at the University of Liverpool. They used **computed tomography** (CT scan) to show that the shape of the semicircular canals of the inner ear differed between Neanderthals and modern humans.

Using this diagnostic feature, they were able to establish that the Grotte du Renne cranium was from a young Neanderthal child. This means that bone tools and ornaments were used at a site *occupied* by Neanderthals. It was still possible, however, that the tools and ornaments were manufactured by anatomically modern humans and subsequently acquired by Neanderthals, perhaps by trade or by scrounging.

This possibility was effectively eliminated by the evidence provided by a French team that completed a detailed reanalysis of the Grotte du Renne record. They discovered that the tools and ornaments were manufactured at the site while Neanderthals were living there. This conclusion is based partly on one piece of bone that was clearly the product of an unsuccessful effort to make an awl (a sharp tool used to punch holes in leather), the most common bone tool found at the site. There were also several ornaments made from the ends of swan bones, along with the matching piece of bone from which the ornament was cut. Taken together, these data show that the tools were made by the people who lived at the site, and these people were Neanderthals.

Thus we are reasonably certain that both Neanderthals and anatomically modern humans carved ornaments and tools out of bone. This conclusion affects our thinking

FIGURE 14.26 These personal ornaments found at Arcy-sur-Cure were likely manufactured by Neanderthals living at the site about 33 kya. The fact that Neanderthals could learn to make and use personal ornaments suggests that the absence of ornaments at most Mousterian sites is not due to some biologically transmitted, cognitive deficit among Neanderthals.

about the causes of the human revolution. The fact that the Neanderthals who lived at Arcy-sur-Cure made and used both bone artifacts and personal adornments is not consistent with the idea that modern behavior is caused by a cognitive innovation that Neanderthals lacked. It seems unlikely that Neanderthals would have been able to borrow ideas from their neighbors if they themselves lacked symbolic abilities. On the other hand, it is plausible that they might have adopted more sophisticated technology from neighboring peoples. The ethnographic record tells us that technological accomplishments do not provide a good measure of social or cognitive complexity. There are many cases in the historical record of technologically more advanced peoples coming in contact with technologically less advanced peoples. In such cases, it is common for the technologically less advanced to adopt ideas and techniques from the more advanced without completely abandoning their own way of life.

chance events. As we explained in Chapters 7 and 8, behavior is subject to the same kinds of evolutionary forces that shape morphology and physiology.

Klein, a strong advocate of the human revolution, has argued that the difficulty with this gradualist account is that it provides no ready explanation of why modern humans waited until 40 kya to leave Africa. If, as we have seen, most of the package of modern behavior was already assembled 100 kya, why didn't modern humans move north and east then, when the climate was relatively warm and moist? Instead, they waited until Europe was cold and dry and they had to cross something like the modern Sahara to get there.

FURTHER READING

Bahn, P. 1998. Neanderthals emancipated. *Nature* 394: 719–721.

Carrol, S. 2003. Genetics and the making of *Homo sapiens*. *Nature* 422: 849–857.

Jobling, M. A., M. E. Hurles, and C. Tyler-Smith. 2004. *Human Evolutionary Genetics*. Garland, New York.

Klein, R. 1999. *The Human Career* (2nd ed.). University of Chicago Press, Chicago.

McBrearty, S., and A. Brooks. 2000. The revolution that wasn't: A new interpretation of modern human behavior. *Journal of Human Evolution* 39: 453–563.

Stringer, C. 1997. *African Exodus: The Origins of Modern Humanity*. Holt, New York.

STUDY QUESTIONS

1. What derived anatomical features distinguish modern humans from other hominins?
2. Describe what the archaeological record tells us about the pattern of human behavior 100 kya and 30 kya. What facts are widely accepted? Which are in dispute?
3. Describe the main differences between the tools of Upper Paleolithic peoples and those of their predecessors.

4. What evidence suggests that Upper Paleolithic peoples were better able to cope with their environments?

5. Explain why it is possible to make phylogenetic trees for mitochondrial DNA and the Y chromosome, but not for other chromosomes.

6. Explain why the mitochondrial DNA and Y chromosome trees are consistent with the hypothesis that modern humans evolved in Africa and then later spread across the rest of the globe.

7. Suppose that you were able to choose three more fossils to extract DNA from and that your goal was to test the hypothesis that humans evolved in Africa. Which three would you pick? Explain why.

8. Describe the two extreme hypotheses that have been put forward to explain the evolution of modern human behavior. Describe and evaluate the evidence and arguments for and against both.

9. How do we know that Neanderthals made the personal ornaments found at Arcy-sur-Cure? How does this fact affect different hypotheses about the evolution of modern human behavior?

KEY TERMS

Upper Paleolithic	Middle Stone Age/MSA
blades	Later Stone Age/LSA
Mode 4	Mode 5
mitochrondria	synonymous substitutions
mitochrondrial DNA/mtDNA	nonsynonymous substitutions
Châttelperronian	

CONTENT FROM *HUMAN EVOLUTION: A MULTI-MEDIA GUIDE TO THE FOSSIL RECORD*

Modern Homo sapiens
  Overview
  Transition to Modernity
  Genetic Evidence
  Upper Paleolithic Cultures

EVOLUTION AND MODERN HUMANS

HUMAN GENETIC DIVERSITY

EXPLAINING HUMAN VARIATION

Human beings vary in myriad ways. In any sizable group of people, there is variation in height, weight, hair color, eye color, food preferences, hobbies, musical tastes, skills, interests, and so on. Some people you know are tall enough to dunk a basketball, some have to roll up the hems of all their pants; some have blue eyes and freckle in the sun, others have dark eyes and can get a terrific tan; some people have perfect pitch, others can't tell a flat from a sharp. Your friends may include heavy drinkers and teetotalers, great cooks and people who can't microwave popcorn, skilled gardeners and some who can't keep a geranium alive, some who play classical music and others who prefer heavy metal.

If we look around the world, we encounter an even wider range of variation. Some of the variation is easy to observe. Language, fashions, customs, religion, technology, architecture, and other aspects of behavior differ among societies. People in different parts of the world also look very different. For example, most of the people in northern Europe have blond hair and pale skin, and most of the people in southern Asia have dark hair and dark skin. As we described in Chapter 13, Arctic peoples are generally shorter and stockier than people who live in the savannas of East Africa. Groups also differ in ways that cannot be detected so readily. For instance, the peoples of the world vary in blood type and the incidence of many genetically transmitted diseases. Figure 15.1 shows that the distributions of three debilitating diseases—PKU (phenylketonuria, introduced in Chapter 3), cystic fibrosis, and Tay-Sachs disease—vary considerably among populations. Tay-Sachs, for example, is nearly 10 times more common among Ashkenazi Jews in New York than among other New Yorkers.

In this chapter we will consider how much of this variation is due to genetic differences among people. We want to know how people vary genetically within and

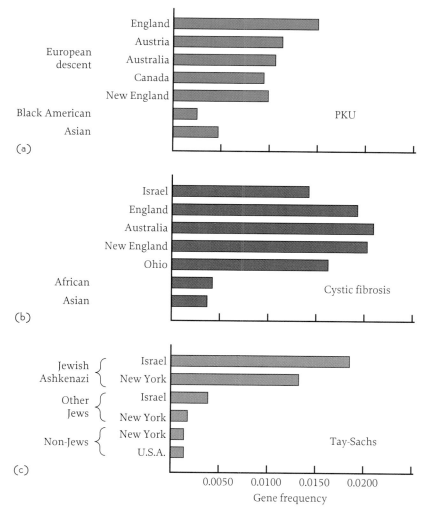

FIGURE 15.1 The distributions of three genetic diseases—(a) PKU, (b) cystic fibrosis, and (c) Tay-Sachs disease—illustrate the existence of variation among human groups.

among societies, and to understand the processes that create and sustain this variation. We'll begin by describing the nature of variation in traits that are influenced by single genes with large effects. Next we'll consider variation in traits that are influenced by many genes. As you will see, the methods that are used to assess variation in traits caused by single genes and multiple genes are quite different. In both cases, we will consider the processes that give rise to variation within and among populations. Finally, we will use our understanding of human genetic diversity to explore the significance and meaning of a concept that plays an important, albeit often negative, role in modern society: race. We will argue that a clear understanding of the nature and source of human genetic variation demonstrates that race is not a valid scientific construct.

Scientists distinguish two sources of human variation: genetic and environmental.

Scientists conventionally divide the causes of human variation into two categories: **Genetic variation** refers to differences between individuals that are caused by the genes they inherited from their parents. **Environmental variation** refers to differences between individuals caused by environmental factors (such as climate, habitat, and competing species) on the organisms' phenotypes. For humans, culture is an important source of environmental variation.

A practical example—variation in body weight—will clarify this distinction. Many environmental factors affect body weight. Some factors, such as the availability of food, have an obvious and direct impact on body weight. The majority of people living under siege in Sarajevo in the mid-1990s were undoubtedly leaner than they were a decade earlier when Sarajevo was a rich, cosmopolitan city. Other environmental effects are more subtle. For example, culture can affect body weight because it influences our ideas about what constitutes an appropriate diet and shapes our standards of physical beauty. In the United States, many young women adopt strict diets and rigorous exercise regimens in order to maintain a slim figure because thinness is considered desirable. But in a number of West African societies, young women are secluded and force-fed large meals several times a day for the express purpose of gaining weight and becoming fat. In these societies, obesity is extremely desirable, and fat women are thought to be very beautiful. Body weight also appears to have an important genetic component. Recent research has shown that individuals with some genotypes are predisposed to be heavier than others, even when diet and levels of activity are controlled.

Genetic and environmental causes of variation may also interact in complicated ways. Consider, for example, two people who have inherited quite different genes affecting body weight. One is easily sated while the other craves food constantly. Both individuals may be thin if they have to subsist on one cup of porridge a day, but only the one who is unconcerned about being thin will gain weight when Big Macs and fries are readily available.

It is difficult to determine the relative importance of genetic and environmental influences for particular phenotypic traits.

It is often difficult to separate the genetic and environmental causes of human variation in real situations. The problem is that both genetic transmission and shared environments cause parents and offspring to be similar. For example, suppose we were to measure the weights of parents and offspring in a series of families living in a range of environments. It is likely that the weight of parents and offspring (corrected for age) would be closely related. However, we would not know whether the association was an effect of genes or environment. Children might resemble their parents because they inherited genes that affect fat metabolism or because they learned eating habits and acquired food preferences from their parents.

Quite different processes create and maintain genetic and environmental variation among groups, and identifying the source of human differences will help us understand why people are the way they are. Genetic variation is governed by the processes of organic evolution: mutation, drift, recombination, and selection. Biologists and anthropologists know a great deal about how the various processes work to shape the living world and how evolutionary processes explain genetic differences among contemporary humans in particular cases.

It is important to distinguish variation within human groups from variation among human groups.

Variation within groups refers to differences between individuals within a given group of people. In the National Basketball Association (NBA), for example, 1.65-m (5-ft 5-in.) Earl Boykins of the Denver Nuggets competes against much larger players, like 2.3-m (7-ft 6-in.) Yao Ming of the Houston Rockets (Figure 15.2). **Variation among groups** refers to differences between entire groups of people. For instance, the average height of NBA players [exemplified in Figure 15.3 by the late, great 2.15-m (7-ft 1-in.) Wilt Chamberlain] is much greater than the average height of professional jockeys [exemplified in the figure by the equally late and equally great 1.5-m (4-ft 11-in.) Willy Shoemaker]. It is important to distinguish these two levels of variation because, as we will see, the causes of the variation within groups can be very different from the causes of variation among groups.

FIGURE 15.2 Variation in stature within the NBA is illustrated by the difference in height between Yao Ming (2.3 m, or 7 ft 6 in.) and Earl Boykins (1.65 m, or 5 ft 5 in.).

FIGURE 15.3 Variation in stature between professional basketball players and professional jockeys is illustrated here by the difference in height between Willy Shoemaker (1.5 m, or 4 ft 11 in.) and Wilt Chamberlain (2.15 m, or 7 ft 1 in.).

VARIATION IN TRAITS INFLUENCED BY SINGLE GENES

By establishing the connection between particular DNA sequences and specific traits, scientists have shown that variation in some traits is genetic.

Although it is sometimes difficult to establish the source of variation in human traits, in some cases we can be certain that variation arises from genetic differences between individuals. For example, recall from Chapter 2 that in West Africa many people suffer from sickle-cell anemia, a disease that causes their red blood cells to have a sickle shape instead of the more typical rounded shape. We know that people with this debilitating disease are homozygous for a gene that codes for one variant of hemoglobin, the protein that transports oxygen molecules in red blood cells. Hemoglobin is made up of two different protein subunits, labeled α (the Greek letter alpha) and β (the Greek letter beta). The DNA sequence of the most common hemoglobin allele, hemoglobin A, specifies the amino acid glutamic acid in the sixth position of the protein chain of the β subunit. But there is another hemoglobin allele, hemoglobin S, which specifies the amino acid valine at this position. People who suffer from sickle-cell anemia are homozygous for the hemoglobin S allele.

We can prove that traits are controlled by genes at a single genetic locus by showing that their patterns of inheritance conform to Mendel's principles.

Although it is often difficult to distinguish between genetic and environmental sources of variation, we can sometimes make this distinction when traits are affected by genes at a single genetic locus. In such cases, Mendel's laws make very detailed predictions about the patterns of inheritance (see Chapter 2). If scientists suspect that a trait is controlled by genes at a single genetic locus, they can test this idea by collecting data on the occurrence of the trait in families. If the pattern of inheritance shows a close fit to the pattern predicted by Mendel's principles (conventionally called "laws"), then we can be confident that the trait is affected by a single genetic locus.

Research on the genetic basis of a language disorder called **specific language impairment** (**SLI**) illustrates the strengths and weaknesses of this approach. Children with SLI have difficulty learning to speak, and in some cases they have small vocabularies and make frequent grammatical errors as adults. SLI is known to run in families, but the genetic basis of the condition is unclear in most cases.

The pattern of inheritance of SLI in one family suggests that at least some cases of SLI are caused by a dominant allele at a single genetic locus. A group of researchers at the Wellcome Trust Centre for Human Genetics in Oxford, England, has studied the expression of SLI in three generations of one family (known as the "KE family"; Figure 15.4). The members of this family who suffer from SLI have severe problems learning grammatical rules, and they also have difficulty with fine motor control of the tongue and jaws. The grandmother (shown as a blue circle) had SLI, but her husband (a yellow triangle) did not. Four of her five children and 11 of her 24 grandchildren also had SLI. Suppose that SLI is caused by a dominant gene. Then because SLI is rare in the population as a whole, the Hardy–Weinberg equations tell us that almost all SLI sufferers will be heterozygotes. Of course, anyone not having SLI must be a homozygote for the normal allele at this locus. From Mendel's laws, on average half of the offspring of a mating between a person with SLI and one without will have SLI and half will have normal linguistic skills. The KE family fits this prediction very well. Both

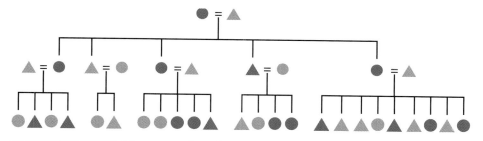

FIGURE 15.4 The pattern of specific language impairment (SLI) in the KE family tree suggests that some cases of SLI are caused by a single dominant gene. Circles represent women, triangles represent men, and blue symbols represent people with SLI. If SLI is caused by a dominant gene, then because SLI is rare in the population as a whole, we know from the Hardy–Weinberg equations that almost all people with SLI will be heterozygotes. Thus, Mendel's principles tell us that, on average, half of the offspring of a mating between a person with SLI and a person without it will have SLI, and half of the offspring will have normal linguistic skills. Notice how well the family shown in this tree fits this prediction.

of the children of the son without SLI are normal, and the rest of the matings produced approximately equal numbers of normal and language-impaired children.

Although the pattern in the KE family is consistent with the idea that SLI is caused by a single dominant gene, it is possible that an environmental factor causes SLI to run in families and that the observed pattern arose by chance. Scientists search for two kinds of data to clinch the case. First, they collect data on more families. The larger the number of families that fit the pattern associated with the inheritance of a single-locus dominant gene, the more confident researchers can be that this pattern did not occur by chance. Second, researchers search for genetic markers (genes whose location in the genome is known) that show the same pattern of inheritance. Thus, if every individual who has SLI also has a specific marker on a particular chromosome, we can be confident that the gene that causes SLI lies close to that genetic marker.

In 1998, the Wellcome Trust researchers demonstrated that SLI in the KE family is closely linked to a genetic marker on chromosome 7, and so it seems likely that SLI in this family is controlled by a gene closely linked to this marker. Even more remarkably, the subsequent discovery of an unrelated person with the same symptoms allowed the researchers to identify the specific gene that causes the disorder in the KE family. The same allele of this gene, named *FOXP2*, was found in all affected members of the KE family and not in 364 unrelated people without SLI. Interestingly, molecular evidence suggests that the *FOXP2* gene has undergone strong directional selection since the divergence of humans and chimpanzees.

The fact that SLI is caused by a single gene does not mean that the gene is responsible for all of the psychological machinery in the human brain that gives rise to language. It means only that damage to the *FOXP2* gene prevents the normal development of some of the psychological machinery necessary for language. If you cut the wire connecting the hard disk to the power supply in your computer, the hard disk will stop working, but that does not mean the wire contains all of the machinery necessary for operation of the hard disk. This argument is supported by the fact that the *FOXP2* gene codes for a transcription factor belonging to a family of genes that play an important role in regulating gene expression during development. *FOXP2* itself is strongly expressed in the brains of developing fetuses. By the same reasoning, SLI in other families may be caused by other genes whose expression is necessary for normal brain development. Just as there are many ways to wreck your hard disk, there may be many mutants at many loci that damage the parts of the brain necessary for language.

Causes of Genetic Variation within Groups

Mutation can maintain deleterious genes in populations, but only at a low frequency.

Many diseases are caused by recessive genes. For example, only people who are homozygous for hemoglobin S are afflicted with sickle-cell anemia. Other diseases caused by recessive alleles include PKU, Tay-Sachs disease, and cystic fibrosis. All of these diseases are caused by mutant genes that code for proteins that do not serve their normal function, and all produce severe debilitation and sometimes death. Why haven't such deleterious genes been eliminated by natural selection?

One answer to this question is that natural selection steadily removes such genes, but they are constantly being reintroduced by mutation. Very low rates of mutation can maintain these deleterious genes because they are recessive traits, so most individuals who carry the gene are heterozygous for the deleterious genes and do not suffer the disastrous consequences that homozygotes suffer. The observed frequency of many deleterious recessive genes is about 1 in 1000. According to the Hardy–Weinberg equations, then, the frequency of newborns homozygous for the recessive allele will be $0.001 \times 0.001 = 0.000001$! Thus, only one in a million babies will carry the disease. This means that even if the disease is fatal, selection will remove only two copies of the deleterious gene for every 1 million people born. Because mutation rates for such deleterious genes are estimated to be a few mutations per million gametes produced, mutation will introduce enough new mutants to maintain a constant frequency of the gene. When this is true, we say that there is **selection–mutation balance**.

Selection can maintain variation within populations if heterozygotes have higher fitness than either of the two homozygotes.

Some lethal genes are too common to be the result of selection–mutation balance. In West African populations, for example, the frequency of the hemoglobin S allele is typically about 1 in 10. How can we account for this? The answer in the case of hemoglobin S is that this allele increases the fitness of heterozygotes. It turns out that individuals who carry one copy of the sickling allele, *S*, and one copy of the normal allele, *A*, are partially protected against the most dangerous form of malaria, called **falciparum malaria** (Figure 15.5). As a consequence, where falciparum malaria is prevalent, heterozygous *AS* newborns are about 15% more likely to reach adulthood than *AA* infants.

When heterozygotes have a higher fitness than either homozygote has, natural selection will maintain a **balanced polymorphism**, a steady state in which both alleles persist in the population. To see why balanced polymorphisms exist, first consider what happens when the *S* allele is introduced into a population and is very rare. Suppose that its frequency is 0.001. The frequency of *SS* individuals will be 0.001×0.001, or about 1 in 1 million, and the frequency of *AS* individuals will be $2 \times 0.001 \times 0.999$, or about 2 in 1000. This means that, for every individual who suffers the debilitating effects of sickle-cell anemia, there will be about 2000 heterozygotes who are partially immune to malaria. Thus, when the *S* allele is rare, most *S* alleles will occur in heterozygotes, and the *S* allele will increase in frequency. However, this trend will not lead to the elimination of the *A* allele. To see why, let's consider what happens when the *S* allele is common and the *A* allele is rare. Now almost all of the *A* alleles will occur in *AS* heterozygotes and be partially resistant to malaria, but almost all of the *S* alleles will occur in *SS* homozygotes and suffer debilitating anemia. The *A* allele has higher fitness than the *S* allele when the *S* allele is common (Box 15.1). The balance between these two processes depends on the fitness advantage of the heterozygotes and the disadvantage of the homozygotes. In this case, the equilibrium frequency for the hemoglobin S allele is about 0.1, approximately the frequency actually observed in West Africa.

Scientists suspect that the relatively high frequencies of genes that cause a number of other genetic diseases may also be the result of heterozygote advantage. For example, the gene that causes Tay-Sachs disease has a frequency as high as 0.05 in some eastern European Jewish populations. Children who are homozygous for this gene seem normal for about the first six months of life. Then, over the next few years a gradual deterioration takes place, leading to blindness, convulsions, and finally death, usually by age four. There is some evidence that individuals who are heterozygous for the Tay-Sachs allele are partially resistant to tuberculosis. Jared Diamond of the University of California Los Angeles points out that tuberculosis was much more prevalent in cities than in rural areas of Europe over the last 400 years. Confined to the crowded urban ghettos of eastern Europe, Jews may have benefited more from increased resistance to tuberculosis than did other Europeans, most of whom lived in rural settings.

Variation may exist because environments have recently changed and genes that were previously beneficial have not yet been eliminated.

Some genetic diseases may be common because the symptoms they create have not always been deleterious. One form of diabetes, **non-insulin-dependent diabetes (NIDD)**, may be an example of such a disease. **Insulin** is a protein that controls the uptake of blood sugar by cells. In NIDD sufferers, blood-sugar levels rise above normal levels because the cells of the body do not respond properly to insulin in the blood. High blood-sugar levels cause a number of problems, including heart disease, kidney damage, and impaired vision. NIDD is also known to have a genetic basis. (The other form of diabetes, insulin-dependent diabetes, occurs because the insulin-producing cells in the pancreas have been destroyed by the body's own immune system. It is unlikely that insulin-dependent diabetes was ever adaptive.)

In some contemporary populations the occurrence of NIDD is very high. On the Micronesian island of Nauru, for example, more than 30% of people over 15 years old now have the disease. Such high rates of NIDD are a recent phenomenon, although the genes that cause the disease are not new. The late human geneticist J. V. Neel of the University of Michigan suggested that the genes now leading to NIDD were beneficial in the past because they caused a rapid buildup of fat reserves during periods of plenty—fat reserves that would help people survive periods of famine in harsh environments. Traditionally, life on Nauru was very difficult. The inhabitants subsisted by fishing and farming, and famine was common. NIDD was virtually unknown during this period. However, Nauru was colonized by Britain, Australia, and New Zealand in recent times, and these influxes brought many changes in the residents'

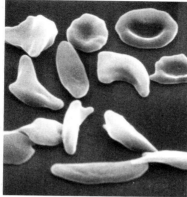

(a)

(b)

FIGURE 15.5 (a) Sufferers of sickle-cell anemia have abnormal red blood cells with a sickle shape. (b) Normal red blood cells are round. Sickle-cell anemia partially protects against falciparum malaria.

BOX 15.1

Calculating Gene Frequencies for a Balanced Polymorphism

It is easy to calculate the frequency of hemoglobin S when selection has reached a stable, balanced polymorphism. Suppose the fitness of AA homozygotes is 1.0, the fitness of AS heterozygotes is 1.15, and the fitness of SS homozygotes is 0, and let p be the equilibrium frequency of the allele S. If individuals mate at random, a fraction p of the S alleles will unite with another S allele to form an SS heterozygote, and a fraction $1 - p$ will unite with an A allele to form an AS heterozygote. Thus the average fitness of the S allele will be

$$0p + 1.15(1 - p)$$

By the same reasoning, the average fitness of the A allele will be

$$1.15p + 1(1 - p)$$

The relationship between the average fitness of each allele and the frequency of hemoglobin S, shown in Figure 15.6, confirms the reasoning given in the text. When S is common, so that p is close to 1, the average fitness of the S allele is close to zero; but when S is rare, its average fitness is almost 1.15. If one gene has a higher fitness than the other, natural selection will increase the frequency of that gene. Thus a steady state will occur when the average fitnesses of the two alleles are equal—that is, when

$$1.15(1 - p) = 1.15p + (1 - p)$$

If you solve for p, you will find that

$$p = \frac{1.15 - 1.10}{1.15 + 1.15 - 1.0}$$

$$= \frac{0.15}{1.30} \approx 0.1$$

which is about the observed frequency of the sickle-cell allele in West Africa.

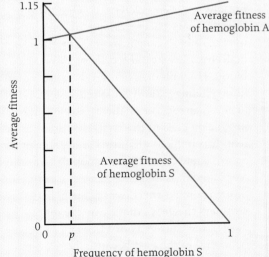

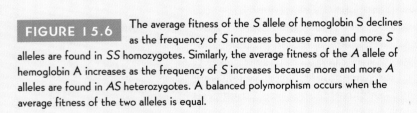

FIGURE 15.6 The average fitness of the S allele of hemoglobin S declines as the frequency of S increases because more and more S alleles are found in SS homozygotes. Similarly, the average fitness of the A allele of hemoglobin A increases as the frequency of S increases because more and more A alleles are found in AS heterozygotes. A balanced polymorphism occurs when the average fitness of the two alleles is equal.

lives. They obtained access to Western food, and prosperity derived from the island's phosphate deposits allowed them to adopt a sedentary lifestyle. NIDD became common. Genes that formerly conferred an advantage on the residents of Nauru now lead to NIDD, the primary cause of nonaccidental death there.

Causes of Genetic Variation among Groups

There are many genetic differences between groups of people living in different parts of the world. The existence of genetic variation among groups is intriguing because we know that all living people are members of a single species, and, as we saw in Chapter 4, gene flow between different populations within a single species tends to make them genetically uniform. In this section we consider several processes that oppose the homogenizing effects of such gene flow and thereby create and maintain genetic variation among human populations.

Selection that favors different genes in different environments creates and maintains variation among groups.

The human species inhabits a wider range of environments than any other mammal. We know that natural selection in different environments may favor different genes, and that natural selection can maintain genetic differences in the face of the homogenizing influence of gene flow if selection is strong enough. Variation in the distribution of hemoglobin genes provides a good example of this process. Hemoglobin S is most common in tropical Africa, around the Mediterranean Sea, and in southern India (Figure 15.7a). Elsewhere it is almost unknown. Generally, hemoglobin S is prevalent where falciparum malaria is common, and hemoglobin A is prevalent where this form of malaria is absent (Figure 15.7b). Southeast Asia represents an exception to this pattern, and it is possible that hemoglobin E, a hemoglobin allele that is common in that region, also provides resistance to malaria.

The gene that controls the digestion of **lactose**, a sugar found in mammalian milk, provides another interesting example of genetic variation maintained by natural selection. Lactose is synthesized in the mammary glands and occurs in large amounts only in mammalian milk. Most mammals have the ability to digest lactose as infants but gradually lose this ability after they are weaned. The vast majority of humans follow the mammalian pattern and cannot digest lactose after the age of five. Such people are said to lack lactase persistence. When people who lack lactase persistence drink more than about half a liter (just over 1 pint) of fresh milk at once, they suffer gastric distress that ranges from mild discomfort to quite severe pain. However, most northern Europeans and members of a number of North African and Arabian populations retain the ability to digest lactose as adults and are said to have lactase persistence. Evidence from family studies indicates that the ability to digest lactose as an adult is controlled by a single dominant gene, labeled **LTC*P**. People who have one copy of LTC*P continue to synthesize lactase as adults, but people who are homozygous for the alternative allele, labeled **LTC*R**, do not. The relationship between genotype and phenotype for the trait of lactase persistence is summarized in Table 15.1. Recent molecular studies indicate that the difference between the two alleles lies in a regulatory region close to the structural gene that codes for lactase.

The prevalence of LTC*P in the deserts of North Africa and Arabia is probably the result of natural selection for the ability to digest large amounts of fresh milk.

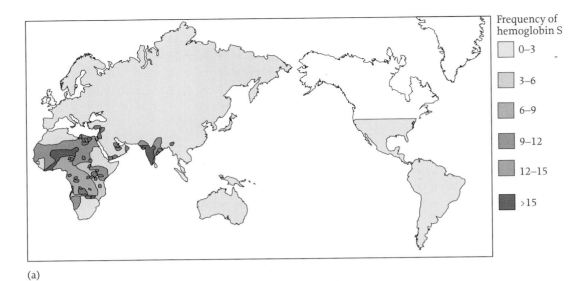

(a)

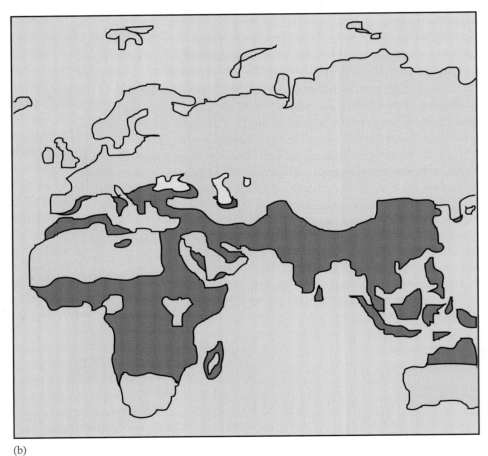

(b)

FIGURE 15.7   Hemoglobin S is common only in areas of the world in which falciparum malaria is prevalent. (a) The colors show the frequency of hemoglobin S throughout the world. (b) The regions of the Old World in which falciparum malaria is prevalent are in red.

TABLE 15.1 The ability to digest lactose is controlled by a single genetic locus with two alleles, *LTC*P* and *LTC*R*.

Genotype	Phenotype
*LTC*P/LTC*P*	Synthesize lactase and are able to digest lactose as adults
*LTC*P/LTC*R*	Synthesize lactase and are able to digest lactose as adults
*LTC*R/LTC*R*	Do not synthesize lactase and are unable to digest lactose as adults

Traditionally, nomadic pastoralism has been the main mode of subsistence in this part of the world, and fresh milk plays a crucial role in this way of life. Gebhard Flatz of the Medizinische Hochschule of Hannover, Germany, studied the Beja, a people who wander with their herds of camels and goats in the desert lands between the Nile and the Red Sea (Figure 15.8). During the nine-month dry season, the Beja rely almost entirely on the milk of their camels and goats. They drink about 3 liters of fresh milk a day, and they obtain virtually all of their energy, protein, and water from milk. Moreover, the milk must be consumed when it is fresh. High desert temperatures prevent the Beja from storing milk, and their nomadic lifestyle makes it difficult to produce dairy products, such as cheese and yogurt, that lack lactose. The Beja probably could not survive without the ability to digest fresh milk. Thus it comes as no surprise that *LTC*P* is common among them; 84% of the Beja are able to digest lactose.

Scientists are less sure about why *LTC*P* is common in northern Europe. Frequencies of *LTC*P* are highest in Scandinavia and Great Britain, decline steadily to the south, and reach lower values around the Mediterranean Sea. Traditionally, large amounts of fresh milk have not been an important part of the diet of the peoples in this region. Low temperatures make milk storage practical, and northern Europeans have long made use of cheese and other processed-milk products. Thus it does not seem likely that the ability to digest fresh milk would provide a big enough advantage to account for the high frequency of *LTC*P*. One possibility is that lactose

FIGURE 15.8 Pastoralists in northern Africa, such as the Beja, herd camels, and during some parts of the year they obtain virtually all of their nourishment from fresh milk. A high proportion of the Beja are able to digest lactose as adults.

digestion is favored in northern Europe because it enhances absorption of vitamin D. When sunlight penetrates the fatty tissue in human skin, vitamin D is synthesized. Consequently, in sunny environments humans rarely suffer vitamin D deficiency. In northern Europe, long, dark, cloudy winters create a potential for vitamin D deficiency. Laboratory studies suggest that the ability to digest lactose increases vitamin D absorption. Thus the ability to digest lactose might be common in northern Europe because it helps prevent vitamin D deficiency.

Many people think of evolution by natural selection as a glacially slow process that acts only over millions of years. The correlation of lactase persistence with dairying, however, suggests that the ability to digest lactose has evolved since people began domesticating livestock in the Middle East and North Africa about 7 kya. This hypothesis is supported by molecular evidence that indicates that the lactase gene has been subjected to recent strong selection. Joel Hirschhorn and his colleagues at Harvard Medical School have identified 101 **single nucleotide polymorphisms (SNPs)** in the lactase gene and surrounding noncoding DNA. SNPs are single nucleotide sites in which the base varies within a population. It turns out that if you identify the base at one site for an individual, then you can predict the base that the same individual will have at the other 100 sites. This is the pattern that we would expect to find if there had been recent strong selection on the lactase gene because selection will tend to increase the frequency of both the beneficial DNA segment and all of the closely linked DNA, even if the latter are not directly beneficial. Over time, though, recombination will tend to destroy these patterns. Thus the pattern of variation in these SNPs suggests that the ability to digest lactose in adulthood evolved relatively recently. We don't need to imagine that selection has huge effects on fitness in order to account for the rapid evolution of this trait. Figure 15.9 shows how fast the frequency of lactase persistence could increase if the ability to digest lactose increased fitness by just 3%. As you can see, even this relatively small benefit could easily explain the high frequencies of persistence that we now find in some parts of the world.

Genetic drift creates variation among isolated populations.

In Chapter 3 we saw that genetic drift causes random changes in gene frequencies. This means that if two populations become isolated from each other, both will change randomly, and over time the two populations will become genetically distinct. Because drift occurs more rapidly in small populations than in large ones, small populations will diverge from one another faster than large ones will. Genetic drift caused by the expansion of a small founding population is sometimes called the **founder effect**.

Genetic differences among members of three religious communities in North America demonstrate how this process can create variation among human groups. Two of these groups are Anabaptist sects—the Old Order Amish (Figure 15.10) and Hutterites—and the third group is the Utah Mormons. Each of these groups forms a well-defined population. About 2000 Mormons first arrived in the area of what is now Salt Lake City in 1847, but members of the church continued to arrive until 1890. Virtually all of the immigrants were of northern European descent. At the turn of the twentieth century, there were about 250,000 people in this area, and about 70% of them belonged to the Mormon church. In contrast, the two Anabaptist groups were much smaller. The founding population of the Old Order Amish was only 200 people, and gene flow from

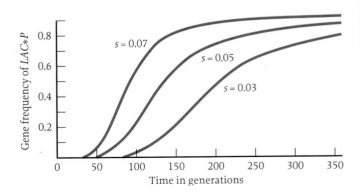

FIGURE 15.9 If the ability to digest lactose as an adult leads to even as little as a 3% increase in fitness (*s*), then it is possible that the *LTC·P* gene has spread in the 7000 years (300 to 350 generations) since the origin of dairying.

outside the group has been very limited. Contemporary Hutterites are all descended from a population of only 443 people and, like the Amish, have been almost closed to immigration.

Researchers studying the genetic composition of each of these populations have found that Mormons are genetically similar to other European populations. Thus, even though Mormon populations have been partly isolated from other European populations for over 150 years, genetic drift has led to very little change. This is just what we would expect, given the size of the Mormon population. In contrast, the two Anabaptist populations are quite distinct from other European populations. Because their founding populations were small and the communities were genetically isolated, drift has created substantial genetic changes in the same period of time.

Genetic drift can also explain why certain genetic diseases are common in some populations

FIGURE 15.10 The Old Order Amish were founded by a group of 200 people. The Amish dress plainly and shun most forms of modern technology, including motor vehicles.

but not in others. For example, Afrikaners in what is now the Republic of South Africa are the descendants of Dutch immigrants who arrived in the seventeenth century. By chance, this small group of early immigrants carried a number of rare genetic diseases, and these genes occurred at much higher frequency among members of the colonizing population than in the Dutch populations from which the immigrants were originally drawn. The Afrikaner population grew very rapidly and preserved these initially high frequencies, causing these genes to occur in higher frequencies among modern Afrikaners than in other populations. For example, sufferers of the genetic disease **porphyria variegata** develop a severe reaction to certain anesthetics. About 30,000 Afrikaners now carry the dominant gene that causes this disease, and every one of these people is descended from a single couple who arrived from Holland in the 1680s.

Current patterns of genetic variation reflect the history of migration and population growth in the human species.

Some of the genetic variation among human groups reflects the history of the peoples of the Earth. In Chapter 14 we explained that the pattern of genetic variation in mitochondrial DNA indicates that the human species underwent a worldwide population expansion about 100,000 years ago. This was only one of several expansions of the world's population. The invention of agriculture led to expansions of farming peoples from the Middle East into Europe, from Southeast Asia into **Oceania** (the Pacific island groups of Polynesia, Melanesia, and Micronesia), and from west-cental Africa to most of the rest of the continent between 4000 and 1000 years ago; the domestication of the horse and associated military innovations led to several expansions of peoples living in the steppes of central Asia between 3000 and 500 years ago; and improvements in ships, navigation, and military organization led to the expansion of European populations during the last 500 years.

L. L. Cavalli-Sforza, a human geneticist at Stanford University, has argued that patterns of genetic variation preserve a record of these expansions. As human populations grow, they often expand into new geographic regions. As the distance grows between parts of the population, they become genetically isolated from one another and begin to accumulate genetic differences. If the population expansion continues, the initial "parent" population may eventually be split into several isolated and genetically differentiated "daughter" populations. If the daughter populations remained completely isolated from all other human populations, they would be just like new species, and we could use the techniques of phylogenetic reconstruction described in Chapter 4 to reconstruct the demographic history of the daughter populations. Such complete isolation is very rare, though. Instead, daughter populations usually experience gene flow with each other and with other human populations they encounter during their expansion, and such gene flow tends to obscure the history of the populations. Nonetheless, Cavalli-Sforza argues, if there is not too much gene flow, the present patterns of genetic variation can help us reconstruct the pattern of past migrations.

Cavalli-Sforza and his colleagues believe that the worldwide pattern of genetic variation results from the expansion of modern humans about 100,000 years ago. Recall from Chapter 14 that supporters of the replacement model believe this expansion began in Africa, and then spread north and east to the rest of Eurasia and Australasia, and finally reached the New World. Cavalli-Sforza and his colleagues have collected data on gene frequencies in populations across the world. Figure 15.11 shows a tree based on the frequencies of 120 genes in 42 populations from every part of the globe. These data are based on what geneticists call "classical markers" because they have been detectable by geneticists for decades. This result was constructed using the genetic-distance methods described in Chapter 4, and it assumes that the rate of genetic change along each branch of the tree has been constant. It is consistent with the view that modern humans originated in Africa and spread to the rest of the world, and that the patterns of genetic variation we see in today's world are the result of that expansion. A simplified version of the tree (Figure 15.12) tells us that African populations first expanded into southern Asia. Then southern Asian populations expanded into northern Asia and Oceania. Finally, northern Asian populations spread into Europe.

The higher resolution allowed by analysis of Y chromosomes and mitochondrial DNA leads to a more complicated picture (Figure 15.13). As with the classical mark-

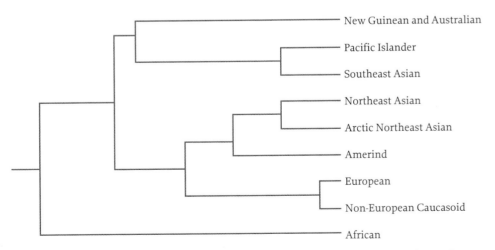

FIGURE 15.11 This tree, based on the frequencies of 120 genes in 42 populations from every part of the globe, is consistent with the hypothesis that humans originated in Africa and spread from there to the rest of the globe.

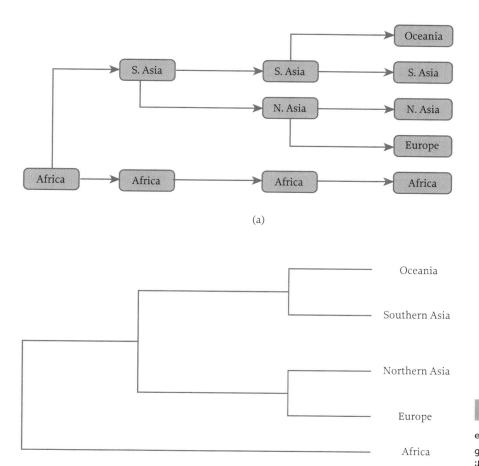

(a)

(b)

FIGURE 15.12 (a) This model of the expansion of early anatomically modern human populations generates the phylogenetic tree shown in (b) if geographically separate populations remain genetically isolated.

ers, the geographic pattern suggests that modern humans are all descended from African populations that moved out of African into southern Asia and subsequently to Europe, northern Asia, and the New World. However, now the results of later population movements can also be seen. Notice for example that Y chromosomes from group C are common in Australia but not elsewhere in Southeast Asia. A more detailed examination of the Y chromosome types common among aboriginal Australians indicates that these C-type chromosomes are most similar to those found in India, and that individuals bearing these chromosomes probably arrived in Australia less than 10,000 years ago. This date coincides with the introduction of the dingo to Australia and with archaeological evidence for new plant-processing techniques. The genetic and archaeological evidence suggests that a new group of foragers arrived in Australia about 10,000 years ago and brought domesticated dogs with them.

VARIATION IN COMPLEX PHENOTYPIC TRAITS

As we saw in Chapter 3, the vast majority of human traits are influenced by many genes, each alone having a relatively small effect. For such traits, it is usually impossible to detect the effects of single genes or to trace the patterns of Mendelian inheri-

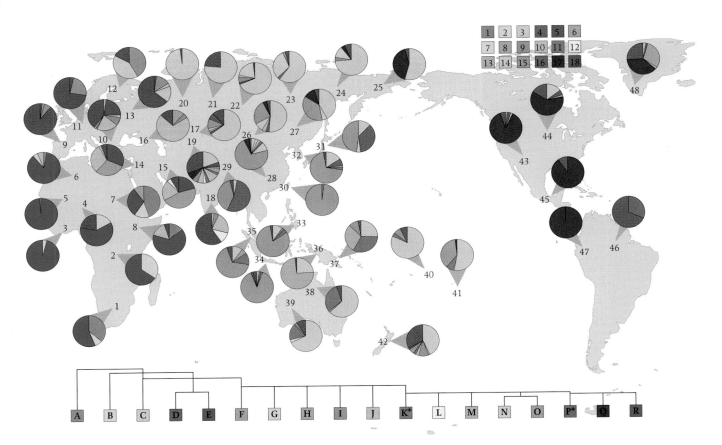

FIGURE 15.13   The geographic pattern of variation in the human Y chromosome is the result of the spread of humans from Africa to Asia and then Australia, the Pacific, and the New World. The tree at the bottom is a simplified version of Figure 14.18, in which the main labeled groups of terminal branches are color-coded. With three exceptions, these groups are all the Y chromosome genotypes that share a particular mutation. The exceptions are groups F*, K*, and P*, which share a mutation with other groups descended from F, K, and P, respectively. The pie graphs indicate the fraction of the Y chromosomes at each site that belong to the various groups labeled on the tree.

tance. Nonetheless, it is still possible to assess the influence of genes on such traits. Geneticists have developed statistical methods that enable them to estimate the relative importance of genetic and environmental components of variation within groups. Because the relative importance of genetic variation and environmental variation will affect the resemblance between parents and offspring, the measure that computes the proportion of variation due to the effects of genes is referred to as the **heritability** of phenotypic traits.

This material will be easier to understand if we have a concrete example in mind. Height is an ideal trait: it is quite easy to measure, it is quite variable within and between populations, it is relatively stable once individuals reach adulthood, and there is a wealth of data on height of individuals in different populations. In any moderately large sample of people there will be a wide range of variation in height. For example, the data in Figure 15.14 are taken from men who joined the British Army in 1939. Some of these young recruits were more than 2 m (around 7 ft) tall; others were less than 1.5 m (5 ft) tall.

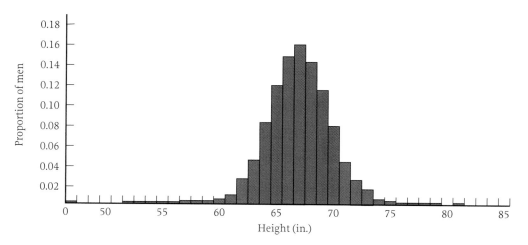

FIGURE 15.14 The heights of men joining the British rmy in 1939 varied considerably, illustrating the range of variation in morphological characters within populations. The tallest men joining the army were more than 2 m (around 84 in., or 7 ft) tall, and others were less than 1.5 m (60 in., or 5 ft) tall.

Genetic Variation within Groups

Under certain conditions, measuring the phenotypic similarities among relatives such as twins allows us to estimate the fraction of the variation within the population that is due to genes.

In Part One we saw that the transmission of genes from parents to offspring causes children and parents to be phenotypically similar. If parents who are taller than average tend to have offspring who are taller than average, and parents who are shorter than average tend to have children who are shorter than average, you might think that height is determined by genes. In contrast, if parents and offspring are no more similar to each other than to other individuals in the population, you might think that genes have little effect on height. The problem with this reasoning is that nongenetic factors may also cause parents and offspring to be similar. It is known that many environmental factors, such as nutritional levels and the prevalence of infectious diseases, also affect height. Human parents directly affect their offspring's environment in many ways: they provide food and shelter, arrange for their children to be inoculated against diseases, and shape their children's beliefs about nutrition. Similarity between the environments of parents and their offspring is called **environmental covariation** and is a serious complication in computing heritability.

Data from studies of twins can be used to separate the effects of genetic transmission from environmental covariation. The technique involves comparing the similarity between monozygotic and dizygotic twins. **Monozygotic** (identical) **twins** begin life when the union of a sperm and an egg produces a single zygote. Then, very early in development, this embryo divides to form two separate, genetically identical individuals. **Dizygotic** (fraternal) **twins** begin life when two different eggs are fertilized by two different sperm to form two independent zygotes. Dizygotic twins share approximately one-half of their genetic material, and they are just like other pairs of full siblings except that they were conceived at the same time. Both monozygotic and dizygotic twins share a womb and experience the same intrauterine environment. After they are born, most twins grow up together in the same family. Thus, if most of the variation in stature has a genetic origin, monozygotic twins are likely to be more similar to one another than dizygotic twins are because monozygotic twins are genetically identical. On the other hand, if most of the variation is due to the environment, and the similarity between parents and offspring is due to having a common family environment, then monozygotic twins will be no more similar to one another than dizygotic twins are. Population genetic theory provides a way to use comparisons of the similarity among dizygotic and monozygotic twins to adjust estimates of heritability for the effects of correlated environments.

Twin studies are useful in trying to estimate the relative magnitude of the effects of genetic variation and environmental variation on phenotypic characters, but the data may be biased in certain ways. For example, twin studies will overestimate heritability if the environments of monozygotic twins are more similar than the environments of dizygotic twins. There are several reasons why this may be the case. In the uterus, some monozygotic twins are more intimately associated than dizygotic twins are. Monozygotic twins are always the same sex. After they are born, monozygotic twins may be treated differently by their parents, family, teachers, and friends than dizygotic twins are treated. It is not uncommon to see monozygotic twins dressed in identical outfits or given rhyming names, and it is inevitable that their physical similarities to one another will be pointed out to them over and over (Figure 15.15).

Studies of monozygotic and dizygotic twins suggest that somewhat more than half of the variation in height in most human populations is due to genetic similarities between parents and their children.

FIGURE 15.15 Identical, or monozygotic, twins are produced when an embryo splits at an early stage and produces two genetically identical individuals.

Genetic Variation among Groups

Stature varies among human populations.

Just as people within groups vary in many ways, groups of people collectively vary in certain characteristics. For example, there is a considerable amount of variation in average height among populations. People from northwestern Europe are tall, averaging about 1.75 m (5 ft 9 in.) in height. People in Italy and other parts of southern Europe are about 12 cm (5 in.) shorter on the average. African populations include very tall peoples like the Nuer and the Maasai, and very short peoples like the !Kung. Within the Western Hemisphere, Native Americans living on the Great Plains of North America and in Patagonia are relatively tall, and peoples in tropical regions of both continents are relatively short.

Some of the variation in body size among human groups appears to be adaptive.

In Chapter 13 we learned that larger body size is favored by natural selection in colder climates. Tall peoples, such as the indigenous people of Patagonia in South America and of the Great Plains in North America, usually live in relatively cold parts of the world, but shorter peoples typically live in warmer areas, like southern Europe or the New World tropics. This pattern suggests that variation in body size among groups may be adaptive, a conjecture that is borne out by data on the relationship between body size and climate from a large number of human groups (Figure 15.16). Thus, at least some fraction of the variation in body size among groups is adaptive.

The fact that variation within populations has a genetic component does not mean that differences between groups are caused solely by genetic differences.

We have seen that a significant proportion of the variation in height and body size within some populations reflects genetic variation. We also know that the variation in height and body shape among populations seems to be adaptive. From these two facts, it might seem logical to conclude that the variation in body size among populations is genetic, and that this variation represents a response to natural selection.

However, this logic is wrong! All of the variation in height within populations could be genetic *and* all of the variation in height among populations could be adaptive, but this would not mean that there is variation among populations in the distribution of genes that influence height. In fact, all of the observed variation in height among groups could be due solely to differences in environmental conditions.

This point is frequently misunderstood, and this misunderstanding leads to serious misconceptions about the nature of genetic variation among human groups. A simple example may help you see why the existence of genetic variation within groups does not imply the existence of genetic variation among groups. Suppose Rob and his neighbor Pete both set out to plant a new lawn. They go to the garden store together, buy a big bag of seed, and divide the contents evenly. Pete, an avid gardener, goes home and plants the seed with great care. He fertilizes, balances the soil acidity, and provides just the right amount of water at just the right times. Rob, who lives next door, scatters the seed in his backyard, waters it infrequently and inadequately, and never even considers fertilizing it. After a few months, the two gardens are very different. Pete's lawn is thick, green, and vigorous, but Rob's lawn hardly justifies the name (Figure 15.17). We know that the difference between the two lawns cannot be due to the genetic characteristics of the grass seed, because Rob and Pete used seed from the same bag. Nonetheless, variation in the height and greenness of the grass within each lawn might be largely genetic because the seeds within each lawn experience very similar environmental conditions. All of the seeds in Pete's lawn get regular water and fertilizer, while all of the seeds in Rob's lawn are neglected to an equal extent. Thus, if the seed company has sold genetically

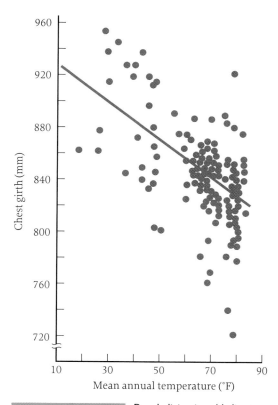

FIGURE 15.16 People living in cold climates have larger bodies than those who live in warm climates. The vertical axis plots mean chest girth for numerous human groups, and the horizontal axis plots the mean yearly temperature in the regions in which each group lives. Because chest girth is a measure of overall size, these data show that people living in colder climates have larger bodies.

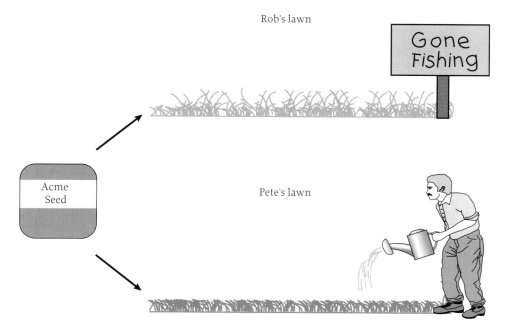

FIGURE 15.17 The differences between two separate lawns planted from the same bag of seed must be environmental. However, if the seed used was genetically variable, the differences within each lawn could be genetic.

variable seed, differences between individual plants within each lawn could be due mainly to genetic differences between the seeds themselves.

Exactly the same argument applies to variation in human stature. The fact that much of the variation in stature among Americans has a genetic component does not mean that stature is determined entirely by genes. It means there is genetic variation that affects stature, and these effects are relatively large in comparison with the effects of environmental differences among Americans. It does not follow that the differences in stature between Americans and other peoples are the result of genetic differences between them. For example, although Americans are taller on average than citizens of Japan, these differences in height are not necessarily genetic. That would be true only if two quite different conditions held: First, there would have to be a difference between Americans and Japanese in the distribution of genes affecting stature. Second, this genetic difference would have to be large compared with the differences in culture and environment between the two groups. The fact that there is genetic variation among Americans does not tell us whether or not Americans are genetically different from Japanese. The relatively small effect of environmental and cultural variation on height among Americans tells us nothing about the average difference in environment between Americans and Japanese.

The increase in stature that coincided with modernization is evidence for the influence of environmental variation on stature.

For height, there is good reason to believe that variation among groups is at least partly environmental. There has been a striking effect of modernization on the average heights of many peoples. For example, Figure 15.18 plots the heights of several groups of English boys between the ages of 5 and 21 in the nineteenth and twentieth centuries. In 1833, 19-year-old factory workers averaged about 160 cm (about 5 ft 3 in.). In 1874, laborers of the same age averaged about 167 cm (5 ft 6 in.). In 1958 the

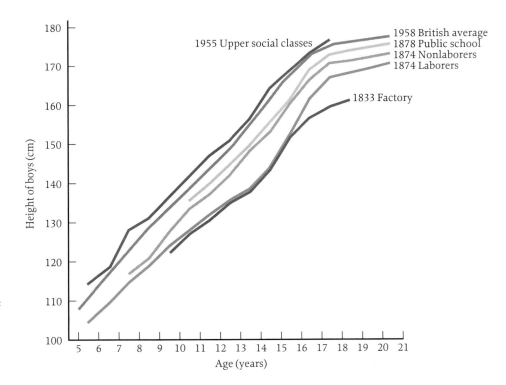

FIGURE 15.18 Height increases with time in English populations, but at any given time richer people are taller. Note that English "public schools" are the equivalent of American private schools.

| TABLE 15.2 | Japanese men who immigrated to Hawaii during the first part of the twenti-eth century were shorter than their children who had been born and raised in Hawaii. The immigrants were similar in height to the Japanese who remained in Japan, which indicates that the immigrants were a representative sample of the Japanese populations from which they came. The fact that the children of the immigrants were taller than their parents shows that environmental factors play an important role in creating variation in stature. (Data from Table 12.3 in G. A. Harrison, J. M. Tanner, D. R. Pilbeam, and P. T. Baker, 1988, Human Biology: An Introduction to Human Evolution, Variation, Growth, and Adaptability, 3rd ed., Oxford Science Publications, Oxford, England.) |

	Average Height (cm)	Sample Size
Japanese immigrants to Hawaii	158.7	171
People from same regions of Japan who remained in Japan	158.4	178
The immigrants' children born in Hawaii	162.8	188

average British 19-year-old stood about 177 cm (5 ft 9 in.). Similar increases in height over the last hundred years can be seen among Swedish, German, Polish, and North American children. These changes have occurred very rapidly, probably too fast to be due solely to natural selection.

There also have been substantial changes in height among immigrants to the United States in the course of a few generations. During the first part of the twentieth century, when Japan had only begun to modernize, many Japanese came to Hawaii to work as laborers on sugar plantations. The immigrants were considerably shorter than their descendants who had been born in Hawaii (Table 15.2). This change in height among immigrants and their children was so rapid that it cannot be the result of genetic change. Instead, it must be due to some environmental difference between Japan and Hawaii in the early twentieth century. The underlying cause of this kind of environmental effect is not completely understood. In 1870s England, poverty was involved to some extent, because relatively wealthy public school boys were taller than less affluent nonlaborers, and nonlaborers were taller than poorer laborers. Observations like this have led some anthropologists to hypothesize that increases in the standard of living associated with modernization improve early childhood nutrition and increase children's growth rates. However, this cannot be the full explanation. Notice that even the richest people in England 120 years ago were shorter than the average person in England is today. Since it seems unlikely that wealthy Britons were malnourished in the 1870s, other factors must have contributed to the increase in stature during the last 120 years. Some authorities think that the control of childhood diseases may have played an important role in these changes.

THE RACE CONCEPT

Race plays an important role in modern life, but the popular conception of race is flawed.

Race is part of everyday life. For better or worse, our race affects how we see the world and how the world sees us; it affects our social relationships, our choice of marriage partners, our educational opportunities, and our employment prospects. We may decry discrimination, but we cannot deny that race plays a major role in many aspects of our lives.

Like any widely used word, *race* means different things to different people. However, the understanding of race held by many North Americans is based on three fundamentally flawed propositions:

1. *The human species can be naturally divided into a small number of distinct races.* According to this view, almost every person is a member of exactly one race; the only exceptions are the offspring of the members of different races. For example, many people in the United States think that people belong to one of three races: descendants of people from Europe, North Africa, and western Asia; descendants of people from sub-Saharan Africa; and descendants of people from eastern Asia.

2. *Members of different races are different in important ways, so knowing a person's race gives you important information about what he or she is like.* For example, biomedical researchers sometimes suggest that race predicts susceptibility to diseases like high blood pressure, heart disease, and infant mortality. Less benignly, some people believe that knowing a person's race reveals something about his or her intelligence or character.

3. *The differences between races are due to biological heritage.* Members of each race are genetically similar to each other, and genetically different from members of other races. Most Americans view African Americans and European Americans as members of different races because they are marked by genetically transmitted characters like skin color. In contrast, the Serbs and Croats of the former Yugoslavia are seen by most Americans as ethnic groups, rather than racial groups, because the differences between them are cultural rather than genetic.

Although many people believe these propositions to be true, they are not consistent with scientific knowledge about human variation. There are genetic differences between groups of people living in different parts of the world, but as we will see, these differences do not mean that the human species can be meaningfully divided into a set of nonoverlapping categories called races. *The common view of race is bad biology.*

Because people vary, it is possible to create classification schemes in which similar peoples are grouped together.

People tend to be more genetically and phenotypically similar to people who live near them than to people who live farther away. As we have seen already in this chapter, this is true for many genes. For example, the sickle-cell gene is common in central Africa and in India but is rare elsewhere; the *LTC*P* gene, which allows adults to digest lactose, predominates in northern Europe and North Africa but is uncommon in the rest of the world. In general, overall genetic similarity is strongly associated with geographic proximity (Figure 15.19). Morphological similarities also link neighboring peoples. For example, most of the people who live near the equator have dark skin (Figure 15.20), and most of the people who live at high latitudes are stocky.

One consequence of these similarities between people is that it is possible to group people into geographically based categories on the basis of their genetic or phenotypic similarities. For example, a classification scheme based on the ability to digest lactose would place northern Europeans and North Africans in one group and the rest of the world in a second group. On the other hand, the overall genetic similarity shown in Figure 15.11 could be used to create a classification system that puts Africans in one

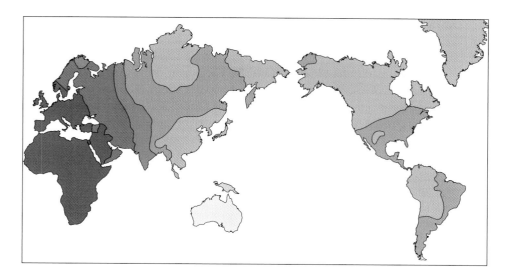

FIGURE 15.19 This contour map of overall genetic similarity is based on a sample of 120 genes from 42 populations assembled by Cavalli-Sforza and his colleagues. The fact that the contours of equal genetic similarity are roughly evenly spaced indicates that there is a smooth east-to-west gradient of overall genetic similarity; there are no sharp boundaries between groups. Sharp boundaries would produce a map in which many contour lines were positioned close together. This map is drawn from the same data used to construct the tree shown in Figure 15.11.

group; Europeans, northern Asians, and Native Americans in a second group; and southern Asians and people from Oceania in a third group.

However, such classification schemes do not support two properties required by the common concept of race. First, classification schemes based on different characters lead to radically different groupings, and the placement of individuals within any single category is often arbitrary. Second, classification schemes are not very informative. The average difference between groups of people living in different parts of the world is much smaller than the differences among individuals within each group.

There is no single natural classification of the human species.

In Chapter 4 we argued that species are distinctive entities that can be unambiguously identified in nature. Racial classifications for humans are quite different, and there is no natural classification scheme for categorizing us. To see what is meant by a natural classification scheme, consider the following analogy. Suppose you are a clerk

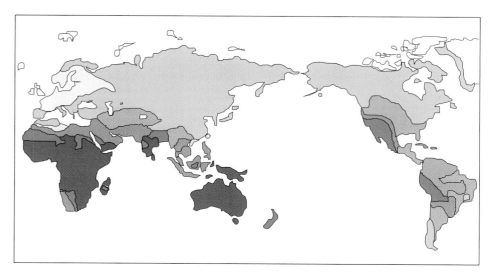

FIGURE 15.20 This map shows contours in skin color. Notice that there are smooth gradients away from the equator.

in a hardware store. Your boss assigns you to classify the contents of two large cabinets. The first cabinet contains power drills made by a number of different manufacturers; there are many drills but only one model per manufacturer. The second cabinet holds various different kinds of screws. The power drills vary in many ways: they have different colors, shapes, weights, and power ratings. The screws also vary: they have different lengths, diameters, pitches, and heads. You will have little trouble sorting the drills into piles according to manufacturer because all the drills made by a single manufacturer are similar in all of their dimensions: they are the same color, shape, and weight and have the same power rating. Moreover, each model is distinctly different; there are no intermediate types. This is a natural classification system. You will have a much harder time classifying the screws. Using length will produce one set of piles; diameter, a second set of piles; and screw pitch, a third set of piles. Moreover, even a classification based on a single characteristic like length will require arbitrary distinctions: Should there be three or four piles? Should the 1.5-inch screws be put in the pile with the smallest screws, or with the next bigger sizes? There is no natural way to classify the screws.

Racial classification schemes based on different sets of characters don't result in the same groupings for all characters. For example, a classification scheme based on the ability to digest lactose would yield very different groupings from one based on resistance to malaria. A classification based on skin color would produce a different grouping from one based on height. This problem cannot be solved by evaluating many different characters at the same time. As Figure 15.21 demonstrates, trees based on a large number of genes and trees based on a large number of morphological characters organize local groups in very different ways. The tree based on genetic traits groups aboriginal peoples from Australia and New Guinea together with people from Southeast Asia, but the one based on morphological features groups these aboriginal peoples with African Pygmies and the !Kung. Also notice that the genetic tree says that people from northern China are more similar to Europeans than they are to Southeast Asians.

This means that folk classification schemes based on skin color and facial morphology are not reliable predictors of overall genetic similarity. In Brazil, for example,

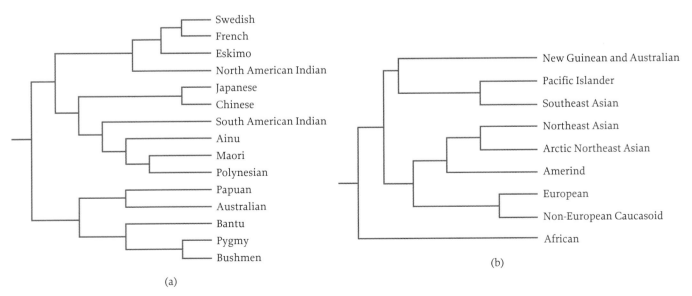

(a)

(b)

FIGURE 15.21 Evolutionary trees for human populations based on morphological similarity (a) look very different from trees based on genetic similarity (b).

people are classified according to what is called *cor* in Portugese. Although the literal translaton of *cor* is "color," the Brazilians' classification is based on more than skin color; it includes the morphology of the lips and eyes, and hair type. *Cor* plays a role in Brazilian society analogous to race in North America: there is substantial prejudice against those classified as "black," and people classified as black earn less money than other Brazilians on average. If the Brazilian folk classification is biologically meaningful, *cor* should be a good predictor of ancestry: "whites" should be of mainly European ancestry, and "blacks" of mainly African ancestry.

To test this idea, Flavia Parra and his colleagues at the Universidade Federal de Minas Gerais in Brazil assembled samples from three populations. They took blood samples from people on São Tomé, an island near the coast of Africa. This is near the area where the ancestors of most African Brazilians lived before being captured and transported to Brazil as slaves. Parra's group also took blood samples from people in Portugal, the area where the ancestors of most European Brazilians came from. It turns out that a small number of genetic loci differ between West African and European populations, and this pattern can be used to compute an index of African ancestry. Finally, these researchers collected two kinds of information from contemporary Brazilians. They assigned each of the individuals to one of three common Brazilian categories— "black," "white," and "intermediate"—on the basis of their phenotypes. They also took blood from each individual, extracted DNA, and typed each individual for genetic loci that differentiate West African and European populations. This analysis allowed them to compute the extent of African ancestry for each individual. The results are quite clear: there is very little correlation between phenotype and ancestry (Figure 15.22). People classified as "black" and people classified as "white" on the basis of their phenotypes are more or less indistinguishable genetically.

Racial classification schemes explain very little of the world's human genetic variation.

Geneticists have tried to account for the patterns of variation across the globe in biochemically detectable characters, such as blood-group loci and other genetic

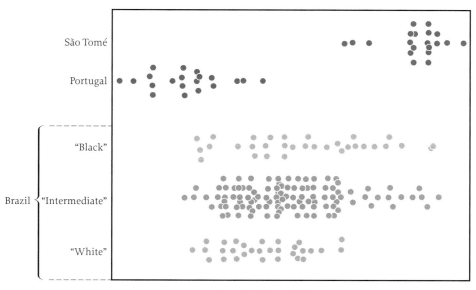

Genetic index of African ancestry

FIGURE 15.22 The genetic index of African ancestry for contemporary Africans living on São Tomé (an island just off the coast of Gabon that served as the entry point for the West African slave trade), contemporary Portuguese, and contemporary Brazilians classifed as "black," "white," and "intermediate" on the basis of skin color, facial morphology, and hair texture. These data indicate that the Brazilian folk classification system predicts little about overall genetic similarity.

polymorphisms. In these studies the human species was categorized first into local groups of people belonging to the same ethnic group, linguistic group, or nationality. Then local groups were collected into larger, geographically based categories that correspond roughly to the usual races. Geneticists computed the amount of variation in these characters within each local group, among groups within each race, and among races. They found that there is much more genetic variation *within* local groups than there is *among* local groups or among races themselves. Differences within local groups account for about 85% of all the variation in the human species. To put this another way, suppose that a malevolent extraterrestrial wiped out the entire human species except for one local group, which it preserved in an extraterrestrial zoo. The alien could pick any local group at random—the Efe, the Inuit, the citizens of Ames, Iowa, or the people of Patagonia—and then wipe out the rest of the humans on the planet. This group would still contain, on average, 85% of the genetic variation that exists in the entire human species. This means that standard racial classifications do not correspond to variation in genetic traits among the peoples of the world.

Racial classification schemes don't represent natural biological categories.

The bottom line is that people can be classified, and such classifications are not necessarily arbitrary, but they do not reflect any *natural* subdivision of the human species into biologically distinct groups. Nor does knowing a person's position in a classification reveal very much about what that person is like.

This conclusion is consistent with what we have learned about human evolution. It seems likely that anatomically modern humans are a very recently evolved species; the evidence from both mitochrondrial DNA and fossils suggests an age of less than 200,000 years. This means that it is less likely that natural selection and genetic drift have produced larger genetic differences within the human species than in other, older species. Recall that the genetic differences among different species of chimpanzees are much greater than among different groups of humans, probably because chimpanzees are a considerably older genus. Moreover, we know that gene flow tends to eliminate genetic differences between groups, and there has been extensive gene flow in human history.

Races represent cultural categories that play an important, but unfortunate, role in society.

Many people find the conclusion that races aren't real completely implausible. They "know" from experience that race is real. The late Martin Luther King Jr.'s dream that "my four little children will one day live in a nation where they will not be judged by the color of their skin, but by the content of their character" has not yet been realized. This is because racism is real and remains a pervasive problem in our society. But it's important to separate the reality of racism from the scientific understanding of the concept of race.

Many view the claim that race is not a valid scientific construct as another example of the "political correctness" that is now common in academia. This view has some support within anthropology. For example, in the November 1994 issue of *Discover* magazine, University of Colorado anthropologist Alice Brues stated,

> A popular political statement now is, "There is no such thing as race." I wonder what people think when they hear this. They would have to suppose that the speaker if he were dropped by parachute into downtown Nairobi, would

be unable to tell, by looking around him, whether he was in Nairobi or Stock-holm. This could only damage his credibility. The visible differences between different populations tell everyone that *there is something here.* (p. 60)

But our intuitions sometimes lead us astray. Our eyes tell us that the Earth is flat, but it is really a sphere. Our intuition tells us that a bullet fired horizontally from a rifle will hit the ground long after a bullet dropped from the muzzle at the same instant, yet in reality they will hit the ground at the same time. Our mind tells us that it is impossible for an elephant to have descended from a shrewlike insectivore, even though that is exactly what happened. The intuition that race is real is also an illusion.

Professor Brues is undoubtedly correct that a parachute jumper who landed in Nairobi would be unlikely to think that he was in Stockholm, Tokyo, Bombay, or Honolulu. But would he be certain that he was in Nairobi, and not in Johannesburg, or Khartoum? Probably not. And what if we imagine an intrepid soul who attempts to bicycle from Nairobi to Stockholm? As she pedaled from Nairobi through Kinshasa, Khartoum, Cairo, Istanbul, and Budapest and finally reached Stockholm, she would observe that people change in many ways, but the change would always be subtle; there would be no sharp boundaries with one kind of people on one side and a dif-ferent kind of people on the other side. The cyclist would see that people vary but this variation does not sort itself out into neat, discrete categories. Thus, our perception of race depends on which of these two metaphors we think captures the essential fea-tures of human variation. The scientific evidence demonstrates quite conclusively that there is considerably more variation within human groups than among them, and this is why our transcontinental bicyclist would not note any sharply defined changes as she made her way slowly north.

In our opinion, race is a culturally constructed category, not a meaningful biolog-ical concept. The characters that North Americans conventionally use to sort people into racial categories tend to be biologically transmitted traits, like skin color and facial features. But we could just as well use culturally transmitted traits like religion, dialect, or class. Classifications based on a small set of biologically transmitted traits have no more scientific validity than do classifications based on religion, language, or politi-cal affiliation. Although skin color and facial features are salient characteristics in the United States, in some societies other types of characters matter much more. Religion is the basis of bitter animosity in Northern Ireland, although the genetic differences between Protestant and Catholic Irish are even more microscopic than are the genetic differences between African Americans and European Americans. In the ethnic con-flicts that have seared the globe in the last decade it is literally a matter of life and death whether you are Azeri or Armenian in Nagorno-Karabakh; Hutu or Tutsi in Rwanda; or Serb or Muslim in Bosnia. Yet in each case, people are divided by culture, not genes.

FURTHER READING

Bodmer, W., and L. L. Cavalli-Sforza. 1976. *Genetics, Evolution, and Man.* Freeman, New York.

Cavalli-Sforza, L. L., P. Menozzi, and A. Piazza. 1994. *The History and Geography of Human Genes.* Princeton University Press, Princeton, NJ.

Falconer, D. R. 1981. *Introduction to Quantitative Genetics.* Longman, London.

Flatz, G. 1987. The genetics of lactose digestion in humans. *Advances in Human Genetics* 16: 1–77.

Harrison, G. A., J. M. Tanner, D. R. Pilbeam, and P. T. Baker. 1988. *Human Biology: An Introduction to Human Evolution, Variation, Growth, and Adaptability* (3rd ed.). Oxford Science, Oxford, England.

Jobling, M. A., M. E. Hurles, and C. Tyler-Smith. 2004. *Human Evolutionary Genetics, Origins, Peoples & Disease*. Garland, New York.

STUDY QUESTIONS

1. What sources of human variation are described in this chapter? Why is it important to distinguish among them?
2. Consider the phenotype of human finger number. Most people have exactly five fingers on each hand, but some people have fewer. What is the source of variation in finger number?
3. Why is it hard to determine the source of variation in human phenotypes? Why might it be easier to determine the source of variation for other animals?
4. Explain how studies of human twins allow researchers to estimate the effects of genetic and environmental differences on phenotypic traits.
5. What kind of body size and shape is best in a humid climate? Why?
6. Suppose you are told that differences in IQ scores among white Americans have a genetic basis. What would that tell you about the differences in average scores between white Americans and Americans belonging to other ethnic groups? Why?
7. Explain why race is not a biologically meaningful category of classification.

KEY TERMS

genetic variation
environmental variation
variation within groups
variation among groups
specific language impairment/SLI
selection-mutation balance
falciparum malaria
balanced polymorphism
non-insulin-dependent diabetes/NIDD
insulin
lactose

LCT∗P
LTC∗R
single nucleotide polymorphisms/SNPs
founder effect
porphyria variegata
Oceania
heritability
environmental covariation
monozygotic twins
dizygotic twins

CHAPTER 16

EVOLUTION AND HUMAN BEHAVIOR

WHY EVOLUTION IS RELEVANT TO HUMAN BEHAVIOR

The application of evolutionary principles to understanding human behavior is controversial.

The theory of evolution is at the core of our understanding of the natural world. By studying how natural selection, recombination, mutation, genetic drift, and other evolutionary processes interact to produce evolutionary change, we come to understand why organisms are the way they are. Of course, our understanding of evolution is far from perfect, and other disciplines, most notably chemistry and physics, contribute greatly to our understanding of life. As the great geneticist Theodosius Dobzhansky once said, however, "Nothing in biology makes sense except in the light of evolution."

So far, the way we have applied evolutionary theory in this book is not controversial. We are principally interested in the evolutionary history of our own species, *Homo sapiens*, but we began by using evolutionary theory to understand the behavior of our closest relatives, the nonhuman primates. Twenty years ago, when evolutionary theory was new to primatology, this approach generated some controversy, but now most primatologists are committed to evolutionary explanations of behavior. In Part Three we used evolutionary theory to develop models of the patterns of behavior that might have characterized early hominins. Although some researchers might debate the fine points of this analysis, there is little disagreement about the value of adaptive reasoning in this context. Perhaps this is because the early hominins were simply "bipedal apes," with brains the size of modern chimpanzee brains. Not many people object to evolutionary analyses of physiological traits, such as lactose tolerance, but they may debate the merits of particular explanations for those traits. Similarly, most people accept evolutionary explanations about why we live so long and mature so slowly. These traits are clearly part of human biology, and there is a broad consensus that evolutionary theory provides an essential key for understanding them.

The consensus evaporates when we enter the the domain of contemporary human behavior. Most social scientists acknowledge that evolution has shaped our bodies, our minds, and our behavior to a limited extent. However, many social scientists have been very critical of attempts to apply evolutionary theory to contemporary human behavior because they think evolutionary analyses imply that behavior is genetically determined. Genetic determinism of behavior in humans seems inconsistent with the fact that so much of our behavior is acquired through learning, and that so much of our behavior and beliefs is strongly influenced by our culture and environment. The notion that evolutionary explanations imply genetic determinism is based on a fundamental misunderstanding about how the natural world works.

All phenotypic traits, including behavioral traits, reflect the interactions between genes and the environment.

Many people have the mistaken view that genetic transmission and learning are mutually exclusive. That is, they believe that behaviors are either genetic and therefore unchangeable, or learned and thereby controlled entirely by environmental contingencies. This assumption lies at the heart of the "nature–nurture question," a debate that has plagued the social sciences for many years.

The nature–nurture debate is based on a false dichotomy. It assumes that there is a clear distinction between the effects of genes (nature) and the effects of the environment (nurture). People often think that genes are like engineering drawings for a finished machine, and that individuals vary simply because their genes carry different specifications. For example, they imagine that basketball player Yao Ming is tall because his genes specified an adult height of 2.3 m (7 ft 6 in.), and Earl Boykins is short because his genes specified an adult height of 1.65 m (5 ft 5 in.).

However, genes are not like blueprints that specify phenotype. Every trait results from the *interaction* of a genetic program with the environment. Thus, genes are more like recipes in the hands of a creative cook, sets of instructions for the construction of an organism using materials available in the environment. At each step, this very complex process depends on the nature of local conditions. The expression of any genotype always depends on the environment. A person's adult height is shaped by the genes they inherited from their parents, how well nourished they were in childhood, and the nature of the diseases they were exposed to when they were growing up.

The expression of behavioral traits is usually more sensitive to environmental conditions than is the expression of morphological and physiological traits. As we saw in Chapter 3, traits that develop uniformly in a wide range of environments, such as finger number, are said to be "canalized." Traits that vary in response to environmental cues, such as subsistence strategies, are said to be "plastic." Every trait, however, whether plastic or canalized, results from the unfolding of a developmental program in a particular environment. Even highly canalized characters can be modified by environmental factors, such as fetal exposure to mutagenic agents or accidents.

Natural selection can shape developmental processes so that organisms develop different adaptive behaviors in different environments.

Some people understand that all traits are influenced by a combination of genes and environment, but they reject evolutionary explanations of human behavior because they have fallen prey to a second, more subtle, misunderstanding. Namely, they believe that natural selection cannot create adaptations unless behavioral differences between individuals are caused by genetic differences. If this were true, it would follow that adaptive explanations of human behavior must be invalid because there is no doubt that most of the variation in behavioral traits, such as foraging strategies, marriage practices, and values, is not due to genetic differences but is instead the product of learning and culture.

This belief is false, however, because natural selection shapes learning mechanisms so that organisms adjust their behavior to local conditions in an adaptive way. Recall from Chapter 3 that this is exactly what happens with soapberry bugs. Male soapberry bugs in Oklahoma guard their mates when females are scarce, but not when females are abundant. Individual males vary their behavior adaptively in response to the local sex ratio. In order for this kind of flexibility in male behavior to evolve, there had to be small genetic differences in the male propensity to guard a mated female, and small genetic differences in how mate guarding is influenced by the local sex ratio. If such variation exists, then natural selection can mold the responses of males so that they are locally adaptive. In any given population, however, most of the observed behavioral variation is due to the fact that individual males respond adaptively to environmental cues.

Behavior in the soapberry bug is relatively simple. Human learning and decision making are immensely more complex and flexible. We know much less about the mechanisms that produce behavioral flexibility in humans than we do about the mechanisms that produce flexibility in mate guarding among soapberry bugs. Nonetheless, such mechanisms must exist, and it is reasonable to assume that they have been shaped by natural selection. (Note that we are not arguing that all behavioral variation in human societies is adaptive. We know that evolution does not produce adaptation in every case, as we discussed in Chapter 3.) The crucial point here is that evolutionary approaches do not imply that differences in behavior among humans are the product of genetic differences between individuals.

In the remainder of this chapter we will consider two ways that evolutionary theory can be used to understand the minds and behavior of modern humans. As you will see, researchers from different academic disciplines have followed different approaches in their efforts to understand how evolution has shaped human behavior. Some have focused on how natural selection has shaped the design of the human brain. Others have tried to understand how the human capacity for culture, and the ability to acquire ideas, beliefs, and values from other group members, have influenced the evolution of human behavior. In the remainder of this chapter and the next we will illustrate how these approaches can be used to gain insights about the behavior of contemporary humans.

UNDERSTANDING HOW WE THINK

Evolutionary analyses provide important insights about how our brains are designed.

The adaptation that most clearly distinguishes humans from other primates is our large and very complex brain. Natural selection hasn't just made our brains big; it has shaped our cognitive abilities in very specific ways and molded the way we think.

Even the most flexible strategies are based on special-purpose psychological mechanisms.

Psychologists once thought that people and other animals had a few general-purpose learning mechanisms that allowed them to modify any aspect of their pheno-types adaptively. However, a considerable body of empirical evidence indicates that animals are predisposed to learn some things and not others. For example, rats quickly learn to avoid novel foods that make them ill. Moreover, rats' food aversions are based solely on the taste of a food that has made them sick, not the food's size, shape, color, or other attributes. This learning rule makes sense because rats live in a very wide range of environments, where they frequently encounter new foods, and usually forage at night when it is dark. In order to determine whether a new food is edible, they taste a small amount first and then wait several hours. If it is poisonous, they soon become ill, and they do not eat it again. Rats may pay attention to the taste of foods, instead of other attributes, because it is often too dark to see what they are eating (Figure 16.1). However, there are limits to the flexibility of this learning mechanism. There are certain items that rats will never sample, and in this way their diet is rigidly controlled by genes. Moreover, the learning process is not affected equally by all environmental contingencies. For example, rats are affected more by the association of novel tastes with gastric distress than they are with other possible associations.

Natural selection determines the kinds of problems that the brains of particular species are good at solving. To understand the psychology of any species, we must know what kinds of problems its members need to solve in nature.

FIGURE 16.1 Rats initially sample small amounts of unfamiliar foods, and if they become ill soon after eating something, they will not eat it again.

Our brains may be designed to solve the kinds of problems that our ancestors faced when they lived in small foraging bands.

We know that people lived in small-scale foraging societies for the vast majority of human history (Figure 16.2); stratified societies with agriculture and high population density have existed for only a few thousand years (Figure 16.3). John Tooby and

Leda Cosmides of the University of California Santa Barbara argue that complex adaptations like the brain evolve slowly, so our brains are designed for life in foraging societies. They use the term **environment of evolutionary adaptedness** (EEA) to refer to the social, technological, and ecological conditions under which human mental abilities evolved. Tooby and Cosmides and their colleagues envision the EEA as being much like the world of contemporary hunter-gatherers.

People living in foraging groups face certain kinds of problems that affect their fitness. For example, food sharing is an essential part of life in modern foraging groups. Although vegetable foods are typically distributed only to family members, meat is nearly always shared more widely. Food sharing is a form of reciprocal altruism. The big problem with reciprocal altruism is that it is costly to interact with individuals who do not reciprocate. Thus, Cosmides and Tooby hypothesized that human cognition should be finely tuned to detect cheaters, and they have accumulated a convincing body of experimental data suggesting that people are very attentive to imbalances in social exchange and violations of social contracts.

As we have seen, there is quite a bit of uncertainty about how early humans lived, and this adds ambiguity to predictions about human psychology based on evolutionary reasoning. Some authorities believe that early members of the genus *Homo* were much like contemporary human foragers. That is, they lived in small bands and subsisted by hunting and gathering. They controlled fire, had home bases, and shared food. They could talk to one another, and they shared cultural beliefs, ideas, and traditions. Other authorities think that the lives of the earliest species of *Homo* were completely unlike those of modern hunter-gatherers. They think that these hominins didn't hunt large game, share food, or have home bases. If early hominins lived like contemporary foragers, then it is reasonable to think that the human brain has evolved to solve the kinds of problems that confront modern foragers, such as detecting freeloaders in social exchange. On the other hand, if lifeways that characterize contemporary foragers did not emerge until 40 kya, then there might not have been enough time for selection to assemble specialized psychological mechanisms to manage the challenges that foragers face, such as food sharing.

FIGURE 16.2 Evolutionary psychologists believe that the human mind has evolved to solve the adaptive challenges that confront food foragers because this is the subsistence strategy that humans have practiced for most of our evolutionary history.

Evolved psychological mechanisms cause human societies to share many universal characteristics.

Much of anthropology (and other social sciences) is based on the assumption that human behavior is not effectively constrained by biology. People have to obtain food, shelter, and other resources necessary for their survival and reproduction. But beyond that, human behavior is not constrained by biology.

As we have just seen, however, this assumption is not very plausible from an evolutionary perspective. It is likely that evolved mechanisms in the human brain channel the evolution of human societies and human culture, making some outcomes much more likely than others. So the right question is, What kinds of mental mechanisms do humans have? We are likely to share some mental mechanisms with other animals, but we may also have certain mental mechanisms that differentiate us from other creatures. In the discussion that follows, we examine two examples of cognitive

FIGURE 16.3 Indigenous peoples of the Mississippi Delta constructed these mound structures about 2500 to 1300 years ago. Monumental architecture like this is based on the ability of one group of people to control the labor of others, a signal of social stratification.

mechanisms that are found in all human societies: one mechanism—inbreeding avoidance—that we may have inherited from our primate ancestors; and another—the capacity for language—that is derived.

Inbreeding Avoidance

The offspring of genetically related parents have lower fitness than the offspring of unrelated parents do.

Geneticists refer to matings between relatives as **inbred matings** and contrast them with **outbred matings** between unrelated individuals. The offspring of inbred matings are much more likely to be homozygous for deleterious recessive alleles than are the offspring of outbred matings. As a consequence, inbred offspring are less robust and have higher mortality than the offspring of outbred matings. In Chapter 15 we discussed a number of genetic diseases, such as PKU, Tay-Sachs disease, and cystic fibrosis, that are caused by a recessive gene. People who are heterozygous for such deleterious recessive alleles are completely normal, but people who are homozygous suffer severe, often fatal consequences. Recall that such alleles occur at low frequencies in most human populations. However, there are many loci in the human genome. Thus, even if the frequency of deleterious recessives at each locus is very small, there is a good chance that everybody has at least one lethal recessive gene somewhere in their genome. Geneticists have estimated that each person carries the equivalent of two to five lethal recessives. Mating with close relatives is deleterious because it greatly increases the chance that both partners will carry a deleterious recessive allele at the same locus. Inbreeding leads to substantial reductions in fitness (Box 16.1). This, in turn, suggests that natural selection should favor behavioral adaptations that reduce the chance of inbreeding.

Matings between close relatives are very rare among nonhuman primates.

Remember from Chapter 7 that in all species of nonhuman primates, members of one or both sexes leave their natal groups near the time of puberty. Adult males do not often remain in groups long enough to be able to mate with their own daughters. It is very likely that dispersal is an adaptation to prevent inbreeding. In principle, primates could remain in natal groups and simply avoid mating with close kin. However, this would limit the number of potential mates and might be unreliable if there were much uncertainty about paternity.

Natural selection has provided at least some primates with another form of protection against inbreeding: a strong inhibition against mating with close kin. In matrilineal macaque groups, some males acquire high rank and mate with adult females before they emigrate. However, matings among maternal kin are extremely uncommon.

Experimental studies conducted by Wendy Saltzman of the University of California Riverside and her colleagues suggest that reproductive inhibition in callitrichids is due partly to inbreeding aversion. Young females housed with their mothers and fathers do not reproduce; but when fathers are replaced with unrelated males, both mothers and daughters breed. Adult female chimpanzees often have opportunities to mate with their fathers (Figure 16.5). More than 30 years of research at Gombe Stream National Park indicates that, in fact, they rarely do. Female chimpanzees seem to have a general aversion to mating with males much older than they are, and males seem to

Why Inbred Matings Are Bad News

The following simple example illustrates why inbreeding increases the chance that offspring will be homozygous for a deleterious recessive gene. Suppose that Cleo has exactly one lethal recessive somewhere in her genome. If she mates with a nonrelative, Mark, the chance that he carries the same deleterious recessive is simply equal to the frequency of heterozygotes carrying that gene in the population. (We don't need to consider the frequency of individuals who are homozygous for the recessive allele, because the allele is lethal in homozygotes.) Let p be the frequency of deleterious recessives in the population. Using the Hardy–Weinberg law, we know that the frequency of heterozygotes is $2p(1 - p)$. The frequency of lethal recessives is typically about 0.001. Thus there is only about 1 chance in 500 that Mark and Cleo carry the same deleterious recessive.

If Cleo mates with her brother, Ptolemy, the story is quite different. Remember from Chapter 8 that r, the coefficient of relatedness, gives the probability that two individuals will inherit the same gene through common descent. For full siblings, $r = 0.5$. Thus if Cleo mates with her brother, there is 1 chance in 2 that he will carry the same lethal recessive that she does, a value that is about 100 times greater than for an outbred mating. Because Cleo and Ptolemy are both heterozygotes for the lethal recessive, Mendel's laws say that 25% of their offspring will be homozygotes and therefore will die. Thus, on average, Cleo will experience a 12.5% decrease in fitness by mating with her brother Ptolemy compared with an outbred mating with Mark.

This simple example actually understates the cost of inbreeding, because people often carry more than one deleterious recessive. Oxford University biologist Robert May derived the estimates in Figure 16.4 using a more realistic model. As you can see, inbreeding is quite bad. Matings between siblings or between parents and their offspring ($r = 0.5$) produce about 40% fewer offspring than do outbred matings. The effects for other inbred matings are smaller but still serious.

Empirical studies strongly support these predictions. Data from 38 species of captive mammals indicate that the offspring of matings among siblings and among parents and their offspring are 33% less likely to survive to adulthood than the offspring of outbred matings. Studies in wild populations, though less conclusive, tell the same story. Several studies suggest that humans are affected by inbreeding in the same way. For example, the 161 children of father–daughter or brother–sister matings studied by geneticist Eva Seemanová, of Charles University in the Czech Republic, were twice as likely to die during their first year as their maternal half-siblings were, and they were 10 times as likely to suffer serious congenital defects. Other evidence comes from studies of Moroccan Jews living in Israel who do not consider marriages between men and their nieces to be incestuous. A study of 131 children from such marriages indicates that they suffer a 20% reduction in fitness compared with a control group from the same population.

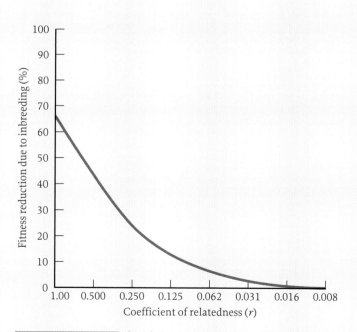

FIGURE 16.4 Population genetic theory predicts that close inbreeding can lead to serious reductions in fitness. The vertical axis plots the percentage of reduction in fitness due to inbreeding, and the horizontal axis plots the coefficient of relatedness among spouses. The curve is based on the conservative assumption that people carry the equivalent of 2.2 lethal recessives.

be generally uninterested in females much younger than themselves. These mechanisms may protect females from mating with their fathers, and vice versa.

Humans rarely mate with close relatives.

During the first half of the twentieth century, cultural anthropologists fanned out across the world to study the lives of exotic peoples. Their hard and sometimes dangerous work has given us an enormous trove of information about the spectacular variety of human lifeways. They found that domestic arrangements vary greatly across cultures: some groups are polygynous, some monogamous, and a few polyandrous. Some people reckon descent through the female line and are subject to the authority of their mother's brother. In some societies, married couples live with the husband's kin, in others they live with the wife's kin, and in some they set up their own households. Some people must marry their mother's brothers' children; others are not allowed to do so.

In all of this variety of domestic arrangements, there is not a *single* ethnographically documented case of a society in which brothers and sisters regularly marry, or one in which parents regularly mate with their own children. The only known case of regular brother–sister mating comes from census data collected by Roman governors of Egypt from A.D. 20 to 258. From the 172 census returns that have survived, it is possible to reconstruct the composition of 113 marriages: 12 were between full siblings and 8 between half-siblings. These marriages seem to have been both legal and socially approved, as both prenuptial agreements and wedding invitations survive.

The pattern for more distant kin is much more variable. Some societies permit both sex and marriage with nieces and nephews or between first cousins; other societies prohibit sex and marriage among even distant relatives. Moreover, the pattern of incest prohibitions in many societies does not conform to genetic categories. For example, even distant kin on the father's side may be taboo in a given society, while maternal cousins may be the most desirable marriage partners in the same society. Sometimes the rules about who can have sex are different from the rules governing who can marry.

Adults are not sexually attracted to the people with whom they grew up.

The fact that inbreeding avoidance is very common among primates suggests that our human ancestors probably also had psychological mechanisms preventing them from mating with close kin. These psychological mechanisms would disappear during human evolution only if they were selected against. However, mating with close relatives is highly deleterious in humans, as it is in other primates. Thus, both theory and data predict that modern humans will have psychological mechanisms that reduce the chance of close inbreeding, at least in the small-scale societies in which human psychology was shaped.

There is evidence that such psychological mechanisms exist. In the late nineteenth century the Finnish sociologist Edward Westermarck speculated that childhood propinquity stifles desire. By this he meant that people who live in intimate association as small children do not find each other sexually attractive as adults. A number of lines of evidence provide support for Westermarck's hypothesis:

FIGURE 16.5 Female chimpanzees avoid mating with closely related males. Although mothers have close and affectionate relationships with their adult sons, matings between mothers and sons are quite uncommon.

- *Taiwanese minor marriage.* Until recently, an unusual form of marriage was widespread in China. In **minor marriages**, children were betrothed and the prospective bride was adopted into the family of her future husband during infancy. There the betrothed couple grew up together like brother and sister. According to Taiwanese informants interviewed by Stanford University anthropologist Arthur Wolf, the partners in minor marriages found each other sexually unexciting. Sexual disinterest was so great that fathers-in-law sometimes had to beat the newlyweds to convince them to consummate their marriage. Wolf's data indicate that minor marriages produced about 30% fewer children than did other arranged marriages (Figure 16.6a) and were much more likely to end in separation or divorce (Figure 16.6b). Infidelity was also more common in minor marriages. When modernization reduced parental authority, many young men and women who were betrothed in minor marriages broke their engagements and married others.

- *Kibbutz age-mates.* Before World War II, many Jewish immigrants to Israel organized themselves into utopian communities called *kibbutzim* (plural of **kibbutz**). In these communities, children were raised in communal nurseries, and they lived intimately with a small group of unrelated age-mates from infancy to adulthood. The ideology of the kibbutzim did not discourage sexual experimentation or marriage by children in such peer groups, but neither behavior occurred. Israeli sociologist Joseph Sepher, himself a kibbutznik, collected data on 2769 marriages in 211 kibbutzim. Only 14 of them were between members of the same peer group, and in all of these cases one partner joined the peer group after the age of six. From data collected in his own kibbutz, Sepher found no instances of premarital sex among members of the same peer group.

- *Third-party attitudes toward incest.* As you may have realized already, aversions to inbreeding extend beyond our attitudes toward our own mating behavior to include strong beliefs about appropriate mating behavior by other individuals. We are disgusted not only by the idea of having sex with our parents or our own

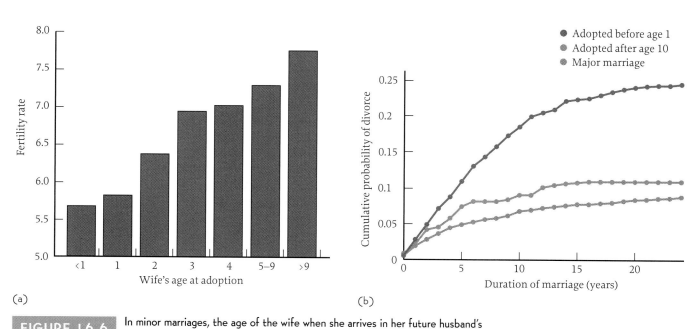

(a)   (b)

FIGURE 16.6 In minor marriages, the age of the wife when she arrives in her future husband's household (age at adoption) affects both fertility and the likelihood of divorce. (a) The fertility of women adopted at young ages is depressed. (b) The younger a woman is when she arrives in her husband's household, the less likely it is that the marriage will survive.

children, but also by the idea of other people having sex with their children. Daniel Fessler and Carlos Navarette of the University of California Los Angeles think that these kinds of "third-party" aversions are a form of "egocentric empathy." Westermarck hypothesized that coresidence during childhood generates sexual aversions to particular partners. If that is the case, then the extent of exposure to siblings of the opposite sex during childhood might also be linked to the strength of feelings about one's own behavior and the strength of feelings about the behavior of others. Debra Lieberman, John Tooby, and Leda Cosmides at the University of California Santa Barbara, along with Fessler and Navarette, have tested these predictions by conducting experiments in which they asked subjects (university undergraduates) to contemplate consensual sibling incest involving hypothetical adults. The results from both studies largely confirmed Westermarck's hypothesis. Those who had grown up with opposite-sex siblings had stronger negative responses to the hypothetical scenario than those who had not. Moreover, women generally had stronger aversive responses to the hypothetical scenario than men did.

Evolutionary interpretations of inbreeding avoidance differ sharply from influential theories about incest and inbreeding avoidance in psychology and cultural anthropology.

Incest and inbreeding avoidance play a central role in many influential theories of human society. Thinkers as diverse as Sigmund Freud (the founder of psychoanalysis) and Claude Lévi-Strauss (the father of structuralist anthropology) have asserted that people harbor a deep desire to have sex with members of their immediate family. According to this view, the existence of culturally imposed rules against incest is all that saves society from these destructive passions. This view is not very plausible from an evolutionary perspective. There are compelling theoretical reasons to expect that natural selection will erect psychological barriers to incest, and good evidence that it has done so in humans and other primates. Both theory and observation suggest that the family is not the focus of desire; it's a tiny island of sexual indifference.

However, the evolutionary analysis we have outlined here is not quite complete. For example, we might expect the Westermarck effect and egocentric empathy to produce an aversion to minor marriage in China. Yet this practice has persisted for a long time. It is possible that psychological mechanisms are supplemented or perhaps superseded by conscious reasoning. People in many societies believe that incest leads to sickness and deformity, and their beliefs may guide their behavior and shape their cultural practices. Finally, it seems clear that attitudes about incest are not based solely on the deleterious effects of inbreeding. If they were, then all societies would have the same kinds of rules about who can have sexual relationships. Instead, we find considerable variation. For example, some societies encourage first cousins to marry, while others prohibit them from doing so.

Human Language

The cognitive capacities that give rise to language are universal.

Languages are clearly part of human culture; children learn to speak the language of the people around them. Although people in different parts of the world speak different languages, they share the morphological adaptations that make spoken language possible and the cognitive capacities that allow them to learn and to use language. The abilities to decode words and to extract meaning from combinations of words are based on highly specialized and derived cognitive mechanisms.

Humans can perceive speech sounds at a much more rapid rate than other sounds.

Phonemes are the basic unit of speech perception, the smallest bits of sound that we recognize as meaningful elements of language. Not all languages recognize the same phonetic distinctions. For example, English speakers hear the difference between "l" and "r" sounds, but Swahili and Japanese speakers lump them together in a single phoneme. Interestingly, infants can initially distinguish the full range of phonemes, but during the first year of life they become insensitive to distinctions that are not part of the language spoken around them.

We are able to decode very rapid streams of phonemes. When we speak normally, we utter about 10 phonemes per second. If we concentrate, we can understand people who bombard us with 25 phonemes per second. By contrast, even the best Morse code operators cannot keep up once the transmission rate exceeds about 5 clicks per second. In fact, if the rate of transmission of a simple sound like a Morse code click exceeds 20 sounds per second, we hear a continuous buzz rather than a sequence of discrete sounds.

It's not hard to imagine why the ability to communicate rapidly was favored by natural selection. The early human who could decode the hurried shout, "Look out, a lion is behind you," would have a substantial edge over the poor fellow who could only say, "Speak more slowly, I can't . . ."

We perceive speech as a sequence of discrete words, even though this is an auditory illusion.

We perceive speech the same way we write it down: as a sequence of discrete words separated by brief silences. However, this is not really the case. There are no periods of silence in ordinary speech. When researchers use an instrument that plots sound level against time, they note that sound level does not drop to zero between spoken words, and in fact, distinct words do not appear on the plot. You may have experienced this phenomenon yourself while listening to people speaking a language you don't understand. You probably couldn't tell when one word ended and another began. This is because there are no gaps between words.

Language is based on grammatical properties that allow us to interpret sequences of words.

Consider the following two sentences: *Org killed your brother* and *Your brother killed Org.* The words are the same, but the meanings produced in your brain are very different. One situation calls for revenge, and the other calls for keeping a close watch on Org's relatives. We hear speech as a sequence of words, like cars coming down an off-ramp. In order to create meaning, our brain interprets this information using rules that specify meaning. These rules are what we call **grammar**. These are not the obscure rules that modern Western people struggle to master in school, but unconscious rules that allow almost every person to speak and understand spoken language effortlessly.

The world's languages share a number of grammatical rules that allow people to use sequences of words to express useful ideas. Some of these rules were listed in a paper written by Steven Pinker, now at Harvard University, and Paul Bloom, now at Yale:

- Words must belong to categories such as "noun," "adjective," and "verb," which correspond to basic features of the world that people need to talk about—in this case, things, the qualities of things, and actions, respectively.
- Some rules allow the listener to figure out what the speaker intended to say about the relationship between things, qualities, and actions. These kinds of rules tell us that Org is a killer in the first sentence above and a corpse in the second. In English these rules are based on the order of the words: the actor comes before the verb, and the recipient of the act comes after the verb. In other languages these distinctions are made by modification of the nouns to indicate, for example, which person is the killer and which is the victim. Linguists refer to this part of grammar as **syntax**.
- Words are grouped into phrases that also belong to categories such as "noun phrase" and "verb phrase." By combining words in this way, language can express a wide range of thoughts with a small number of words. The noun *brother* by itself is not very informative. By combining *brother* with other words in noun phrases, however, you can refer to *my brother, your brother, your older brother, your husband's brother,* and so on.
- Some rules allow a listener to determine which words the speaker meant to group together in phrases. These rules allow us to distinguish between the sentences *Your brother killed my father* and *My brother killed your father* by telling us that *your* is to be grouped with *brother* in the first sentence, and with *father* in the second sentence.
- Verbs are modified to indicate when events occurred. For example, we can distinguish between the statements *Org killed your brother* and *Org will kill your brother.* Verbs are also modified to indicate the relative timing of two events. Thus, *Org had killed your brother before Thag showed up* tells the listener that Thag did not witness the killing.

There are other grammatical rules for forming questions, for indicating states of mind, and for reducing the amount of memory necessary to process a sentence.

This system is not perfect. Even with all the grammatical rules in action and with help from context, many sentences allow for more than one meaning. Wordplay such as double entendre relies on such ambiguity. Nonetheless, by combining discrete words using grammatical rules, humans can communicate an incredible range of useful information (and make very bad puns).

The psychological mechanisms that structure language are different from other cognitive abilities.

Many researchers think that language is a manifestation of a powerful general-purpose learning mechanism. Humans learn many complicated rule-bound skills, like playing the piano or long division. However, no one would argue that our brains evolved because of the selective advantages of learning to play the piano or doing long division. Instead, we must have a powerful and very general learning mechanism that allows us to acquire diverse skills. According to this view, language is not a specialized adaptation; it is just another complex skill acquired by the same general learning mechanism.

This point of view does not take into account several lines of evidence that suggest that we have special-purpose cognitive mechanisms specifically shaped by natural selection to allow people to learn and to use language. First, we would expect to find a direct relationship between cognitive abilities and linguistic skills if the ability to learn and to use language were based solely on general cognitive ability. However, this is not the case. Some stroke victims lose nearly all of their linguistic ability, but their intelligence is not diminished. These people are not fluent language users, and they have particular difficulty with syntax, but they do not show any other cognitive deficiencies. There are also people who speak grammatically but are otherwise mentally impaired. For example, many children suffering from spina bifida are severely retarded, and they cannot learn to read, write, or do simple arithmetic. Nonetheless, some of them speak fluently and at length, using grammatically correct sentences, but the content of their speech is nonsensical.

Second, if the ability to use language were like other complex skills, children would need considerable tutoring to learn grammatical rules. But children seem to know some universal grammatical principles without having to learn them. Most children begin to acquire the grammar of their native language when they are about two years old, and by the time they are three, most children make relatively few grammatical errors. Many linguists believe that children have to learn the surface grammar of their own language—how to speak properly in English, Spanish, or Farsi, for example—but they do not need to learn the deep, universal grammar that underlies all languages. This knowledge is innate. In fact, these linguists think that without some innate knowledge of the structure of language, children could not learn language at all.

Comparative evidence suggests that animals can associate concepts with arbitrary symbols but have only a limited mastery of syntax. Thus, human language capacities are derived traits.

Language is based on the ability to attach meaning to arbitrary sounds, and this ability allows us to communicate about things outside of our immediate experience. When one person smells smoke and yells "Fire!" in a crowded theater, the rest of the audience knows it's time to head for the exits, even if they can't smell the smoke themselves. Moreover, the audience knows that there is a fire in the theater—not a flood or an earthquake or a sighting of Elvis.

Robert Seyfarth and Dorothy Cheney of the University of Pennsylvania have demonstrated that alarm calls in vervet monkeys work in much the same way (Figure 16.7). Vervets give acoustically distinct calls when they encounter snakes, eagles, and leopards, and they also respond differently to each of these predators. When they see a leopard, they rush up into a tree; when they see an eagle, they dive into dense bushes; and when they see a snake, they look down at the ground around them. Cleverly designed playback experiments demonstrate that the monkeys associate particular kinds of calls with particular predators. When they hear a tape-recorded eagle call, for example, they dive into the bushes, even though there is no eagle to be seen (Table 16.1).

Great apes also demonstrate these kinds of abilities in laboratory studies. For example, a gorilla named Koko was taught to use American Sign Language to

FIGURE 16.7 Vervet monkeys associate particular calls with specific predators.

TABLE 16.1	Responses of vervet monkeys to playbacks of alarm calls. The data indicate that the monkeys recognize the meaning of different calls. (From Table 1 in R. M. Seyfarth, D. L. Cheney, and P. Marler, 1980, Monkey responses to three different alarm calls: Evidence of predator classification and semantic communication, Science, 210: 801–803.)

| | RESPONSE TO PLAYBACK | | | |
Type of Alarm Call	Climb into Tree	Run into Bushes	Look Up	Look Down
Leopard	8	2	4	1
Eagle	2	6	7	4
Snake	2	2	2	14

FIGURE 16.8 Koko was trained by psychologist Francine Patterson, of the Gorilla Foundation, to use American Sign Language.

FIGURE 16.9 Kanzi, a bonobo trained by Sue Savage-Rumbaugh and Duane Rumbaugh (now at the Great Ape Trust in Des Moines, Iowa) uses abstract lexigrams to communicate.

communicate (Figure 16.8); and a bonobo named Kanzi points to abstract written symbols, or lexigrams, to tell his trainers when he wants a banana or a tickle (Figure 16.9). Surprisingly, these skills are not limited to primates. A gray parrot named Alex has learned to count, label, and categorize objects as "the same" or "different" in spoken English. These individuals, and several others, have extensive vocabularies and seem to grasp the relationship between objects and arbitrary symbols.

None of the language-trained animals have mastered the full complexity of human syntax. They can string symbols together to make requests ("Sue tickle"), create novel terms ("water bird"), and express thoughts ("surprise hide"). However, nearly all of the strings are either short or repetitive. For example, they might sign "give orange give give orange give orange orange." Chimpanzees have been taught to use simple rules that specify the order between elements, such as action–agent (for example, "hug Roger"). Kanzi, the bonobo trained by Sue Savage-Rumbaugh at Georgia State University, has gone one step further. Some of Kanzi's comments involve two lexigrams, or a lexigram and a gesture, and he has spontaneously generated simple rules to order these elements. For example, Kanzi is more likely to name an action ("keep away") first, and the agent or object second ("balloon"), than vice versa. He is also more likely to name something and then point to the object, than he is to point to the object and then name it.

However, syntax is more than word order. Kanzi's combinations of elements and those of the most fluent signing apes both fail to meet several of the criteria for syntax that were defined earlier in the chapter. For example, the elements are not grouped into phrases, they provide no information about the timing of events, and much of the meaning must be derived from context. What do we make of the apes' achievements and limitations? Patricia Greenfield of the

University of California Los Angeles emphasizes that ape abilities are comparable to those of very young children just beginning to acquire language, and she suggests that there is continuity between the linguistic abilities of apes and humans. Others, including Steven Pinker, emphasize differences between apes and young children in the ease with which they learn symbols, acquire grammatical rules, and naturally grasp the complexities of syntax. Those that emphasize this gap conclude that linguistic abilities in humans are derived traits.

EVOLUTION AND HUMAN CULTURE

For many anthropologists, culture is what makes us human. Each of us is immersed in a cultural milieu that influences the way we see the world, shapes our beliefs about right and wrong, and endows us with the knowledge and technical skills to get along in our own environment. Despite the central importance of culture in anthropology, there is little consensus about how or why culture arose in the evolution of the human lineage. In the discussion that follows, we present a view of the evolution of human culture developed by one of us (R. B.) and Peter Richerson at the University of California Davis. Although we believe strongly in this approach to understanding the evolution of culture, there is not a broad consensus among anthropologists that this, or any other particular view of the origins of culture, is correct.

Culture is information acquired by individuals by imitation, teaching, and other forms of social learning.

Before we begin, we must provide some definitions of terms. We define **culture** as information acquired by individuals through some form of social learning. Cultural variation is the product of differences among individuals that exist because they have acquired different behaviors as a result of some form of social learning. The properties of culture are sometimes quite different from the properties of other forms of environmental variation. If people acquire behavior from others through teaching or imitation, then culturally transmitted adaptations can gradually accumulate over many generations (Figures 16.10 and 16.11). For example, one man may learn to fletch his arrows and teach this skill to his sons. One of his sons may learn from his neighbor to dip his arrows in poison, and his sons may learn both of these behaviors by watching their father. To understand culture, we need to take its cumulative nature into account.

Culture Is a Derived Trait in Humans

Culture is common among other animals.

In the last few years, primatologists have documented a range of behavioral variation across groups in chimpanzees, orangutans, and *Cebus* monkeys. For example, chimpanzees living on the western shores of Lake Tanganyika raise their arms and clasp hands while they groom, but chimpanzees living on the eastern shores of the lake don't. Orangutans in some areas use sticks to pry seeds out of fruits, but at other sites, orangutans are unable to extract the seeds because they have not mastered this technique.

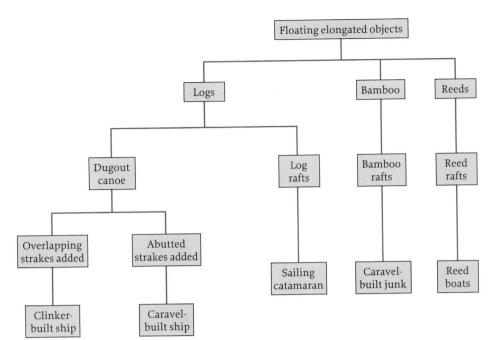

FIGURE 16.10 We can trace the development of certain technological innovations through time. In China, the first boats were elongated structures that floated on the water. Some were made of logs, others of bamboo or reeds. Floating logs were transformed into canoes with the addition of a keel. Other types of ships, including Chinese junks, have a square hull and no keel.

Cebus monkeys show considerable variation in foraging techniques and social conventions. For example, capuchins at some sites participate in long bouts of mutual hand sniffing (Figure 16.12), but capuchins at other sites never display this behavior. In some cases, scientists have documented the appearance, diffusion, and eventual extinction of behavioral variants.

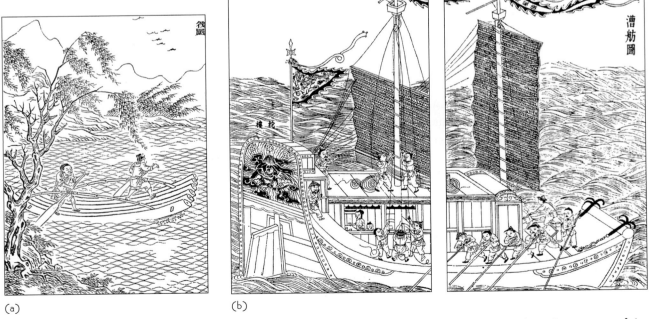

(a)   (b)

FIGURE 16.11   Bamboo rafts like the one shown in (a) may have been the precursors of the great Chinese junks (b).

Cumulative cultural change is rare in nature.

These compendiums of behavioral traditions are important because they demonstrate that behavioral innovations can arise and spread through primate groups, but they neglect a question that is very important for understanding the evolution of the human capacity for culture: How are these behaviors acquired and transmitted? Do individuals learn these behaviors on their own, or do they learn them by observing others? There are at least two nongenetic transmission mechanisms that can account for persistent behavioral differences between populations.

FIGURE 16.12 Capuchin monkeys display a variety of behaviors that seem to vary from group to group. Here a capuchin monkey at Lomas Barbudal Biological Reserve in Costa Rica sniffs the hands of its partner. (Photograph courtesy of Susan Perry and Sharon Kessler.)

1. **Social facilitation** occurs when the activity of one animal indirectly increases the chance that other animals will learn the behavior on their own. Social facilitation would account for the persistence of tool use in the following scenario: Nuts are a greatly desired food, and young chimpanzees find eating nut meats highly reinforcing. Young chimpanzees accompany their mothers while they are foraging. This means that young animals spend a lot of time in proximity to nuts and hammer stones in populations in which females use stones to break open nuts. Young chimpanzees fool around with stone hammers and anvils until they master the skill of opening the nuts. They do not learn the skill by watching their mothers. In populations in which chimpanzees do not use stones to open nuts, young chimpanzees never spend enough time in proximity to both nuts and hammer stones to acquire the skill.

2. **Observational learning** occurs when "naïve" animals learn how to perform an action by watching the behavior of experienced, skilled animals. If this were the case, the tool tradition would be preserved because young chimpanzees actually imitate the behavior of older ones. (As we will see, evidence for imitation in apes is scant.)

Social facilitation and observational learning can both can lead to persistent behavioral differences between populations. However, there is an important distinction between the two processes. Social facilitation can only preserve variation in behavior that organisms can learn on their own. Observational learning allows cumulative cultural change. To see the difference, consider the following hypothetical example: Suppose that, in especially favorable circumstances, an early hominin learns to strike rocks together to make useful flakes. Her companions, who spend time near her, will be exposed to the same kinds of conditions, and some of them might learn to make flakes too, entirely on their own. This behavior will be preserved by social facilitation because groups in which tools are used will spend more time in proximity to the appropriate stones. However, that will be as far as it goes. If an especially talented individual finds a way to improve the flakes, this innovation will not spread to other members of the group; rather, each individual will have to learn the behavior by himself. With observational learning, on the other hand, innovations can persist as long as younger individuals are able to acquire the modified behavior by observing the actions of others. As a result, observational learning can lead to the cumulative evolution of behaviors that no single individual could invent on its own.

There is some debate about whether observational learning occurs in nonhuman primates. Many of the behavioral traditions described by primatologists are relatively simple and could be learned independently by individuals in each generation. Some behaviors, such as potato washing by Japanese macaques (Figure 16.13), were once thought to have been spread through imitation but are now thought to be the product of social facilitation and individual learning. But claims that nonhuman primates have "culture" continue to be debated.

In the field, it is always very difficult to know how behaviors are acquired and to distinguish between social facilitation and observational learning (Box 16.2). To sort this out, researchers turn to the laboratory, where they can carefully control animals' experience and their opportunities to learn novel skills. One of the most influential experiments was conducted by Elisabetta Visalberghi, of the Consiglio Nazionale delle Ricerche in Rome. Capuchins are extremely clever monkeys, well known for their manipulative abilities. In captivity, they are skilled tool users. Visalberghi gave several capuchins the opportunity to learn to use a stick to push a peanut out of a horizontal, clear plastic tube. She then allowed a second set of monkeys to watch the skilled monkeys get the peanut out of the tube. If capuchins learn by observing conspecifics, then we would expect monkeys housed with the skilled monkeys to learn faster than monkeys who did not have a chance to watch skilled models. However, exposure to skilled models didn't make any difference. So Visalberghi and her colleague Dorothy Fragaszy of the University of Georgia wryly concluded that "monkeys don't ape."

FIGURE 16.13   A young Japanese macaque learned to wash her sweet potatoes, and many members of her group subsequently began to wash their potatoes as well. For many years, this was thought to be a good example of observational learning. However, we now think each monkey learned the behavior by trial and error.

This experiment inspired many new experiments based on the same basic design: allow naïve individuals to observe an expert perform a rewarding task using a particular technique, and then find out whether the naïve individuals can master the task and if they use the same technique that they observed. For example, Andrew Whiten of St. Andrews University and his colleagues constructed an "artificial fruit." The fruit is really a box that can be opened in two different ways. Naïve chimpanzees in one group saw the box opened one way; naïve chimpanzees in another group saw the box opened the other way. The researchers observed that naïve chimpanzees tend to use the technique that they have seen demonstrated. On the other hand, a body of experiments conducted by Michael Tomasello and his colleagues suggest the opposite conclusion. That is, chimpanzees are quick to learn that tools can be used to obtain rewards, but they do not learn the same techniques that models use. Thus, important questions remain unresolved about how chimpanzees and other primates learn novel skills and behaviors.

But even if apes can learn new skills by watching others, there is still a huge gap between humans and other primates. No other primate relies on observational learning to the same extent that humans do, other primates have not elaborated behavioral traditions to the same extent as humans have, and the behavior of other primates is much less variable from group to group or from region to region than the behavior of humans is.

Understanding the Sources of Behavioral Variation

BOX 16.2

In the early 1960s, Jane Goodall reported that chimpanzees at Gombe stripped the leaves from slender twigs and used the twigs as probes to fish for termites. As Goodall continued her work, she documented a number of additional forms of tool use. For example, chimpanzees at Gombe also use long, slender twigs to collect ants, crumple leaves and use them as "sponges" to soak up water from tree hollows, and use leaves to wipe debris from their bodies. These observations made a huge impact in anthropology and the social sciences because tool use was thought to be a defining feature of the human species. Darwin himself thought that bipedal evolution had been favored by natural selection because it enabled early hominins to carry their tools from one place to another. But the discovery that tool use occurs in a wide range of species—from Caledonian crows that fashion hooks out of twigs (Figure 16.14), to sea otters that use rocks to bash open abalone shells, and *Cebus* monkeys that use large stones to crack hard-shelled nuts (Figure 16.15)—demonstrates that humans do not have a monopoly on technology.

When researchers began to study chimpanzees at other sites in Africa, they discovered new forms of tool use. Moreover, they began to realize that the behavioral repertoire of chimpanzees varied across groups. For example, in the Mahale Mountains, just miles away from Gombe, chimpanzees raise their arms and clasp hands while they groom (Figure 16.16), a position never adopted at Gombe. As more chimpanzee communities were habituated and studied, more and more examples of behavioral differences were accumulated.

In 1999, a group of researchers led by Andrew Whiten of St. Andrews University published an exhaustive catalogue of behavioral variants observed in seven well-studied wild

FIGURE 16.15 In northeastern Brazil, *Cebus* monkeys use large stones to smash open hard-shelled nuts on sandstone slabs. This behavior is now being studied by a group of researchers including Dorothy Fragaszy, Elisabetta Visalberghi, Patricia Iza, and Eduardo Ottoni of the Universidade de São Paulo, and Marino Gomes de Oliveira of Fundação BioBrasil. (Photo by Peter Oxford.)

FIGURE 16.14 Caledonian crows make and use tools in nature. Here a female has bent the end of a twig and is using the hooked end to retrieve a food reward. (Photo courtesy of the Behavioral Ecology Research Group, Department of Zoology, University of Oxford.)

Chimpanzees in the Mahale Mountains in Tanzania often hold their hands above their heads and clasp their partner's hands as they groom. This grooming posture has never been seen at Gombe, just 135 km (about 90 miles) away. (Photograph courtesy of William C. McGrew.)

der twigs (or wands) for ant dipping, but chimpanzees in the Ivory Coast's Taï Forest typically choose shorter sticks. Gombe chimpanzees hold the wand in one hand and pull it slowly through the other hand, and then pop the ants into the mouth. Taï chimpanzees hold the wand in one hand, then use their lips to wipe the ants directly off the wand.

Whiten and his colleagues note that "it is difficult to see how such behaviour patterns could be perpetuated by social learning processes simpler than imitation." According to this view, chimpanzees at Taï and Gombe use different techniques because they see others in their communities using those techniques, and they imitate the behaviors that they have observed. Close observation also seems to play a role in the development of termite fishing in chimpanzees. Elizabeth Lonsdorf and her colleagues at the University of Minnesota videotaped young chimpanzees while their mothers were fishing for termites. She found that young females watched their mothers carefully as they fished for termites, but young males were considerably less attentive (Figure 16.17). Lonsdorf also discovered that mothers varied in their fishing techniques, some consistently using longer twigs than others. Daughters tended to use the same kinds of tools that their mothers used, but sons did not.

You might be convinced that observational learning is the only plausible explanation for differences in ant-dipping techniques at Taï and Gombe. After all, it seems implausible that all the chimpanzees at Gombe would adopt one method, while all the chimpanzees at Taï would adopt a different method, if each chimpanzee figured out the technique for ant dipping on its own. This makes sense as long as there

chimpanzee communities. Their list includes 39 behaviors that are common in some communities but absent in others. These behaviors include various forms of tool use, distinctive grooming styles, and courtship gestures. Some behaviors, such as fishing for algae, occur in only one community; other behaviors, such as dipping for ants, occur in multiple communities. In some cases, behavioral variants seem to be regionally specific. Thus, chimpanzees use stones to hammer open hard-shelled nuts in West Africa but not in East Africa. Other behaviors, such as rain displays that are performed at the beginning of rainstorms, occur in widely separated locations.

In some cases, chimpanzees in different groups perform similar tasks in different ways. For example, chimpanzees use different techniques to collect safari ants. Safari ants live in subterranean nests and often migrate through the underbrush in dense swarms. Columns of marching ants, numbering hundreds of thousands of individuals, provide a mobile snack for chimpanzees. Safari ants aggressively defend themselves and have a painful bite. Chimpanzees apparently find ants tasty but don't like being bitten, so they use tools to gather them. Chimpanzees at Gombe choose long, slen-

Young female chimpanzees carefully watch their mothers termite fishing, and they tend to acquire the same kinds of techniques that their mothers use. Males are much less attentive to their mothers, and do not match their mother's techniques.

is no functional reason why one technique works better than another technique. However, Tatyana Humle, now at the University of Wisconsin, and Tesuro Matsuzawa of Kyoto University, have discovered that functional distinctions do underlie differences in ant-dipping techniques in a small community of chimpanzees in Boussou, Guinea. The Boussou chimpanzees are accomplished tool users. They use stones to hammer open nuts, they use sticks to scoop algae off the water surface, and they also dip for ants.

The Boussou chimpanzees' ant-dipping tools are intermediate in length between the tools used at Gombe and those used at Taï. Sometimes Boussou chimpanzees use the two-handed pull-and-wipe method like the Gombe chimpanzees, and sometimes they wipe the tool with their mouths like the chimpanzees at Taï do (Figure 16.18). Humle and Matsuzawa discovered that the chimpanzees at Boussou eat two different kinds of ants: red ones and black ones. The black ants are considerably more aggressive than the red ants, and their bite is more painful; both kinds of ants are more aggressive at their subterranean nests than when they are migrating. The chimpanzees use longer tools to dip for black ants than to dip for red ones, and they use longer tools when they dip at the entrance to the nest than when they dip into the migrating swarms. They are also more likely to dip from a position of safety off the ground when they dip at the nest site than when they dip on migrating ants. The chimpanzees used the pull-through technique mainly when they used long tools at ant nests. They tended to use the mouthing technique when they were feeding on migrating ants with shorter sticks. Thus the chimpanzees at Boussou apparently use safer techniques for more dangerous prey. Humle and Matsuzawa think that there is good reason to believe that the ants that chimpanzees eat at Gombe are like the black ants at Boussou, and the ants that

FIGURE 16.18 A chimpanzee in Boussou uses its mouth to collect ants from a wand.

the chimpanzees eat at Taï are like the milder red ants at Boussou.

Humle and Matsuzawa's observations are important because they remind us just how hard it is to figure out how behaviors are acquired and transmitted in natural settings. It is possible that behavioral differences across chimpanzee communities represent cultural traditions passed from one generation to the next as naïve youngsters observe skilled adults and imitate their behavior. However, it is also possible that behavioral differences arise as individuals within groups encounter similar conditions, gain exposure to the artifacts that others use, and then learn the skill on their own. Many different learning mechanisms could produce the kinds of stable differences in behavior that researchers have documented among different chimpanzee groups, and it takes careful work under controlled conditions to discriminate between them.

Culture Is an Adaptation

Observational learning is not a by-product of intelligence and social life.

Chimpanzees and capuchins are among the world's cleverest creatures. In nature, they use tools and perform many complex behaviors; in captivity, they can be taught extremely demanding tasks. Chimpanzees and capuchins live in social groups and have ample opportunity to observe the behavior of other individuals, and yet the best evidence suggests that neither chimpanzees nor capuchins make much use of observational learning. Thus, observational learning appears not to be simply a by-product

FIGURE 16.19 Infants are prone to spontaneous imitation of the behaviors they observe. Here a 13-month-old infant flosses her two teeth.

of intelligence and of having opportunities for observation. Instead, observational learning seems to require special psychological mechanisms.

This conclusion suggests, in turn, that the psychological mechanisms that enable humans to learn by observation are adaptations that have been shaped by natural selection because culture is beneficial (Figure 16.19). Of course, this need not be the case. Observational learning could be a by-product of some other adaptation that is unique to humans, such as language. But given the great importance of culture in human affairs, it is important to think about the possible adaptive advantages of culture.

Culture allows humans to exploit a wide range of environments using a universal set of mental mechanisms.

The archaeological record suggests that during the Pleistocene, humans occupied virtually all of Africa, Eurasia, and Australia. Modern hunter-gatherers developed an astounding variety of subsistence practices and social systems. Consider just a few examples. The Copper Eskimo lived in the high Arctic, spending their summers hunting near the mouth of the Mackenzie River and the long dark months of the winter living on the sea ice and hunting seals. Groups were small and heavily dependent on hunting. The !Xo lived in the central Kalahari Desert of Botswana, collecting seeds, tubers, and melons, hunting impala and gemsbok, enduring fierce heat, and living without surface water for months at a time. Their small, nomadic bands were linked together in large clusters organized along male kinship lines. The Chumash lived on the southern California coast, gathering shellfish and seeds and fishing the Pacific from great plank boats. They lived in large permanent villages with pronounced division of labor and extensive social stratification.

The fact that the !Xo could acquire the knowledge, tools, and skills necessary to survive the rigors of the Kalahari is not so surprising; many other species live there too. What is amazing is that the same brain that allowed the !Xo to survive in the Kalahari also permitted the Copper Eskimo to acquire the very different knowledge, tools, and skills necessary to live on the tundra and ice north of the Arctic Circle, and the Chumash to acquire the skills necessary to cope with life in crowded, hierarchical settlements. No other animal occupies a comparable range of habitats or utilizes a comparable range of subsistence techniques and social structures. For example, savanna baboons, the most widespread primate species, are limited to Africa and Arabia, and the diet, group size, and social systems of these far-flung baboon populations are much more similar to one another than the diet, group size, and social systems of human hunter-gatherers are.

Humans can live in a wider range of environments than other primates because culture allows us to accumulate better strategies for exploiting local environments much more rapidly than genetic inheritance can produce adaptive modifications. Animals like baboons adapt to different environments using various learning mechanisms. For example, they learn how to acquire and process the food they eat. Baboons in the lush wetlands of the Okavango Delta of Botswana learn how to harvest roots of water plants and how to hunt young antelope during the birth season. Baboons living in the harsh desert of nearby Namibia must learn how to find water and process desert foods. All such learning mechanisms require prior knowledge about the environment: where to search for food, what strategies can be used to process the food, which flavors are reinforcing, and so on. More detailed and more accurate knowledge allows more accurate adaptation because it allows animals to avoid errors and acquire a more specialized adaptation.

In most animals, this knowledge is stored in the genes. Imagine that you captured a group of baboons from the Okavango Delta and moved them to the Namibian desert. It's a very good bet that the first few months would be tough for the baboons, but after a relatively short time the transplanted group of baboons would probably be quite similar to their neighbors. They would eat the same foods, have the same activity patterns, and use the same grooming techniques. They might even adjust the size of their groups and the size of their social networks. The transplanted baboons would become similar to the local baboons because they acquire a great deal of information about how to be a baboon genetically; it is hardwired. To be sure, the transplanted baboons would have to learn where things are, where to sleep, which foods are desirable, and which foods to avoid, but they would be able to do this without contact with local baboons because they have the basic knowledge built in.

Human culture allows accurate adaptation to a wider range of environments because *cumulative* cultural adaptation provides more accurate and more detailed information about the local environment than genetic inheritance systems can provide. People are smart, but individual humans can't figure out how to live in the Arctic, the Kalahari, or anywhere else without drawing on the knowledge of local people. Think about being plunked down on an Arctic beach with a pile of driftwood and seal skins and trying to make a kayak (Figure 16.20). You already know a lot: what a kayak looks like, roughly how big it is, and something about its construction. Nonetheless, you would almost certainly fail. And, supposing you did make a passable kayak, you would still have dozens of other skills to master before you could make a contribution to the Inuit economy. The Inuit could make kayaks, and do all the other things they needed to do to stay alive in a challenging environment, because they could make use of a vast pool of useful information stored in the minds of other people in their population.

FIGURE 16.20 The people of Nunivak Island in the Bering Sea hunted seals from kayaks like this one. These kayaks weighed less than 15 kg (less than about 30 lb) and could be paddled at speeds of up to 11 km (7 miles) per hour.

The information contained in this pool is accurate and adaptive because the combination of individual learning and observational learning leads to rapid, cumulative adaptation. Even if most individuals blindly imitate the behavior of others, some individuals may occasionally come up with a better idea, and this will nudge traditions in an adaptive direction. Observational learning preserves the many small nudges and exposes the modified traditions to another round of nudging. This process generates adaptation more quickly than genetic inheritance does. The complexity of cultural traditions can explode to the limits of our capacity to learn them, far past our ability to make careful, detailed decisions about them.

Culture can lead to evolutionary outcomes not predicted by ordinary evolutionary theory.

The importance of culture in human affairs has led many anthropologists to conclude that evolutionary thinking has little to contribute to understanding human behavior. They argue that because evolution shapes genetically determined behaviors, but not behaviors that are learned, culture is independent of biology. This argument is a manifestation of the nature–nurture controversy, and we explained why this reasoning is flawed at the beginning of the chapter. Although many anthropologists have rejected evolutionary thinking about culture, many evolutionists have made the opposite mistake. They reject the idea that culture makes any *fundamental* difference in the way that evolution has shaped human behavior and psychology. Some evolutionists argue that the genes underlying the psychological machinery that gives rise to human behavior were shaped by natural selection, so, at least in ancestral environments, the machinery must have led to fitness-enhancing behavior. If the adaptation doesn't enhance fitness in modern environments, that's because our evolved psychology is designed for life in a different kind of world.

We think both sides in this argument are wrong. We can't understand humans without understanding the complex interplay between biology and culture. This is because cumulative cultural evolution is rooted in a novel evolutionary trade-off between benefits and costs. Observational learning is beneficial because it allows human populations to accumulate vast reservoirs of adaptive information over many generations, leading to the cumulative cultural evolution of highly adaptive behaviors and technology. Because this process is much faster than genetic evolution, it allows human populations to evolve cultural adaptations to local environments: kayaks in the Arctic and blowguns in the Amazon. The ability to adjust rapidly to local conditions was highly adaptive for early humans because the Pleistocene was a time of extremely rapid fluctuations in world climates. However, the psychological mechanisms that create this benefit come with a built-in cost. Remember that the advantage of observational learning is that it avoids the need for all individuals to figure out everything for themselves. They can just do what others do. But to get the benefits of social learning, people have to be credulous, generally accepting that other people are doing things in a sensible and proper way.

This credulity opens up human minds to the spread of maladaptive beliefs and behaviors. If our neighbors believe that it's beneficial to bleed sick people and that it's a good idea to treat corn with lye, we believe that too. This is how we get wondrous adaptations like kayaks and blowguns. But we have little protection against the perpetuation of maladaptions that somehow arise. Even though the capacities that give rise to culture and shape its content must be (or at least must have been) adaptive on average, the behavior observed in any particular society at any particular time may reflect evolved maladaptations. Examples of these sorts of maladaptations are not hard to find.

Maladaptive beliefs can spread because culture is not acquired just from parents.

There are good reasons to expect maladaptive beliefs to spread. As you know, we acquire all our genes from our parents, but we acquire culturally transmitted beliefs and values from our parents, friends, teachers, religious figures, the media, and other sources. Biologist Richard Dawkins of Oxford University refers to culturally transmitted beliefs and values as **memes**. The logic of natural selection applies to memes in the same way that it does to genes. Memes compete for our memory and for our attention, and not all memes can survive. Some memes are more likely to survive and be transmitted than others. Memes are heritable, often passing from one individual to another without major change. As a result, some memes spread and others are lost.

Even though memes are subject to natural selection, it is not necessarily true that only memes that enhance our genetic fitness will spread. The rules of cultural transmission are different from the rules of genetic transmission, so the outcome of selection on memes can be different from the outcome of selection on genes. The basic rules of genetic transmission are simple. With some exceptions, every gene that an individual carries in her body is equally likely to be incorporated into her gametes, and the only way that those genes can be transmitted is through her offspring. Thus, only genes that increase reproductive success will spread. Cultural transmission is much more complicated. Memes are acquired and transmitted throughout an individual's life, and they can be acquired from grandparents, siblings, friends, coworkers, teachers, and even completely impersonal sources such as books, television, and now the Internet.

Genes are transmitted directly from parents to offspring, but memes may spread along many different pathways. This means that memes can spread even if they do not enhance reproductive fitness. If ideas about dangerous hobbies like rock climbing or heroin use spread from friend to friend, these ideas can persist even though they reduce survival and individual reproductive success (Figure 16.21). Beliefs about heaven and hell can spread from priest to parishioner, even if the priest is celibate. Moreover, cultural variants may accumulate and be transmitted within groups of people who form clans, fraternities, business firms, religious sects, or political parties. This process can generate groups that are defined by cultural values and traditions, not by genetic relatedness.

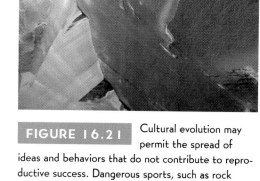

FIGURE 16.21 Cultural evolution may permit the spread of ideas and behaviors that do not contribute to reproductive success. Dangerous sports, such as rock climbing, may be examples of such behaviors.

Culture is part of human biology, but culture makes human evolution qualitatively different from that of other organisms.

The fact that culture can lead to outcomes not predicted by conventional evolutionary theory does not mean that human behavior has somehow transcended biology. The idea that culture is separate from biology is a popular misconception that cannot withstand scrutiny. Culture is generated from organic structures in the brain that were produced by the processes of evolution. However, cultural transmission leads to novel evolutionary processes. Thus, to understand the whole of human behavior, evolutionary theory must be modified to account for the complexities introduced by these poorly understood processes.

The fact that culture can lead to outcomes that would not be predicted by conventional evolutionary theory does not mean that ordinary evolutionary reasoning is useless. The fact that there are processes that lead to the spread of risky behaviors like rock climbing does not mean that these are the only processes that influence cultural behavior. Earlier in this chapter we saw that it is likely that many aspects of human psychology have been shaped by natural selection so that people learn to behave adaptively. We love our children, have strong aversions to mating with close relatives, and have an uncanny facility for learning language. There is every reason to suspect that these predispositions play an important role in determining which memes spread and which don't. As long as this is the case, ordinary evolutionary reasoning will be useful for understanding human behavior.

FURTHER READING

Barkow, J., L. Cosmides, and J. Tooby, eds. 1992. *The Adapted Mind, Evolutionary Psychology and the Generation of Culture.* Oxford University Press, New York.

Fragaszy, D. M., and S. Perry, eds. 2003. *The Biology of Traditions.* Cambridge University Press, Cambridge, England.

Richerson, P. J., and R. Boyd. 2004. *Not by Genes Alone: How Culture Transformed Human Evolution.* University of Chicago Press, Chicago.

Whiten, A., V. Horner, and S. Marshall-Pescini. 2003. Cultural panthropology. *Evolutionary Anthropology* 12: 92–105.

STUDY QUESTIONS

1. Much of the behavior of all primates is learned. Nonetheless, we have suggested many times that primate behavior has been shaped by natural selection. How can natural selection shape behaviors that are learned?

2. Many of the things that we do are consistent with general predictions derived from evolutionary theory. We love our children, help our relatives, and avoid sex with close kin. But there are also many aspects of the behavior of members of our own society that seem unlikely to increase individual fitness. What are some of these behaviors?

3. In some species of primates there seems to be an aversion to mating with close kin. The aversion seems to be stronger for females than for males. Why do you think this might be the case? Under what conditions would you expect this gender difference to disappear?

4. What evidence suggests that language is based on special-purpose cognitive mechanisms?

5. Human speech is composed of many words that can be rearranged to make different meanings. What are the advantages of this system?

6. The verb *to ape* means "to copy or imitate." Is its meaning consistent with what we now know about the learning processes of other primates?

7. Primatologists have documented many examples of behaviors that vary across populations, and some have concluded that this variation is a form of culture. Explain why this is or is not a reasonable conclusion.

8. Some things that we do seem to be maladaptive (think about skydiving, drug abuse, and collecting classic cars). Some people would argue that these behaviors provide evidence that natural selection has no important impact on modern humans. Is this a reasonable argument? Why or why not?

KEY TERMS

environment of evolutionary
 adaptedness/EEA
inbred matings
outbred matings
kibbutz

phonemes
grammar
syntax
social facilitation
observational learning

CHAPTER 17

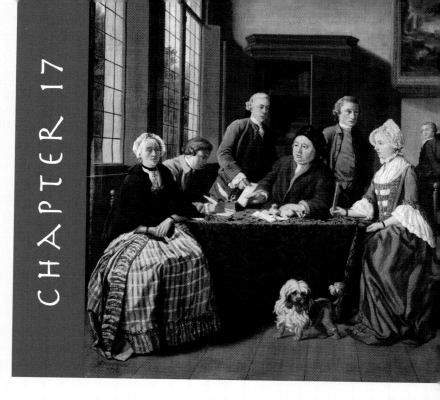

HUMAN MATE CHOICE AND PARENTING

You should be convinced by now that every aspect of the human phenotype is the product of evolutionary processes. Over the last 10 million years, natural selection, mutation, and genetic drift have made us into large, nearly hairless, bipedal primates, with grasping hands and large brains. The same evolutionary processes have molded the psychological mechanisms that influence human behavior, causing us to behave differently from other primates in some contexts. There is simply no other explanation.

However, the fact that modern humans are the product of evolutionary processes does not necessarily mean that evolutionary theory will help us understand contemporary human behavior. Constraints on adaptation, which we discussed in Chapter 3, may have predominated in the evolution of human behavior, limiting the usefulness of adaptive reasoning. In Chapter 16, however, we saw that evolutionary theory has provided useful insights into certain aspects of human psychology, social organization, and culture. In this chapter we will present several more examples that show how evolutionary thinking can help us understand the day-to-day behavior of humans in modern societies.

In choosing these examples, we have focused on reproductive behavior because mating and parenting strongly affect fitness. The unfortunate person who chooses a mate that is lazy, infertile, or unfaithful is likely to have many fewer children than the person whose spouse is healthy, hardworking, and faithful. Neglectful parents who fail to provide adequate care for their offspring will raise fewer children than will attentive parents who carefully nurture their offspring. Because reproductive decisions are likely to have a marked effect on fitness, there is good reason to expect that our psychology has been shaped by natural selection to improve the chances of making good choices about who we mate with, who we marry, and how we raise our children. We do not have enough space to provide a comprehensive analysis of these topics. Instead, we focus on certain examples in which evolutionary theory provides novel and fundamental insights into human behavior and physiology.

THE PSYCHOLOGY OF HUMAN MATE PREFERENCES

Marry

Children—(if it Please God)—Constant companion, (& friend in old age) who will feel interested in one,—object *to be* beloved and played with. better than a dog anyhow.—Home, & someone to take care of house—Charms of music & female chit-chat.—These things good for one's health.—*but terrible loss of time.*—

My God, it is intolerable to think of spending one's whole life, like a neuter bee, working, working, & nothing after all.—No, no won't do.—Imagine living all one's day solitary in smoky dirty London house.—Only picture to yourself nice soft wife on a sofa with good fire, & books, & music perhaps—Compare this vision with the dingy reality of Grt. Marlbro St.

Marry—Mary—Marry Q.E.D.

Not Marry

Freedom to go where one liked—choice of Society & *little of it.*—Conversation of clever men at clubs—Not forced to visit relatives, & to bend in every trifle.—to have the expense & anxiety of children—perhaps quarrelling —**Loss of time.**—cannot read in the Evenings—fatness & idleness—Anxiety & responsibility—less money for books & c—if many children forced to gain one's bread.—(But then it is very bad for ones health to work too much)

Perhaps my wife wont like London; then the sentence is banishment & degradation into indolent, idle fool.

From p. 444 in F. Burkhardt and S. Smith, eds., 1986, *The Correspondence of Charles Darwin*, Vol. 2, 1837–1843, Cambridge University Press, Cambridge, England.

These are the thoughts of 29-year-old Charles Darwin, recently returned from his five-year voyage on the HMS *Beagle*. Soon after writing these words, Darwin married his cousin Emma, the daughter of Josiah Wedgwood, the progressive and immensely wealthy manufacturer of Wedgwood china (Figure 17.1). By all accounts, Charles and Emma were a devoted couple. Emma bore Charles 10 children and nursed him through countless bouts of illness. Charles toiled over his work and astutely managed his investments, parlaying his modest inheritance and his wife's more substantial one into a considerable fortune.

Darwin's frank reflections on advantages and disadvantages of marriage were very much those of a conventional, upper-class, Victorian gentleman. But people of every culture, class, and gender have faced the problem of choosing mates. Sometimes people choose their own mates, and other times parents arrange their children's marriages. But everywhere, people care about the kind of person they will marry.

Evolutionary theory generates some testable predictions about the psychology of human mate preferences.

For much of their evolutionary history, humans have lived in foraging societies. The adaptive challenges that men and women face in these kinds of societies are likely to have shaped their mating strategies. For women, this might have meant choosing men who would provide them with access to resources. Recall from Chapter 12 that there is considerable interdependence between men and women in foraging societies. Women mainly gather plant foods, and men are mainly responsible for hunting. Women's consumption exceeds their production for much of their reproductive life. Children do not begin to provide substantial amounts of their own food until they reach adolescence. Thus, for women, it might be important to choose a mate who will be a good provider.

Men's reproductive success depends largely on the fertility of their partners, so it is plausible that selection favored men who focused on this attribute. Women's fertility is highest when they are in their 20s and declines to zero when women reach menopause, at about 50 years of age (Figure 17.2). Thus, selection should have favored men who chose young and healthy mates. Because picture IDs were scarce in the Pleistocene, selection may have shaped men's psychology so that they are attracted to cues that reliably predict youth and health, such as smooth skin, good muscle tone, symmetrical features, and shiny hair.

For both men and women, it is important to find mates that they can get along with. Human children are dependent on their parents for a remarkably long time. Dur-

FIGURE 17.1 Charles Darwin courted and married his cousin Emma Wedgwood. This portrait was painted when Emma was 32, just after the birth of her first child.

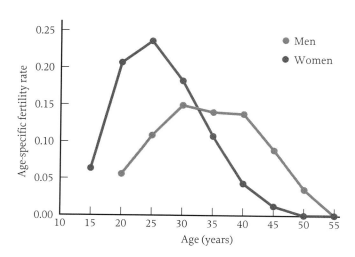

FIGURE 17.2 Age-specific fertility rates give the probability of producing a child at particular ages. !Kung women have their first child between the ages of 15 and 19 and the highest fertility rate in their 20s. Women's fertility falls to zero by age 50. !Kung men do not begin to reproduce until their early 20s, and their fertility rates are fairly stable in their 30s and 40s, dropping to low levels in their 50s.

FIGURE 17.3 People in 33 countries (red) were surveyed about the qualities of an ideal mate.

ing this period, both parents provide food and shelter for their children. Because parental investment lasts for many years, adaptive thinking predicts that both men and women will value traits in their partners that help them sustain their relationships. Both are likely to value personal qualities like compatibility, agreeableness, reliability, and tolerance.

If evolution has shaped the psychology of human mating strategies, then we would expect to find common patterns across societies.

David Buss, a psychologist at the University of Texas, was among the first to test the evolutionary logic underlying human mating preferences and tactics. Buss enlisted colleagues in 33 countries to administer standardized questionnaires about the qualities of desirable mates to more than 10,000 men and women. Most of the data were collected in Western industrialized nations, and most samples represent university students in urban populations within those countries (Figure 17.3). In the questionnaires, people were asked to rate a number of traits of potential mates—good looks, good financial prospects, compatibility, and so on—according to their desirability. Respondents were also asked about their preferred age at marriage and the preferred age difference between themselves and their spouse.

People generally care most about the personal qualities of their mates.

Men and women around the world rate mutual attraction or love above all other traits (Table 17.1). The next most highly desired traits for both men and women are personal attributes, such as dependability, emotional stability and maturity, and a pleasing disposition. Good health is the fifth most highly rated trait for men and the seventh for women. Good financial prospect is the thirteenth most highly rated trait by men and the twelfth by women. Good looks are rated tenth by men and thirteenth by women. It is interesting, and somewhat surprising, that neither sex seems to value chastity highly. Perhaps this is because people were asked to evaluate the desirability of sexual experience before marriage (that is, virginity), not fidelity during their marriage.

TABLE 17.1 Men and women from more than 33 countries around the world (shown in Figure 17.3) were asked to rate the desirability of a variety of traits in prospective mates. The rankings of the values assigned to each trait, on average, are given here. Subjects were asked to rate each trait from 0 (irrelevant or unimportant) to 3 (indispensable). Thus, high ranks (low numbers) represent traits that were generally thought to be important. (From Table 4 in D. M. Buss et al., 1990, International preferences in selecting mates: A study of 37 cultures, *Journal of Cross-Cultural Psychology,* 21: 5–47.)

| | RANKING OF TRAIT BY: | |
Trait	Males	Females
Mutual attraction/love	1	1
Dependable character	2	2
Emotional stability and maturity	3	3
Pleasing disposition	4	4
Good health	5	7
Education and intelligence	6	5
Sociability	7	6
Desire for home and children	8	8
Refinement, neatness	9	10
Good looks	10	13
Ambition and industriousness	11	9
Good cook and housekeeper	12	15
Good financial prospect	13	12
Similar education	14	11
Favorable social status or rating	15	14
Chastity*	16	18
Similar religious background	17	16
Similar political background	18	17

*Chastity was defined in this study as having no sexual experience before marriage.

Men and women show the differences in mate preferences predicted by parental investment theory.

Even though the ranking of the scores assigned to these traits is similar for men and women, there are consistent differences between men and women in how desirable each gender thinks these traits are. Buss found that people's sex had the greatest effect on their ratings of the following traits: "good financial prospect," "good looks," "good cook and housekeeper," "ambition and industriousness." As the evolutionary model predicts, women value good financial prospects and ambition more than men do, and men value good looks more than women do. Gender has a smaller and somewhat less uniform effect on ratings of chastity. In 23 populations, men value chastity significantly more than women do; in the remaining populations, men and women value chastity equally. There are no populations in which women value chastity significantly more than men do.

Men and women differ about the preferred ages of their partners.

Evolutionary reasoning suggests that men's mate preferences will be strongly influenced by the reproductive potential of prospective mates. Therefore, we would predict men to choose mates with high fertility or high reproductive value. By the same token, we would expect women to be less concerned about their partner's age than about their partner's ability to provide resources for them and their offspring.

A considerable amount of evidence suggests that men consistently seek and marry partners who are younger than they are and that women seek and marry partners somewhat older than themselves. Douglas Kenrick and Richard Keefe of the University of Arizona surveyed marriage records in two cities in the United States and in a small Philippine village, as well as personal advertisements in the United States, northern Europe, and India. In all of these cases, they found a similar pattern. As men get older, the age difference between them and their wives increases. Thus, young men marry women slightly younger than themselves, and older men marry partners considerably younger than themselves (Figure 17.4a). As women get older, there is little change in the age difference between them and their husbands (Figure 17.4b). Newspaper advertisements in which advertisers specify the range of ages for prospective mates demonstrate a very similar pattern. Henry Harpending of the University of Utah has found very similar patterns among the Herero, a pastoralist group in the northern Kalahari Desert of Botswana, in which marriages are unstable, divorce is common, and women have a considerable amount of financial independence.

Although men advertise for and marry progressively younger women, the actual age of their partners does not seem to fit the prediction that males will choose fertile mates. Older men seek and marry women who are considerably younger than they themselves are, but not young enough to produce children. Men's choices about who to date and who to marry may be driven by multiple factors, not just women's fertility. Older men may desire younger women, but they may also want to find someone who shares their taste in music, has similar goals in life, and so on. Furthermore, men's preferences and their marriages may reflect their own attractiveness in the mating market. Older men may want young women but know they will have to settle for partners closer to their own age. Together, the data taken from personal advertisements and from marriage records reflect individual desires tempered with reality.

Men and women vary about the preferred number of partners.

In addition to having different criteria for the ideal mate, men and women may also have different mating tactics. Because women devote nine months to each pregnancy and nurse their children for even longer, selection is likely to have favored a psychology that makes them cautious about involvement in sexual relationships that would expose them to the risks of pregnancy. (Of course, nowadays birth control reduces the risk of pregnancy for women, but effective methods of contraception are a recent innovation. Human mating tactics evolved in a world without such technology.) Women are likely to prefer stable, committed relationships with men who are willing and able to help care for them and their offspring. Because the costs of conception are borne mainly by women, men can afford to be more flexible in their mat-

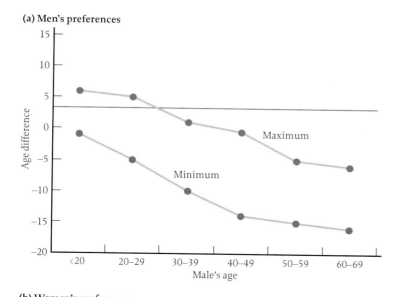

(a) Men's preferences

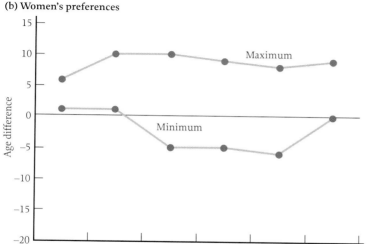

(b) Women's preferences

FIGURE 17.4 Mate preferences by sex. (a) In personal advertisements, all but the youngest men state preferences for women who are younger than themselves. As men get older, the age difference between themselves and preferred mates increases. (b) Women typically prefer men who are somewhat older than themselves, and these preferences remain the same as women get older.

ing tactics, and to have a psychology that makes them more open to mating opportunities that do not involve long-term commitments. However, we would expect men to form committed long-term relationships because children that receive care from both parents are more likely to thrive.

David Schmitt of Bradley University has coordinated a comprehensive cross-cultural study of human sexuality, sampling people, mainly university students, in 62 countries around the world. In this survey, people were asked about the traits that they valued in potential mates and were also asked about various aspects of their mating tactics. For example, they were asked the number of sexual partners they would like to have over various time intervals ranging from 6 months to 30 years. For all time intervals, men reported preferring a larger number of sexual partners than women did (Figure 17.5). This difference seems to be common cross-culturally (Figure 17.6), although the magnitude of the sex difference and the number of partners desired vary considerably.

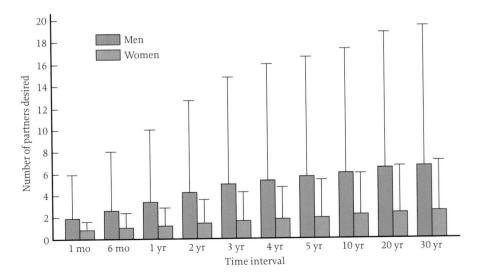

FIGURE 17.5 Men typically prefer a larger number of sexual partners across all time intervals than women do. Note, however, that there is also more variability in men's preferences than in women's preferences. This means that some men prefer large numbers of partners, while others prefer many fewer partners.

Differences in mating tactics may contribute to misunderstandings between men and women.

When you meet someone you're attracted to, you probably feel excitement and some degree of uncertainty. Some of this uncertainty arises because you are not sure whether the other person is as attracted to you as you are to him or her. And some of this uncertainty arises because you don't know what the other person's intentions are. Martie Haselton of the University of California Los Angeles, and David Buss, have pointed out that this uncertainty generates different kinds of problems for men and women. To understand the logic of their argument, let's think about what kinds of mistakes you could make in deciding whether someone was attracted to you. A false positive arises if you think the other person is attracted to you, when in fact that's not the case. A false negative occurs if you think the other person doesn't like you, when they

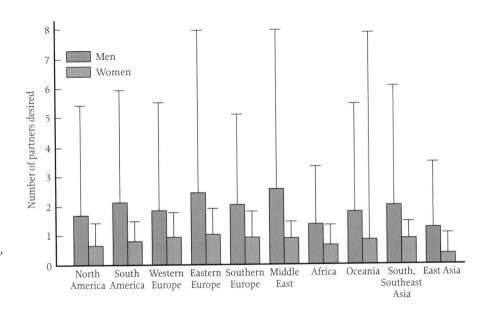

FIGURE 17.6 In every world region, men prefer significantly more sexual partners than women do over the next month.

really do. Both kinds of errors are costly: a false positive could lead you to make an overture that would be rejected ("Sorry, I need to wash my hair"); a false negative could prevent you from making any overture at all (and you will have nothing better to do than wash your hair).

Now think about the kinds of errors that can arise when there is uncertainty about the other person's intentions about the relationship. Haselton and Buss hypothesize that natural selection will predispose men and women to bias their judgments about new partners' sexual intentions and commitment in differ-ent ways. Women, who could become pregnant, are expected to be cautious about their partner's intentions, and as a result they will make more false negative errors than false positive errors. Put another way, evolutionary reasoning predicts that women are more likely to under-estimate men's commitment than to overestimate it. Men, who are interested in pursuing both short-term and long-term relationships, are expected to minimize the chance of missing sexual opportunities, and as a result they will make more false positive errors than false negative errors. That is, they are likely to overestimate women's sexual interest more often than they underestimate it.

Haselton and Buss have conducted a number of dif-ferent studies on college students in the United States to test this hypothesis. They have asked men and women to evaluate sexual intent and commitment in members of their own sex and the opposite sex; to imagine how they would interpret various kinds of signals directed to them-selves (for example, held hands, declared love); and to recall instances when their own intentions were misunder-stood by members of the opposite sex. The results con-form to Haselton and Buss's predictions: men tend to overestimate women's sexual intent, and women tend to underestimate men's interest in commitment (Figure 17.7).

Culture predicts people's mate preferences better than gender does.

Both of the large cross-cultural data sets reveal con-siderable variation from country to country. Buss and his colleagues found that the country of residence has a greater effect than gender does on variation in all of the 18 traits in Table 17.1, except for "good financial prospect." This means that knowing where a person lives tells you more about what he or she values in a mate than knowing the person's gender does. Of the 18 traits, chastity (defined in Buss's study as no sexual experience before marriage) shows the greatest variability among populations. In Sweden, men rate chastity at 0.25, and women rate it at 0.28 on a scale of 0 (irrelevant) to 3 (indispensable). In contrast, Chinese men rate chastity

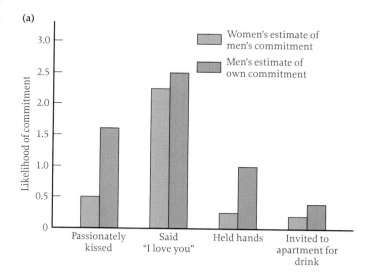

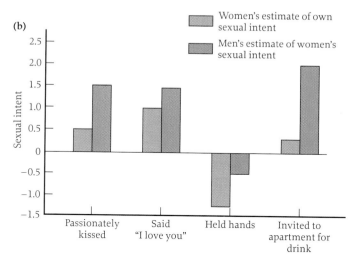

FIGURE 17.7 Men and women were asked to estimate how they would interpret particular courtship signals by members of the opposite sex and how they would rate the same sig-nals by members of their own sex, using a scale from +3 (very likely) to –3 (very unlikely). (a) Men and women estimated the extent of commit-ment implied by a number of different courtship signals. Women tended to underestimate men's interest in commitment relative to men's own esti-mates of their interest in commitment. (b) Men and women estimated the likelihood of sexual intent implied by the same signals. Men tended to overestimate women's sexual intent relative to women's perceptions of their own intent.

	Current mate is child's mother	Current mate is not child's mother
TABLE 17.2 When men invest time, energy, or other resources in children, they may be enhancing their parenting effort or their mating effort.		
Man is child's father	Parenting effort + mating effort	Parenting effort
Man is not child's father	Mating Effort	Neither

2.54, and Chinese women rate it 2.61 (Figure 17.8). This means that there is more similarity between men and women from the same population than there is among members of each sex from different populations.

This result illustrates an important point: evolutionary explanations that invoke an evolved psychology and cultural explanations that are based on the social and cultural milieu are not mutually exclusive. The cross-cultural data suggest there are some uniformities in people's mate preferences that are the result of evolved psychological mechanisms. People everywhere want to marry kind, caring, trustworthy people. Men want to marry young women and women want to marry older men. But this is not the whole story. The cross-cultural data also suggest that human mate preferences are strongly influenced by the cultural and economic environment in which we live. Ultimately, culture also arises out of our evolved psychology, and the cultural variation in mate preferences that the cross-cultural data reveal must therefore also be explicable in evolutionary terms. However, the way our evolved psychology shapes the cul-

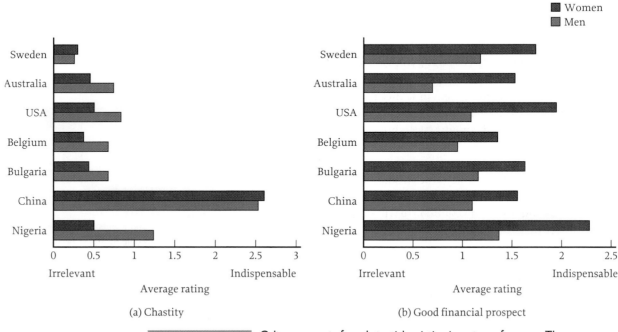

(a) Chastity

(b) Good financial prospect

FIGURE 17.8 Culture accounts for substantial variation in mate preferences. The average ratings given by men and women in several countries surveyed are shown for (a) the trait with the highest interpopulation variability ("chastity") and (b) the trait with the lowest interpopulation variability ("good financial prospect").

tures in which we live is complicated and poorly understood, and many interesting questions remain unresolved. Evolutionary theory does not yet explain, for example, why chastity is considered essential in China but undesirable in Sweden.

Evolutionary analyses of mate choice have generated considerable controversy over at least two different issues.

As we explained in Chapter 16, evolutionary analyses of human behavior generate considerable controversy. Work on human mate choice is no exception. Many critics have complained that evolutionary analyses simply reflect and reinforce Western cultural values, which celebrate women's youth and beauty and men's wealth and power. They argue that researchers are not studying evolved preferences, but rather are learning about cultural values and beliefs. In response, advocates of evolutionary analyses argue that the cross-cultural uniformity of mate preferences and mating tactics reflects evolved psychological predispositions that are modified, but not created, by culture.

Other critics have accepted the general logic of evolutionary reasoning but have questioned the methods used to assess mate preferences and mating tactics. Most of the early work was based on pencil-and-paper tests in which people, often undergraduates, were asked about their preferences and their personal experiences. These kinds of data may be biased in a number of ways. For example, cultural norms may lead men to exaggerate their sexual experience and women to understate their desire for sexual variety, even on anonymous surveys.

Moreover, for reasons that are not fully understood, results derived from "forced choice" tests, in which subjects are asked which of two alternatives they would choose, sometimes deviate from results in which subjects are asked to rate the same alternatives. Although this may seem like an arcane methodological issue, it may be important if different kinds of data generate different conclusions about the psychology underlying mating tactics.

This problem arises in studies of sex differences in jealousy. Buss and his colleagues have hypothesized that there will be sex differences in responses to emotional and sexual infidelity. Men, who are concerned about being cuckolded, will be more upset about their partners' sexual infidelity than about their emotional infidelity. On the other hand, women, who are concerned about access to resources for themselves and their offspring, will be more upset about their partners' emotional infidelity than about their sexual infidelity. When subjects are asked which form of infidelity would upset them more, men typically report that they would be more upset by sexual infidelity than emotional infidelity, and women report the opposite pattern. However, when Christine Harris of the University of California San Diego asked men and women *how much* they focused on each type of infidelity, these sex differences disappeared.

Harris suggests that the forced-choice paradigm provides a misleading picture of the sex differences underlying jealousy. It's possible, she argues, that men find sexual infidelity more upsetting than emotional infidelity in the forced-choice paradigm because they assume that their partners would not sleep with someone else unless they were emotionally involved with them; thus sexual infidelity is more upsetting because it implies both emotional and sexual betrayal. Similarly, women may be more upset by emotional infidelity than by sexual infidelity because they assume that if their partners are emotionally involved with someone else, they are also sleeping with them, but not vice versa. Although this explanation is cogent, we don't know if it is correct.

It is not yet clear whether sex differences in sexual jealousy are robust, or which type of methodology generates the most meaningful information. However, researchers are developing a battery of new methods that will allow them to investigate sexual jealousy and other aspects of mating preferences more fully.

SOME SOCIAL CONSEQUENCES OF MATE PREFERENCES

You might wonder how people's mate preferences influence their actual decisions and choices about marriage partners. In this section we describe the findings from one ethnographic study suggesting that these kinds of preferences actually influence people's behavior in social situations, and consequently shape the societies in which they live.

Kipsigis Bridewealth

Evolutionary theory explains marriage patterns among the Kipsigis, a group of East African pastoralists.

Among the Kipsigis, a group of Kalenjin-speaking people who live in the Rift Valley province of Kenya, women usually marry in their late teens (Figure 17.9), men usually marry for the first time when they are in their early 20s, and it is common for men to have several wives—a practice called **polygyny**. As in many other societies, the groom's father makes a **bridewealth** payment to the father of the bride at the time of marriage. The payment, tendered in livestock and cash, compensates the bride's family for the loss of her labor and gives the groom rights to her labor and the children she bears during her marriage. The amount of the payment is settled through protracted negotiations between the father of the groom and the father of the bride. The average bridewealth consists of six cows, six goats or sheep, and 800 Kenyan shillings. This is about one-third of the average man's cattle holdings, one-half of his goat and sheep herd, and two months' wages for men who hold salaried positions. Because men marry polygynously, there is competition over eligible women. Often the bride's father entertains several competing marriage offers before he chooses a groom for his daughter. The prospective bride and groom have little voice in the decisions their fathers make.

Anthropologist Monique Borgerhoff Mulder at the University of California Davis reasoned that Kipsigis bridewealth payments provide a concrete index of the qualities that each party values in prospective spouses. The groom's father is likely to prefer a bride who will bear his son many healthy children. His bridewealth offer is expected to reflect the potential reproductive value of the prospective bride. The groom's father is also expected to prefer that his son marry a woman who will devote her labor to his household. Kipsigis women who remain near their own family's households are likely to be called on to help their mothers with the harvest and to

FIGURE 17.9 These Kipsigis women are eligible for marriage. Their fathers will negotiate with the fathers of their prospective husbands over bridewealth. (Photograph courtesy of Monique Borgerhoff Mulder.)

assist their mothers in childbirth. Thus, the groom's father may prefer a woman whose natal family is distant from his son's household. The bride's father is likely to have a different perspective on the negotiations. Because wealthy men can provide their wives with larger plots of land to farm and more resources, we would expect the bride's father to prefer that his daughter marry a relatively wealthy man. At the same time, because the bride's family will be deprived of her labor and assistance if she moves far away from their land, the bride's father is likely to prefer a groom who lives nearby. The fathers of the bride and groom are expected to weigh the costs and benefits of prospective unions in their negotiations over bridewealth payments. For example, although the bride's father may prefer a high bridewealth payment, he may settle for a lower payment if the groom is particularly desirable. In order to determine whether these preferences affected bridewealth payments, Borgerhoff Mulder recorded the number of cows, sheep, and goats, as well as the amount of money, that each groom's family paid to the bride's family.

Plump women whose menarche occurred at an early age fetched the highest bridewealth payments.

Borgerhoff Mulder found that bridewealth increased as the bride's age at the time of **menarche** (her first menstruation) decreased (Figure 17.10). That is, the highest bridewealths were paid for the women who were youngest when they first menstruated. Among the Kipsigis, age at menarche is a reliable index of women's reproductive potential. Kipsigis women who reach menarche early have longer reproductive life spans, higher annual fertility, and higher survivorship among their offspring than do women who mature at later ages.

Borgerhoff Mulder wondered how a man assessed his prospective bride's reproductive potential, since men often do not know their bride's exact age, nor her age at reaching menarche. One way to be sure of a woman's ability to produce children would be to select one who had already demonstrated her fertility by becoming pregnant or

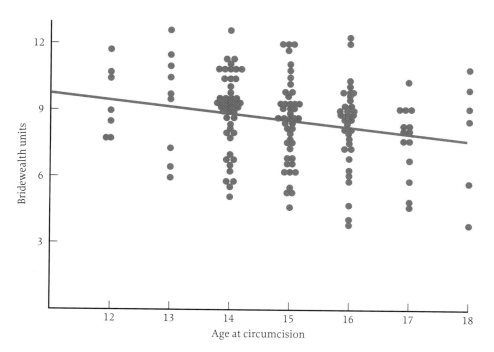

FIGURE 17.10 Kipsigis girls who mature early fetch larger bridewealth payments than older girls do. Among the Kipsigis, girls undergo circumcision (removal of the clitoris) within a year of menarche. The largest bridewealths were paid for the girls who underwent menarche and circumcision at the youngest ages. Bridewealth is transformed into standardized units to account for the fact that the value of livestock varies over time.

producing a child. However, bridewealths for such women were typically lower than bridewealths for women who had never conceived. Instead, bridewealth payments were associated with the physical attributes of women. The fathers of brides who were considered by the Kipsigis to be plump received significantly higher bridewealth payments than did the fathers of brides considered to be skinny. The plumpness of prospective brides may be a reliable correlate of the age at menarche, since menarcheal age is determined partly by body weight. Plumpness may also be valued because a woman's ability to conceive is determined partly by her nutritional status.

Bridewealth payments are also related to the distance between the bride's home and the groom's home; the farther she moves, the less likely she is to provide help to her mother, and the higher the bridewealth payment will be. However, there is no relationship between the wealth of the father of the groom and the bridewealth payment. The bride's father does not lower the bridewealth payment to secure a wealthy husband for his daughter. Although this finding was unexpected, Borgerhoff Mulder suggested that it may be related to the fact that differences in wealth among the Kipsigis are unstable over time. A wealthy man who has large livestock herds may become relatively poor if his herds are raided, decimated by disease, or diverted to pay for another wife. Although land is not subject to these vicissitudes, the Kipsigis traditionally have not held legal title to their lands.

Nyinba Polyandry

Polyandry is rare in humans and other mammals.

Because mammalian females are generally the limiting resource for males, males usually compete for access to females. As a result, among mammalian species polygyny is much more common than polyandry. As you may recall from Chapter 5, **polyandry** is a mating system in which one female is paired with two or more males. Polyandry is rare because, all other things being equal, males that share access to a single female produce fewer offspring than do males that maintain exclusive access to one or more females. The reproductive costs of polyandry may be somewhat reduced if brothers share access to a female—an arrangement called **fraternal polyandry**. Thus, in accordance with Hamilton's rule (see Chapter 8), it is plausible to assume that natural selection has shaped human psychology so that a man is more willing to care for his brother's offspring than for children who are unrelated to him.

Polyandrous marriage systems are as rare among human societies as they are in other mammalian species. Anthropologist George Peter Murdock compiled information about marriage practices in a sample of 862 societies around the world. In 83% of these societies, polygyny is the preferred form of marriage. In 16% of the societies, marriages are exclusively monogamous. Polyandry is practiced in only 0.5% of Murdock's sample.

Polyandrous marriage occurs in several societies in the Himalayas.

Although polyandry is generally rare among human societies, there are several societies in the highlands of the Himalayas in which fraternal polyandry is the preferred form of marriage. In these societies, several brothers marry one woman and establish a communal household. Because polyandry seems to limit a man's reproductive opportunities, we might ask why men tolerate polyandrous marriages. Nancy

Levine of the University of California Los Angeles studied one of these polyandrous societies: the Nyinba of northwestern Nepal (Figure 17.11). Her data shed some interesting light on this question.

The Nyinba live in four prosperous villages, nestled between 2850 and 3300 m (about 9500 and 11,000 ft) on gently sloping hillsides. Nyinba families support themselves through a combination of agriculture, herding, and long-distance trade. The Nyinba believe that marriages of three brothers are most desirable, allowing one husband to farm, another to herd livestock, and the third to engage in trade. In practice, however, all brothers marry jointly, no matter how many there are. The Nyinba are quite concerned with the paternity of their children. Women identify the fathers of each of their children, a task made easier by the fact that one or more of their husbands is often away from home for lengthy periods tending to family business. Although it is impossible to be certain how accurate these paternity assessments are, it is clear that the Nyinba place great stock in them. Men develop particularly close relationships with the children they have fathered, and fathers bequeath their share in the family's landholdings to their own sons.

FIGURE 17.11 Polyandry is the preferred form of marriage among the Nyinba. In this family, three brothers are married to one woman. The Nyinba consider this to be the ideal number of cohusbands. (Photograph courtesy of Nancy Levine.)

Evolutionary theory helps explain why some polyandrous marriages are successful and others fail.

Polyandry is the ideal form of marriage among the Nyinba, but reality does not always conform to this ideal. Although all men marry jointly, some marriages break down when one or more of the brothers brings another wife into the household or leaves to set up an independent household. When households break up, each man receives a share of the estate and takes with him all the children he has fathered during the marriage. Levine reasoned that a comparison of marriages that remain intact and marriages that break up would provide clues about the kinds of problems that arise in polyandrous marriages. Together with one of us (J. B. S.), Levine has used her data to determine whether male decisions conform to predictions derived from evolutionary theory.

All other things being equal, the more men who are married to a single woman, the lower each man's reproductive success is likely to be. Thus we might expect marriages with many cohusbands to be less stable than marriages with fewer cohusbands. This turns out to be the case; few of the largest communal Nyinba marriages remained intact. Only 10% of the marriages that began with two cohusbands partitioned (split apart), but 58% of the marriages that began with four cohusbands did (Figure 17.12).

Some Nyinba marriages are arranged by parents; others are set up by the oldest brother. In either case, the wife is usually a few years younger than the oldest cohusband. In some marriages, the youngest cohusbands are considerably younger than their wives. Remember that Buss's cross-cultural data suggest that men prefer to marry younger women. Thus we might expect that men would be dissatisfied with marriages to women much older than themselves. In fact, the Nyinba men who were most junior to their wives were the ones most likely to initiate partitions. When these men

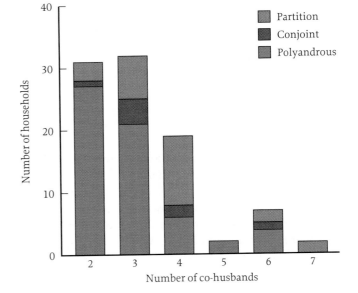

FIGURE 17.12   Among the Nyinba, marriages that contain more than three brothers are more likely to break up than marriages that include only two or three cohusbands. "Polyandrous" households are those that maintain intact polyandrous marriages. "Conjoint" households are those that added a second wife, which often effectively created two separate marriages within a single household. "Partitioned" households are those in which one or more men left the fraternal marriage and formed a new household.

remarried, they invariably chose women who were younger than themselves and who were younger than their first wives.

If the basic problem with polyandry is that it limits male reproductive opportunities, then we should expect males' satisfaction with their marriages to be linked to their reproductive performance. In most marriages, the oldest brother fathers the wife's first child. The second-oldest brother often fathers the second child, and so on. However, disparities in reproductive success among cohusbands often persist. A man's place in the birth order is directly related to his reproductive success, so the oldest brothers have more children than the youngest brothers have. Men's decisions to terminate their marriages are associated with their reproductive performance: men who remained in stable polyandrous marriages fathered an average of 0.1 children per year; men who terminated their marriages had produced, on average, only 0.04 children per year during the relationship.

Finally, kin selection theory would lead us to predict that close kinship among cohusbands would help stabilize polyandrous marriages because men should be more willing to invest in their brothers' children. We can test this prediction because there is a considerable amount of variation in the degrees of relatedness among cohusbands within households. Some households are composed entirely of full siblings. In others, the cohusbands may have multiple fathers, who are related themselves, and multiple mothers, who may also be related. The result is a complicated web of relationships among cohusbands. Contrary to predictions based on kin selection, there is no difference in the degree of relatedness among cohusbands in households that remained intact and households that dissolved. Moreover, when men remarried, they showed no inclination to align themselves with the cohusbands to whom they were more closely related.

Evolutionary theory explains certain aspects of polyandry but not others.

Polyandry is rare among human societies, as we would expect from an evolutionary perspective. When marriages dissolve, they seem to do so in ways that are consistent with predictions derived from evolutionary theory. Men leave marriages in which the women are much older, and they leave marriages when they have not fathered many children. In their decisions about whether to maintain their marriages, however, men seem to ignore kinship to their cohusbands—a surprising result, from an evolutionary perspective.

RAISING CHILDREN

Children are a lot of work. In addition to the maternal burdens of pregnancy and lactation, children need to be fed, carried, protected, clothed, and educated. In most human societies, this job is done mainly by parents. All this seems so normal and so obvious that it is easy to overlook the possibility that societies could be organized in many other ways. For example, babies could be bought and sold or traded back and forth like pets or domesticated animals. Such a system would have lots of practical advantages. It would be easy to adjust the size of your family, to select the ratio or order of boys and girls, and to regulate the spacing between children. Women who preferred to avoid pregnancy and delivery, or who were unable to bear children, could have a family just as easily as anyone else. Some people might even opt to skip the dubious pleasures of frequent diaper changes and midnight feedings, and acquire children who were already two or three years old. But this is not what happens. Instead, in virtually every society, most people raise their own children. It seems likely that people do this because their evolved psychology causes them to value their own children very differently from other people's children.

There is, however, considerable cross-cultural variation in human parenting behavior. In some societies, most men are devoted fathers; in others, men take little interest in children and rarely interact with their own offspring. In some societies, parents supervise their children's every waking moment; in others, children are reared mainly by older siblings and cousins. Moreover, some aspects of parenting behavior in some societies seem hard to reconcile with the idea that human psychology has been shaped by natural selection. There are cultures in which parents sometimes kill or physically abuse their own infant children, and cultures in which people regularly raise children who are not their own. Because such behaviors seem inconsistent with the idea that human behavior is controlled by evolved predispositions, their existence has been cited as evidence that evolutionary reasoning has little relevance to understanding modern human behavior.

In the remainder of this chapter we will consider data suggesting that observed patterns of parental investment, forms of neglect and abuse of children, and adoption are consistent with the idea that human behavior is influenced by evolved predispositions. Moreover, evolutionary reasoning can give fresh and useful insights about when and why such behaviors occur. However, we will also see that there are aspects of contemporary parenting behavior that evolutionary reasoning has not yet explained.

Parenting Effort and Mating Effort

Some human paternal investment is parenting effort, but some is mating effort.

Human children require care from their parents long after they are weaned. The lengthy period of parental investment means that some marriages will end before children are grown. Although death, separation, and divorce divide families, remarriage can create new families that include children from the partners' previous marriages and children from their present marriage. The theory of kin selection generally predicts that men and women will direct care selectively to kin, and this might lead them to favor genetic offspring over stepchildren. At the same time, if men and women value their mates' investment in their children, then care for children might also be a form of mating effort.

Kermyt Anderson, Hillard Kaplan, and Jane Lancaster have pointed out that, in humans, parental care may function as both parenting effort and mating effort. Thus, when husbands and wives invest in their genetic offspring, their efforts may enhance their inclusive fitness and the quality and durability of their marital relationship. Although parenting effort and mating are confounded when we think about husbands and wives investing in their joint progeny, it is possible to tease mating effort and parenting effort apart, as we show in Table 17.2. As you can see, investment in stepchildren reflects mating effort but not parenting effort. On the other hand, investment in one's own offspring from a previous marriage reflects parenting effort but not mating effort. Investment in a mate's stepchildren from a previous relationship does not constitute parenting effort or mating effort.

Anderson and his colleagues interviewed a large number of men in Albuquerque, New Mexico, and asked them about their reproductive histories. They gathered information about a number of different measures of parental care, including whether men provided any financial support for children attending college and how much time they spent with children in the previous year. The data in Figure 17.13 show that both mating effort and parenting effort seem to influence the extent of men's investment in children. Men are much more likely to provide money for their college-age children when they are still married to the mothers of those children than they are to provide money for their own children after the marriage has ended or for stepchildren. The magnitude of the effects of mating effort and parenting effort seem to be about the same; men invest nearly as much in their stepchildren as they do in their own offspring from a previous relationship.

There is some evidence that similar patterns influence men in other societies. Anderson and his colleagues surveyed Xhosa high school students in Cape Town,

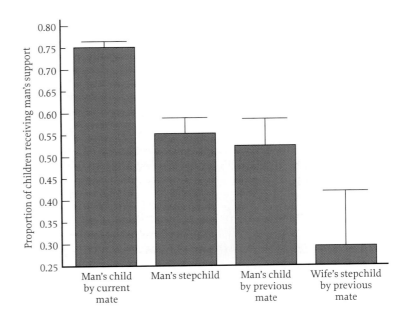

FIGURE 17.13 The proportion of children attending college who receive support from men varies considerably. Men are significantly more likely to contribute to the costs of college for their own children when they are still with the child's mother than they are to contribute to the costs of college for their stepchildren. However, men are equally likely to provide funds for stepchildren and their own children by a previous mate.

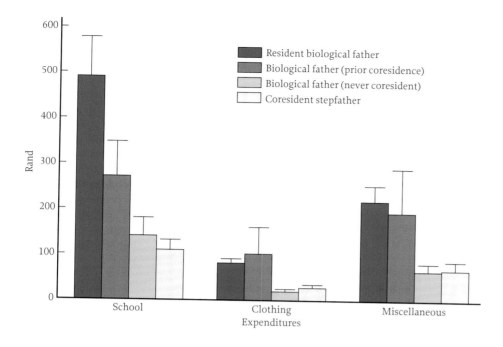

FIGURE 17.14 In South Africa, Xhosa men make bigger financial expenditures on behalf of their resident offspring than on behalf of stepchildren. Note that some children had never lived with their genetic fathers, while others had lived with them sometime in the past. Coresidence with genetic fathers has a sizable impact on the amount of investment children receive. At the time that these data were collected, one U.S. dollar was worth about 5 rand.

South Africa, and asked them about the amount and kinds of care they obtained from their fathers and stepfathers. These students spent more time with their resident genetic fathers than with their resident stepfathers, and they spent more time with their resident stepfathers than with nonresident genetic fathers (Figure 17.14). At the same time, the students generally received less financial support from their stepfathers than from their genetic fathers; they received the most support from resident genetic fathers.

Frank Marlowe of Harvard University has investigated the trade-off between mating effort and parenting effort among the Hadza, a foraging group in Tanzania. Marriages among the Hadza are relatively impermanent, so men may find themselves living with the mothers of their children or with women who have children from previous relationships. Hadza men hunt and bring food back to camp, where it is distributed. Although they are out of camp for about the same amounts of time, men who live with only biological offspring bring substantially more food back to camp than do men who live with at least one stepchild.

Hadza men also spend about 2% to 5% of every day caring directly for children. This care takes various forms: they may hold, clean, feed, comfort, or feed children; play with them; talk with them; and stay close to them (Figure 17.15). Marlowe predicted that men would provide more care for genetic offspring than for stepchildren, but he also predicted that men would make trade-offs between the extent of their parenting effort and that of their mating effort. In addition, Marlowe postulated that when men's mating opportunities were limited, they would invest more time in caring for their genetic offspring; and that when mating opportunities were abundant, men would increase mating effort and decrease parenting effort. As he expected, Marlowe found that Hadza men provide more care for their genetic offspring than for stepchildren, but they do spend some time caring for stepchildren. Caring for stepchildren might be a form of mating

FIGURE 17.15 A Hadza man holds his child on his lap. Men provide more care for genetic offspring than for stepchildren, and they seem to balance mating effort against parenting effort. (Photo courtesy of Frank Marlowe.)

effort. Moreover, when men were living in larger camps with more eligible women, they provided less direct care for children.

Grandparental Care

Grandparents are closely related to their grandchildren and are often in a position to contribute to their care.

In human societies, men and women often live long enough to have grandchildren. Grandparents may play an active role in their grandchildren's lives, particularly if all their own children are grown. Grandparental care is likely to be a product of kin selection because grandparents are closely related to their grandchildren.

Uncertainty about paternity favors matrilateral biases in grandparental investment.

In humans, there is more certainty about maternity than about paternity. There is no doubt who the mother of a given infant is, but there is sometimes some uncertainty about the identity of the father. This asymmetry in certainty about parentage skews estimates of relatedness. Let's begin with the simplest case to review how we calculate relatedness between parents, offspring, and grandoffspring. The coefficient of relatedness (r) is 0.5 for women and their daughters, and 0.5 again for those daughters and their own daughters; so the relatedness between women and their daughter's children is equal to 0.5×0.5, or 0.25. For mothers of sons, however, the calculations need to take into account the probability, p, that the son is actually the father of his mate's children. When $p = 1$, there is no uncertainty about paternity; when $p < 1$, there is some uncertainty about who is the father. Thus the degree of relatedness between mothers and their son's children is devalued by p, or $0.5 \times 0.5p = 0.25p$. The same logic tells us that men are related to their sons and daughters by $0.5p$, to their daughters' children by $0.25p$, and to their sons' children by $0.25p^2$. Whenever $p < 1$, women will be more closely related to their children than their husbands are, and both parents will be more closely related to their daughters' children than to their sons' children.

This asymmetry in relatedness to grandchildren led Harald Euler and Barbara Weitzel of the University of Kassel to predict that there would be matrilateral biases in grandparental investment. Euler and Weitzel surveyed nearly 2000 German adults and asked them how much care they received from their grandparents (Figure 17.16). In this population, both maternal grandparents provide more care than their paternal counterparts do. And on both sides of the family, grandmothers provide more care for children than grandfathers do. However, maternal

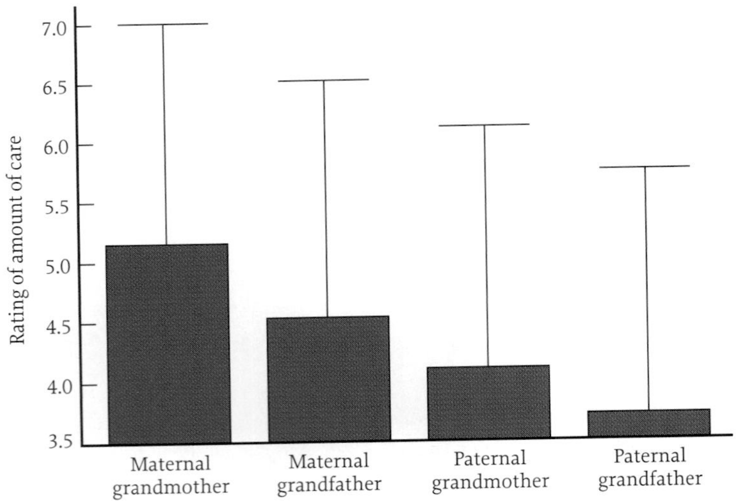

FIGURE 17.16 In Germany, maternal grandparents provide more care on average than paternal grandparents, perhaps because parents are more confident of their daughters' maternity than their sons' paternity. Adults were asked to rate the amount of care that they received from each of their grandparents on a seven-point scale. High scores correspond to higher amounts of care. Here the mean and the standard deviations are plotted.

grandfathers provide significantly more care than paternal grandmothers do, even though both are related to their grandchildren by 0.25p. Of course, the matrilateral bias in care for grandchildren might arise if there were a tendency for children to live closer to their maternal grandparents than to their paternal grandparents. Euler and Weitzel's data indicate that this is not the case, however. Moreover, the pattern held when the authors controlled for grandparents' age and availability.

Discriminative Parental Solicitude

Evolutionary theory predicts that parents should terminate investment in offspring if their prospects are poor.

Parents' ability to invest in their offspring is limited, so investment in one offspring limits their ability to invest in other offspring. If a child's prospects for survival are poor or the costs of raising a particular child are very high, then investment will not enhance parental fitness. Thus, natural selection is expected to favor mechanisms that cause parents to adjust investment in offspring in relation to the offspring's impact on parental fitness. Before birth, maternal physiology is expected to terminate pregnancies when fetal prospects are poor or the costs of carrying the fetus to term threaten maternal health. After birth, we expect parental psychology to be sensitive to the child's condition, the parents' economic circumstances, and the parents' alternative opportunities to pass on their genetic material. This phenomenon has been labeled **discriminative parental solicitude** by psychologists Martin Daly and Margo Wilson of McMaster University in Ontario, Canada.

Prenatal Investment

Selection sometimes favors the termination of pregnancy when either the mother or the fetus is in poor condition.

The costs of pregnancy must be weighed against the potential benefits. For women, every pregnancy is a costly investment. Pregnancy lasts nine months, and during their pregnancies, women need more food and work somewhat less efficiently. Childbirth is also a major source of mortality among women. The magnitude of these costs, and their impact on the mother's future reproductive success, will depend partly on the mother's condition at conception. Women who are in poor health or badly nourished will be less able to meet the demands of pregnancy and will take longer to recover from their pregnancies than will healthy, well-nourished women.

The fitness benefit that the mother will gain depends largely on the quality of the fetus that she carries. If the fetus is sickly or suffers from a genetic defect, it will be unlikely to survive. If the mother is in poor condition, she may be unable to provide the resources that her fetus or infant needs to survive. In such cases, the costs of the pregnancy will outweigh the benefits. Thus we would expect natural selection to favor physiological mechanisms that terminate a pregnancy when there is something wrong with the fetus or the mother is in very poor condition.

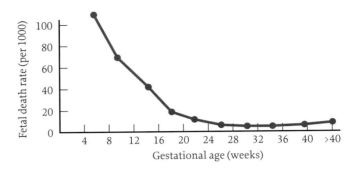

FIGURE 17.17 Most fetal deaths occur during the first trimester of pregnancy.

In some cases, miscarriages appear to provide a maternal mechanism for maintaining quality control over pregnancy.

It has been estimated that only 22% of all conceptions are carried to term. The rest are miscarried, usually before the twelfth week of pregnancy (Figure 17.17). In the majority of cases, miscarriages are associated with overt chromosomal abnormalities in the fetus. There are good reasons to expect that natural selection acting on mothers would favor miscarriages under these circumstances.

Most people think of miscarriages as pathologies—maladaptive and unpredictable aberrations that cause the fetus to die and the pregnancy to be terminated. For those who suffer miscarriages, the loss of a fetus is often deeply distressing. In such cases, parents may never know why their fetus died or whether their next one is likely to survive. As upsetting as miscarriages are for parents, there is good reason to believe that some of the mechanisms that produce miscarriages may be adaptive responses that have been favored by natural selection over the course of human history.

There is conflict between mother and fetus over whether a given pregnancy should be sustained.

To this point we have considered miscarriages from the point of view of selection acting on the mother. The picture looks different from the perspective of the fetus. As we explained in Chapter 7, the fetus has a greater interest in its own survival (and a weaker interest in any future sibling's survival) than does its mother. Thus there will be a certain scope for conflict between the mother and her fetus over whether a given pregnancy should continue. Natural selection will favor genes expressed in the fetus that prevent those spontaneous abortions that are induced by genes expressed in the mother. The fetus plays an active role in promoting its own survival.

In humans, pregnancy is maintained by the production of **luteinizing hormone** (**LH**) by the anterior pituitary gland. Luteinizing hormone, in turn, stimulates the production of another hormone, **progesterone**, by the **corpus luteum**, a mass of endocrine cells formed in the mother's ovary after ovulation. Progesterone inhibits uterine contractions that would terminate the pregnancy.

From very early in pregnancy, the fetus influences the levels of progesterone in its mother's bloodstream. Initially the fetus releases **human chorionic gonadotropin** (**hCG**), which stimulates the production of progesterone by the mother's corpus luteum, directly into the mother's bloodstream. By the eighth week of pregnancy, however, the fetus can produce enough progesterone to sustain the pregnancy, even if the corpus luteum is removed from the mother. Thus, the eight-week-old fetus is able to manipulate the maternal endocrine system to enhance its own viability (Figure 17.18). It seems quite plausible that the ability of the fetus to manipulate the level of progesterone in its mother's bloodstream has been favored by natural selection because this ability allows the fetus to protect itself from premature termination of the pregnancy.

Infanticide

Cross-cultural analyses indicate that infanticide occurs when a child is unlikely to survive, when parents cannot care for the child, or when the child is not sired by the mother's husband.

Children have been deliberately killed, fatally neglected, or abandoned by their parents through recorded history and across the world's cultures. It is tempting to dismiss infanticide by parents as a pathological side effect of modern life. After all, a child is a parent's main vehicle for perpetuating his or her own genetic material. However, evolutionary analyses suggest that parents who kill their own children may, under some circumstances, have higher fitness than parents who do not do so.

Daly and Wilson used the Human Relations Area Files, a vast compendium of ethnographic information from societies around the world, to compile a list of the reasons why parents commit infanticide. They based their study on a standard, randomly selected sample of 60 societies that represent traditional societies around the world. In 39 of these societies, infanticide is mentioned in ethnographic accounts, and in 35 of the societies Daly and Wilson found information about the circumstances under which infanticide occurs. There are three main classes of reasons why parents commit infanticide: (1) the child is seriously ill or deformed, (2) the parents' circumstances do not allow them to raise the child, or (3) the child is not sired by the mother's husband (Table 17.3).

In 21 societies, children are killed if they are born with major deformities or if they are very ill. In traditional societies, children with major deformities or serious illnesses require substantial amounts of care and impose a considerable burden on their families. They are unlikely to survive, and if they do, they are unlikely to be able to support themselves, marry, or produce children of their own. Thus it is plausible that selection has shaped human psychology so that we are predisposed to terminate investment in such children and to reserve resources for healthy children.

In some societies, children are killed when the parents' present circumstances make it too difficult or dangerous for them to raise the children. This includes cases in which births are too closely spaced, twins are born, the mother has died, there is no male present to support the children, or the mother is unmarried. It is easy to see why these factors might make it very difficult to raise children. Twins, for example, strain a mother's abilities to nurse her children and to provide adequate care for them. Moreover, twins often are born prematurely and are of below-average birth weight, making them less likely to survive than single infants. In traditional societies, it is not always possible to obtain substitutes for breast milk. This problem dooms some children whose mothers die when they are very young. If a child's father dies before it is born or when it is an infant, its mother may have great difficulty supporting herself and her older children. The child's presence might also reduce her chances of remarrying and producing additional children in the future. When circumstances are not favorable, parents who attempt to rear children may squander precious resources that might better be reserved for older children or for children born at more favorable times.

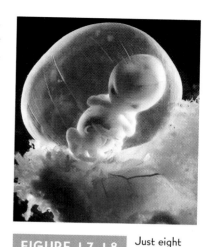

FIGURE 17.18 Just eight weeks after conception, the human fetus is able to produce enough progesterone to sustain pregnancy.

TABLE 17.3   In a representative cross-cultural sample of societies, the rationales for infanticide are consistent with predictions derived from evolutionary theory. Parents commit infanticide mainly when the child is unlikely to thrive, when their own circumstances make it difficult to raise the child, or when the child was not fathered by the mother's spouse. (Data from Table 3.1 in M. Daly and M. Wilson, 1988, *Homicide*, Aldine de Gruyter, New York.)

Rationale	Number of Societies
Child is of poor quality	
Deformed or ill	21
Parents' circumstances do not allow them to raise child	
Twins born	14
Children too closely spaced or numerous	11
No male support for child	6
Mother died	6
Mother unmarried	14
Economic hardship	3
Born in wrong season	1
Quarrel with husband	1
Paternity is not assigned to wife's husband	
Adulterous conception	15
Nontribal sire	3
Sired by mother's previous husband	2
Other reasons	15

In 20 societies, infanticide occurs when a child is sired during an extramarital relationship, by a previous husband, or by a member of a different tribe who is not married to the mother. Recall from Chapter 8 that the theory of kin selection predicts that altruism will be restricted to kin. According to this theory, males should not be predisposed to support other men's children.

Before you condemn those who kill their children, keep in mind that the options for parents in traditional societies are very different from the options available to people in our own society, who have access to modern health care systems, social services, and so on. Decisions by parents to kill their children are often reluctant responses to the painful realities of life. Whether these considerations are sufficient to absolve the parents in your eyes depends, of course, on your moral beliefs.

Adoption

In many societies, children are adopted and raised by adults other than their biological parents.

Adoption is the flip side of infanticide. In some regions of the world, particularly Oceania and the Arctic, a substantial fraction of all children are raised in adoptive households. Adoption can be thought of as a form of altruism because it takes considerable time, energy, and resources to raise children. In other animal species, including other primates, voluntary extended care of others' offspring is uncommon. There are, however, examples of *in*voluntary extended care: some species of parasitic birds, such as cuckoos, routinely lay their eggs in other birds' nests, tricking the unwitting hosts into rearing cuckoo young as their own (Figure 17.19). The fact that we label such behavior **nest parasitism** reflects the fact that this is a form of exploitation, enhancing the fitness of the parasitic cuckoo at the expense of its host. In human societies, there is no need to trick adoptive parents into caring for children, since they are typically eager to assume responsibility for other people's children.

ADOPTION IN OCEANIA

Adoption is very common in the societies of the Pacific Islands, and the pattern is consistent with the predictions of evolutionary theory.

FIGURE 17.19 Cuckoos lay their eggs in the nests of other species of birds. The hosts unwittingly rear the alien chicks as their own.

At first glance, adoption seems inconsistent with evolutionary theory, which predicts that people should be reluctant to invest time and energy in unrelated children. This apparent inconsistency prompted Marshall Sahlins, a cultural anthropologist at the University of Chicago, to cite adoption as an example of a human behavior that contradicts the logic of evolutionary theory. In *The Use and Abuse of Biology*, Sahlins pointed out that adoption is very common in the societies of the Pacific Islands. In fact, it is so common that the majority of households in many of these island societies include at least one adopted child. According to Sahlins, such widespread altruism toward nonkin shows that evolutionary reasoning does not apply to contemporary humans, and demonstrates that human societies are free to invent almost any social arrangement.

Sahlins's challenge prompted one of us (J. B. S.) to take a closer look at the pattern of adoption in a number of the societies of Oceania. The data tell a very interesting story. Adoption seems to provide an adaptive means for birth parents to regulate the sizes of their families and to enhance the quality of care they give to each of their offspring. Moreover, many of the features of adoption transactions in Oceania are consistent with predictions derived from kin selection theory:

- Adoptive parents are usually close kin, such as grandparents, aunts, or uncles (Figure 17.20). Close kin participate in adoptive transactions much more often than would be expected if children were adopted at random, without respect to kinship.
- Natural parents give up children when they cannot afford to raise them, and they rarely give up firstborn children for adoption.
- Adoptive parents generally have no dependent children; they are usually childless or the parents of grown children, but sometimes they are simply wealthy enough to be able to raise an additional child.
- Natural parents are often reluctant to give up their children for adoption, and regret the need to do so.

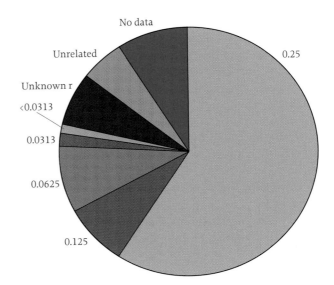

FIGURE 17.20   People in Oceania adopt mainly kin. Information about the numbers of adoptive parents and their relationships with their adoptive children in a number of communities in Oceania have been collected by ethnographers. The numbers here represent the degree of relatedness between the child and its adoptive parents. Most adoptions involve close kin ($r \leq .25$), such as aunts, uncles, and grandparents.

- Natural parents maintain contact with their children after adoption, and terminate adoptions if children are neglected, mistreated, or unhappy.
- Natural parents prefer to have their children adopted by well-to-do individuals who can provide adequately for the children's daily needs or bestow property on them.
- Sometimes asymmetries in investment exist between natural and adoptive children; adopted children often inherit less property from their adoptive parents than biological offspring do.

The same pattern characterizes adoptions in some traditional societies elsewhere.

A number of of the same features characterize adoption in the North American Arctic, where many children are adopted. Moreover, adoption transactions in Oceania and the North American Arctic bear a striking similarity to the more temporary fosterage arrangements that are common in West Africa. In each of these cases, birth parents seem to act in ways that increase the health, security, and welfare of their children, and care for children is preferentially delegated to close kin.

Adoption transactions seem to fit predictions derived from the theory of kin selection, but this does not mean that we have fully explained human adoption transactions. Not all adoption transactions in Oceania fit the evolutionary model. We have not accounted for the fact that adoption is more common in some societies than in others, or that there are some societies, including our own, in which adoption is not restricted to relatives. It is to this problem that we now turn.

ADOPTION IN INDUSTRIALIZED SOCIETIES

In contemporary industrial societies, adoption often involves strangers.

As you read the preceding description of adoption in the traditional societies of Oceania, you may have been struck by how different these transactions are from adoptions

TABLE 17.4	The patterns of adoption in the United States and Oceania show both similarities and differences.		

Feature	Oceania	United States
Economic difficulties figure in the adoption decision.	Yes	Yes
Adoptive parents are usually childless or wealthy enough to raise another child.	Yes	Yes
Children are adopted by close kin.	Usually	Sometimes*
Biological parents often regret the need for adoption.	Yes	Yes
Legal authority for the child is transferred exclusively to the adoptive parents.	No	Yes
The identities of the natural parents are known to the adoptee.	Yes	No
Contact is maintained between biological parents and child after adoption.	Yes	No
Biological parents want children to be adopted by well-to-do people.	Yes	Yes
Asymmetries exist in the care of adoptive children and biological children.	Yes	?

*In the United States from 1952 to 1971, approximately one-fourth of adoptions involved kin, another fourth involved stepparents, and one-half involved unrelated individuals.

in the United States. In the United States and in other industrialized nations, adoption is often a formal, legal process involving strangers. But as you compare the features of adoption transactions in Oceania and the United States in Table 17.4, you will see that there are similarities as well as differences between them.

In Oceania and in the United States, children are given up for adoption mainly when their parents are unable to raise them, and they are frequently adopted by people who have no dependent children of their own or by people who can afford to raise additional children. In both Oceania and the United States, biological parents often give up their children with considerable reluctance and hope to place their children in homes where the children's prospects will be improved.

The main difference between adoption in Oceania and in the United States is that in Oceania, adoption is a relatively informal, open transaction between close kin, while in the United States, adoption is often a formal, legal transaction between strangers. The anonymity of the U.S. model has a number of ramifications. When the identity of the biological parents is not disclosed to the adoptive parents or the adopted child, contact between biological parents and their children is broken. In Oceania, adopted children's interests are protected by their biological parents, who maintain contact with them and reserve the right to terminate the adoption if the children are mistreated or unhappy. In the United States, government agencies are responsible for protecting the interests of adopted children. In recent years, some of these differences have been eroded as open adoptions have become more common, and the courts have given adoptees the right to find out the identities of their birth parents.

The different patterns of adoption in Oceania and the United States may result from the same basic evolved psychological motivations.

Overall, the similarities in adoption seem to be related to people's motivations about children; the differences are related to how the transactions are organized in each culture. In Oceania and the United States, people seem to share concerns about their children. We are deeply concerned about the welfare of our own children, and many people have deep desires to raise children. These feelings are likely to be the product of evolved psychological predispositions that motivate us to cherish and protect children. Although people in Oceania and the United States may have very similar feelings about children, the adoption process is clearly different. Nepotism (favoring relatives over nonrelatives) is a central element in adoption transactions in Oceania but not in the United States. People in Oceania rely on their families to help them find homes for their children and to obtain children; people in the United States turn to adoption agencies, private attorneys, and strangers.

We are not sure why kinship plays such a fundamental role in adoption transactions in Oceania but not in industrialized societies like the United States and Canada. We can speculate that this difference is due to differences in the availability of children who are eligible for adoption, or to the tendency of family members to be dispersed geographically. However, nepotism generally seems to play a more important role in the societies of Oceania than it does in industrialized societies.

IS HUMAN EVOLUTION OVER?

This question is often raised by students in our courses, and it seems a sensible question to consider as we come to the end of the story of human evolution. As we have seen, modern humans are the product of millions of years of evolutionary change. Of course, so are cockroaches, peacocks, and orchids. All of the organisms that we see around us, including people, are the products of evolution, but they are not finished products. They are simply works in progress.

In one sense, however, human evolution is over. Because cultural change is much faster than genetic change, most of the changes in human societies since at least the origin of agriculture, almost 10 kya, have been the result of cultural, not genetic, evolution. Most of the evolution of human behavior and human societies is not driven by natural selection and the other processes of organic evolution; rather, it is driven by learning and other psychological mechanisms that shape cultural evolution. However, this fact does not mean that evolutionary theory or human evolutionary history is irrelevant to understanding contemporary human behavior. Natural selection has shaped the physiological mechanisms and psychological machinery that govern learning and other mechanisms of cultural change, and an understanding of human evolution can provide important insights into human nature and the behavior of modern peoples.

FURTHER READING

Barrett, L., R. I. M. Dunbar, and J. Lycett. 2002. *Human Evolutionary Psychology.* Princeton University Press, Princeton, NJ.

Betzig, L., M. Borgerhoff Mulder, and P. Turke, eds. 1988. *Human Reproductive Behavior.* Cambridge University Press, Cambridge, England.

Buss, D. 1994. *The Evolution of Desire.* Basic Books, New York.

Daly, M., and M. Wilson. 1988. *Homicide.* Aldine de Gruyter, Hawthorne, NY.

Silk, J. B. 1990. Human adoption in evolutionary perspective. *Human Nature* 1: 25–52.

STUDY QUESTIONS

1. In Chapter 7 we said that the reproductive success of most male primates depends on the number of females with which they mate. Here we discussed Buss's argument that a man's reproductive success will depend mainly on the health and fertility of his mate. Why are humans different from most primates? Among what other primate species should we expect males to attend to the physical characteristics of females when choosing mates?

2. Why should men value fidelity in prospective mates more than women should?

3. In Buss's cross-cultural survey, what was the most important attribute in a mate for both men and women? Does this result falsify his evolutionary reasoning?

4. Are Borgerhoff Mulder's observations about the Kipsigis consistent with Buss's cross-cultural results? Explain why or why not.

5. Explain why polyandry is so rare among mammalian species. Some bird species are polyandrous. What kinds of behaviors would you expect to find among females in these species?

6. Explain why individual organisms may sometimes increase their fitness by killing their own offspring. What does your reasoning predict about the contexts in which infanticide will occur?

7. Newborn babies attract an incredible amount of attention. Complete strangers stop to chuck them under the chin, coo at them, and comment on how cute they are. If natural selection has favored discriminative parental solicitude, as the data on infanticide seem to suggest, how can you explain the unselective attraction to newborn infants?

8. Why is adoption altruistic from an evolutionary perspective? How can adopting other people's babies be adaptive?

KEY TERMS

polygyny	discriminative parental solicitude
bridewealth	luteinizing hormone/LH
menarche	progesterone
polyandry	corpus luteum
fraternal polyandry	human chorionic Gonadotropin/hCG

THERE IS GRANDEUR IN THIS VIEW OF LIFE...

Here we end our account of how humans evolved. As we promised in the Prologue, the story has not been a simple one. We began, in Part One, by explaining how evolution works: how evolutionary processes create the exquisite complexity of organic design, and how these processes give rise to the stunning diversity of life. Next we used these ideas in Part Two to understand the ecology and behavior of nonhuman primates: why they live in groups, why the behavior of males and females differs, why animals compete and cooperate, and why primates are so smart compared with other kinds of animals. Then, in Part Three, we combined our understanding of how evolution works and our knowledge of the behavior of other primates with information gleaned from the fossil record to reconstruct the history of the human lineage. We traced each step in the transformation from a shrewlike insectivore living at the time of dinosaurs; to a monkeylike creature inhabiting the Oligocene swamps of northern Africa; to an apelike creature living in the canopy of the Miocene forests; to the small-brained, bipedal hominins who ranged over Pliocene woodlands and savannas; to the large-brained and technically more skilled early members of the genus *Homo*, who migrated to most of the Old World; and finally, to creatures much like ourselves who created spectacular art, constructed simple structures, and hunted large and dangerous game just 100 kya. Finally, in Part Four, we turned to look at ourselves—to assess the magnitude and significance of genetic variation in the human species, to ask how culture influences human evolution, and to consider how we choose our mates and how we raise our children.

Evolutionary analyses of human behavior are not always well received. In Darwin's day, many were deeply troubled by the implications of his theory. One Victorian matron, informed that Darwin believed humans to be descended from apes, is reported to have said, "Let us hope that it is not true, and if it is true, that it does not become widely known." Darwin's theory profoundly changed the way we see ourselves. Before Darwin, most people believed that humans were fundamentally different from other animals. Human uniqueness and human superiority were unquestioned. But we now know that all aspects of the human phenotype are products of organic evolution—exactly the same processes that create the diversity of life around us. Nonetheless, many people still feel that we diminish ourselves by explaining human behavior in the same terms that we use to explain the behavior of chimpanzees or soapberry bugs or finches.

In contrast, we think the story of human evolution is breathtaking in its grandeur. With a few simple processes, we can explain how we arose, why we are the way we are, and how we relate to the rest of the universe. It is an amazing story. But perhaps Darwin himself (Figure 1) put it best in the final passage of *On the Origin of Species*:

> It is interesting to contemplate an entangled bank, clothed with many plants of many kinds, with birds singing on the bushes, with various insects flitting about, and with worms crawling through the damp earth, and to reflect that these elaborately constructed forms, so different from each other, and dependent on each other in so complex a manner, have all been produced by laws acting around us. These laws, taken in the largest sense, being Growth with Reproduction; Inheritance which is almost implied by reproduction; Variability from the indirect and direct action of the external conditions of life, and from use and disuse; a Ratio of Increase so high as to lead to a Struggle for Life, and as a consequence, Natural Selection, entailing Divergence of Character and the Extinction of less-improved forms. Thus, from the war of nature, from famine and death, the most exalted object which we are capable of conceiving, namely, the production of the higher animals, directly follows. There is grandeur in this view of life, with it several powers having been originally breathed into a few forms or only one; and that, whilst this planet has gone cycling on according to the fixed law of gravity, from so simple a beginning endless forms most beautiful and most wonderful have been, and are being evolved. [From C. Darwin, 1964 (1859), *On the Origin of Species*, 1st ed., Harvard University Press, Cambridge, MA, p. 490.]

FIGURE 1 Charles Darwin died in 1882 and was buried in Westminster Abbey beneath the monument to Isaac Newton.

APPENDIX

THE SKELETAL ANATOMY OF PRIMATES

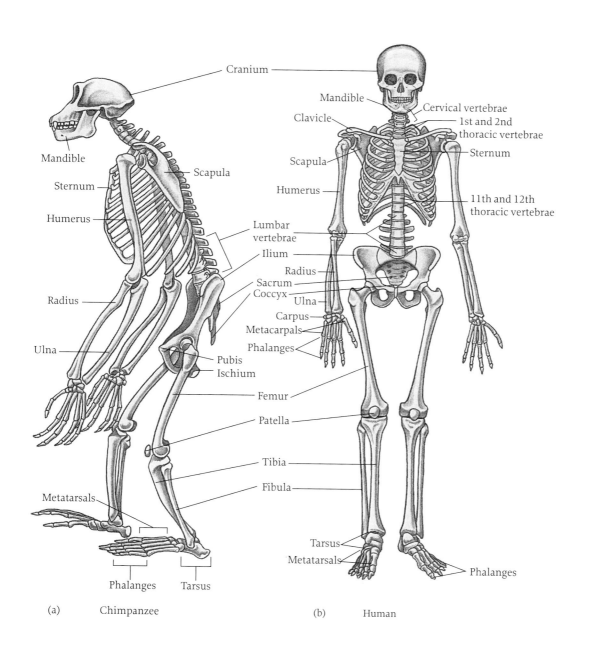

Cranium

Mandible

Sternum

Humerus

Scapula

Radius

Ulna

Metatarsals

Phalanges   Tarsus

(a)   Chimpanzee

Mandible

Clavicle

Scapula

Humerus

Lumbar
vertebrae

Ilium

Radius

Sacrum

Coccyx

Ulna

Carpus

Metacarpals

Phalanges

Pubis

Ischium

Femur

Patella

Tibia

Fibula

Cervical vertebrae

1st and 2nd
thoracic vertebrae

Sternum

11th and 12th
thoracic vertebrae

Tarsus

Metatarsals

Phalanges

(b)   Human

GLOSSARY

abductor A muscle whose contraction moves a limb away from the midline of the body. The abductors that connect the pelvis to the femur act to keep the body upright during bipedal walking.

Acheulean A tool industry found at sites dated at 1.6 to 0.3 mya and associated with *Homo ergaster* and some archaic *Homo sapiens*. Named after the French village of Saint-Acheul, where it was first discovered, the Acheulean industry is dominated by teardrop-shaped hand axes and blunt cleavers.

achondroplasia A genetic disease caused by a dominant gene that leads to the development of short stature and disproportionately short arms and legs.

activator A protein that increases transcription of a regulated gene. Compare *repressor*.

actor The individual performing a given behavior. Compare *recipient*.

adaptation A feature of an organism created by the process of natural selection.

adaptive grade The basic way that an animal makes a living. Distantly related animals can belong to the same adaptive grade.

adaptive radiation The process in which a single lineage diversifies into a number of species, each characterized by distinctive adaptations. The diversification of the mammals at the beginning of the Cenozoic era is an example of an adaptive radiation.

adenine One of the four bases of the DNA molecule. The complementary base of adenine is thymine.

affiliative Friendly.

alkaloids Secondary compounds produced and kept in plant tissues to make the plant distasteful or even poisonous to herbivores.

allele One of two or more alternative forms of a gene. For example, the *A* and *S* alleles are two forms of the gene controlling the amino acid sequence of one of the subunits of hemoglobin.

allele frequency See *gene frequency*.

alliance An interaction in which two or more animals jointly initiate aggression against, or respond to aggression from, one or more other animals. Also called *coalition*.

allopatric speciation Speciation that occurs when two or more populations of a single species are geographically isolated from each other and then diverge to form two or more new species. Compare *parapatric speciation* and *sympatric speciation*.

altruism (altruistic, adj.) Behavior that reduces the fitness of the individual performing the behavior (the actor) but increases the fitness of the individual affected by the behavior (the recipient). Compare *mutualism*. See also *selfish* and *spiteful*.

amino acids Molecules that are linked in a chain to form proteins. There are 20 different amino acids, all of which share the same molecular backbone but have a different side chain.

aminoacyl-tRNA synthetases Enzymes that link amino acids to the appropriate varieties of transfer RNA as part of protein synthesis.

analogy (analogous, adj.) Similarity between traits that is due to convergent evolution, not common descent. For example, the fact that humans and kangaroos are both bipedal is an analogy. Compare *homology*.

ancestral trait A trait that appears earlier in the evolution of a lineage or clade. Ancestral traits are contrasted with *derived traits*, which appear later in the evolution of a lineage or clade. For example, the presence of a tail is ancestral in the primate lineage, and the absence of a tail is derived. Systematists must avoid using ancestral similarities when constructing phylogenies.

angiosperms The flowering plants. The radiation of the angiosperms during the Cretaceous period may have played an important role in the evolution of the primates. Compare *gymnosperms*.

anthropoid Any member of the primate suborder that includes the monkeys and apes. The only other primate suborder (the *prosimians*) includes lemurs, lorises, and tarsiers.

anticodon The sequence of bases on a transfer RNA molecule that binds complementarily to a particular *codon*. For example, for the codon ATC the corresponding anticodon is TAG because A binds to T, and G to C.

apatite crystal A crystalline material found in tooth enamel.

arboreal Active predominantly in trees. Compare *terrestrial*.

archaic *Homo sapiens* An older term for hominins with larger brains and more modern crania that appear in the fossil record about 500 kya in Africa and Europe, and somewhat later in eastern Asia.

argon–argon dating A sophisticated variant of the potassium–argon dating method that allows very small samples to be dated accurately.

australopithecine Any member of the genus *Australopithecus*, consisting of extinct hominins that lived from 4.2 mya to about 1.8 mya. Australopithecines were characterized by bipedal locomotion, robust teeth and jaws, and ape-sized brains. (In this text we recognize six species: *A. anamensis*, *A. afarensis*, *A. africanus*, *A. gahri*, *A. habilis*, and *A. rudolfensis*.)

bachelor male A male that has not been able to establish residence in a bisexual group. Bachelor males may live alone or reside in all-male groups.

balanced polymorphism A steady state in which two or more alleles coexist in a population. This state occurs when heterozygotes have a higher fitness than any homozygote.

basal metabolic rate The rate of energy use required to maintain life when an animal is at rest.

base One of four molecules—adenine, guanine, cytosine, and thymine—that are bound to the DNA backbone. Different sequences of bases encode the information necessary for protein synthesis.

basicranium (basicrania, pl.) The base or underside of the cranium.

biface A flat stone tool made by working both sides of a core until there is an edge along the entire circumference. See also *hand ax*.

bilaterally symmetrical Describing an animal whose morphology on one side of the midline is a mirror image of the morphology on the other side.

binocular vision Vision in which both eyes can focus together on a distant object to produce three-dimensional images. See also *stereoscopic vision*.

biochemical pathway Any of the chains of chemical reactions by which organisms regulate their structure and chemistry.

biological species concept The concept that species are defined as a group of organisms that cannot interbreed in nature. Adherents of the biological species concept believe that the resulting lack of gene flow is necessary to maintain differences between closely related species. Compare *ecological species concept*.

bipedal Describing locomotion in which the animal walks upright on two (hind) legs. Compare *quadrupedal*.

blade A stone tool, made from a flake, that is at least twice as long as it is wide. Blades dominate the tool traditions of the Upper Paleolithic.

blending inheritance A model of inheritance, widely held during the nineteenth century, in which the hereditary material of the mother and father was thought to combine irreversibly in the offspring.

brachiation A form of movement in which the body is propelled by the arms alone with a phase of free flight between handholds. Only gibbons and siamangs are true brachiators.

brainstem The portion of the brain that lies between the cerebrum and the spinal cord and provides the major route for communication between the forebrain, the spinal cord, and the peripheral nerves.

bridewealth The collection of valuable items that is transferred from the groom's family to the bride's family at the time of marriage.

camera-type eye An eye in which light passes through a transparent opening and is then focused by a lens on photosensitive tissue. Camera-type eyes are found in vertebrates, mollusks, and some arthropods.

canalized Describing traits that are very insensitive to environmental conditions during development, resulting in similar phenotypes in a wide range of environments. Compare *plastic*.

canine The sharp, pointed tooth that lies between the incisors and the premolars in primates.

carbohydrates Certain organic molecules with the formula $C_nH_{2n}O_n$, including common sugars and starches.

carbon-14 dating A dating method based on an unstable isotope of carbon with atomic weight of 14. Carbon-14 is produced in the atmosphere by cosmic radiation and is taken up by living organisms. After organisms die, the carbon-14 present in their bodies decays to a stable isotope (nitrogen-14) at a constant rate. By measuring the ratio of carbon-14 to the stable isotope of carbon (carbon-12) in organic remains, scientists can estimate the length of time that has passed since the organism died. The carbon-14 method is useful for dating specimens that are younger than about 40,000 years old. Also called *radiocarbon dating*.

cathemeral Active both during the day and at night.

character A trait or attribute of the phenotype of an organism.

character displacement The result of competition between two species that causes the members of different species to become morphologically or behaviorally more different from each other.

Châtelperronian An Upper Paleolithic tool industry found in France and Spain that dates from 36 to 32 kya and is associated with Neanderthal fossil remains.

chromosome A linear body in the cell nucleus that carries genes and appears during cell division. Staining cells with dyes reveals that different chromosomes are marked by different banding patterns.

cladistic taxonomy A system for classifying organisms in which patterns of descent are the only criteria used. Compare *evolutionary taxonomy*.

cleaver A biface stone tool with a broad, flat edge. Cleavers are common at Acheulean sites.

coalition See *alliance*.

coding sequence A DNA sequence that encodes the amino acid sequence of a protein.

codon A sequence of three DNA bases on a DNA molecule that constitutes one "word" in the message used to create a specific protein. There are 64 different codons. Compare *anticodon*.

coefficient of relatedness (r) An index measuring the degree of genetic closeness between two individuals. The index ranges from 0 (for no relation) to 1 (which occurs only between an individual and itself, or between identical twins). For example, the coefficient of relatedness between an individual and its parents or its siblings is 0.5.

cognitive map A mental representation of the location of objects in space and time that allows for efficient navigation.

collected food Type of food resource, such as a leaf or fruit, that can be gathered and eaten directly. Compare *extracted food* and *hunted food*.

combinatorial control The control of gene expression in which more than one regulatory protein is used and expression is allowed only in a specific combination of conditions.

comparative method A method for establishing the function of a phenotypic trait by comparing different species.

compound eye An eye in which the image is formed by a large number of discrete photoreceptors. Compound eyes are found in insects and other arthropods.

computed tomography (**CT**) An X-ray technique that generates three-dimensional images.

conspecifics Members of the same species.

contest competition A form of competition that occurs when resources are clumped in space and worth defending. Compare *scramble competition*.

continental drift The movement over the surface of the globe of the immense plates of relatively light material that make up the continents.

continuous variation Phenotypic variation in which there is a continuum of types. Height in humans is an example of continuous variation. Compare *discontinuous variation*.

convergence The evolution of similar adaptations in unrelated species. The evolution of camera-type eyes in both vertebrates and mollusks is an example of convergence. See also *analogy*.

core A piece of stone from which smaller flakes are removed. Cores and/or flakes may themselves be useful tools.

corpus luteum A mass of cells that forms in a woman's ovary after ovulation and produces the hormone progesterone.

correlated characters Traits that are statistically associated in a population. For example, arm length and leg length are correlated if people with long arms also tend to have long legs, and vice versa. Correlated characters arise when particular genes affect multiple characters. See also *pleiotropic effects*.

correlated response An evolutionary change in one character caused by selection on a second, correlated character. For example, selection favoring only long legs will also increase arm length if arm length and leg length are positively correlated.

cortex The original, unmodified surface of a stone used to make stone tools.

cross In genetics, a mating between chosen parents.

crossing over The exchange of genetic material between homologous chromosomes during meiosis. Crossing over causes recombination of genes carried on the same chromosome.

crural index The ratio of the length of the shin bone (tibia) to the length of the thigh bone (femur).

CT See *computed tomography*.

culture Information stored in human brains that is acquired by imitation, teaching, or some other form of social learning and that is capable of affecting behavior or some other aspect of the individual's phenotype.

cusp A projection on the biting surface of a molar or premolar.

cytoplasm The material that is inside the cell but outside the nucleus.

cytosine One of the four bases of the DNA molecule. The complementary base of cytosine is guanine.

dental formula The number of incisors, canines, premolars, and molars in the upper and lower jaws.

deoxyribonucleic acid See *DNA*.

derived trait A trait that appears later in the evolution of a lineage or clade. Derived traits are contrasted with *ancestral traits*, which appear earlier in the evolution of a lineage or clade. For example, the absence of a tail is derived in the hominin lineage, and the presence of a tail is ancestral. Systematists seek to use derived similarities when constructing phylogenies.

development All of the processes by which the single-celled zygote is transformed into a multicellular adult.

diastema (diastemata, pl.) A gap between adjacent teeth.

diploid Referring to cells containing pairs of homologous chromosomes, in which one chromosome of each pair is inherited from each parent. Also referring to organisms whose somatic (body) cells are diploid; all primates are diploid. Compare *haploid*.

discontinuous variation Phenotypic variation in which there are a discrete number of phenotypes with no intermediate types. Pea color in Mendel's experiments is an example of discontinuous variation. Compare *continuous variation*.

discriminative parental solicitude A tendency among parents to adjust their investment in offspring according to cues that predict (or did predict in past environments) the likelihood that the offspring will survive and reproduce.

diurnal Active only during the day. Compare *nocturnal*. See also *cathemeral*.

dizygotic twins Twins that result from the fertilization of two separate eggs by two separate sperm. Dizygotic twins are no more closely related than other full siblings. Compare *monozygotic twins*.

DNA Deoxyribonucleic acid, the molecule that carries hereditary information in almost all living organisms. DNA consists of two very long phosphate–sugar backbones (called "strands") to which the bases adenine, cytosine, guanine, and thymine are bound. Hydrogen bonds between the bases bind the two strands together.

dominance The ability of one individual to intimidate or defeat another individual in a pairwise (dyadic) encounter. In some cases, dominance is assessed from the outcome of aggressive encounters; in other cases, dominance is assessed from the outcome of competitive encounters.

dominance hierarchy A ranking of individuals in a group that reflects their relative dominance.

dominance matrix A square table constructed to keep track of dominance interactions among a group of individuals. Usually winners are listed down the left side and losers are listed across the top, and the number of times each individual defeats another is entered in the cells of the matrix. Individuals are ordered in the matrix so as to minimize the number of entries below the diagonal. This ordering is then used to construct the dominance hierarchy.

dominant Describing an allele that results in the same phenotype whether in the homozygous or the heterozygous state. Compare *recessive*.

dyadic Describing an interaction that involves two individuals. Also called *pairwise*.

ecological species concept The concept that natural selection plays an important role in maintaining the differences between species, and that the absence of interbreeding between two populations is not a necessary condition for defining them as separate species. Compare *biological species concept*.

EEA See *environment of evolutionary adaptedness*.

electron-spin-resonance dating A technique used to date fossil teeth by measuring the density of electrons trapped in apatite crystals in teeth. This method is important for sites that are too young to be dated by potassium–argon dating (less than 500 kya) and too old to be dated with carbon-14 dating (more than 40 kya).

endocranial volume The volume inside the braincase.

environment of evolutionary adaptedness (EEA) The past environment(s) in which currently observed adaptations were shaped. For example, the psychological mechanisms that cause contemporary humans to overeat were likely shaped in an environment of evolutionary adaptedness in which overeating was rarely a problem.

environmental covariation The effect on phenotypes that occurs when the environments of parents and offspring are similar. Because environmental covariation causes the phenotypes of parents and offspring to be similar, it can falsely increase estimates of heritability.

environmental variation Phenotypic differences between individuals that exist because those individuals developed in different environments. Compare *genetic variation*.

enzyme A protein that serves as a catalyst, increasing the rate at which particular chemical reactions occur at a given temperature. Enzymes can control the chemical composition of cells by causing some chemical reactions to occur much faster than others.

equilibrium A steady state in which neither gene frequencies nor genotypic frequencies change.

estrus A period during the reproductive cycle of most mammals (and most primates) when the female is receptive to mating and is capable of conceiving.

eukaryotes Organisms whose cells have cellular organelles, cell nuclei, and chromosomes. All plants and animals are eukaryotes. Compare *prokaryotes*.

evolutionary taxonomy A system for classifying organisms that uses both patterns of descent and patterns of overall similarity. Compare *cladistic taxonomy*.

executive brain Composed of the neocortex and the striatum, a structure in the basal ganglia that is functionally linked to the neocortex.

exon A segment of the DNA in eukaryotes that is translated into protein. Compare *intron*.

extracted food Food that is embedded in a matrix, encased in a hard shell, or otherwise difficult to extract. Extracted foods require complicated, carefully coordinated techniques to process. Compare *collected food* and *hunted food*.

F$_0$, F$_1$, and F$_2$ generations A system for keeping track of generations in breeding experiments. The initial generation is called the F$_0$ generation, the offspring of the F$_0$ generation constitute the F$_1$ generation, and the offspring of the F$_1$ generation constitute the F$_2$ generation.

falciparum malaria A severe form of malaria. The sickle-cell allele for hemoglobin is common in West Africa because it confers resistance to falciparum malaria in the heterozygous state.

family A taxonomic level above genus but below order. A family may contain several genera, and an order may contain several families. Humans belong to the family Hominidae, and the other great apes belong to the family Pongidae.

fecundity The biological capacity to reproduce. In humans, fecundity may be greater than fertility (the actual number of children produced) when people limit family size.

femur The thigh bone.

fixation A state that occurs when all of the individuals in a population are homozygous for the same allele at a particular locus.

flake A smaller chip of stone knocked from a larger stone core.

folivore (folivorous, adj.) An animal whose diet consists mostly of leaves.

food sharing The practice of sharing food within a social group. Food sharing has been widely observed among contemporary human foragers but is relatively uncommon in other primate species.

foramen magnum The large hole in the bottom of the cranium through which the spinal cord passes.

fossil A trace of life more than 10,000 years old preserved in rock. Fossils can be mineralized bones, plant parts, impressions of soft body parts, or tracks.

founder effect A form of genetic drift that occurs when a small population colonizes a new habitat and subsequently greatly increases in number. Random genetic changes due to the small size of the initial population are amplified by subsequent population growth.

fraternal polyandry A marriage system in which two or more brothers marry a single wife. Fraternal polyandry is practiced among the Nyinba, a Tibetan-speaking group living in western Nepal.

frugivore (frugivorous, adj.) An animal whose diet consists mostly of fruit.

gametes In animals, eggs and sperm.

gene A segment of the chromosome that produces a recognizable effect on phenotype and segregates as a unit during gamete formation.

gene flow The movement of genes from one population to another, or from one part of a population to another, as the result of interbreeding.

gene frequency The fraction of the genes at a genetic locus that are a particular allele (therefore also called *allele frequency*). For example, a population that contains 250 *AA* individuals, 200 *AS* individuals, and 50 *SS* individuals has 700 copies of the *A* allele and 300 copies of the *S* allele; therefore the frequency of the *S* allele is 0.3.

gene tree A phylogenetic tree tracing the pattern of descent for a particular gene.

genetic distance A measure of the overall genetic similarity of individuals or species. The best estimates of genetic distance utilize large numbers of genes.

genetic drift Random change in gene frequencies due to sampling variation that occurs in any finite population. Genetic drift is more rapid in small populations than in large populations.

genetic variation Phenotypic differences between individuals that result from the fact that those individuals have inherited different genes from their parents. Compare *environmental variation*.

genome All of the genetic information carried by an organism.

genotype The combination of alleles that characterizes an individual at some set of genetic loci. For example, in populations with only the *A* and *S* alleles at the hemoglobin locus, that locus has only three possible genotypes: *AA*, *AS*, and *SS*. (*SA* is the same as *AS*.) Compare *phenotype*.

genotypic frequency The fraction of individuals in a population that have a particular genotype.

genus (genera, pl.) A taxonomic category below family and above species. There may be several species in a genus, and several genera in a family.

Gondwanaland The more southerly of the two supercontinents that existed from about 180 mya to roughly 140 mya. Gondwanaland included the continental plates that now make up Africa, South America, Antarctica, Australia, New Guinea, Madagascar, and the Indian subcontinent. Compare *Laurasia*.

grammar All of the rules of meaning that our brains use to interpret information provided by language.

grooming The process of picking through hair to remove dirt, dead skin, ectoparasites, and other material. Grooming is a common form of affiliative behavior among primates.

guanine One of the four bases of the DNA molecule. The complementary base of guanine is cytosine.

gum A sticky carbohydrate produced by some trees in response to physical damage. Gum is an important food for many primates.

gummivore (gummivorous, adj.) An animal whose diet consists mostly of gum.

gymnosperms A group of plants that reproduce without flowering. Modern gymnosperms include pines, redwoods, and firs. Compare *angiosperms*.

haft To attach a spear point, ax head, or similar implement to a handle. Hafting greatly increases the force that can be applied to the tool.

Hamilton's rule A rule predicting that altruistic behavior among relatives will be favored by natural selection if $rb > c$, where r is the *coefficient of relatedness* between actor and recipient, b is the sum of the benefits of performing the behavior on the fitness of the recipient(s), and c is the cost, in decreased fitness of the donor, of performing the behavior. See also *kin selection*.

hand ax The most common type of biface stone tool found in Acheulean sites. It is flat and teardrop-shaped, with a sharp point at the narrow end.

haploid A cell with only one copy of each chromosome. Gametes are haploid, as are the cells of some asexual organisms. Compare *diploid*.

haplorhine Any member of the group containing tarsiers and anthropoid primates. The system that classifies primates into haplorhines and strepsirhines is a cladistic alternative to the system used in this text, in which

primates are divided into prosimians and anthropoids, and tarsiers are grouped with prosimians. Compare *strepsirhine*.

Hardy–Weinberg equilibrium The unchanging frequency of genotypes that results from sexual reproduction and occurs in the absence of other evolutionary forces such as natural selection, mutation, or genetic drift.

hCG See *human chorionic gonadotropin*.

hemoglobin A protein in blood that carries oxygen, including two α (alpha) and two β (beta) subunits.

heritability The fraction of the phenotypic variation in the population that is the result of genetic variation.

heterozygous Referring to a diploid organism whose cells carry two different alleles for a particular genetic locus. Organisms that are heterozygous are called "heterozygotes." Compare *homozygous*.

hindlimb-dominated Describing an animal that habitually carries most of its weight on its hindlimbs.

home base A temporary camp that members of a group return to each day. At the home base, food is shared, processed, cooked, and eaten; subsistence tools are manufactured and repaired; and social life is conducted.

home range The area in which an individual or a group of animals travels, feeds, rests, and socializes. Territorial species actively defend the borders of their home ranges.

hominin Any member of the family Hominidae, including all species of *Australopithecus* and *Homo*.

hominoid Any member of the superfamily Hominoidea, which includes humans, all the living apes, and numerous extinct apelike and humanlike species from the Miocene, Pliocene, and Pleistocene epochs.

Homo heidelbergensis Middle Pleistocene hominins from Africa and western Eurasia. These hominins had large brains and very robust skulls and postcrania.

homologous chromosomes A pair of chromosomes in a diploid cell in which one member of the pair is derived from the father and one is derived from the mother.

homologous pair See *homologous chromosomes*.

homology (homologous, adj.) Similarity between traits that is due to common ancestry, not convergence. For example, the reason that gorillas and baboons are both quadrupedal is that they are both descended from a quadrupedal ancestor. Compare *analogy*.

homozygous Referring to a diploid organism whose chromosomes carry two copies of the same allele at a single genetic locus. Organisms that are homozygous are called "homozygotes." Compare *heterozygous*.

human chorionic gonadotropin (hCG) A hormone, secreted by the embryo during the early part of pregnancy, that stimulates the mother's body to produce progesterone.

humerus The bone in the upper part of the forelimb (arm).

hunted food Live animal prey captured by human foragers or nonhuman primates. Compare *collected food* and *extracted food*.

hybrid zone A geographic region where two or more populations of the same species or two different species overlap and interbreed. Hybrid zones usually occur at the habitat margins of the respective populations.

ilium (ilia, pl.) One of the three bones in the pelvis.

inbred mating Mating between closely related individuals. Also called *inbreeding*. Compare *outbred mating*.

inbreeding See *inbred mating*.

incisors The front teeth in mammals. In anthropoid primates, incisors are used for cutting, and there are two on each side of the upper and lower jaw.

independent assortment The principle, discovered by Mendel, that each of the genes at a single locus on a pair of homologous chromosomes is equally likely to be transmitted when gametes (eggs and sperm) are formed. This happens because during meiosis the probability that a particular chromosome will enter a gamete is 0.5 and is independent of whether other nonhomologous chromosomes enter the same gamete. Thus, knowing that an individual received a particular chromosome from its mother (and thus a particular allele) tells you nothing about the probability that it received other, nonhomologous chromosomes from its mother.

infraorder The taxonomic level between order and superfamily. An order may contain several infraorders, and an infraorder may contain several superfamilies.

insectivore (insectivorous, adj.) An animal whose diet consists mostly of insects.

insulin A substance that is created by the pancreas and is involved in the regulation of blood sugar.

intersexual selection A form of sexual selection in which females choose who they mate with. The result is that traits making males more attractive to females are selected for. Compare *intrasexual selection*.

intrasexual selection A form of sexual selection in which males compete with other males for access to females. The result is that traits making males more successful in such competition, like large body size or large canines, are selected for. Compare *intersexual selection*.

intron A segment of the DNA in eukaryotes that is not translated into protein. Compare *exon*.

isotope A chemical element with the same atomic number and properties as another element but having a differ-

ent atomic weight. Unstable isotopes spontaneously change into more stable isotopes. See also *radioactive decay*.

kibbutz (kibbutzim, pl.) An agricultural settlement in Israel, usually organized according to collectivist principles.

kin selection A theory stating that altruistic acts will be favored by selection if the product of the benefit to the recipient and the degree of relatedness (*r*) between the actor and recipient exceeds the cost to the actor. See also *Hamilton's rule*.

knapping The process of manufacturing stone tools.

knuckle walking A form of quadrupedal locomotion in which, in the forelimbs, weight is supported by the knuckles, rather than by the palm or outstretched fingers. Chimpanzees and gorillas are knuckle walkers.

lactation (lactate, v.) Production of milk by the mammary glands in females; also, the period during which milk is produced for nursing offspring. Lactation is a characteristic feature of mammals.

lactose A sugar present in mammalian milk. Most mammals—including most humans—lose the ability to digest lactose as adults.

Later Stone Age (LSA) The tool industries found in Africa that correspond in age and type to the Upper Paleolithic industries in Europe.

Laurasia The more northerly of the two supercontinents that existed from roughly 180 mya to 140 mya. Laurasia included what is now North America, Greenland, Europe, and parts of Asia. Compare *Gondwanaland*.

Levallois technique A three-step toolmaking method used by Neanderthals. The knapper first makes a core having a precisely shaped convex surface, then makes a striking platform at one end of the core, and finally knocks a flake off the striking platform.

LH See *luteinizing hormone*.

linked Referring to genes located on the same chromosome. The closer together two loci are, the more likely they are to be linked. Compare *unlinked*.

locus (loci, pl.) The position on a chromosome that is occupied by a particular gene.

LSA See *Later Stone Age*.

luteinizing hormone (LH) A hormone secreted by the pituitary gland. In females, LH stimulates ovulation, formation and maintenance of the corpus luteum, and production of the hormone progesterone. In males, LH stimulates production of the hormone androgen.

macroevolution Evolution of new species, families, and higher taxa. Compare *microevolution*.

maladaptive Detrimental to fitness.

mandible The lower jaw. Compare *maxilla*.

marsupial A mammal that gives birth to live young that continue their development in a pouch equipped with mammary glands. Marsupials include kangaroos and opossums.

mate guarding A form of mating in which the male defends his mate after copulation to prevent other males from mating with her.

mating efforts Efforts made to secure access to a mate, such as courtship displays or provision of resources to the prospective mate.

mating system The form of courtship, mating, and parenting behavior that characterizes a particular species or population. An example is polygyny.

matrilineage Individuals related through the maternal line.

maxilla The upper jaw. Compare *mandible*.

meiosis The process of cell division in which haploid gametes (eggs and sperm) are created. Compare *mitosis*.

meme A term coined by Richard Dawkins to refer to a unit of cultural information (belief or value) transmitted by imitation and teaching.

menarche First menstruation.

messenger RNA (mRNA) A form of RNA that carries specifications for protein synthesis from DNA to the ribosomes.

microevolution Evolution of populations within a species. Compare *macroevolution*.

microlith A very small stone flake. Typical of African Later Stone Age industries, microliths were probably hafted onto wood handles to make spears and axes.

Middle Stone Age (MSA) The stone tool industries of sub-Saharan Africa and southern and eastern Asia that existed 250 to 40 kya. The MSA is the counterpart of the Middle Paleolithic (Mousterian) in Europe. The MSA industries varied, but flake tools were manufactured in all of them.

mineralization (mineralized, adj.) The process by which organic material in the bones of dead animals is replaced by minerals from the surrounding rock, creating fossils.

minor marriage A form of marriage, formerly widespread in China, in which children were betrothed in infancy and then raised together in the household of the prospective groom.

mitochondrial DNA (mtDNA) DNA in the mitochondria that is particularly useful for evolutionary analyses, for two reasons: (1) mitochondria are inherited only from the mother, and thus there is no recombination, and (2) mtDNA accumulates mutations at relatively high rates, thus serving as a more accurate molecular clock for recent changes (last few million years).

mitochondrion (mitochondria, pl.) A cellular organelle that is involved in basic energy processing.

mitosis The process of division of somatic (normal body) cells through which new diploid cells are created. Compare *meiosis*.

Mode 1 A category of simple stone tools made by removing flakes from cores without any systematic shaping of the core. Both the flakes and the cores were probably used as tools themselves. Tools in the Oldowan industry are Mode 1 tools.

Mode 2 A category of stone tools in which cores are shaped into symmetrical bifaces by the removal of flakes. The Acheulean industry is typified by Mode 2 tools.

Mode 3 A category of stone tools made by striking large symmetrical flakes from carefully prepared stone cores using the Levallois technique. The Mousterian industry in Europe and the Middle Stone Age industries in Africa are typified by Mode 3 tools.

Mode 4 A category of stone tools in which blades are common. Mode 4 tools are found in some Middle Stone Age industries in Africa, and they predominate in the Upper Paleolithic industries of Europe.

Mode 5 A category of stone tools in which microliths are common. The African Later Stone Age industries are typified by Mode 5 tools.

modern synthesis An explanation for the evolution of continuously varying traits that combines the theory and empirical evidence of both Mendelian genetics and Darwinism.

molars The broad, square back teeth that are generally adapted for crushing and grinding in primates. Anthropoid primates have three molars on each side of the upper and lower jaw.

molecular clock The hypothesis that genetic change occurs at a constant rate and thus can be used to measure the time elapsed since two species shared a common ancestor. The molecular clock is based on observed regularities in the rate of genetic change along different phylogenetic lines.

monozygotic twins Twins that result from the fertilization of one egg by a single sperm. Early in development the fertilized egg splits to create two zygotes. Compare *dizygotic twins*.

morphology The form and structure of an organism; also a field of study that focuses on the form and structure of organisms.

Mousterian A stone tool industry characterized by points, side scrapers, and denticulates (tools with small tooth-like notches on the working edge), but an absence of hand axes. The Mousterian is generally associated with Neanderthals in Europe.

mRNA See *messenger RNA*.

MSA See *Middle Stone Age*.

mtDNA See *mitochondrial DNA*.

mutation A spontaneous change in the chemical structure of DNA.

mutualism (mutualistic, adj.) Behavior that increases the fitness of both actor and recipient. Compare *altruism*. See also *selfish* and *spiteful*.

natal group The group into which an individual is born. In many primate species the females remain in their natal groups throughout their lives, while the males emigrate and join new groups.

natural selection The process that produces adaptation. Natural selection is based on three postulates: (1) the availability of resources is limited; (2) organisms vary in the ability to survive and reproduce; and (3) traits that influence survival and reproduction are transmitted from parents to offspring. When these three postulates hold, natural selection produces adaptation.

Neanderthal A form of archaic *Homo sapiens* found in western Eurasia from about 127 kya to about 30 kya. Neanderthals had large brains and elongated skulls with very large faces. They were also characterized by very robust bodies.

negatively correlated Describing a statistical relationship between two variables in which larger values of one variable tend to co-occur with smaller values of the other variable. For example, the size and number of seeds produced by an individual plant are negatively correlated in some plant populations. Compare *positively correlated*.

neocortex Part of the cerebral cortex; generally thought to be most closely associated with problem solving and behavioral flexibility. In mammals, the neocortex covers virtually the entire surface of the forebrain.

neocortex ratio The size of the neocortex in relation to the rest of the brain.

nest parasitism A behavior seen in some bird species in which females lay eggs in the nests of other birds, which then unwittingly raise the alien chicks as their own.

neutral theory A theory postulating that genetic change is caused only by mutation and drift.

niche The way of life, or "trade," of a particular species— what foods it eats and how the food is acquired.

NIDD See *non-insulin-dependent diabetes*.

nocturnal Active only during the night. Compare *diurnal*. See also *cathemeral*.

non-insulin-dependent diabetes (NIDD) A form of diabetes in which cells of the body do not respond properly to levels of insulin in the blood. NIDD is known to have a genetic basis.

nonsynonymous substitution Substitution of one nucleotide for another in a DNA sequence that changes the amino acid coded for. Compare *synonymous substitution*.

nucleus (nuclei, pl.) The distinct part of the cell that contains the chromosomes. Eukaryotes (fungi, protozoans, plants, and animals) all have nucleated cells; prokaryotes (bacteria) do not.

observational learning A form of learning in which animals observe the behavior of other individuals and thereby learn to perform a new behavior. Compare *social facilitation*.

occipital torus A horizontal ridge at the back of the skull in *Homo erectus* and archaic *Homo sapiens*.

Oceania A region of the South Pacific that includes Polynesia, Melanesia, and Micronesia.

Oldowan A set of simple stone tools made by removing flakes from cores without any systematic shaping of the core. Both the flakes and the cores were probably used as tools. This industry is found in Africa at sites that date from about 2.5 mya.

olfaction (olfactory, adj.) The sense of smell.

opposable Describing the property of the thumb or big toe that enables some primates (Old World monkeys and apes) to touch the tips of the thumb and forefinger together.

organelle A portion of the cell that is enclosed in a membrane and has a specific function; examples are mitochondria and the nucleus.

out-group A taxonomic group that is related to a group of interest and can be used to determine which traits are ancestral and which are derived.

outbred mating Mating between unrelated individuals. Compare *inbred mating*.

pairwise Describing an interaction that involves two individuals. Also called *dyadic*.

paleontologist A scientist who studies fossilized remains of plant and animal species.

Pangaea The massive single continent that contained all of the Earth's dry land until about 125 mya.

parapatric speciation A two-step process of speciation in which (1) selection causes the differentiation of geographically separate, partially isolated populations of a species and (2) subsequently the populations become reproductively isolated as a result of reinforcement. Compare *allopatric speciation* and *sympatric speciation*.

parent–offspring conflict Conflict that arises between parents and their offspring over how much the parents will invest in the offspring. These conflicts stem from the opposing genetic interests of parents and offspring.

parenting efforts Efforts made to enhance the survivorship of offspring, such as lactation and infant care.

phenotype The observable characteristics of organisms. Individuals with the same phenotype may have different genotypes. Compare *genotype*.

phenotypic matching A mechanism for kin recognition in which animals assess similarities between themselves and others.

philopatry The tendency in some animals to remain in their natal (birth) groups throughout their lives. In many Old World monkey species, females are philopatric.

phoneme The basic unit of speech perception. Phonemes are the smallest bits of sound that we recognize as meaningful elements of language. People who speak different languages recognize slightly different sets of phonemes.

phylogeny The evolutionary relationships among a group of species, usually diagrammed as a "family tree."

pick A triangular-shaped biface stone tool found in Acheulean sites.

placental mammal A mammal that gives birth to live young that developed for a period of time in the uterus and were nourished by blood delivered to a placenta.

plastic Describing traits that are very sensitive to environmental conditions during development, resulting in different phenotypes in different environments. Compare *canalized*.

pleiotropic effects Phenotypic effects created by genes that influence multiple characters. See also *correlated characters*.

plesiadapiform Any member of a group of primatelike mammals that lived during the Paleocene (65 to 55 mya). Although many paleontologists do not consider them to have been primates, the plesiadapiforms probably were similar to the earliest primates, who lived around the same time.

pneumatized Containing cavities filled with air.

polyandry A mating system in which a single female forms a stable pair-bond with two different males at the same time. Polyandry is generally rare among mammals, but it is thought to occur in some species of marmosets and tamarins. Compare *polygyny*.

polygyny A mating system in which a single male mates with many females. Polygyny is the most common mating system among primate species. Compare *polyandry*.

population genetics The branch of biology dealing with the processes that change the genetic composition of populations through time.

porphyria variegata A genetic disease caused by a dominant gene in which carriers of the gene develop a severe reaction to certain anesthetics.

positively correlated Describing a statistical relationship between two variables in which larger values of one variable tend to co-occur with larger values of the other variable. For example, the height and weight of individuals are positively correlated in human populations. Compare *negatively correlated*.

postcranium (postcrania, pl.; postcranial, adj.) The skeleton excluding the skull.

potassium–argon dating A radiometric method of dating the age of a rock or mineral by measuring the rate at which potassium-40, an unstable isotope of potassium, is transformed into argon. This method can be used to date volcanic rocks that are at least 500,000 years old.

prehensile Describing the ability of hands, feet, or tails to grasp objects, such as food items or branches.

premolars The teeth that lie between the canines and molars.

primary structure The sequence of amino acids that make up a protein.

proconsulid Any member of a group of early-Miocene hominoids that includes the genus *Proconsul*.

progesterone An ovarian steroid that plays an important role in preparing the fetus to sustain a pregnancy.

prognathic See *subnasal prognathism*.

prokaryotes Organisms that lack a cell nucleus or separate chromosomes. Bacteria are prokaryotes. Compare *eukaryotes*.

prosimian Any member of the primate suborder that contains the lemurs, lorises, and tarsiers. Compare *anthropoid*.

protein A large molecule that consists of a long chain of amino acids. Many proteins are enzyme catalysts; others perform structural functions.

Punnett square A diagram that uses gene (or allele) frequencies to calculate the genotypic frequencies for the next generation.

quadrupedal Describing locomotion in which the animal moves on all four limbs. Compare *bipedal*.

radioactive decay Spontaneous change from one isotope of an element to another isotope of the same element or to an entirely different element. Radioactive decay occurs at a constant rate that can be measured precisely in the laboratory.

radiocarbon dating See *carbon-14 dating*.

radiometric method Any dating method that takes advantage of the fact that isotopes of certain elements change spontaneously from one isotope to another at a constant rate.

rain shadow An area of reduced rainfall found on the lee (downwind) side of large mountains and mountain ranges.

recessive Describing an allele that is expressed in the phenotype only when it is in the homozygous state. Compare *dominant*.

recipient The individual to whom a particular behavior is directed. Compare *actor*.

reciprocal altruism A theory that altruism can evolve if pairs of individuals take turns giving and receiving altruism over the course of many encounters.

recombination The creation of novel genotypes as a result of the random segregation of chromosomes and of crossing over.

redirected aggression A behavior in which the recipient of aggression threatens or attacks a previously uninvolved party. For instance, if A attacks B and B then attacks C, B's attacks are an example of redirected aggression.

regulatory gene A DNA sequence that regulates the expression of a structural gene often by binding to an activator or repressor.

regulatory sequence A regulatory gene.

reinforcement The process in which selection acts against the likelihood of hybrids occurring between members of two phenotypically distinctive populations, leading to the evolution of mechanisms that prevent interbreeding.

repressor A protein that decreases transcription of a regulated gene. Compare *activator*.

reproductive isolation A relationship between two populations in which there is no gene flow between them.

reproductive skew The extent to which reproduction is evenly distributed among individuals of one sex within a group. If reproductive skew is high, reproduction is monopolized by a small number of individuals. If it is low, reproduction is egalitarian.

ribonucleic acid See *RNA*.

ribosome A small organelle composed of protein and nucleic acid that temporarily holds together the messenger RNA and transfer RNAs during protein synthesis.

RNA Ribonucleic acid, a long molecule that plays several different important roles in protein synthesis. RNA differs from DNA in that it has a slightly different chemical backbone and it contains the base thymine instead of uracil.

rock shelter A site sheltered by an overhang of rock.

sagittal crest A sharp fin of bone that runs along the midline of the skull that increases the area available for the attachment of chewing muscles.

sagittal keel A feature running along the midline of the skull shaped like a shallow, upside-down V. The sagittal keel is a derived characteristic of *Homo erectus*.

sampling variation The variation in the composition of small samples drawn from a large population.

scapula (scapulae, pl.) Shoulder blade.

scramble competition A form of competition that occurs when resources are distributed evenly through space and not worth defending. Compare *contest competition*.

secondary compounds Toxic (poisonous) chemical compounds produced by plants and concentrated in plant tissues to prevent animals from eating the plant.

selection–mutation balance An equilibrium that occurs when the rate at which selection removes a deleterious gene is balanced by the rate at which mutation introduces that gene. The frequency of genes at selection–mutation balance is typically quite low.

selfish Describing behavior that increases the fitness of the actor and decreases the fitness of the recipient. Compare *spiteful*. See also *altruism* and *mutualism*.

sex ratio The number of individuals of one sex in relation to the number of the opposite sex. By convention, sex ratios are generally expressed as the number of males to the number of females.

sexual dimorphism Differences between sexually mature males and females in body size or morphology.

sexual selection A form of natural selection that results from differential mating success in one gender. In mammals, sexual selection usually occurs in males and may be due to male–male competition.

sexual selection infanticide hypothesis A hypothesis postulating that infanticide has been favored by sexual selection because males who kill unweaned infants are able to enhance their own reproductive prospects if they (1) kill infants whose deaths hasten their mothers' resumption of cycling, (2) do not kill their own infants, and (3) are able to mate with the mothers of the infants that they kill.

sickle-cell anemia A severe form of anemia that afflicts people who are homozygous for the sickle-cell gene.

silverback A mature male gorilla. The term derives from the fact that the hair on the back and shoulders turns silvery gray in mature males.

single nucleotide polymorphism (SNP) A site in the genome where DNA varies from one person to the next by a single nucleotide.

SLI See *specific language impairment*.

SNP See *single nucleotide polymorphism*.

social facilitation The situation that occurs when the performance of a behavior by older individuals increases the probability that younger individuals will acquire that behavior on their own. Social facilitation does not mean that young individuals copy the behavior of older individuals. For example, the feeding behavior of older individuals may bring younger individuals in contact with the foods that adults are eating and, therefore, increase the chance that they acquire a preference for those foods. Compare *observational learning*.

social intelligence hypothesis The hypothesis that the relatively sophisticated cognitive abilities of higher primates are the outcome of selective pressures that favored intelligence as a means to gain advantages in social groups.

socioecology The study of how social structure is influenced by ecological conditions. Socioecological models posit that the distribution of resources influences competitive regimes, which in turn influence the distribution of females, dispersal patterns, the nature of dominance relationships, and the quality of social bonds among females.

species (sing. and pl.) A group of organisms classified together at the lowest level of the taxonomic hierarchy. Biologists disagree about how to define a species (see *biological species concept* and *ecological species concept*).

specific language impairment (SLI) A family of language disorders in which the affected person experiences difficulty using language but is of otherwise normal intelligence. Evidence suggests that at least some cases of SLI are hereditary.

spiteful Describing behavior that is costly both to the actor and to the recipient. Compare *selfish*. See also *altruism* and *mutualism*.

stabilizing selection Selection pressures that favor average phenotypes. Stabilizing selection reduces the amount of variation in the population but does not alter the mean value of the trait.

stereoscopic vision Vision in which three-dimensional images are produced because each eye sends a signal of the visual image to both hemispheres in the brain. Stereoscopic vision requires binocular vision.

strategy A complex of behaviors deployed in a specific functional context, such as mating, parenting, or foraging.

stratum (strata, pl.) A geological layer.

strepsirhine Any member of the group containing lemurs and lorises. The system classifying primates into haplorhines and strepsirhines is a cladistic alternative to the system used in this text, in which primates are divided into prosimians and anthropoids, and tarsiers are grouped with prosimians. Compare *haplorhine*.

striatum A structure composed of two of the basal ganglia of the forebrain: the caudate nucleus and the putamen.

structural gene A DNA sequence that encodes the amino acid sequence of a protein.

subnasal prognathism The condition in which the part of the face below the nose is pushed out.

superfamily The taxonomic level that lies between infraorder and family. An infraorder may contain several superfamilies, and a superfamily may contain several families. For example, humans are a member of the superfamily Hominoidea, which contains the families Hominidae and Pongidae.

sutures Wavy joints between bones that mesh together and are separated by fibrous tissue.

sympatric speciation A hypothesis that speciation can result from selective pressures favoring different phenotypes within a population, without positing geographic isolation as a factor. Compare *allopatric speciation* and *parapatric speciation*.

synonymous substitution Substitution of one nucleotide for another in a DNA sequence that does not change the amino acid coded for. Compare *nonsynonymous substitution*.

syntax The grammatical rules that allow people to assign meaning to strings of words.

systematics A branch of biology that is concerned with the procedures for constructing phylogenies. Compare *taxonomy*.

tactical deception The use of normal parts of an animal's behavioral repertoire in an unusual context to achieve specific objectives that are beneficial to the actor.

tapetum (tapeta, pl.) A layer behind the retina that reflects light in some organisms.

taphonomy The study of the processes that affect the state of the remains of an organism from the time the organism dies until it is fossilized.

taurodont root A single broad tooth root in molars, resulting from the fusion of three roots. Taurodont roots were characteristic of Neanderthals.

taxonomy A branch of biology that is concerned with the use of phylogenies for naming and classifying organisms. Compare *systematics*.

temporalis muscle A large muscle involved in chewing. The temporalis muscles attach to the side of the cranium and to the mandible.

terrestrial Active predominantly on the ground. Compare *arboreal*.

territory A fixed area occupied by animals that defend the boundaries against intrusion by other individuals or groups of the same species.

tertiary structure The three-dimensional folded shape of a protein.

testes (testis, sing.) The male organs responsible for sperm production.

theory of mind The capacity to be aware of the thoughts, knowledge, or perceptions of other individuals. A theory of mind may be a prerequisite for deception, imitation, teaching, and empathy. It is generally thought that humans, and possibly chimpanzees, are the only primates to possess a theory of mind.

thermoluminescence dating A technique used to date crystalline materials by measuring the density of trapped electrons in the crystal lattice. This method is important for sites that are too young to be dated using potassium–argon dating (less than 500,000 years old) and too old to be dated with carbon-14 dating (more than 40,000 years old).

third-party relationships Relationships among other individuals. For example, monkeys and apes are believed to understand something about the nature of kinship relationships among other group members.

thymine One of the four bases of the DNA molecule. The complementary base of thymine is adenine.

tibia The larger of the two long bones in the lower leg.

torque A twisting force that generates rotary motion.

toxin A chemical compound that is poisonous or toxic.

trait A characteristic of an organism.

transfer RNA (tRNA) A form of RNA that facilitates protein synthesis by first binding to amino acids in the cytoplasm and then binding to the appropriate site on the mRNA molecule. There is at least one distinct form of tRNA for each amino acid.

transitive Describing a property of triadic (three-way) relationships in which the relationships between the first and second elements and the second and third elements automatically determine the relationship between the first and third elements. For example, if A is greater than B and B is greater than C, then A is greater than C. In many primate species, dominance relationships are transitive.

tRNA See *transfer RNA*.

unlinked Referring to genes on different chromosomes. Compare *linked*.

Upper Paleolithic The period from about 45 kya to about 10 kya in Europe, North Africa, and parts of Asia. The tool kits from this period are dominated by blades.

uracil One of the four bases of the RNA molecule. Uracil corresponds to the base thymine in DNA; as with thymine, its complementary base is adenine.

variant The particular form of a trait. For example, blue eyes, brown eyes, and gray eyes are variants of the trait eye color.

variation among groups Differences in the average phenotype or genotype between groups.

variation within groups Differences in phenotype or genotype between individuals in a group.

vertical clinging and leaping A form of locomotion in which the animal clings to vertical supports and moves by leaping from one vertical support to another.

vestibular system A system of tubes that are embedded in the inner ear (inside the cranium) and are part of the system that animals use to maintain balance.

viviparity Giving birth to live young.

zygomatic arch A cheekbone.

zygote The cell formed by the union of an egg and sperm.

CREDITS

Part II: Baboon footprint (Joan B. Silk).

CHAPTER 5

5.0: Photograph by John Haskew, Mondika Research Center; Courtesy of Diane Doran; 5.1a :ardea.com/Liz Bomford; 5.1b: © Grospas/Nature; 5.1c: Ivan Polunin/NHPA; 5.1d: © Jan Lindblad/Photo Researchers, Inc.; 5.1e: TomVezo.com; 5.2a: © Dr. U. Nebelsiek/Peter Arnold, Inc.; 5.2b: © Karl & Kay Ammann/Bruce Coleman, Inc.; 5.2c: Oxford Scientific/photolibrary; 5.3: Courtesy of Carola Borries; 5.4: © 2006 Norbert Wu, www.norbertwu.com; 5.5: From Figure 3.1 in J. G. Fleagle, 1988, Primate Adaptation and Evolution, Academic Press, San Diego, CA; 5.6: Robert Boyd; 5.7: Oxford Scientific/photolibrary; 5.8: Premaphotos Wildlife; 5.9a: Courtesy of Carlão Limeira; 5.9b: Robert Boyd and Joan B. Silk; 5.9c: Courtesy of Susan Perry; 5.10a: © Art Wolfe/artwolfe.com; 5.10b: Courtesy of Carola Borries; 5.11a: Courtesy of Kathy West; 5.11b: Robert Boyd and Joan B. Silk; 5.11c: Courtesy of Marina Cords; 5.12: From Figures 6.21 and 6.23 in R. D. Martin, 1990, Primate Origins and Evolution, Princeton University Press, Princeton, NJ; 5.14a: © Karl & Kay Ammann/Bruce Coleman, Inc.; 5.14b: © Michael J. Doolittle/Peter Arnold, Inc.; 5.15a-b: Courtesy of John Mitani; 5.16a: Joan B. Silk; 5.16b-c: Courtesy of John Mitani; 5.17: Courtesy of Sue Boinski; 5.20: Photograph by Colin Chapman; 5.21: Photograph by Colin Chapman.

CHAPTER 6

6.0: Timothy G. Laman/Getty Images; 6.1: Joan B. Silk; 6.3: Joan B. Silk; 6.5: From Figure 5.13 in A. Richard, 1985, Primates in Nature, W. H. Freeman, New York; 6.6: From Figure 8.9 in J. G. Fleagle, 1988, Primate Adaptation and Evolution, Academic Press, San Diego, CA; 6.7: © Norbert Wu/www.norbertwu.com; 6.8a: Courtesy of Lynne Isbell; 6.8b: Courtesy of Susan Perry; 6.8c: Courtesy of John Mitani; 6.8d: Joan B. Silk; 6.8e: Courtesy of Carlão Limeira; 6.8f: Courtesy of Carola Borries; 6.9: From Figure 4.1 in J. Terborgh, 1983, Five New World Primates, Princeton University Press, Princeton, NJ; 6.10: Courtesy of Susan Perry; 6.11: Courtesy of Lynne Isbell; 6.12: Oxford Scientific/photolibrary; 6.13: Joan B. Silk; 6.15: Norman Tomalin/Bruce Coleman, Inc.; 6.16: Courtesy of Carola Borries; 6.17a-b: Joan B. Silk; 6.18a-b, d-e: Robert Boyd; 6.18c: © Wolfgang Kaehler, www.wkaehlerphoto.com; 6.19a: Courtesy of Sue Boinski; 6.19b: Robert Boyd; 6.20a-c: Robert Boyd; 6.21: Courtesy of Lysa Leland; 6.22a: Joan B. Silk; 6.22b: © Norman Meyers/Bruce Coleman, Inc.; 6.24: Joan B. Silk; 6.25: Joan B. Silk; 6.26: Courtesy of Charles Jansen; 6.28: © BIOS/Peter Arnold, Inc.; 6.31a-b: Courtesy of Lynne Isbell; 6.32: Courtesy of Carola Borries.

CHAPTER 7

7.0: Nick Gordon/Nature Picture Library; 7.1: Joan B. Silk; 7.2: henryholdsworth.com; 7.3: Photograph by Kathy West; 7.4: R. Kuiter/OSF/Animals Animals; 7.6: © Ardella Reed Stock Pho-

tography; 7.5: Courtesy of Kathy West; 7.7: Adapted from data in A. Mori, 1979, Analysis of population changes by measurement of body weight in the Koshima troop of Japanese monkeys, Primates 20:371–397; 7.8: © Sarah Blaffer Hrdy/Anthro-Photo; 7.10a–b: From Figures 1 and 2 in A. Pusey, J. Williams, and J. Goodall, 1997, The influence of dominance rank on the reproductive success of female chimpanzees, Science 277:828–831; 7.11: Courtesy of Carola Borries; 7.13: Joan B. Silk; 7.14: Robert Boyd and Joan B. Silk; 7.15: Adapted from Appendix 1 in P. C. Lee, P. Majluf, and I. J. Gordon, 1991, Growth, weaning, and maternal investment from a comparative perspective, Journal of Zoology, London 225:99–114; 7.17: Premaphotos Wildlife; 7.18a: © Art Wolfe/artwolfe.com; 7.18b: © BIOS/Peter Arnold, Inc.; 7.19a–b: Robert Boyd; 7.19c: From Figure 23.2 in C. Packer et al., Reproductive success of lions, pp. 363–383 in Reproductive Success, ed. by T. H. Clutton-Brock, University of Chicago Press, Chicago; 7.20: Joan B. Silk; 7.21: From Figure 2 in P. H. Harvey and A. H. Harcourt, 1984, Sperm competition, testes size, and breeding systems in primates, pp. 589–599 in Sperm Competition and the Evolution of Animal Mating Systems, ed. by R. L. Smith, Academic Press, New York; 7.22: From Figure 2 in P. H. Harvey and A. H. Harcourt, 1984, Sperm competition, testes size, and breeding systems in primates, pp. 589–599 in Sperm Competition and the Evolution of Animal Mating Systems, ed. by R. L. Smith, Academic Press, New York; 7.23: Courtesy of Mark Chappell; 7.24: © Schafer & Hill/Peter Arnold, Inc.; 7.25: © Charles Sleicher; 7.26: © Jean-Paul Ferrero/AUSCAPE International; 7.27: From Figure 2 in R. Palombit, Infanticide and the evolution of pair bonds in nonhuman primates, Evolutionary Anthropology, 7:117-129; 7.28: © Erwin & Peggy Bauer; 7.29: Data from Figure 2 in P. A. Garber, 1997, One for all and breeding for one: Cooperation and competition as a tamarin reproductive strategy, Evolutionary Anthropology 5:187–199; 7.30: © Erwin & Peggy Bauer; 7.31: Tom Brakefield/Corbis; 7.32: Joan B. Silk; 7.33a: Robert Boyd and Joan B. Silk; 7.33b: Robert Boyd; 7.35a: © Sarah Blaffer Hrdy/Anthro-Photo; 7.35b: Courtesy of Carola Borries; 7.36: Photograph by Ryne A. Palombit; 7.37a: From Figure 2 in C. M. Crockett and R. Sekulic, 1984, Infanticide: Comparative and Evolutionary Perspectives, ed. by G. Hausfater and S. B. Hrdy, Aldine, New York; 7.37b: Courtesy of Carola Borries; 7.38: Premaphotos Wildlife; 7.39: Photograph by Charles Janson; 7.40: Joan B. Silk; 7.42: Joan B. Silk.

CHAPTER 8

8.0: Courtesy of Susan Alberts; 8.1: Premaphotos Wildlife; 8.2: Photograph by Susan Perry; 8.7a: Courtesy of Kathy West; 8.7b: Joan B. Silk; 8.8a–b: Joan B. Silk; 8.9: Joan B. Silk; 8.10: Joan B. Silk; 8.11: Joan B. Silk; 8.12a: Courtesy of Susan Perry; 8.12b: Courtesy of Marina Cords; 8.12c: Joan B. Silk; 8.12d: Courtesy of John Mitani; 8.13: From Figure 4 in E. Kapsalis and C. M. Berman, 1996, Models of affiliative relationships among free-ranging rhesus monkeys (Macaca mulatta) I. Criteria for kinship, Behaviour 133:1209–1234; 8.14: From Figure 2 in F. Aureli, C. P. van Schaik, and J.A.R.A.M. van

Hooff, 1989, Functional aspects of reconciliation among captive long-tailed macaques (Macaca fasicularis), American Journal of Primatology 19:39–51; 8.15: Joan B. Silk; 8.16a: From Table 1 in J.A. Johnson, 1987, Dominance rank in juvenile olive baboons, Papio anubis: the influence of gender, size, maternal rank and orphaning, Animal Behaviour 35:1694–1708; 8.16b: Joan B. Silk; 8.17: Joan B. Silk; 8.18: From Figure 1 in R. M. Seyfarth and D. L. Cheney, 1984, Grooming, alliances, and reciprocal altruism in vervet monkeys, Nature 308:541–543; 8.19: From Figure 2 in F. B. M. de Waal, 1997, The chimpanzee's service economy: Food for grooming, Evolution & Human Behavior 18:375–386.

CHAPTER 9

9.0: Courtesy of Susan Perry; 9.1: Cover: Time Life/Getty Images; Photo: AP/Wide World Photos; 9.2: W. Perry Conway/Corbis; 9.3: Joan B. Silk; 9.4: Gallo Images/Corbis; 9.5: Yann Arthus-Bertrand/Corbis; 9.6a :Joan B. Silk; 9.6b: Courtesy of Susan Perry; 9.7: Wolfgang Kohler, The Mentality of Apes. Routledge & Kegan Paul, Ltd., London, 1927. Reproduced with permission from the publisher; 9.8a: © Steve Bloom; 9.8b: Photograph courtesy of Tetsuro Matsuzawa; 9.9a–c: Redrawn from p. 109 in S.J. Jones, R. D. Martin, and D. A. Pilbeam, 1992, The Cambridge Encyclopedia of Human Evolution, Cambridge University Press, Cambridge; 9.10: From figures 1a and 1c in Reader, S.M., & Laland, K.N. 2002. Social intelligence, innovation, and enhanced brain size in primates. Proc. Nat. Acad. Sci.,99:4436-4441; 9.11: From Fig 2 in Byrne, R.W. and Corp, N. 2004. Neocortex size predicts deception rate in primates. Proc. R. Soc. Lond. B 271: 1693-1699; 9.12: Kevin Schafer/Corbis; 9.13: Robert Boyd and Joan B. Silk; 9.14: Joan B. Silk; 9.15a: Joan B. Silk; 9.15b: Courtesy of Susan Perry; 9.17: Courtesy of E. Menzel; 9.18: Courtesy of the Goodall Institute; 9.19: Photo courtesy of the Cognitive Evolution Group, University of Louisiana at Lafayette.

Part III: Bipedal footprints dating from approx. 3.5 mya discovered in Laetoli, Tanzania. (John Reader, 1999/Science Photo Library/Photo Researchers, Inc.).

CHAPTER 10

10.1a-b: Figure from pp. 38–39 in R. J. G. Savage and M. R. Long, 1986, Mammal Evolution, Facts on File and the British Museum (Natural History), Oxford, England; 10.2: From Figure 5.3 in R. D. Martin, 1990, Primate Origins and Evolution: A Phylogenetic Reconstruction, Princeton University Press, Princeton, NJ; 10.3: © Art Wolfe/artwolfe.com; 10.5: Adapted from Zachos, J. et al., Trends, rhythms, and aberrations in global climate 65 mya to present, Science, 292:688 2001; 10.7: From p. 200 in S. J. Jones, R. D. Martin, and D. A. Pilbeam, 1992, The Cambridge Encyclopedia of Human Evolution, Cambridge University Press, Cambridge; 10.8: Sargis 2002, Science 298:1564, November; 10.9: From J. G. Fleagle, 1998, Primate Adaptation and Evolution, 2nd Ed., Academic Press, San Diego, p. 353; 10.10a–b: From Figure 3 in K. D. Rose, 1994/5, The earliest primates, Evolutionary Anthropology 3:159–173; 10.11: From Figure 11.9 in J. G. Fleagle, 1988, Primate Adaptation and Evolution, Academic Press, San Diego; 10.12: From Figure 12.5 in J. G. Fleagle, 1988, Primate Adaptation and Evolution, Academic Press, San Diego; 10.13: From Figure 12.6 in J. G. Fleagle, 1988, Primate Adaptation and Evolution, Academic Press, San Diego; 10.14a–c: From Figure 1.10 in G. Conroy, 1990, Primate Evolution, W. W. Norton, New York; 10.15: From Figure 12.13 in J. G. Fleagle, 1988, Primate Adaptation and Evolution, Academic Press, San Diego; 10.16: From Figure 14.11 in J. G. Fleagle, 1998, Primate Adaptation and Evolution, 2nd Ed., Academic Press, San Diego; 10.17: Robert Boyd; 10.18a–c: From Figure 2 in R. D. Martin, 1993, Primate origins: Plugging the gaps, Nature 363:223–234;10.20: Christophe Ratier/NHPA.UK; 10.21: Photograph courtesy of Laura MacLatchy. 10.22: From Figure 13.7 in J. G. Fleagle, 1988, Primate Adaptation and Evolution, Academic Press, San Diego. 10.24: From Figure 2.23 in R. D. Martin, 1990, Primate Origins and Evolution: A Phylogenetic Reconstruction, Princeton University Press, Princeton, NJ; 10.25: Photo courtesy of Salvador Moya-Sola. Reproduced with permission of Nature, 379: 156-159.

CHAPTER 11

11.0: Courtesy M.P.F.T. 11.1: Robert Boyd; 11.2: B. Senut et al./C.R. Acad. Sci. Paris, Sciences de la Terre et des planetes/Earth and Planetary Sciences 332 (2001) 141; 11.3: © M.P.F.T. Photo courtesy of Michel Brunet; 11.4: From Figure 1 in T. D. White, D. C. Johanson, and W. H. Kimbel, 1981, Australopithecus africanus: Its phyletic position reconsidered. South African Journal of Science 77:445–470; 11.5: © National Museum of Kenya; 11.6: From Figure 1 in M. G. Leakey, C. S. Feibel, I. McDougall, and A. Walker, 1995, New four-million-year-old hominid species from Kanapoi and Allia Bay, Kenya, Nature 376:565–571; 11.7: © Institute of Human Origins/Nanci Kahn; 11.8: John Reader/Science Photo Library/Photo Researchers, Inc; 11.9: Redrawn from p. 237 in S. Jones, R. Martin, and D. Pilbeam, 1992, The Cambridge Encyclopedia of Human Evolution, Cambridge University Press, Cambridge; 11.10: Courtesy of Richard Klein; 11.11a–c: Courtesy of Richard Klein; 11.12: From Figures 20.15 and 20.20 in L. Aeillo and C. Dean, 1990, An Introduction to Evolutionary Human Anatomy, Academic Press, New York; 11.14: From p. 67 in R. Lewin, 1989, Human Evolution, Blackwell Scientific, Boston; 11.15: © 1999 John Reader; 11.16: Joan B. Silk; 11.17: Courtesy of Richard Klein; 11.18: From Cean, C. et al. Nature 2001, 414:695, Figure in Australopithecine-GrowthDean.eps; 11.20: Courtesy of Alan Walker; 11.21: © National Museum of Kenya; 11.24: © National Museum of Kenya; 11.25a–c: From p. 80 in R. Lewin, 1989, Human Evolution, Blackwell Scientific, Boston; 11.27: Lee Berger, University of Witswatersrand; 11.28: From Figure 1 in A. Sillen, 1992, Strontium-calcium ratios (Sr/Ca) of Australopithecus robustus and associated fauna from Swartkrans, Journal of Human Evolution 23:495–516; 11.29: © Wolfgang Kaehler, www.wkaehlerphoto.com; 11.30: Meave Leakey; 11.34: Robert Boyd; 11.36a–b: Joan B. Silk; 11.37a–b: Joan B. Silk; 11.38: Joan B. Silk; 11.39: From Figure 3.1 in R. I. M. Dunbar, 1988, Primate Social Systems, Cornell University Press, Ithaca, NY; 11.40a–b: Courtesy of Craig Stanford; 11.41: From Figure 1 in C. Stanford, J. Wallis, H. Matama, and J. Goodall, 1994, Patterns of predation by chimpanzees on red colobus monkeys at Gombe National Park, 1982–1991, American Journal of Physical Anthropology 94:213–228; 11.42: Robert Boyd; 11.43: Courtesy of William McGrew; 11.44: Adapted from Table 13 in C. Boesch and H. Boesch, 1989, Hunting behavior of wild chimpanzees in the Tai National Park, American Journal of Physical Anthropology 78: 547–573; 11.45: Courtesy of Craig Stanford.

CHAPTER 12

12.0: Peter Johnson/Corbis; 12.1: From p. 113 in K. Schick and N. Tooth, 1993, Making Silent Stones Speak, Simon and Schuster, New York; 12.2: Redrawn from p. 141 in K. Schick and N. Tooth, 1993, Making Silent Stones Speak, Simon and Schuster, New York; 12.4: Courtesy of Kim Hill; 12.:5 Courtesy of Nicholas Blurton Jones;12.9: Francesco d'Errico; 12.10: From Figure 6.1 in R. Potts, 1988, Early Hominid Activities in Olduvai, Aldine de Gruyter, New York; 12.11: From Figure 2 in R. Potts, 1984, Home bases and early hominids, American Scientist 72:338–347; 12.12: From Figure 3 in R. Potts, 1884, Home bases and early hominids, American Scientist 72:338–347; 12.13: © Art Wolfe/artwolfe.com; 12.14a–c: National Museum of Natural History, Smithsonian Institution, Washington, DC; 12.15a–b: © Y. Arthus-Bertrand/Peter Arnold, Inc.; 12.16: Robert Boyd and Joan B. Silk; 12.17: Joan B. Silk; 12.18: Robert Boyd; 12.19: Robert Boyd; 12.20: Drawing by Christian Gurney from Figure 2 in Henry T. Bunn and Ellen M. Kroll, 1986, Systematic butchery by Plio/Pleistocene homonids at Olduvai Gorge, Tanzania, Current Anthropology, 27:432–452; 12.21: Courtesy of Robert Bailey; 12.23: From Figure 4 in R. Potts, 1984, Home bases and early hominids, American Scientist, 72:338–347.

CHAPTER 13

13.0: Photo courtesy Prof. Peter Brown, The University of New England, Armidale, Australia. Permission also granted by Nature Publishing Group (http://www.nature.com); 13.1: Redrawn from Updike, Neil D. 1995. Mammalian migration and climate over the last seven million years. In Paleoclimate and Evolution, With Emphasis on Human Origins, E.S. Vrba, G.H. Denton, T.C.Partridge and L.H.Burckle, eds. New Haven, Yale University Press; 13.3: David Lordkipanidze; 13.:5 © Alan Walker; 13.6: Photograph courtesy of Alan Walker, reprinted with permission of the National Museums of Kenya; 13.8: From p. 232 in K. Schick and N. Tooth, 1993, Making Silent Stones Speak, Simon and Schuster, New York; 13.10: National Museum of Kenya; 13.11: National Museum of

Kenya; 13.12: From Figure 17 in C. Stringer and C. Gamble, 1993, In Search of the Neanderthals, Thames & Hudson, New York; 13.14: National Anthropological Archive; 13.16: From p. 224 in S. Jones, R. Martin, and D. Pilbeam, 1992, The Cambridge Encyclopedia of Human Evolution, Cambridge University Press, Cambridge; 13.17: R. Potts, Smithsonian Institution; 13.21: Courtesy of Richard Klein; 13.23: Photo courtesy Prof. Peter Brown, The University of New England, Armidale, Australia. Permission also granted by Nature Publishing Group (http://www.nature.com); 13.24: Javier Trueba/Madrid Scientific Films; 13.25: From Figure 1, Greenland Ice-Core Project, 1993, Climate instability during the last interglacial period recorded in the GRIP ice core, Nature 364:203–207; 13.28a–b: From Figure 34 in C. Stringer and C. Gamble, 1993, In Search of the Neanderthals, Thames & Hudson, New York; 13.29: From Figure 3.2 in C. Stringer, 1984, Adaptation in the Pleistocene, pp.55–83 in Hominid Evolution and Community Ecology, ed. by R. Foley, Academic Press, New York; 13.30a: © BIOS/Peter Arnold, Inc.; 13.30b: Maurizio Lanini/Corbis; 13.31: ©Martin Harvey; 13.32: © Erik Trinkaus; 13.33: Joan B. Silk; 13.34: Eric Delson; 13.35: Photograph by Michael Day.

CHAPTER 14

14.0: © Jean Clottes/DRAC Rhône-Alpes; 14.4: Courtesy of Science; 14.5: © Jean-Michel Labat/AUSCAPE International; 14.6: From Figure 5.6 in F. Bordes, 1968, La Paleolithique dans le monde, Hachette, Paris; 14.7: From Figure 7.9 in R. Klein, 1989, The Human Career, University of Chicago Press, Chicago; 14.8: From Figure 34.8 in O. Soffer, 1989, The Middle to Upper Paleolithic transition on the Russian Plain, pp. 714–742 in The Human Revolution, ed. by P. Mellars and C. Stringer, Princeton University Press, Princeton, NJ; 14.10: From Figure 13.4 in P. Mellars, 1996, The Neanderthal Legacy, Princeton University Press, Princeton, NJ; 14.11: © Jean Clottes/DRAC Rhône-Alpes; 14.12: Ulmer Museum, Ulm, Germany; 14.20: Housed in National Museum of Ethiopia, Addis Ababa. Photo © 2001 David L. Brill\Brill, Atlanta; 14.21: From Figure 15 in P. Mellars, 1989, Major issues in the emergence of modern humans, Current Anthropology, 30:349–385; 14.22: S. McBrearty and A. Brooks; 14.23: © Chip Clark: Museum of Natural History, Smithsonian; 14.24: © Centre for Development Studies, University of Bergen.

Part IV: Footprint on the moon from the Apollo landing (NASA).

CHAPTER 15

15.0: Ariel Skelley/Corbis; 15.1a–c: From Figure 20.12 in W. Bodmer and L. L. Cavalli-Sforza, 1976, Genetics, Evolution, and Man, W. H. Freeman, New York; 15.2: AP/Wide World Photos. 15.3a: George Long/Sports Illustrated; 15.3b: Richard Mackson/Sports Illustrated; 15.4: Adapted from p. 4 of M. Gopnik and M. B. Crago, 1991, Familial aggregation of a developmental language disorder, Cognition 39:1–50; 15.5a-b: Courtesy of Marion I. Barnhart, Wayne State University Medical School, Detroit, MI; 15.7: From Figures 2.14.1B and 2.14.9 in L. L. Cavalli-Sforza, P. Menozzi, and A. Piazza, 1994, The History and Geography of Human Genes, Princeton University Press, Princeton, NJ; 15.8: Sarah Errington/Panos Pictures; 15.9: From Figure 100 in G. Flatz, 1987, The genetics of lactose digestion in humans, Advances in Human Genetics 16:60; 15.10: © Jean Higgins/Unicorn Stock; 15.11: From Figure 2.3.3 in L. L. Cavalli-Sforza, P. Menozzi, and A. Piazza, 1994, The History and Geography of Human Genes, Princeton University Press, Princeton, NJ; 15.14: From Figure 1 (p. 241) in D. Hatl, 1980, Principles of Population Genetics, Sinauer, Sunderland, MA; 15.15: © John W. McDonough; 15.16: From Figure 19.12 in W. F. Bodmer and L. L. Cavalli-Sforza, 1976, Genetics, Evolution, and Man, Freeman, New York; 15.18: From Figure 16.6 in G. A. Harrison, J. M. Tanner, D. R. Pilbeam, and P. T. Baker, 1988, Human Biology: An Introduction to Human Evolution, Variation, Growth, and Adaptability, 3rd ed., Oxford Science Publications, Oxford; 15.19: From Figure 2.11.1 in L. L. Cavalli-Sforza, P. Menozzi, and A. Piazza, 1994, The History and Geography of Human Genes, Princeton University Press, Princeton, NJ; 15.20: From Figure 2.13.4 in L. L. Cavalli-Sforza, P. Menozzi, and A. Piazza, 1994, The History and Geography of Human Genes, Princeton University Press, Princeton, NJ; 15.21a–b: From Figures 2.2.3 and 2.3.3 in L. L. Cavalli-Sforza, P. Menozzi, and A. Piazza, 1994, The History and Geography of Human Genes, Princeton University Press, Princeton, NJ.

CHAPTER 16

16.0: © Perry van Duijnhoven; 16.1: © Stephen Dalton/Photo Researchers, Inc.; 16.2: © Joe Cavanaugh/DDB Stock Photo; 16.3: Oil painting by Martin Pate. Courtesy Cultural Resources, National Park Service; 16.4: From Figure 6.6 in W. H. Durham, 1992, Coevolution, Stanford University Press, Stanford, CA; 16.5: Anup Shah/Nature Picture Library; 16.6a: From Table 12.6 in A. P. Wolf, 1995, Sexual Attraction and Childhood Association, Stanford University Press, Stanford, CA; 16.6b: From Tables 11.1, 11.2, and 11.3 in A.P. Wolf, 1995, Sexual Attraction and Childhood Association, Stanford University Press, Stanford, CA; 16.7: Premaphotos Wildlife; 16.8: Dr. Ronald H. Cohn/The Gorilla Foundation; 16.9: Steve Winters/Courtesy of the Language Research Center; 16.10: Adapted from Table 71 in J Needham, 1988, Science and Civilization in China, vol. 4, pt. 3, Cambridge University Press, Cambridge, UK; 16.11a–b: From Figures 929 and 943 in J. Needham, 1988, Science and Civilization in China, vol. 4, pt. 3, Cambridge University Press, Cambridge, UK; 16.12: Courtesy of Susan Perry; 16.13: © Kennosuke Tsuda/Nature Production, Tokyo; 16.14: Behavioral Ecology Research Group, University of Oxford; 16.15: © Pete Oxford/Minden Pictures; 16.16: Courtesy of William C. McGrew; 16.17: Anup Shah/Nature Picture Library; 16.18: Photograph by Gaku Ohashi, Primate Research Institute, Kyoto University; 16.19: Robert Boyd; 16.20: Corbis; 16.21: © Kennan Harvey.

CHAPTER 17

17.0: The Marriage Contract, Jan Jozef II Horemans, 1768. Rockoxhouse, Antwerp (inv. nr. 77.9); 17.1: English Heritage Photo Library/G. P. Darwin on behalf of Darwin Heirlooms Trust; 17.2: From Tables 7.1 and 13.3 in N. Howell, 1979, Demography of the Dobe !Kung, Academic Press, New York; 17.3: From Figure 6.10 in J. Wood, 1994, Dynamics of Human Reproduction, Aldine de Gruyter, New York 17.4a-b: Redrawn from Kenrick & Keefe, 1997; 17.5. Redrawn from Table 2 in Schmitt et al. 2003; 17.6: Redrawn from Table 3 in Schmitt et al. 2003; 17.7a-b: Data redrawn from Haselton and Buss 2000; 17.8: From Tables 2 and 6 in D. Buss, 1989, Sex differences in human mate preferences: Evolutionary hypotheses tested in 37 cultures, Behavioral and Brain Sciences 12:1–49; 17.9: Photograph courtesy of Monique Borgerhoff Mulder; 17.10: From Figure 3.2 in M. Borgerhoff Mulder, 1988, Kipsigis bridewealth payments, pp. 65–82 in Human Reproductive Behavior, ed. by L. Betzig, M. Borgerhoff Mulder, and P. Turke, Cambridge University Press, Cambridge; 17.11: Photograph courtesy of Nancy Levine; 17.12: From N. L. Levine and J. B. Silk, 1997, When polyandry fails: Sources of instability in polyandrous marriages, Current Anthropology 38:375–398; 17.13: From Table 2 in Anderson et al. 1999. Evolution and Human Behavior 20:405-431; 17.14: From Figure 2 in Anderson et al. 1999. Evolution and Human Behavior 20:433-451;17.15: Courtesy of Frank Marlowe; 17.16: From Table 1 in Euler & Weitzel, 1996; 17.18: Photograph by Lennart Nilsson/Albert Bonniers Forlag AB, Behold Man, Little, Brown and Company; 17.19: Stephen Dalton/Photo Researchers, Inc.; 17.20: Adapted from Table III in J. B. Silk, 1980, Adoption in Oceania, American Anthropologist 82:799–820.

EPILOGUE

1: Courtesy of the National Portrait Gallery, London.

APPENDIX

1: From Figure B.1.b–c in G. Conroy, 1990, Primate Evolution, W. W. Norton, New York.

INDEX

Page numbers in *italics* refer to illustrations, tables and figures.